The History of Science Fiction

Palgrave Histories of Literature

Titles include:

Adam Roberts
THE HISTORY OF SCIENCE FICTION

Forthcoming titles in the series:

Andrew Pepper
THE HISTORY OF CRIME WRITING

Adeline Johns-Putra
THE HISTORY OF THE EPIC

Palgrave Histories of Literature
Series Standing Order ISBN 1–4039–1196–7 (hardback) 1–4039–1197–5 (paperback)
(*outside North America only*)

You can receive future titles in this series as they are published by placing a standing order.
Please contact your bookseller or, in case of difficulty, write to us at the address below with
your name and address, the title of the series and the ISBN quoted above.

Customer Services Department, Macmillan Distribution Ltd, Houndmills, Basingstoke,
Hampshire RG21 6XS, England

The History of Science Fiction

Adam Roberts

First published 2006 by
PALGRAVE MACMILLAN
Houndmills, Basingstoke, Hampshire RG21 6XS and
175 Fifth Avenue, New York, N.Y. 10010
Companies and representatives throughout the world.

PALGRAVE MACMILLAN is the global academic imprint of the Palgrave
Macmillan division of St. Martin's Press, LLC and of Palgrave Macmillan Ltd.
Macmillan® is a registered trademark in the United States, United Kingdom
and other countries. Palgrave is a registered trademark in the European
Union and other countries.

ISBN-13: 978–0–333–97022–5 hardback
ISBN-10: 0–333–97022–5 hardback

This book is printed on paper suitable for recycling and made from fully
managed and sustained forest sources.

A catalogue record for this book is available from the British Library.

A catalog record for this book is available from the Library of Congress.

10 9 8 7 6 5 4 3 2
15 14 13 12 11 10 09 08 07 06

Printed and bound in Great Britain by
Antony Rowe Ltd, Chippenham and Eastbourne

Contents

Preface

Science fiction is too large a subject to be represented exhaustively in a Critical History, even in one such as this, which has been allowed fairly ample dimensions. The reader seeking greater comprehensiveness is advised to consult John Clute and Peter Nicholls' astonishingly full *Encyclopedia of Science Fiction*. The present study is not a complete account of the genre, but rather an attempt to trace a line that connects one specific mode of 'fantastic' literature, which we now call science fiction, from earliest times through to the present day. The majority of texts examined are novels, short or long, and these remain the dominant form of SF, although the 'short story' (a slightly different thing from 'short novels'), cinema, television, comic books and other forms of cultural production play an increasingly large part in the later stages. As a Critical History, this work also has a certain case to make. I hope to avoid tendentiousness, but my argument is not neutral – even if such a thing as a purely neutral critical argument could exist – and I sketch it out here so that readers can be forewarned and prepare themselves to read what follows in a sympathetic or hostile frame of mind, whichever suits them better.

I argue that the roots of what we now call science fiction are found in the fantastic voyages of the Ancient Greek novel; and I use the Vernean phrase *voyages extraordinaires*, which I find to be the most supple and useful descriptor for these sorts of texts. Narratives of travel and adventure, often with fantastic (which is to say, impossible or fantastical) interludes, were among the most popular modes of ancient culture: the epic provides many examples, such as Odysseus' encounter with the Cyclops or with the sorceress Circe, who turned his men into pigs. Nor is this a surprising cultural development, given that the Greeks had a culture in which actual travel and exploration played major roles. I argue that among these many accounts of lengthy and entertaining sea voyages, or treks by land, is a category of *voyages extraordinaires* of a different sort: voyages into the sky, and especially voyages to other planets. It was possible for a Greek, in theory, to charter a ship and travel to the Gates of Herakles, or to India, or even (we can hypothesise, although there is no evidence that such voyages took place) beyond the setting sun to the Fortunate Islands, to America or the Arctic wastes. This is merely to say that the technologies of travel available to the Greeks enabled such voyages. But it was of course *not* possible for these people to travel to the Moon, as Antonius Diogenes' protagonists do. Moving the journey upwards represents a radical departure in the mode of traveller's tales.

In other words, I am arguing that the ur-form of the SF text is 'a story about interplanetary travel'. It still seems to me that stories of journeying through space form the core of the genre, although many critics would disagree. Travels 'upwards' through space, or sometimes 'downwards' into hollow-earth marvels (distinguished from more conventional 'ordinary' travels over the surface of the globe), are the trunk, as it were, from which the various other modes of SF branch off. Speaking broadly, these other branches are twofold. First, there are 'travels through time' as a corollary of 'travels through space'. It is not, I think, coincidental that this sub-genre comes into being and rapidly becomes vigorous in the late nineteenth century and throughout the twentieth – which is to say, at the time that 'science' was establishing the intimate interrelations between these quantities of 'time' and 'space'. A second branch, a major bough in fact (to continue the metaphor), is 'stories about technology'. Because long-distance travel already involves a range of complex technologies – as (for example) sail ships, technologies of long-term life support away from land, navigation, military ordnance and various others – it is again not surprising that 'tales of technology' begin as embedded elements in extraordinary voyages. The android woman, fluent in all the world's languages, encountered in Charles Sorel's *Gazettes et nouvelles ordinaires de divers pays lointains* ('Gazettes and News from Various Faraway Countries', 1632) represents, perhaps, the first instance of this sort of SF; although through the nineteenth century this mode can be seen separating itself off from the voyage to become a much more distinct sub-genre. Jules Verne is the first genius of 'technology fiction', and by the twentieth century techno-fiction had arguably become the dominant form of SF. In my first chapter I discuss 'science fiction' and 'technology fiction' as deserving equivalent theoretical attention.

These three forms, broadly conceived, delineate my rough sense of what SF 'is': stories of travel through space (to other worlds, planets, stars), stories of travel through time (into the past or into the future) and stories of imaginary technologies (machinery, robots, computers, cyborgs and cyber-culture). There is a fourth form, utopian fiction, which is often discussed by critics of science fiction as belonging to any reasonable definition of the form. My assumption in this study is that utopian fiction is indeed science fiction, although it takes as its starting point philosophy and social theory. We might say that the 'novum' (or new thing), the nub of Darko Suvin's definition of SF, in *Utopia* is 'a new social organisation', and for many critics this is enough to make the book science-fictional. I argue in this study that utopian fiction must be discussed as a parallel development to SF, including as it often does voyages to distant lands or into different times, and facilitated with different technologies. But it is easy to see why some critics wish to exclude utopias from a discussion of SF: because utopian writing is fundamentally a satiric mode of literature; that takes its force from the implied contrast between the 'ideal' society being described and the imperfect world in which the author and his/her readers actually live.

When it comes to SF I cling, perhaps naively, to the belief that the worlds encountered in the genre's best texts are more than simply modified forms of this world – which is to say that SF embodies a genuine and radical Will to Otherness, a fascination with the outer reaches of imaginative possibility. Not all utopias partake of this alterity, but utopias must be discussed nevertheless. For one thing, many practitioners of SF have regarded utopian fiction as part of their practice and have themselves written utopias. For another, as SF developed it became more and more concerned with the intricacies of 'world-building', in which writers construe alternative but self-consistent societies. World-building has now become one of the most prized things that SF provides its readers, and much of the grammar of such constructions derives from utopian fiction. So, although developing independently of the genre, utopian writing becomes a sort of para-SF, entwining itself round the genre in the nineteenth and twentieth centuries.[1]

This thumbnail sketch already implies one of the major questions that must be addressed by any historian of science fiction. We may take the form as commencing with the interplanetary *voyages extraordinaires* of the Ancient Greek novel; we may then trace the development of these ideas through seventeenth-century works such as Kepler's *Somnium* (1634), Godwin's *The Man in the Moone, or a Discourse of a Voyage Thither by Domingo Gonsales, the Speedy Messenger* (1638) and Cyrano de Bergerac's *Voyage à la lune* (1655). From this period it is possible to identify an effectively unbroken line of continual textual production in the science-fictional mode. The question, then, is: Why is there so pronounced and so lengthy a gap in the record between the Greeks and the Renaissance? Over a thousand years passes between these two periods in which science fiction is not written. Why?

While no answer to this question could be definitive, several possibilities suggest themselves. In the present work I argue a particular line, and it happens to have important consequences for my definition of science fiction. Accordingly, it is worth rehearsing here. Stating the case briefly, I argue that the re-emergence of science fiction is correlative to the Protestant Reformation. During the late sixteenth and early seventeenth centuries the balance of scientific enquiry shifted to Protestant countries, where the sort of speculation that could be perceived as contrary to biblical revelation could be undertaken with more (although not total) freedom. René Descartes, for instance, settled in Holland in 1629, in part because his native French Catholic culture was proving to be hostile to his scientific enquiries. In Descartes' mind was the recent condemnation of Galileo's astronomical work by the Catholic Church, a shocking development for many scientific thinkers at that time. Indeed, there were more troubling developments than this, especially for the more imaginatively minded (which is to say, science-fictional) thinkers.

A little-known but none the less key development in the history of the genre, I would argue, occurred in 1600 when Giordano Bruno the Nolan was burned at the stake by the Catholic Inquisition in Rome. Bruno's crime had been to argue that the universe was infinite and contained innumerable worlds – an example of speculative rather than empirical science, and accordingly science-fictional. Bruno was condemned to death for contradicting the teaching of the Catholic Church, but it takes a moment's consideration to realise why the idea of innumerable inhabited worlds was deemed so shocking. Dante, for instance, postulated inhabitants on the various worlds of his cosmos (although, of course, his poem is situated in a Ptolemaic rather than a Copernican solar system), and his writing was considered to be pious rather than otherwise.

The problem, identified and discussed brilliantly (though eccentrically) by William Empson in his posthumously published *Essays on Renaissance Literature* (1993), can be put in these terms: if there are many worlds, with many populations of beings living on them, then this tends to deny the uniqueness of the crucifixion and so devalue Christianity itself, perhaps terminally. The Church teaches that God sent Christ to Earth to save humanity, a race created in His image. This was a singular, miraculous event, a sacrament that connects humanity to God, or at least provides humanity with the possibility of that connection. But if humanity is but one among many populations of beings in the cosmos, what of the others? Have they too been redeemed by their own Christs (a possibility that would fatally degrade the uniqueness of Christ's sacrifice in *this* world) or has God simply omitted to provide them with salvation (which paints God in a very cruel light)? Under the logic of the Ptolemaic system the solar system is a sort of extension of the Earth, inhabited by human souls or angels created directly by God, and the stars are a fixed sphere, making an immense, decorative backdrop. In this cosmos a single Christ can redeem everything. But if the cosmos is infinite, this belief becomes difficult to sustain.

The revolutionary point of the Copernican cosmos is wholly to reconfigure the focus away from Earth and mankind. Either Christ died only once and God has ignored the rest of this vast creation, or he died in every possible world. As Empson puts it, 'either the Father had been totally unjust to the Martians, or Christ was crucified on Mars too; indeed, on all inhabited planets, so that his identity in any one appearance became precarious' (Empson, 1: 130). Half a century later this idea was still so shocking and destabilising to Catholic orthodoxy that Bruno was burned to death even for implying it.[2]

This may seem like an obscure point of theological quibbling, but I suggest that it marks a crucial point of cleavage in the development (or redevelopment) of western science fiction. To an orthodox Catholic imagination a plurality of inhabited worlds becomes an intolerable supposition; other stars and planets become a *theological* rather than a material reality, as they were for Dante – a sort of spiritual window-dressing to God's essentially human-sized creation. But to a Protestant imagination (or to a sceptical humanist Catholic imagination, such as Descartes' or Voltaire's) the cosmos expands before the probing inquiries of empirical science through the seventeenth and eighteenth centuries, and the imaginative-speculative exploration of that universe expands with it. This is the science fiction imagination, and it becomes increasingly a function of western Protestant culture. From this SF develops as an imaginatively expansive and (crucially) *materialist* mode of literature, as opposed to the magical-fantastic, fundamentally religious mode that comes to be known as Fantasy.

This, in turn, connects with another form of definition often applied to the mode: science fiction, in contemporary publishing and bookselling practice, is distinguished from 'Fantasy', the latter involving tales of fantastic or non-realist form in which the mechanism for the fantasy is magic rather than technology. The key text in the development of twentieth-century Fantasy (a genre of astonishing and continuing fertility) is J. R. R. Tolkien's *Lord of the Rings* (1952–53); elves, men and dwarves, aided by a wizard, battle malign orcs and monsters as part of a cosmic war between the forces of 'good' and 'evil', the latter epitomised by Sauron, an evil, yet god-like being. There are very few machines in *Lord of the Rings*, and most of those that do appear are aligned with the forces of evil. Instead, it is magic that operates in the world to good or ill effect. The action revolves around a golden ring imbued with magical powers, a device that renders the wearer invisible but also grants the ring-bearer enormous power-to-command.

The Lord of the Rings is a profoundly Catholic work, not so much in terms of conscious allegory (Tolkien always expressed 'cordial dislike' for allegory) as in its detailed working-out; a drama of fall and redemption, in which a saviour returns to overthrow evil. The use of a sacramental symbol – the gold band symbolising marriage in Christian service – as the core element of the fantasy is also significant. From a Catholic perspective something magical is involved in marriage, 'magical' in the forceful sense that Jesus' miracles can also be described as 'magical'. To a Catholic the communion wafer actually becomes the body of Christ during the mass – transubstantiation is a literal process, another magical sacrament. To a Protestant worshipper it is *symbolic* of Christ, but is otherwise wholly material and bread.

This is to revisit the very rationale of the Protestant Reformation itself. Norman Davies, for instance, summarises the movement in the following terms:

> The Protestant movement contained a very strong impulse to 'take the magic out of religion' ... none the less, in the fifteenth, sixteenth and seventeenth centuries Europe continued to be devoted to every form of magical belief. The landscape was filled with alchemists, astrologers, diviners, conjurers, healers and witches ... Magic held its own through the Reformation period. In this respect, therefore, the Protestant onslaught on magic enjoyed only partial success, even in the countries where Protestantism was to be nominally triumphant. But the intentions of the radicals were unmistakable. After Wycliffe came Luther's attack on the indulgences (1517) and Calvin's dismissal of transubstantiation as 'conjury' (1536) ... Protestant Christianity was supposedly magic-free. (Davies, p. 405)

Davies goes on to point out that magic remained, stubbornly, even in this supposedly 'purged' religion of rational and conscious belief ('it proved virtually impossible to abandon the consecration of church buildings, of battle standards, of food, of ships, and of burial grounds'). But a separation begins here in the broad stream of 'fantastic' or 'non-realist' fiction. 'Catholic' imaginations countenance magic and produce traditional romance, magical-Gothic, horror, Tolkienian fantasy and Marquezian magic realism. Protestant imaginations increasingly replace the instrumental function of magic with technological devices, and produce science fiction. This present history depends, then, on a historicised definition of SF as that form of fantastic romance in which magic has been replaced by the materialist discourses of science.

To say this is not to deny a distinctive Catholic strand in science fiction; on the contrary, I argue that this strand is present in the vast majority of good SF, whether written by Catholic authors or not. If I am asked to condense it into a single sentence, my thesis is that science fiction is determined precisely by the dialectic between 'Protestant' and 'Catholic' (or, if one prefers less sectarian terms, between 'deism' and 'magical pantheism') that emerges out of the seventeenth century. SF texts mediate these cultural determinants with different emphases, some more strictly materialist, some more mystical or magical. Many of the most celebrated works of what is sometimes called 'Catholic' SF are deeply embedded in this sacramental, 'magical' vision – it is this, I would argue, rather than a fascination with theological questions as such, that distinguishes 'Catholic' SF. So, for instance, Walter M. Miller's *A Canticle for Leibowitz* (1959) spreads its narrative over many centuries, from the aftermath of a nuclear war in which society is reduced to primitivism, to the gradual rebirth of civilisation and the growth of technology again, to the point where mankind again plans rocket trips to the stars, and once more threatens to destroy itself with nuclear weapons. The coherence of this long narrative is provided by its focus on a group of monks living in the American desert, and in large part Miller provides a detailed, realist-manner description of their quotidian life. But the novel also depends on several magical turns of events; one is the character of Leibowitz himself, a hermit in the post-nuclear desert who seems immortal (the same character appears in each chapter, although they are separated by centuries), and is indeed specifically identified with the Wandering Jew. As the bombs fall for a second time, a mutant, head-like growth on the shoulder of a woman apparently comes to life. The magical element is not merely dropped into the novel for bizarreness' sake; rather it functions as an endorsement of the supernatural, the presence of God, in a world that had been atom bomb-ruined by societies too purely secular and rational.

Gene Wolfe's *Long Sun* tetralogy (1993–96) features as its hero a humble priest with many Graham Greene-like Catholic trappings, although the religion he serves is not Catholicism: the novels take place on a gigantic spaceship, tube-shaped and revolving to provide gravity to those living within it. This ship is on a generations-long journey to a new world – indeed, the journey has been so lengthy that the passengers have forgotten that they inhabit a spaceship at all. The religion which Silk, Wolfe's priest-hero, follows is one in which 'windows' (advanced TV screens) display the personages of the gods themselves; but over the course of the novels these supposed gods are revealed to be nothing but the downloaded personalities of ancient individuals, several of whom have gone mad in their electronic environment. The overall trajectory of this series presents the unmasking of features of the world that had been taken to be natural and supernatural, revealing them to be nothing but technological – a plotline we might term, utilising this crude binary, 'Protestant'. Yet Catholic Wolfe cannot abandon magic; a mysterious and non-technological numinous being, 'the Outsider', lurks behind Silk's actions throughout the book, and the first novel opens with a revelation experienced by Silk in which – magically – the Outsider reaches through into the world of the long sun and provides insight. Wolfe's most famous tetralogy, *The Book of the New Sun* (1980–87), is also science fiction figured as Fantasy. Severian advances from apprentice Torturer, to Torturer proper, and eventually to world-emperor, travelling through a world so far into the future that its bizarre rituals and paraphernalia seem utterly magical. The fantastic-magical elements predominate, and the far-future technological features can never quite be reduced to materialist explanation: appropriately so because in a manner of speaking the protagonist, Severian the Torturer, is Christ.

The narrative of this Critical History, in other words, sees a nascent form of SF in Ancient Greece that disappears, or becomes suppressed, with the rise to cultural dominance of the Catholic Church; and which re-emerges when the new cosmology of the sixteenth century inflects the theology of Protestant thinkers in the seventeenth. The death of Bruno in 1600 is a short-hand for this crux; and the doctrine of the plurality of worlds that engages seventeenth-century thinkers informs almost all the newer interplanetary texts that SF from this period involves.

I discuss the implications of this in greater detail in chapter 4, but in brief this means that when interplanetary travel romances return to western culture in the seventeenth and eighteenth centuries, they are all vitally concerned with the *theological* implications of the aliens they describe. Where a modern-day astronaut might greet extraterrestrials with some benign liberal platitude ('We come in peace'), star-travellers in stories from this period are all keen to hear the answer to one crucial question: 'Do you believe in Jesus Christ?' When Francis Godwin's hero travels to the Moon and meets aliens there, his first words are 'Jesus Maria':

> No sooner was the word Jesus out of my mouth but young and old fell all down upon their knees, at which I not a little rejoiced, holding up both their hands on high, and repeating certain words which I understood not. (Godwin, *Man in the Moone*, p. 96)

This is important because, for Godwin and his readers, space-aliens are not esoteric curiosities, but crucial proof or disproof of divine truth. In Godwin's work, his Lunites are closely enough associated with the Earth to be able to share in the earthly Christ's redemptive power; but all of the space fantasies of this period in which aliens are encountered worry away at the question of the plurality of worlds and what this means for Christian revelation. Wilkins' *Discovery of a World in the Moone* (1638), for instance, postulates lunar inhabitants, but goes on to question whether such beings 'are the seed of Adam, whether they are in a blessed estate, or else what means there may be for their salvation'. Wilkins quotes Thomas Campanella to the effect that Lunarians must be 'liable to the same misery [of original sin] with us, out of which, perhaps, they were delivered by the same means as we, the death of Christ' (Wilkins, pp. 186–92). The topic re-emerges regularly in SF throughout the nineteenth and twentieth centuries. C. S. Lewis, for instance, sends his space-travellers to an inhabited Mars in *Out of the Silent Planet* (1938); Lewis's trilogy of science fiction novels are all concerned with such theological issues, and his solution to the problem is to argue that Christ is unique to Earth because only Earth has fallen into the clutches of the Devil.

James Blish's *A Case of Conscience* (1958) sees a priest exorcising an entire planet (Lithia) literally out of existence, as a devilish imposture to beguile the faithful with the corrosive conceptual possibilities of a Brunoesque multiplicity of inhabited worlds. Blish's Hugo Award-winning novel is a brilliantly intelligent piece of fiction; but reading it from a Catholic perspective is a different experience from reading it from a non-Catholic one. Under the logic of the former, the novel is an interesting exploration of a theological conundrum; under the logic of the latter, a heartbreaking story of human arrogance and short-sightedness, with the blameless Lithians terrible victims.

More recently, Dan Simmons' highly regarded *Hyperion* (1989) opens with a story about a priest haunted by precisely this question: whether Christ is a universal saviour or merely a parochial, earthbound figure. He travels to a distant corner of the galaxy and finds there a race of apparently idiot aliens who all bear a glowing cross on their torsos. The priest, ecstatic, decides that this is proof of the universality of Christ. In the event, the cross shape turns out to be an especially pernicious parasitic form of alien life, which is only coincidentally cruciform, and the priest is disillusioned. But this theological opening to the *Hyperion* sequence of novels is appropriate. *Hyperion* establishes that a sadistic creature called the shrike is kidnapping and inflicting terrible pain on various inhabitants of the planet Hyperion; in the novel's sequel, *The Fall of Hyperion*, it becomes apparent that malign machine intelligences are using the shrike to summon a numinous spirit of compassion from its hiding place (the bait being a larger and larger quantity of suffering human beings) so that they can destroy it. That there is a mystical principle of compassion behind the events of the cosmos, in conflict with a ruthless opposite principle, casts a pseudo-Christian battle between Christ and Satan onto a galactic stage.

By asserting that this opposition of 'Protestant/humanist' technology and 'Catholic' magic is radically constitutive of science fiction (and necessarily, therefore, by contradistinguishing SF from Fantasy) I may be reminding readers of Arthur C. Clarke's famous dictum: 'Any sufficiently advanced technology is indistinguishable from magic' (Clarke, *Profiles of the Future*, 1969). But it hardly needs pointing out that far from co-opting magic into SF, Clarke's statement reduces all 'magic' to technological reality; what seems at first glance to be miraculous becomes, when properly analysed, only technological, albeit technology of a wonderfully advanced sort. In other words, Clarke in effect denies 'Catholic' SF altogether; for him 'Catholic' SF is always and inevitably 'Protestant-humanist' SF in disguise; something entirely appropriate to Clarke's own SF corpus, where the apparently 'transcendent' (for instance, the ending of *2001: A Space Odyssey*) is later rationalised in materialist, technological terms (in the three sequels to that work).

In other words, I am suggesting here a modification to the crude distinction between 'magical' Fantasy and 'scientific' SF. It is not the fact that Fantasy is magical *as such* that distinguishes it from SF. It is the fact that it is *sacramental*. Fantasy is supernatural, SF extraordinary, and there is the world of difference between the two. Once we accept that a 'wizard' is a form of priest, we see that there is always a priest in Fantasy. This priestly role is almost always taken (in effect) by a technological artefact in SF.

To sum up: it is the argument of the present Critical History that post-1600 SF has been intimately shaped by this dialectic between 'magic' and 'technology'. Indeed, the subdivisions of the field popular among fans in fact map out positions on the line from magical to technological, 'Hard SF' aligning itself closer to the latter term, 'Soft SF' to the former. The later chapters of this history present the case that the dialectic between 'science and magic' (or 'fact and mysticism' or 'rationalism and religion') actively informs all the major classics of twentieth-century SF: that *Metropolis* or *Dune* or *Star Wars* or Stan Robinson's *Mars* books or the *Matrix* films all articulate precisely this dynamic, and do so for deep reasons connected to the determining history of the genre.

I would like, in this preface, to say one further thing in defence of the scope of this Critical History. Brian Aldiss traced the origins of SF to Mary Shelley's *Frankenstein*; Thomas Disch to Edgar Allan Poe; Patrick Parrinder to H. G. Wells and Jules Verne; Samuel Delany thinks 'there is no reason to run SF too much back before 1926 when Hugo Gernsback coined the ... term' – powerful critics all, who find the sort of proto-science fiction given so much attention in this work incongruous to their definitions of the form. Delany, for instance, has argued that 'More, Kepler, Cyrano ... would be absolutely at sea with the codic conventions by which we make sense of the sentences in a contemporary SF text' and branded 'these preposterous and historically insensitive genealogies with Mary Shelley as our grandmother or Lucian of Samosata as our great-great grandfather' as 'just pedagogic snobbery (or insecurity)' (Delany, pp. 25–6). I have been, in my time, persuaded by these arguments; I am no longer. The fact that Phidias would probably be baffled by the work of Henry Moore does not mean that we can usefully deny that both figures were practitioners of sculpture. To say that Lucian would probably not understand Delany's *Dhalgren* is only to say that forms evolve and change, and that full comprehension requires a certain attention to these changes and evolutions. Science fiction has certainly evolved, but the fact does not deprive its earlier manifestations of a place in the tradition of SF; and it is that tradition that this study seeks to explore, along the lines outlined in the previous paragraph.

It seems to me, having read many critical-historical works about SF in preparation for the present study, that critics exclude SF written before 1927 (or 1870, or 1818, as the case may be) not because there is a coherent rationale for limiting SF to works written after such a date, but rather because the individual critics prefer the later writing, and don't really enjoy so-called 'proto-science fiction'. There's no disputing *gustibus* in such matters, of course; but it is worth stressing that the fact that a particular individual doesn't really enjoy reading Kepler's *Somnium*, or Restif's *La Découverte australe* is not *by itself* a reason for excluding such works from a comprehensive history of SF. More, it is the thesis of this study that present-day SF is still being influenced substantively by the determining cultural dialectic out of which it arose four centuries ago.

I have already mentioned my debt to Clute and Nicholl's *Encyclopedia of Science Fiction*, but it is something that bears reiteration: nobody working in this field can afford to be without that near-miraculous scholarly accumulation of fact and interpretation. I cite it fairly often in the pages that follow, and indeed its influence can be taken to be ubiquitous. Many others have assisted in the creation of this work, and I would like to list them here: Ariel and all at the alienonline.net (where some of the material published here was first aired), Tim Armstrong, Jane Brockett, Andrew Butler, Ria Cheyne, Gabe Chouinard, Samuel Delany, Robert Eaglestone, Malcolm Edwards, Brian Green, Julie Green, Gareth Griffiths, Miriam Jones, David Langford, Roger Levy, James Lovegrove, Nick Lowe, Liam McNamara, Aris Mousoutzanis, Patrick Parrinder, Una O'Farrell Tate, Pam Thurshwell, Andy Richards, Rachel Roberts, Nicola Sinclair, Simon Spanton-Walker and Darko Suvin. Mark Bould read the entire typescript in detail and picked up a plethora of errors, omissions and idiocies: I am deeply grateful to him. Dr Abraham Kawa was invaluable to the project, both in discussion and in detailed feedback on the MS, which he read with an expert eye and a brilliant mind. Gillian Redfern also read the whole thing and provided me with extensive, penetrating and intelligent criticisms. I am also grateful to the anonymous reader employed by Palgrave Macmillan for raising several points.

It is conventional at this point to note that any errors or omissions that remain are my own responsibility, and this is of course true; but in this case it bears a double iteration. A certain level of error is almost impossible to purge wholly from scholarship, but my previous criticism on SF has been marred by a too great degree of mistake and sloppiness. I have endeavoured to do things better in this book, but am only too aware of the manifold insufficiencies and impatience in my own critical intelligence. I would like to ask readers uncovering errors in the pages to follow to let me know via my website: adamroberts.com.

And since we're on the subject of gratitude, let me record that I am not in the least grateful to the British Arts and Humanities Research Board, who, when I applied for research funding that would significantly have eased and facilitated the work on this project, turned me down for reasons which (in accordance with their own policy) they did not divulge. A plague on their house. That this book was ever completed owes nothing to them at all.

My own university education training is in English Literature and the Classics, and I am most comfortable discussing these fields; where I talk about areas of speciality further from my main focus (for instance, French science fiction) I have tried to discuss my conclusions with individuals more expert in these matters than I am. Foreign titles are given in their original forms; all translations from Greek, Latin and French are my own unless specifically stated otherwise. I lack the linguistic facility to attempt translation from other tongues, and have relied here on the work of others. I have transliterated Greek names and titles directly, except where a name is so familiar in an Anglicised form that direct transliteration would tend to confuse (so, 'Lucian' not 'Loukianos'). Where works have variant titles, something common in SF, I cite what I consider the title by which it is most widely known.

References

Davies, Norman, *Europe. A History* (Oxford: Oxford University Press 1996)
Delany, Samuel, *Silent Interviews on Language, Race, Sex, Science Fiction, and Some Comics: a Collection of Written Interviews* (Hanover, NH and London: Wesleyan University Press 1994)

Eco, Umberto, *The Island of the Day Before* (1994; transl. William Weaver; London: Minerva 1996)

Empson, William, *Essays on Renaissance Literature*, ed. John Haffenden, 2 vols (Cambridge: Cambridge University Press 1993)

Godwin, Francis, *The Man in the Moone*, with a modern introduction by Andy Johnson and Ron Shoesmith (Herefordshire: Logaston Press 1996)

Wilkins, John, *The Discovery of a World in the Moone* (1638). A facsimile reproduction with an introduction by Barbara Shapiro (Delmar, NY: Scholars' Facsimiles and Reprints 1973)

1
Definitions

Three definitions

The obvious place to begin a Critical History of science fiction is with a definition of its topic, but this is no easy matter. Many critics have offered definitions of SF, and the resulting critical discourse is a divergent and contested field. One particularly influential approach is that of Darko Suvin (b. 1930), who calls SF

> a literary genre or verbal construct whose necessary and sufficient conditions are the *presence and interaction of estrangement and cognition, and whose main device is an imaginative framework alternative to the author's empirical environment.* (Suvin, p. 37)

Suvin goes on, usefully, to isolate what he calls 'the novum' (plural: nova), the fictional device, artefact or premise that focuses the difference between the world the reader inhabits and the fictional world of the SF text. This novum might be something material, such as a spaceship, a time machine or a communications device; or it might be something conceptual, such as a new conception of gender or consciousness. Suvin's 'cognitive estrangement' balances radical alterity and a familiar sameness, such that 'by imagining strange worlds we come to see our own conditions of life in a new and potentially revolutionary perspective' (Parrinder, p. 4).

The critic and novelist Damien Broderick (b. 1944) has developed and refined Suvin's insights. He notes that the flowering of SF in the nineteenth and twentieth centuries reflected the great cultural, scientific and technological upheavals (he calls these 'epistemic changes') of that era, and pins down with more precise language the strategies employed by the bulk of SF texts:

> SF is that species of storytelling native to a culture undergoing the epistemic changes implicated in the rise and supersession of technical-industrial modes of production, distribution, consumption and disposal. It is marked by (i) metaphoric strategies and metonymic tactics, (ii) the foregrounding of icons and interpretive schemata from a collectively constituted generic 'mega-text'

1

[*i.e. all previously published SF*] and the concomitant de-emphasis of 'fine writing' and characterisation, and (iii) certain priorities more often found in scientific and postmodern texts than in literary models: specifically, attention to the object in preference to the subject. (Broderick, p. 155; my addition)

The writer and critic Samuel Delany (b. 1942) has, on the other hand, challenged the validity of defining SF in terms of its subject matter, and suggests instead that it is 'a vast play of codic conventions' which readers can apply to texts at the level of the *sentence* as much as the level of the text. He suggests that sentences such as 'her world exploded' or 'he turned on his left side' *mean* different things, depending on whether a reader approaches them as SF or ordinary fiction. He suggests: 'most of our specific SF expectations will be organized around the question: what in the portrayed world of the story, by statement or implication, must be different from ours in order for this sentence to be normally uttered?' (Delany, pp. 27–8, 31). For Delany, in other words, SF is as much a *reading strategy* as it is anything else.

Other critics who have attempted definitions (and there have been many) have explored different approaches. Brian Stableford (b. 1948), John Clute (b. 1940) and Peter Nicholls (b. 1939), in their lengthy entry 'Definitions of SF' in Clute and Nicholls' *Encyclopedia of Science Fiction* (2nd edn., 1993) quote sixteen separate definitions, from Hugo Gernsback's in 1926 ('a charming romance intermingled with scientific fact and prophetic vision') to Norman Spinrad's more recent 'science fiction is anything published as science fiction' (Clute and Nicholls, pp. 311–14). There is among all these thinkers no single consensus on what SF is, beyond agreement that it is a form of cultural discourse (primarily literary, but latterly increasingly cinematic, televisual, comic book and gaming) that involves a world-view differentiated in one way or another from the actual world in which its readers live. The degree of differentiation (the strangeness of the novum, to use Suvin's term) varies from text to text, but more often than not involves instances of technological hardware that have become, to a degree, reified with use: the spaceship, the alien, the robot, the time-machine, and so on. The *nature* of differentiation, however, remains debated, some critics defining science fiction as that branch of 'fantastic' or 'non-realist' fiction in which difference is located within a *materialist, scientific* discourse, whether or not the science invoked is strictly consonant with science as it is understood today. This means that faster-than-light travel (impossible, according to contemporary scientific orthodoxy) is a staple of science fiction, provided that such travel is rationalised within the text through some device or technology. A tale in which a character travels from Earth to Mars simply by 'wishing' or 'imagining' it might be defined as 'fantastic' or 'magic realist' rather than strictly science-fictional. On the other hand, few SF texts adhere with complete consistency to the scientific, or pseudo-scientific, logics of their conception. It would, for example, be perverse to deny that Edgar Rice Burroughs' *A Princess of Mars* (1912) is a work of science fiction, and yet the protagonist travels from the Earth to Mars precisely by 'wishing' the journey.

Some critics are comfortable defining as SF a range of texts more normally classified as magic realist or fantastic. In part there has been a reaction to the perceived 'ghettoisation' of SF, by which the literary establishment in America and Europe dismisses texts by category, privileging so-called 'literary fiction' over so-called 'genre fiction' as if the category 'literary fiction' were anything other than a genre, and in many cases ranking 'science fiction' as especially juvenile and valueless, below 'historical fiction' and 'crime fiction' in their notional pecking order. This persistent prejudice does real harm by creating a climate in which it is harder for writers to work and gain recognition, thereby damaging literature in general. Polemic is probably out of place in a Critical History, so we can limit ourselves to observing how perniciously ridiculous these notions are, and (perhaps) humanely pitying the blinkered attitude of literary editors, reviewers and the intelligentsia literature that has been infected by them.[1]

This study has been unable to avoid the often tedious debates concerning 'definition': but my aim is to present an historically determined narrative of the genre's evolution rather than offering an apophthegmatic version of the sentence 'SF is such-and-such'. This narrative is outlined in the chapters that follow, and it sees SF as a specific – and, as it happens, dominant – version of *fantastic* (rather than *realist*) literature: texts that adduce qualia that are not to be found in the real world in order to reflect certain effects back on that world. The specificity of this fantasy is determined by the cultural and historical circumstances of the genre's birth: the Protestant Reformation, and a cultural dialectic between 'Protestant' rationalist post-Copernican science on the one hand, and 'Catholic' theology, magic and mysticism, on the other. Those texts where the latter term predominates are often called 'Fantasy'; those largely or wholly within the aegis of the former are called 'Hard SF'. In between – the majority of texts with which we will have to deal – we find 'SF' as it is broadly conceived. But it is one of the theses of this study that pretty much all the classic texts of SF articulate this fundamentally religious dialectic. In asserting this I am not saying, as some critics have done, that SF embodies 'religious myth' or secularises religious themes. SF may, of course, do either or both of these things, but this is not my argument. My argument is that the genre as a whole still bears the imprint of the cultural crisis that gave it birth, and that this crisis happened to be a European religious one. This is, I think, worth stating unambiguously at the beginning of the study, so that the reader (who may well and profitably disagree with the emphases that follow) can position herself with respect to the argument. No critical history of science fiction could be wholly consensual, and nothing I argue here will please all, or perhaps even many, critics in the field.

The remainder of this chapter will be concerned with a more detailed discussion of some key terms in the definition of SF; specifically 'science' and 'technology'.

The scientific

We need to define the term 'science' as it appears as a modifier in the phrase 'science fiction'. For some critics this is the crucial question when it comes to defining the genre. Brian Aldiss's influential argument that SF 'begins' in 1818 with Mary Shelley's *Frankenstein* (although Aldiss himself lists numerous important ancestors) relies on the assumption that SF could not have originated any earlier than the nineteenth century precisely because it is only in the nineteenth century that 'science' as we now understand the term obtained widespread cultural currency. To quote Peter Nicholls: 'SF proper requires a consciousness of the scientific outlook ... a cognitive, scientific way of viewing the world did not emerge until the 17th century, and did not percolate into society at large until the 18th (partly) and the 19th (to a large extent)' (Clute and Nicholls, pp. 567–8).

'Science' as the term is generally understood means, roughly, a discipline which seeks to understand and explain the cosmos in materialist (rather than spiritual or supernatural) terms; a deductive, experimental discourse characterised by what the German philosopher Karl Popper (1902–1994) called 'falsifiability', whereby the accumulation of empirical data can disprove but never actively prove a theory. Because this version of 'science' is instrumental, it aligns the discourse closely to technology, specifically with the enormous technological advances associated with the Industrial Revolution. This sense of 'science' may explain why nineteenth- and twentieth-century SF is so much more fascinated with items of technology than it is with less 'applied' forms of scientific discourse (mathematics, biology, geography, chemistry, psychology, geology and the like) as such. Of course, there are examples of SF that take the term in this proper sense: Abbot's *Flatland* (1884), for instance, stands at the head of a vigorous little tradition of SF based on mathematical premises. But the great majority of SF written in the nineteenth and twentieth centuries is actually 'extrapolated technology fiction'. In an earlier critical study of science fiction (published by Routledge in 2000) I was quite persuaded by the argument that only a nineteenth-century 'scientific' cultural milieu could meld the constituent generic elements (fantastic voyage, utopia, future-tale, satire, and so on) into 'science fiction' as we understand it. I have since changed my mind. To put it simply, I no longer see why a distinctively modern conception of 'science' need underlie 'science fiction', given that 'science' more broadly conceived as a non-theological mode of understanding the natural world goes back a great deal further than the nineteenth century.

Of course, *something* happened to 'science' in the Victorian age. To be precise, with the nineteenth century's conception of science comes a cultural division into arts and sciences, a perceived separation between what C. P. Snow in his influential 1959 lectures called *The Two Cultures*. Stefan Collini, in an introduction to a recent reprint of Snow's study, points out that the term 'scientist' was first proposed in 1834 along the lines of 'artist':

> the lack of a single term to describe 'students of the knowledge of the material world' had bothered meetings of the British Association for the Advancement of Science in the early 1830s, at one of which 'some ingenious gentleman proposed that, by analogy with artist, they might form scientist' (Snow, p. xii)

This is indicative of the sense, growing in culture through the mid-nineteenth century, that art and science form a binary; and with that there inevitably follows 'the economy of the binary':

> Like all binaries art and science needed to be yoked together (yet held apart) in order to accrue the strengths of their polar positions: soft versus hard, intuitive versus analytical, indicative versus deductive, visual versus logical, random versus systematic ... two things seemed clear (in the mid-19th century): art occupied the domain of the creative, intervening mind, and the scientific ethos seemed to demand precisely the suppression of such impulses ... (Jones and Gallison, pp. 2–3)

The drift of modern mind, informed by this cultural tradition, defines 'science' *in opposition to* 'art', such that science becomes inimical to aesthetics, a lamentable state of affairs for an art like SF which seeks precisely to explore the aesthetics of scientific premises. Taking SF out of the ghetto becomes part of the larger project of breaking down this pseudo-distinction. It seems natural to us; it is inscribed in our educational syllabuses from the earliest schooling and is reinforced by many aspects of culture. But we must bear in mind that it is a nineteenth-century cultural construction rather than a 'natural' state of affairs.

A much fuller sense of the possibilities of the genre is unlocked by taking science fiction back past the nineteenth century and exploring ways in which earlier notions of science informed fiction – to deconstruct, in other words, the logic of cultural binarism that wants to make 'science' and 'fiction' mutually exclusive terms. In fact, it can be asserted that science fiction itself, as a broad statement of aesthetic strategy, has always sought to resist the notion of 'the two cultures'. SF is the place where art and science connect. It is empirical proof that arts and science do not constitute a binary economy.

It helps, in working through the implications of this, to understand how notions of 'science' have shifted in the last century or so. Older theories of science assumed, in an unembarrassed way, that science provides systematic generalisations that explain the truth of the material world. For Bertrand Russell (1872–1970), for instance, scientific method involved a straightforward passage from observation to generalisation, although with 'a careful choice of significant facts on the one hand, and, on the other hand, various means of arriving at laws otherwise than by mere generalisation' (Russell, p. 3). That this definition depends on a rather arbitrary consensual sense of what distinguishes 'scientific generalisation' from 'mere generalisation' is one of its flaws. Another is the belief that data lead by accumulation to water-tight generalisations, or 'truths'. But this rather woolly sense of science was challenged by Popper.

Popper's insight was that science does not produce theories that 'explain' or 'determine' the world, because all scientific theorising is empirically contingent. Any theory can never be proved by observation, it can only be falsified. Observing a thousand two-legged penguins does not *prove* that penguins have two legs; on the other hand, observing a single three-legged penguin *falsifies* the theory. What follows from this is the notion that a scientific theory (for instance, 'that penguins have two legs') is not 'the truth', but rather a contingent explanation for the data as they stand. We can think of this as the positivist definition of 'science'. The American philosopher Robert Nozick (1938–2002) neatly summarised this school of thought, which he called 'the standard model of science' in our post-Popperian culture, although he went on to challenge it on a number of grounds:

> Karl Popper presents an appealing picture of science as formulating sharp theories that are open to empirical testing and to empirical refutation. Scientific theories are not induced from the data, but are imaginative creations designed to explain the data. (Nozick, *Invariances*, p. 103)

This notion of science as 'imaginative creation' is of the greatest interest to the critic and historian of SF, since SF is itself a more thoroughgoing mode of imaginative creation allied to science. One of the most appealing consequences of Popper's position is its unstated implication that SF is *a mode of doing science* (or 'philosophy' more generally conceived) as well as a mode of doing fiction. Not all philosophers of science would find this idea acceptable. Popper himself could see no place for imaginative creation, at least in the sense of 'the innovative, ingenious imaginative leap' that is the currency of SF, in his version of 'science':

> The question of how it happens that a new idea occurs to a man—whether it is a musical theme, a dramatic conflict, or a scientific theory—may be of interest to empirical psychology; but it is irrelevant to the logical analysis of scientific knowledge. (Popper, *The Logic of Scientific Discovery*, p. 31)

One objection to the idea that SF might count as a genuine aspect of science as well as a branch of literature is that fiction, and other such cultural-artistic discourses (such as cinema, TV, the graphic novel and the like), operate according to aesthetic rather than logical-deductive processes. The force of this objection depends on a belief that the *process* of fiction, reading and writing, while occasionally deductive, is more frequently intuitive, metaphoric, metonymic, suggestive, psychological and imagistic. Even the hardest of Hard SF will partake of these 'soft' or aesthetic elements to some degree. But other philosophers of science have pointed out that it is a mistake to reduce 'scientific process' purely to logic. Ernest Nagel (1901–1985), for instance, stresses the importance of analogy to scientific practice: his example is 'the kinetic theory of gases' which is often theorised as if the particles acted 'like billiard balls' (Nagel, p. 110). For Nagel, analogies and hypotheses, while having obvious limitations, nevertheless 'can serve as fruitful instruments of systematic research' (p. 108). Similar modular thinking, whereby a model is constructed of a particular system, 'may be intrinsically valuable because it suggests ways of expanding the theory embedded within it' (p. 117). Many critics have seen SF as a modular system, with fictive 'worlds' modelling reality on a range of different levels, from the practical to the symbolic. Gwyneth Jones (b. 1952), SF author and critic, plausibly brings the whole of SF under the rubric of the experiment: 'the business of the [SF] writer is to set up equipment in a laboratory of the mind such that the "what if" in question is at once isolated and provided with the nutrients it needs. This view of SF,' she adds, 'is not new to science fiction writers and critics, but it is worth restating: the essence of SF is the experiment' (Jones, p. 4).

A fuller perspective on the role of science in SF can be obtained via the work of the American philosopher Paul Feyerabend (1924–1994). His book *Against Method* (1975) is a powerful polemic against 'method' in science. The best way to do science, says Feyerabend, is anarchically – 'anarchism, whilst perhaps not the most attractive *political* philosophy, is certainly excellent medicine for the *philosophy of science*', he says. Scientific rules limit possible advances in science: 'the only principle that does not inhibit progress is: *anything goes*'. Feyerabend proposes a free-for-all proliferation of scientific theories, even though some – or perhaps many – will be kooky, mystical, daft or unpalatable. However odd these theories get, Feyerabend is sure that in their interaction better and better models will emerge, better and better science will be practised. The alternative, he says, is to propose uniformity, a situation in which the powers-that-be manipulate consensus by force. This is rather close to the situation that presently obtains in science: scientists that advocate telepathy, alien abduction, the power of crystals and the like are frozen out of the scientific community by a mix of ridicule, cold-shouldering and the financial penalties of being unable to raise funds to pursue their research. Increasingly, the only way to obtain funding is to work within the accepted frameworks. Feyerabend argues that a 'proliferation of theories is beneficial for science, while uniformity impairs its critical power. Uniformity also endangers the free development of the individual' (Feyerabend, p. 5). So, for example, conventional science was not apprised of the environmental dangers of technological advance; awareness of such issues was raised by groups outside science: 'Green' political advocates, New Age enthusiasts and cranks of all sorts. And yet such figures have been vital in broadening useful debate on global warming, the environmental impact of technology, carbon economy; all things that 'science' now takes seriously. Feyerabend says:

> Non-scientific procedures cannot be pushed aside by argument. To say: 'the procedure you used is non-scientific, therefore we cannot trust your results and cannot give you money for research' assumes that 'science' is successful and that it is successful because it uses uniform procedures. The first part of this assertion is not true, if by 'science' we mean things done by scientists – there are lots of failures also. The second part – that successes are due to uniform procedures – is not true because there are no such procedures. Scientists are like architects who build buildings of different sizes and different shapes and who can be judged only after the event, ie after they have finished their structure. It may stand up, it may fall down, nobody knows. (Feyerabend, p. 2)

Against Method is a polemic rather than a manifesto for change in science, and it is perhaps hard to see how his ideas might be put into practice (grant-awarding bodies, after all, do need *some* criteria to determine who gets research money and who doesn't, there being many more applications than money to fund them). And yet it is the case that there does exist a space where the sort of 'science' Feyerabend is proposing already takes place; where brilliantly unorthodox thinkers bounce ideas around regardless of how strange they seem at first; in which experiments are conducted and blue-sky research undertaken. This space is called science fiction. Although he makes no mention of literature, Feyerabend's perspective includes, implicitly, the notion that SF is a crucial component of science as well as of culture. Research councils may rarely give money for the study of interstellar colonisation, time travel, extrasensory perception, mutant cactuses or virtual reality; but publishers *will* give money if the 'research' (which is to say, the novelisation) is good enough. Steven Hawking's *A Brief History of Time: from the Big Bang to Black Holes* (1988) is a dull historical account of things that have already happened in science and some cautious speculation about things for which Hawking lacks empirical data. On the other hand, Will McCarthy's novel *The Collapsium* (2000) is a riveting account of how science might be, or will be, or ought to be. McCarthy imagines black holes not as highly compressed stars, but as very heavy elementary particles. His protagonist manages to assemble these particles into the material after which the novel is named, and from that wonderful Feyerabendian scientific experiment all sorts of fascinating things follow, including but not limited to plausible faster-than-light travel.

A Feyerabendian sense of the genre 'science fiction' would be alive to the fluid possibilities of the genre in a way that the (still widespread) older notion of science as a discourse with a special relationship to 'the truth' does not. To return to Russell's 1931 book on *The Scientific Outlook* for a moment. After elaborating the many advantages of a scientific outlook, Russell moves on to propose 'scientific world government' as a radical solution to the ills of the day. This government, he says, 'will embrace all eminent men of science except a few wrong-headed and anarchical cranks' (Russell, p. 193) (a qualification which speaks to the essentially conformist and coercive nature of 'scientific discourse' as Russell understands it). This scientific government, he continues,

> will possess the sole up-to-date armaments, and will be the repository of all new secrets in the art of war. There will, therefore, be no more war, since resistance by the unscientific will be doomed to obvious failure. The society of experts will control propaganda and education. It will teach loyalty to the world government, and make nationalism high treason. The government, being an oligarchy, will instil submissiveness into the great bulk of the population, confining initiative and the habit of command to its members. (Russell, p. 193)

This distinctly unappealing picture is, although Russell does not say so, science fiction. It owes much to H. G. Wells, and looks forward to Aldous Huxley's *Brave New World*, which was published the following year ('a life of easygoing and frivolous pleasure may be provided for the manual workers ...', Russell, p. 211). Russell's book, in other words, is an example of philosophy *as* SF. Russell is quite aware of the fact that in his vision 'features that everybody would consider desirable are mixed with features that are repulsive' (Russell, p. 214). Indeed, the point of this work, for our purposes, is that it stands as an example of the extrapolation of this older, scientific logic to its ideological conclusions. This is a vision of science as oppressive dogma, a mode of social domination, which frequently finds expression in science fiction. Feyerabend's version of science, which specifically privileges the very 'cranks and anarchists' that Russell dismisses, has by far the greater potential.

The technological

According to the respected SF author and critic Theodore Sturgeon (1918–1985), 'the word "science" derives from the Latin *scientia*, which means not method or system but *knowledge*. The concept of SF as a "knowledge fiction" satisfied me completely' (Sturgeon, p. 73). Sturgeon prefers this phrase, because it allows him to include, for instance, *The Lord of the Flies* in the SF category 'because of its profound investigation of the origins of religion and secular power in a human society'. The oblique snobbery of such redefinition depends on a buried sense that conventional definitions of SF exclude 'proper' literature (*Brave New World, Nineteen Eighty-four, Gravity's Rainbow* and the like), leaving the genre with the dregs of populist, Pulp and adventure yarns – a snobbery common to many SF intellectuals and academics, and not without a rationale. But the roots of it as a prejudice are, philosophically, very revealing. And philosophy is the key context here: 'philosophy' (from the Greek, meaning 'love of wisdom') has had its turn as a word for what we nowadays call 'science', particularly as 'natural philosophy'.

The crucial distinction here is not between 'science' and 'knowledge', but between 'science' and 'technology'. These two words are often taken together, with the latter seen as a specific example of the former. According to *Chambers Dictionary of Science and Technology*, technology is 'the practice, description and terminology of any of the applied sciences which have practical value and/or industrial use' (Walker, p. 1150). But in fact this distinction uncovers a split at the very root of the discourse within which 'science fiction' (among many other things) needs to be oriented. The definition of science evoked in Walker's particular reference work ('the ordered arrangement of ascertained knowledge, including the methods by which such knowledge is extended and the criteria by which its truth is tested', Walker, p. 1021) draws out the emphasis on 'truth', 'knowledge' and 'order'. Which is to say, 'science' becomes a more or less restrictive idealist philosophical framework, restrictive (as most scientists assert) by the nature of things 'out there'. Technology, on the other hand, is the discourse of tools and machines, 'tools' being extensions of the human worker, like hammer and saws, and 'machines' being devices that stand apart from the human worker. Friedrich Engels (1820–1895) was one of the first to make this distinction between the tool and the machine, and did so by way of articulating what he saw as the nature of the Industrial Machine, which tends to 'alienate' humanity from its own labour. But, taken conceptually, we find tools and machines at the core of most science fiction: such that spaceships, robots, time-machines and virtual technology (computers and virtual realities) are the four most commonly occurring tropes of the field: which is to say, Suvin's novum is almost always technological in form. There are nova of a more conceptual or 'scientific' nature, of course; but it is rare for these to be wholly uninvolved with technology. Ursula Le Guin's conceptual novum in *The Left Hand of Darkness* (1969) postulates an alien people without fixed gender, but her novel also includes a series of technological nova, among them the 'ansible' (a faster-than-light communications device) and a spaceship. Christopher Priest's *Inverted World* (1974) presents us with a striking 'science' fictional tale, a case of upended scientific logics, a city whose inhabitants live not (as we do) in a finite world located within an infinite universe, but in an infinite world within a finite universe. Nevertheless the narrative resolves itself back into 'technology fiction' at the end, with the apparent nature of the world revealed as a function of the particular energy technologies that power the motile city at the centre of the book.

Despite the genre's reliance on technology, and despite the many brilliant effects that machines and tools can achieve within the aesthetic framework of an SF text, there remains a certain bias. 'The novel of ideas' has traditionally been privileged over the instrumental novel of the machine, in the same way that 'real fiction' (meaning a particular sub-genre of 'mainstream, literary fiction') is privileged over science fiction by the literary establishment. It is only relatively recently, in philosophical terms, that discourses have been developed to allow us to challenge this prejudice.

One of the most influential philosophical interventions in the question of 'technology' is the 1953 essay 'The Question of Technology' by the German philosopher Martin Heidegger (1889–1976). Heidegger takes the word back to its Greek roots: 'from earliest times until Plato the word *technê* is linked to the word *epistêmê*', but from Plato and Aristotle onwards a distinction begins to be made between them (Heidegger, pp. 318–19). Ἐπιστήμη (*epistêmê*) is the Greek word for 'knowledge' (it is the root of the English word 'epistemology'), and by extension it means 'finding things out about the universe' in an open-ended, dialectical manner; which is to say, it means 'science'. Τέχνη (*technê*), on the other hand, the root of the word 'technology', means 'a specific skill or ability', the knowledge of how to make something, and is used, by extension, to mean 'cunning devices, arts, wiles'. English has a similar complex of implication in the word 'artificial', which means both 'the work of an artificer or artist' (where 'art' has positive implication) and something suspect, *ersatz*, less-than-real. Fifth- and fourth-century BC Greek thinkers divided these two forms of 'knowledge': Plato and Aristotle reserved 'episteme' to themselves, and dismissed 'techne' as the trick of the unethical, rhetoric-rather-than-truth 'Sophists'. According to Bernard Stiegler:

> The separation is determined by a political context, one in which the philosopher accuses the Sophist of instrumentalizing the *logos* ['truth', 'knowledge', 'the underlying order of things'; *logos* also means 'the word'] as rhetoric and logography, that is, both as an instrument of power and a renunciation of knowledge ... It is in the inheritance of this conflict – in which the philosophical *episteme* is pitched against the sophistic *technê*, whereby all technical knowledge

is devalued – that the essence of technical beings in general is conceived. (Stiegler, p. 1; my gloss)

By 'instrumentalizing the *logos*', Stiegler means that the Sophists were accused of turning 'truth' into an instrument, of being amorally concerned with the means rather than ends. As this distinction is traced down the centuries of philosophical tradition, we can see that 'techne' becomes associated with an emptying out of meaning and validity. For example, Stiegler quotes Edmund Husserl's assessment that 'algebra' is the 'emptying of meaning' from 'the actually spatio-temporal ideal-ities' of geometry, constituting 'a mere art of achieving results, through a calculating technique according to technical rules' (Husserl, *The Crisis of the Universal Sciences and Transcendental Phenomenology* (1970), quoted in Stiegler, p. 3).

Heidegger's essay challenges, and indeed overturns, this understanding of 'tech-nics'. For him technology is not an instrument, it is a mode of knowing, 'a mode of revealing ... where *alêtheia*, truth, happens' (Heidegger, p. 319). Far from seeing technology as merely the 'practice of science', Heidegger argues that science is in fact a function of technology. He means this not only in the sense that 'modern physics, as experimental, is dependent upon technical apparatus' (Heidegger, pp. 319–20), although this is of course true. He means rather, in the words of Timothy Clark, that technology 'is not the application of science. There is not the-ory on the one side and its practical implementation on the other. Rather science is one manifestation of the technological stance towards entities' (Clark, p. 37). Heidegger thinks that technology, from windmills to hydroelectric plants, 'enframes' the world in a certain way, allowing or shaping the ways in which we 'know' the world around us.

It may be that technology encourages us to think of the world only as what Heidegger calls 'standing-reserve', a quantity of raw material to be harnessed; and indeed it is possible to take Heidegger's essay as a statement of hostility to the increasing pace of technological change (politically conservative, Heidegger declared his preference for windmills over hydroelectric plants, and indeed felt physically sick in modern cities 'surrounded on every side by mechanization and regimented space', Clark, p. 36). But this is not what 'The Question of Technology' is saying. As a mode of knowing, of enframing, the world, technology is 'not some-thing fundamentally new or even modern. Rather it fulfils Western Philosophy's oldest desire for knowledge of what is real' (Scharff and Dusek, p. 247). Heidegger's undoubted hostility to much modern technology was based not on the fact that it was technology as such, but rather on the peculiarly Heideggerian question of whether it is likely to make us feel 'at home' or not.

Nevertheless, it is Heidegger's insight into the way technology 'enframes' the world for humanity that makes him a crucial (though admittedly unlikely) figure to bring into a discussion of the definition of science fiction. In another essay, 'What Calls for Thinking?' (1954), Heidegger famously, perhaps notoriously, declared 'Science does not think' (Heidegger, p. 373). What he meant by this (and he conceded in the essay that 'this is a shocking statement') was that science does not 'enframe' in the way that technology does. Science fiction, on the other hand,

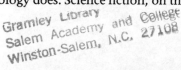

does think: not only in the sense of rehearsing a great many concepts, possibilities, intellectual dramas and the like, but in this deeper sense of textually enframing the world by positing the world's alternatives. We could say (to adopt Heidegger's idiom) that science does not think *except in science fiction*; but this is actually only a way of saying something simpler: that SF is actually technology fiction in this Heideggerean sense.

It seems perverse to say it, but perhaps it is Heidegger who represents the best starting-point for a thoroughgoing theorisation of 'science fiction'. Heidegger's most famous philosophical work centred not on questions of technology, but on the issue of 'Being', the ontological condition of humanity. Bernard Stiegler, in his complex ongoing theoretical study *Technics and Time*, has set out to revise Heidegger's philosophy of *Dasein*, or 'Being', to allow certain technological objects (he is a little obscure as to which precisely) access to the same authentic Being-in-the-World that characterises human beings. Heidegger distinguishes between the existence of a creature like a man (*Dasein*) and the existence of an object which we categorise solely in terms of its use (*Zuhandenheit*). Yet according to Stiegler this denigration of the 'technical object' becomes less and less tenable in a world in which the technological not only interpenetrates human life at almost every level, but in which such objects themselves move further from the sort of dumb instrumentality that characterises a spade or a pair of glasses, and closer to the thinking-machine and the self-aware object. On the other hand, no machine in the present world is truly self-aware. To speak more precisely, the place where Stiegler's technological *Dasein* actually obtains is science fiction itself. One of the key themes of SF for the last half-century has been precisely to delineate and explore the place where the technical object achieves *Dasein*, a Being-in-the-World and a Being-towards-Death. Neither a chair, a typewriter nor a thermostat can have 'authentic' Being in the sense that Heideggerans, or existential philosophers, use the word: but Asimov's robots all possess precisely this quality.

It can be argued, and with some justification, that SF has rarely followed through the possibilities that this philosophical state of affairs has afforded it: that when the 'technical' has been introduced, it has more often than not been to denigrate it. Stiegler considers the newest technologies of genetic manipulation, concluding that 'they make imaginable and possible the fabrication of a "new humanity"':

> without having to dive into science-fiction nightmares, one can see that even their simple current applications destroy the oldest ideas that humanity has of itself – and this, at the very moment when psychoanalysis and anthropology are exhuming the constitutive dimension of these ideas, as much for the psyche as for the social body ... [technology is] for the first time directly confronting the very form of this question: what is the nature of the human. (Stiegler, p. 87)

Donna Harraway is the most famous cultural critic to celebrate the possibilities of this technological reinvention of the category 'human' in terms of its diversity and possibility, as well as insisting on the increasing relevance of talking in terms of 'the inextricable weave of the organic, technical, textual, mythic, economic and political threads that make up the flesh of the world' (Harraway, in Gray, p. xii). Like Harraway, Stiegler argues that 'the human is a technical being that cannot (merely) be characterised physiologically and specifically (in the zoological sense)' (Stiegler, p. 50); although, unlike Harraway, Stiegler's emphasis is on ontology, rather than on the many technical prostheses that augment contemporary life as such. Similarly, with regard to culture and society he is adamant that 'the technical dynamic *precedes* the social dynamic and imposes itself thereupon' (Stiegler, p. 67). In both cases, it is a *'technical* fiction' rather than a 'science fiction' more generally conceived that is able to penetrate to the root of things.

Machines today are redefining the human; and yet the dominant story-thread of twentieth-century mainstream SF has been precisely how machines *return* to humanity, how their developmental trajectory brings them back into discourses of humanity. Asimov's story 'The Bicentennial Man' (1976) is a core fable in this regard. After decades of robot stories in which he used the trope of 'the robot' as a means of exploring aspects of humanity, Asimov finally wrote a story about a robot literally turning himself into a human being (his own assessment of the story is that 'of all the robot stories I ever wrote [it] is my favourite and, I think, the best', Asimov, *The Complete Robot*, p. 603). Andrew Martin begins the story as a metal creature with a positronic brain, whose being is entirely determined by the 'three laws of robotics' for which Asimov is famous. A flaw in his programming makes him creative (a flaw erased by his manufacturers in all subsequent robots), and during his lifetime he accumulates money through royalties earned on his art, enabling him first to buy his freedom, then to have the metal portions of his body replaced with organic ones, and finally to petition the Legislative Establishment to have himself legally recognised as human. Public opinion makes this impossible, despite Andrew Martin's egregious virtue, until he instructs a surgeon to make one last adjustment: 'decades ago, my positronic brain was connected to organic nerves. Now, one last operation has arranged that connection in such a way that slowly – quite slowly – the potential is being drained from my pathways' (Asimov, *The Complete Robot*, p. 680). By dying, the robot sways public opinion; on his 200th birthday he is declared human, and dies. By taking on human weakness the machine is able to take on Being-towards-Death, and this defuses human fear of the machine. We see this same archetypal narrative structure in a great deal of science fiction: the character of the android Data in *Star Trek: the Next Generation* who yearns, Pinocchio-like, to become human is never challenged in his strange desire. Robot stories can be traced back to fables in which automata are mistaken for human beings such as Hoffman's 'Der Sandmann' (1816) or J. Storer Clouston's *Button Brains* (1933), the point of such tales being the transfer from a machinic to a humanitarian ethic and logic.

The demonisation of the machine is a continuing aesthetic SF strategy: Gregory Benford's 'Ocean' series of novels, beginning with *In the Ocean of the Night* (1977), postulates a galactic conflict between organic life and a brutalising inorganic machine race. The narrative arc of the first 100 *Perry Rhodan* novels (1961–71) pit the 'peace lord of the universe' against the malign 'robot regent' of the planet

Arkon. The Star Trek franchise has returned many times to the machinic villains named 'The Borg'. The vastly popular *Matrix* films pit organic life in a massive, violent war against 'the machines'. And so on through a thousand examples, with only a few SF authors of merit positing the opposite line (Greg Egan is, perhaps, the most eminent of these).

Why this bias? In philosophical terms, the machines are seen as inherently less authentic than organic life because they fall under the rubric of *techne* rather than *episteme*; it is this rhetoric that governs the devaluing of the machinic. 'Good' means amenable to humanisation, like Asimov's saintly Bicentennial Man; 'bad' means resistant to this process. More recent SF has been bolder in deconstructing this notion, with a range of cyberpunk and other texts exploring the validities of a technological perspective, but the bulk of the genre reproduces the ancient bias.

A number of more recent theorists and philosophers have published work which provides a way out of this constrictive dilemma, but it has achieved a lesser cultural penetration than it deserves. The thought of the French philosopher Michel Serres (b. 1930) challenges the very notion that two separate cultures, such as 'art and science', actually exist. Serres' focus is always on the connections between the various discourses of science, culture and thought, and although these connections are not straightforward ('from the sciences of man to the exact sciences, or inversely, the path does not cross a homogeneous and empty space ... it follows a path that is difficult to gauge') they are crucial. As he apophthegmatically puts it, 'criticism is a generalised physics' (Serres, p. xi). Neither science, art nor religion is composed of 'facts' or 'predicates'; instead they are determined by the complex dynamics of interrelations. Science fiction is the proper literature for this, and Serres has published a critical monograph on at least one SF author (*Jouvences: Sur Jules Verne*, 1974). A much more influential French thinker, Gilles Deleuze (1925–1995), published a wide range of philosophical texts attacking essentialism and revisioning reality wholly in terms of 'machines' – 'desiring machines' and machines productive of all manner of flow (interrupted by various sorts of 'interruption' machine) as a means of replacing the older atomic or 'monadic' bias of the western scientific tradition. For Deleuze this machinic rhizomatic 'becoming' is something enthusiastically to be celebrated. The opening of one of his most famous books, *L'Anti-Oedipe* ('The Anti-Oedipus', 1972, co-written with Félix Guattari) revels in the machinic cosmos of 'desiring production', the strenuously joyful functioning of desire itself:

> It is at work everywhere, functioning smoothly at times, at other times in fits and starts. It breathes, it heats, it eats. It shits and fucks. What a mistake to have ever said *the* id. Everywhere *it* is machines – real ones, not figurative ones: machines driving other machines, machines being driven by other machines, with all the necessary couplings and connections. An organ-machine is plugged into an energy-source-machine: the one produces a flow that the other interrupts. The breast is a machine that produces milk, and the mouth a machine coupled to it. (Deleuze and Guattari, p. 1)

This is the fluid ecstasy of the modern world; and which literature is better placed to apprehend it than SF?

'In real life' and 'in SF'

My own training and biases as a critic have left me suspicious of binaries, and I worry that precisely such a binary model has emerged from this chapter of definitions. Any distinguishing of 'realist' and 'science-fictional', of course, occurs under the sign of erasure, as it were; and reading texts through these notional categories happens always with a sense of the ways in which the two terms bleed into one another, the ways in which SF writers utilise realist strategies, and 'realism' itself, are always contingent on the sorts of imaginative and speculative constructions that characterise SF. The same is true of the blurred binaries 'art/science', 'romance/the novel' and 'science/technology'; in each case there is no prior term, and the interplay between the categories must be understood as fully dialectic and in process. But in this chapter I have not, I concede, quite shaken off the dust of one of these binaries, and I want to finish by acknowledging my bias. It has to do with the different understandings of the 'science' that underpins science fiction, and the sorts of fictions that result from them.

A shorthand for this binary, although not a very satisfactory one, might be 'Hard SF' versus 'Soft SF', a distinction often made by SF fans themselves. More precisely, we might say, it is the difference between the science in science fiction deriving from the rigid, Russellian notion (with correlatives of 'truth' and 'correctness'), and the science in science fiction deriving from the anarchical Feyerabendian sense of the term (with correlatives of 'imaginative intellectual play' and 'extrapolation'). My preference as a reader and writer is for the latter. However, many SF writers and fans take a particular pleasure in the correctness of the science of science fiction, 'correct' here being understood as 'not transgressing the laws of science as they are presently understood'. Gwyneth Jones asks: 'Does it matter if the science is sound? The fantasy fanciers will say no, the SF faithful will say yes.' She goes on to point out that Larry Niven's Hugo and Nebula Award-winning *Ringworld* (1970), 'one of the great, classic "engineering feat" SF novels, reached print in the first instance with terrible mistakes in its science', and that Niven, 'as free as any SF novelist alive from moral qualms about social verisimilitude or cultural relativity, acquiesced to the helpful advice he received from Dyson Sphere buffs, and obediently corrected his fantasy for later editions' (Jones, p. 16).

The shibboleth here is consistency, and one problem with its application is that fans tend to overlook substantive transgressions of scientific orthodoxy (spacecraft that can travel faster than light) while becoming agitated about minor features (the mechanism by which Niven's 'ringworld', a massive ring of habitable land circling a star, is kept precisely in its orbit). This inconsistency in applying, precisely, criteria of consistency reveals an ideological ground; for only an ideological belief in science as 'truth' can sanction the sort of misprision necessarily perpetuated by this sort of analysis. Another example: Robert Lambourne, Michael Shallis and Michael Shortland analyse various SF texts that deal with centrifugal forces. Space

habitats or spacecraft that are spun to give the illusion of gravity in a free-fall environment are a popular recourse of the SF text, in part because such centrifugal
environments avoid the need for the 'pseudo-science' artificial gravity. Lambourne
et al. discuss the way the Coriolis effect, created by the constant rotation, would
determine life inside such an environment:

> In the short story 'Small World' (1978), by Bob Shaw, for example, a projectile
> is described as travelling across a cylindrical space habitat along an S-shaped
> trajectory. In fact, the reversal of the Coriolis force after the projectile passes the
> midpoint of its course and starts its descent, means that the path is C-shaped
> when viewed from the drum, as shown in figure 5(b). (Lambourne et al., p. 55)

The category error here is the 'in fact'. A story is not 'fact'; nor does fictional entry
into one or other discourse of science render it so. Application of conventional
scientific orthodoxy as a criterion of judgement for an aesthetic object is fundamentally foolish even when applied with absolute consistency; and when applied
inconsistently, as it often is (swallowing the camel of faster-than-light travel but
straining at the gnat of, for instance, S-shaped ballistic trajectories inside spinning
environments) it combines deadness with muddle. Our choice is between a textual
universe run along the oppressive lines of Russell's scientific world government, or
a science fiction that plays anarchically with 'science' along the lines Feyerabend
suggests. This seems to me no choice at all.

And yet there is *something* in Lambourne's 'in fact'. A personal anecdote: I sat in
a cinema audience in Aberdeen when the film *Star Trek III: The Search for Spock* was
first shown in that city in 1985. In the film the Federation starship *Enterprise* has
been stolen by its former captain, Kirk, so that he and a few of his friends can go
on an unauthorised search for his colleague Spock, who is believed dead. This crew
of paunchy old geezers finds Spock's rejuvenated body on an artificially created
'Genesis' planet. But they have been followed through space by a band of maraud-
ing and violent Klingon warriors who challenge the ship to a space-duel, even
though the Klingon craft is a tiny fighter and the *Enterprise* a massive starship. As
it happens, because the *Enterprise* is without its usual complement of crew, it is
extremely vulnerable (though the Klingons do not realise this). The Klingons fire,
and with one shot they disable the *Enterprise*. At this moment in the film, with a
shot on screen of the Klingon ship positioned in space directly in front of the
Enterprise, I heard somebody behind me in the cinema stage-whispering to his
companion: 'Dear me, no, of course that's a Klingon D7 pseudo-fighter; it doesn't
fire disruptor bolts like that. In real life this confrontation couldn't really happen.'

'In real life.' We are familiar with the idea that films reflect 'real life' poorly. On
screen the good guy always wins the girl, we see the threatened disaster averted in
the nick of time, the bad guy getting his comeuppance, and in each case we are
aware of the fact that in the world we actually inhabit these things do not often
happen that way. But the consensus 'real life isn't like that' is usually applied to
films that mimic our actual existence; the sentence 'real life is nothing like *Notting
Hill* or *Die Hard* or *When Harry Met Sally*' is one form of locution. The sentence 'real

life is nothing like this scene in *Star Trek III: The Search for Spock*' is quite another, and the difference between them is instructive.

As a statement to the effect that actual life bears no resemblance to the special effects, future-world space battle of this particular film, the sentence is strictly accurate: 'in real life, the primitiveness of contemporary space technology and the non-existence of alien races means that no such space battle is possible'. But it is clear that the speaker did not mean the words in that sense. He meant 'this cinematic representation of a battle between a Federation cruiser and a Klingon pseudo-fighter does not map accurately onto the *reality* of such a battle'. What might this reality be, as far as this individual is concerned?

To answer this question is to excavate a little the cultural phenomenon of *Star Trek*, and of Fandom more generally. Fans are integral to the way contemporary SF operates: numerous fan-created magazines, websites and conventions generate much of the energy on which the continuing vitality of the genre depends. Yet the 'fan', and especially the 'science fiction fan', has a very low cultural currency today. He or she exists in a cultural climate of low-level ridicule and dismissal; thought of as obsessive cultists, unskilled at social interaction, physically unattractive and unhygienic, outsiders, nerds; to instance a cultural icon with whom many people will be familiar, the comic book store owner in *The Simpsons* cartoon series. Behind all this negative social construction (which, as with any derogatory stereotype, relates less to reality and more to prevalent ideological fascinations and anxieties) is the twofold baseline perception: that fans are *'fanatical'* (the former term, of course, derived originally from the latter) in some dangerous sense; and that fans are *passive* receptacles of consumer culture.

The American critic Henry Jenkins (b. 1958) has done more than anybody else to overturn this cultural stereotype. His breakthrough study *Textual Poachers: Television Fans and Participatory Culture* (1992), working largely with the example of *Star Trek* fans, demonstrated that fans, far from being passive, are often extremely active, both in proselytising for their favourite shows and in terms of textual production – re-appropriating material from those shows, writing their own fiction and producing their own art (often in 'slash zines', in which two favourite characters are placed in erotic congruence, their names separated by a slash: 'Kirk/Spock', for example). Jenkins shows the extent to which fans are *creative*, active participants in the textual universes of their favourite shows.[2] Jenkins' liberating analysis not only critiques the lazy stereotyping tendencies of modern society; it opens up the 'fan' as a crucial category for any analysis of SF. The important thing about fans is that they care, and they care in an active, engaged and creative way. They care (as in the example of *Star Trek III*) about consistency; about production values; about the quality and range of the texts available to them. They champion the works they admire, and often they strive directly to involve themselves in that work. Naturally, this enthusiasm can slide into cliquishness, in which schoolyard shibboleths are used to determine who is 'us' and who 'not us'. Moreover, I suspect that few people who have spent time with fans at conventions and elsewhere will disagree when I describe this sort of siege mentality, which can spill into tribal obstinacy or paranoia. But the fundamental point is that fans love

SF, and love is not an emotion to be treated lightly. Most SF authors working today (I'm tempted to say *all*) began as fans, and many continue as fans. Science fiction is a community, not an elite. Fans more often than not embody a huge, detailed and working knowledge of their genre, and can locate new texts within a framework of intertextual reference and connection with impressive facility. And the trope of 'the fan' embodies not only actual humans who follow SF, but the position of the new SF text (novel, film) in respect of the whole genre, and – as I have been arguing – in an ideal sense the relationship (active, engaged, creative) between 'SF' and science that underpins the definition of the genre this chapter has sought to sketch.

Conclusion

The three definitions of SF cited at the beginning of this chapter remain extremely useful for scholars in the field, despite the tendency of some critics to nibble away at them. This chapter has sought not to replace Suvin's, Delany's and Broderick's definitions, but to go a little deeper into some of the assumptions underlying 'science fiction' as a piece of terminology. My conclusion is that SF is better defined as 'technology fiction' provided we take 'technology' not as a synonym for 'gadgetry' but in a Heideggerean sense as a mode of 'enframing' the world, a manifestation of a fundamentally philosophical outlook. As a genre, therefore, SF textually embodies this 'enframing', taking as its 'standing reserve' not only the discourses of science and technology, but also the whole backlist of SF itself, the intertextual tradition that this study will go on to examine. To the extent that SF enters into the discourse of 'science' (as it very frequently does) the best way of theorising this is as a Feyerabendian proliferation of theories rather than a notional uniformity or 'truth'. The useful shorthand for this is 'Fortean', from the writings of the American journalist and writer Charles Fort (1874–1932). This pluralism, and range of speculative possibility, frees SF from what Heidegger saw as the danger in technological 'enframing', the way in which 'it banishes man into the kind of revealing that is an ordering. Where this ordering holds sway, it drives out every other possibility of revealing' (Heidegger, p. 332). In this philosophical sense, SF must be a disorderly technology fiction.

I should perhaps add that many readers of SF will not recognise the genre from my description here. 'Technology fiction' is most often taken as precisely the bland, gadget-driven narratives I say it should not be: 'Hard SF' either as machine or cosmological fiction – stories about spacecraft, weapons, prostheses, or about the universe as physics presently understands it, in which an iron rule of 'truth' applies. 'Soft SF', on the other hand, is given more leeway by readers. By what strange logic 'techno' fiction finds itself falling back against this untested and ultimately Platonic absolute 'truth', and 'science' fiction finds itself able to explore the imaginative possibilities of human thought untrammelled by such concerns is not immediately clear. My belief, although it is not one I hold dogmatically, is that this division is explicable in the context of the historical development of science fiction itself. As outlined in the Preface (and as elaborated in much greater detail in the whole of this book) I take modern SF to arise at the cleavage of what I call broadly 'Catholic' and 'Protestant' fictive worldviews, a separation I date from around the turn of the seventeenth century.

I should be clear here: I am very specifically *not* saying that science fiction is exclusively a Protestant, and 'Fantasy' exclusively a Catholic, literature. There are very many great Catholic science fiction writers, and many great Protestant Fantasists, and increasingly (although only since the late twentieth century) a very great many excellent SF and Fantasy writers who come from neither cultural milieu. Rather, I am suggesting that, speaking historically, SF expresses a particular dialectic determined originally by the separation of 'Protestant' and 'Catholic' world-views (or if one prefers less sectarianly charged terms, between 'deism' and 'magical pantheism') that emerged in the seventeenth century. SF texts mediate these cultural determinants with different emphases, some more strictly materialist, some more mystical or magical. But without an understanding of the broader historical context many aspects of the tradition of SF are incomprehensible.

I think this explains why a Catholic writer like Jules Verne limits his science fiction to technological devices, where a Protestant writer like H. G. Wells expands his vision in speculative and universal directions. It seems to me (to mention three eminent Catholic writers of the genre) that the 'mystical' turn in SF, the introduction of 'magic' (as in Blish's *Black Sunday*), of God (as 'The Outsider' in Wolfe's *Long Sun* tetralogy) or of miracles (in Miller's *Canticle*) index an impulse to mark out that place where technology ends specifically *as a magic*, mystic area; the God of the Gaps of which the philosophers sometimes speak. Protestant traditions, such as produced writers like Stapledon, Heinlein or Robinson, are less respectful of the veil of the temple, and produce a different more fully scientific or knowledge fiction. In other words, I am suggesting that it is the mystical, the quasi-religious, that 'enframes' in the limiting and 'ordering' manner deplored by Heidegger: a discourse that insists upon one and only one interpretation of the cosmos.

References

Asimov, Isaac, 'The Bicentennial Man' (1976), in *The Complete Robot* (London: Grafton 1982)

Broderick, Damien, *Reading by Starlight: Postmodern Science Fiction* (London: Routledge 1995)

Clark, Timothy, *Martin Heidegger* (London: Routledge 2002)

Clute, John and Peter Nicholls, *Encyclopedia of Science Fiction* (2nd edn., London: Orbit 1993)

Delany, Samuel, *Silent Interviews on Language, Race, Sex, Science Fiction, and Some Comics: a Collection of Written Interviews* (Hanover, NH and London: Wesleyan University Press 1994)

Deleuze, Gilles and Felix Guattari, *The Anti-Oedipus: Capitalism and Schizophrenia* (1972; transl. Robert Hurley, Mark Seem and Helen R. Lane, London: Athlone Press 1984)

Doody, Margaret Anne, *The True Story of the Novel* (New Brunswick, NJ: Rutgers University Press 1996)

Feyerabend, Paul, *Against Method* (1975; 3rd edn., London: Verso 1993)

Frye, Northrop, *The Secular Scripture: a Study of the Structure of Romance* (Cambridge, MA: Harvard University Press 1976)

Gray, Chris Hables (ed.), *The Cyborg Handbook* (London and New York: Routledge 1995)

Heidegger, Martin, 'The Question of Technology' (1953), in David Farrell Krell (ed. and transl.), *Martin Heidegger: Basic Writings* (London: Routledge 1993), pp. 311–41

Jones, Caroline A. and Peter Gallison, with Amy Slaton (eds), *Picturing Science and Producing Art* (New York and London: Routledge 1998)

Jones, Gwyneth, *Deconstructing the Starships: Science, Fiction and Reality* (Liverpool: Liverpool University Press 1999)

Lambourne, Robert, Michael Shallis and Michael Shortland, *Close Encounters? Science and Science Fiction* (Bristol and New York: Adam Hilger 1990)

Nagel, Ernest, *The Structure of Science: Problems in the Logic of Scientific Explanation* (Indianapolis: Hackett 1979)

Nozick, Robert, *Invariances: the Structure of the Objective World* (Cambridge, MA: Belknap Press 2001)

Parrinder, Patrick, 'Revisiting Suvin's Poetics of Science Fiction', in Parrinder (ed.) *Learning from Other Worlds: Estrangement, Cognition and the Politics of Science Fiction and Utopia* (Liverpool: Liverpool University Press 2000), 36–50

Popper, Karl, *The Logic of Scientific Discovery* (in German 1934, in English 1959; London: Routledge 1999)

Russell, Bertrand, *The Scientific Outlook* (1931; London: Routledge 2001)

Scharff, Robert C. and Val Dusek (eds), *Philosophy of Technology: the Technological Condition* (Oxford: Blackwell 2003)

Snow, Charles Percy, *The Two Cultures* (1959; with an introduction by Stefan Collini, Cambridge: Cambridge University Press 1993)

Stiegler, Bernard, *Technics and Time, 1: the Fault of Epimetheus* (transl. Richard Beardsworth and George Collins; Stanford, CA: Stanford University Press 1998)

Sturgeon, Theodore, 'Bookshelf', *Galaxy* 34 (1973) 3: 69–73

Suvin, Darko, *Positions and Suppositions in Science Fiction* (London: Macmillan 1988)

Walker, Peter (ed.), *Chambers Dictionary of Science and Technology* (Edinburgh: Chambers Harrap 1999)

2
Science Fiction and the Ancient Novel

This study began by arguing that the task of defining science fiction resolves itself not into a pseudo-'truth claim', hard-edged definition of the field, but rather into a delineation of the continuum by which SF can be meaningfully separated out as that form of the Fantastic that embodies a technical (materialist) 'enframing', as opposed to the religious (supernatural) approach we would today call 'Fantasy'.

Looking at the origins of the novel (the mode central to SF for much of its life as a genre) crystallises precisely this question. Margaret Anne Doody notes that the 'history of the novel' advanced by mainstream criticism as dating from the late seventeenth or eighteenth century overlooks the fact that 'the Novel as a form of literature in the West has a continuous history of about two thousand years' (Doody, p. 1). As Doody herself points out, this is no secret: classicists have undertaken many studies of 'the Ancient Novel', a form very popular in the first few centuries AD. Between five and eight complete novels (depending on how long one requires a 'novel' to be), two detailed summaries and a large number of fragments have survived; and these a small fraction of the total number of novels written in Greek and Latin during the classical period.

Doody suggests some of the reasons why almost all critics of the English and Continental novel have ignored this vigorous novelist culture, reasons that are extremely instructive to fans of science fiction. Put briefly, since the eighteenth century there has been a tendency to separate out 'the novel' from 'the romance', reserving serious critical attention for the former (conceived as in essence 'realist') and denigrating the latter as fantastical, escapist or vulgar. Doody's account of the reputation of romance will strike a chord of recognition with the SF aficionado: as a genre it 'is despicable, a term reserved for a certain low section of bookstore ... conveying literary pleasure the critic thinks readers would be better off without. It describes work that fails to meet the requirements of realism' (Doody, pp. 15–16). Nowadays, of course, in Doody's words, 'realism has faded away like the Cheshire cat, leaving its smile of reason behind'. It is time, she insists, 'to drop the pretence that the primary demand of a long work of prose fiction is that it should be "realistic" '; a compelling rallying-cry that readers of science fiction have been making, explicitly or implicitly, for decades now.

In polemical mode Doody insists:

> Romance and the Novel are one. The separation between them is part of a problem not part of a solution. ... as the emphasis on that supposed distinction has often done more harm than good, I propose to do without it altogether. I shall call all the works I am dealing with 'novels', as that is the term we feel most positive about. (Doody, pp. 15–16)

This is the practice followed by the present work.

All surviving Ancient novels involve fantastic elements to one degree or another; and the task for a historian of science fiction is to sketch the fluid borders where SF can usefully be contradistinguished from a more supernatural Fantastic. Ancient authors were comfortable with the literary idioms of what today we would call 'fantasy', 'magic realism', 'satire' and even, in a sense, 'surrealism'. For example, the Roman author Apuleius wrote a comic romance under the title *Metamorphoses* (*c*. AD 170, sometimes known as 'The Golden Ass'), in which the protagonist Lucius is transformed into the form of an ass by black magic, undergoes various adventures and is finally returned to human shape by the goddess Isis. Though delightful, the book is difficult to read as SF: it is, in a core sense, a religious and even devotional fable. The science that most frequently informed Ancient SF was either practical and technical (as in the sciences associated with naval navigation and warfare) or philosophical.

Another vital feature is the trope of odyssey, or *voyage extraordinaire*, which has occupied so central a place in discourses of science fiction, and which finds its origin in Ancient Greek literature. From Homer's *Odyssey* (seventh century BC), through epics, plays, histories, dialogues and later prose romances, Greek culture produced many hundreds of examples of fantastic voyages. Some of these were travellers' tales, based on actual or augmented experience (as, for example, voyages to Africa, India or over the Atlantic); some were purely fantastical and imaginative (journeys to the lands of the dead or to the heavens). Some 'liberal' critics of SF are content to classify all such *voyages extraordinaires* as early examples of the genre. A more common strategy is to concentrate critical attention on a small group of these narratives which detail journeys into the atmosphere, or journeys to the Moon and solar system.

The Ancient cosmos

The most important observation to make about Ancient Greek SF is that its proper realm is between the mundane, or earthly, and the divine, or theological, idioms. To have a clearer sense of this, we need to understand the version of the cosmos that Greek science provided imaginative writers. The conceptual model of the solar system altered over time, but was constant in one regard: celestial bodies were thought of as perfect, eternal and divine, and were indeed actively worshipped by many. This presents contemporary perspectives with a certain difficulty in separating out the scientific from the theological perspectives of the heavens, a difficulty made more acute by the fact that, for many Greeks (as for many medieval Europeans), there was no meaningful distinction between these categories in the first place.

The earliest Greek cosmologies postulated that the stars were glimpses of fire seen through holes in a sphere of mist surrounding the Earth. Sometime around 530 BC, Pythagorean thought replaced this model with one in which a mass of fire occupied the central position, and the Earth, Moon, Sun and other celestial bodies revolved around this point at intervals determined by the harmonies of musical scales, all these celestial bodies being thought of as divine. Later models placed the Earth in the centre of the cosmos: in the fourth century BC Aristotle, for instance, postulated that the universe was composed of 59 concentric spheres, with the Earth in the middle. The four inner terrestrial spheres constituted the four elements (Fire, Air, Earth and Water) which Plato and Aristotle believed were the fundamental ingredients of all earthly matter. The remaining 55 spheres, composed of a mysterious 'fifth element' not found on Earth, carried the celestial bodies in a series of circular revolutions around the stationary Earth. This system is known today as the Ptolemaic model, although Ptolemy (more properly Claudius Ptolemaeus, a geographer and astronomer who worked in the second century AD) was only one of many subsequent astronomers who refined this version of cosmological structure. Though satisfying theologically, this geocentric version of the cosmos does not explain all the motions observable in the night sky; in particular, planets seem to wander (the Greek πλανήτης, *planētēs*, means 'wanderer', or more forcefully, 'vagrant', 'tramp') from the path that would be expected if they moved in a clear circle around the Earth. Hipparchus (*c*. 190–120 BC) proposed that the planets moved eccentrically within their spheres. Ptolemy suggested epicycles, circular orbits around a point that itself orbits as a circle around the Earth, to solve this problem. This theory was preferred by many because it preserved the circularity of celestial orbits; the circle was thought of as a more perfect trajectory and therefore more appropriate to divine beings. This model survived for nearly 1,500 years unchallenged.

The important thing to realise about this conceptual set-up is that it makes a distinction, in a manner of speaking, between *secular* and *divine* heavens. Aristotle's treatise on *Meteorology* (Μετεωρολογικόν, *c*. 330 BC) discusses not only aspects of the weather such as thunder and lightning, rain and wind, but also a number of phenomena we would think of as astronomical (comets, for example, and the Milky Way) *as if they were* meteorological. The reason for this is that, because such phenomena could not be reduced to the purity or simplicity required by the theological model of the cosmos, they were assumed to belong to the *sky* rather than the divine heavens.[1]

What is interesting here is the relative fluidity with which 'sky' and 'outer space' relate to one another in the Ancient mind. It misrepresents this sort of ur-SF to see 'journeys to the Moon' as belonging to a category separate from 'journeys into the air' more generally conceived. The Moon is seen by many writers as being 'in the air', in the same realm as the Sun, stars, clouds and birds, and no more or less accessible than any of these things. Accordingly, the much larger corpus of

Ancient Greek fantastical tales about flight and airborne explorations of strange lands has the same relationship to subsequent SF as the lunar adventures of Antonius Diogenes or Lucian. The important differential is religious, not scientific. The sky and aerial phenomena are thought of as akin to earthly phenomena; for some writers even the Moon and Sun fall into this category. But the stars (for instance) are thought of as divine. In the words of Benjamin Farrington:

> Such, in the Pythagorean, Platonic, early Aristotelian and Stoic conceptions was the nature of the universe. The starry heavens were the visible image of the divine. As such they shared the lot of the gods and became the province of the theologian ... To hold other views was not a scientific error but a heresy. (Farrington, 2: 87–8)

It is taken as axiomatic in the present study that science fiction is distinct from theology, being natural and material where the latter is supernatural and spiritual. Ancient SF, including voyages into the air and voyages to the Moon, are conceptually distinct from journeys to the stars; they connect with material, practical discourses such as the science of navigation rather than strictly theological idioms.

The *voyage extraordinaire* into the sky has a lengthy pedigree. Euripides' tragedy Βελλεροφόντς (*Bellerophontēs*), now known to us only in fragmentary form, was produced perhaps in 430 BC. It dramatises Bellerophon's attempt to fly to Olympus on the back of the winged horse Pegasus in order to confront the gods directly with their various injustices. Pegasus unseats him in mid-air and he tumbles back to Earth; he appears at the end of the play crippled by his fall. Aristophanes, the comic poet, satirised Euripides in many plays, and Ειρήνη (*Eirēnē*, 'Peace', 421 BC) ridicules this play in particular. Aristophanes' protagonist is a farmer called Trugaios, a man so wearied of Athens' long war with Sparta that he flies to Heaven on the back of a huge dung-beetle to remonstrate with Zeus. A later Aristophanes play, Ὀρνιθες (*Ornithes*, 'Birds', 414 BC) continues the aerial theme. In it, two Athenians, Euelpidēs and Pisthetairos, disgusted with the degeneracy of their countrymen, pay suit to the King of the Birds, Epops. They persuade him and his avian followers to build a utopian new city in mid-air, called Nephelokokkugia, or 'Cloud-Cuckoo-town'.

The earliest surviving text that takes a strictly cosmic rather than merely aerial view is the *Somnium Scipionis* ('The Dream of Scipio', 51 BC) by Marcus Tullius Cicero (106–43 BC), a brief prose fable placed by Cicero at the end of his political tract *De republica*. In this text the younger Scipio dreams of his deceased and eminent grandfather, the older Scipio. He is shown the dwelling among the stars that is destined for those who follow the path of virtue, and particularly for patriots who defend their country: the text zooms out, as it were, from Carthage to a breathtaking perspective of the whole of the galactic 'Milky Circle' where the dreamer sees 'stars which we never see from Earth'. The fable's sense of wonder (stars 'larger than we have ever imagined ... the Earth seemed to me so small that I was scornful of our Roman empire, which covers only a single point, as it were, upon its surface') rather overpowers the ethical moral of the whole.

Early novels

The works cited so far occupy a variety of literary modes: poetry, drama, philosophical discourse. Of particular interest to the development of SF, however, is the novel; SF has always been (until the very recent significant interventions in film and TV) primarily a novelistic form. It is perhaps for this reason that Lucian of Samosata is so often cited as the first SF author. He certainly did write a proto-SF novel involving a journey to the Moon, but in fact romances of this type were being written 100 years before Lucian, although his is the first to survive in its entirety into modern times. But a number of pre-Lucian SF novels can still be identified. Especially interesting for our purposes are those texts in which a scientific discourse and a speculative-fantastic discourse come together. One author with an interest in both fields was Plutarch (Mestrius Plutarchus, *c.* AD 45–*c.* 125), an essayist and historian from Chaeronea in Boeotia, Greece. A prolific and wide-ranging writer, the author of some 227 books, much of his work has survived to the present day, including 50 biographies of famous figures, and 78 miscellaneous works. These latter, collected under the rather misleading title of *Moralia* ('Moral Works'), include works on philosophy, religion, rhetoric, biology, physics and cosmology.

Plutarch's Περί του εμφαινόμενου προσώπου τω κύκλω της σελήνης (*Peri tou emphainomenou prosōpou to kuklō tēs selēnēs*, 'On the Face Apparent in the Circle of the Moon') dates from about AD 80. Most of this work is a discussion between various individuals about possible explanations for the marks evident on the face of the Moon. The theories rehearsed reflect contemporary scientific inquiry: on the one hand, the belief that the Moon is made of a fiery, clear substance that acts as a mirror, with the marks visible being reflections of terrestrial oceans; on the other, that the Moon is made of earth, or an earthy substance, and its marks are impurities in it, or shadows cast by sunlight. From this, the discussion moves on to the question of whether the moon is inhabited. One of the interlocutors, Sulla, recalls meeting a stranger from a continent on the far side of the Atlantic who had revealed the nature of the heavens to him: that human beings are born with three elements (body, mind and soul), and that terrestrial death destroys only the body, whereupon his mind-soul migrates to the Moon to live there until a second death releases the soul. 'One death reduces a man from three to two, and another reduces him from two to one ... where the goddess here on earth severs soul from body quickly and violently, Phersephonēs (on the Moon) separates mind and soul slowly and gently' (943b). This goddess judges the inhabitants of her realm, 'sweeping off some as they cling to the Moon', although those 'who have found a firm footing walk about there triumphantly crowned with feathers' (943d):

> Just as our Earth contains deep gulfs ... so those features are also present on the moon. The largest of these is called 'Hekatē's Innermost Place', where souls suffer and are punished for the things they have done after they have become spiritual beings, and two lengthy ones are called 'the Gates', because souls pass through them from the side of the moon that faces the heavens to the side that

faces the earth. The side of the moon that faces the heavens is called 'the Elysian Plain', the other side 'the House of Anti-chthonic Phersephonēs'. (Plutarch, *Peri tou prosōpou*, 944c)

Of itself, this eschatological fantasy would be considered by most a work of speculative religion rather than SF; but the juxtaposition of science and fiction creates a whole that works on the boundaries of established science in a speculative fashion. Neither is this mixture of scientific enquiry and fantastical extrapolation an uncomfortable one. As D. A. Russell points out:

> The passage from science to fantasy should not be misinterpreted. It is to miss the tone and purpose of the dialogue to detect a clash between the clarity and acumen of the preceding arguments on astrophysics and the moon-mythology ... Both science and religious myth belong to the same range of elevated 'cosmic' subjects. They demand elaboration and magnificence, not bare factual statement. (Russell, p. 72)

In other words, Plutarch's fantasy is a way of doing science via elaboration and invention, which is to say, it is SF. Something similar happens in another Plutarchian dialogue, Περί του Σωκράτους δαιμονίου (*Peri tou Sōkratous daimoniou*, 'On the "Daimon" of Socrates', written *c*. AD 90). As with the *Peri tou prosōpou* the science fiction element here is only a small portion of the whole – a whole which is, in this case, an historical tale set in 379 BC about a group of conspirators who are planning to overthrow the tyrants of Thebes. One character relates an out-of-body experience:

> When he looked around, the Earth was nowhere to be seen, but he saw islands shining fire softly upon one another with first one colour, then another, like red-hot iron dipped in cold water (*or*, like cloth dipped in dye), as the many-coloured light kept changing. They seemed infinite in number, and absolutely enormous, and though they were not all the same they were nonetheless all circular in shape ... In their midst was a sea or a lake, through whose blue-grey transparency the colours travelled; and of the islands, some sailed out across the flow, and others were swept along with it, the sea drifting round ... (Plutarch, *Sōkratous daimoniou*, 590c–d)

This trope of interplanetary space as a 'sea' and celestial bodies as 'islands' is a common one in the Ancient world, a mode of expression balanced between literal and metaphorical (it still is: we launch our spaceships from capes, call them shuttles and vessels).

Antonius Diogenes' Τα υπέρ Θούλην άπιστα (*Ta huper Thulēn apista*, 'The Wonders beyond Thule') dates from around AD 100. It is available to us now only in condensed form, as one of the books summarised by the ninth century AD Byzantine scholar Photius in his Βιβλιοθήκη (*Bibliothēkē*, 'Library'). As with the Plutarch works mentioned above, the visit to the Moon in this early novel is only one small part of a larger series of romantic adventures. The protagonists travel widely, from

Sicily to Thule (the Greek term for a far northern island, possibly Iceland or Scandinavia), and into the Arctic Circle from where it is possible to pass upwards to the Moon (Photius, 111a). The lunar world is described as a 'γην καθαρώτατην', 'a clean, pure, spotless land' (the phrase might also be translated as 'open' or 'free land'). Returning to Earth, the characters have many more adventures, which ultimately end happily. Much in the story is of fantastic-romantic provenance: the heroes, captured by 'cruel and stupid' Celts, escape on horses that change colour; in Spain they encounter a village where the people see at night but are blind during the day; one of their party, Astraios, has eyes in which the pupils dilate and contract in consonance with the phases of the Moon; characters die, their deaths later revealed as only seeming, and are brought back to life. But what makes the work especially interesting is the extent to which this romance enters into a dialogue with the science of its time. Reyhl, for instance, thinks that Antonius dramatised Pythagorean philosophy in his romance, 'following the Pythagorean belief that the hills and valleys of the moon were inhabited by fantastic creatures' (quoted in Georgiadou and Larmour, p. 39).

The classical author most consistently cited as a 'father of science fiction' is Lucian (Λουκιανός, *c.* AD 120–90), sometimes called Lucian Samosata after his birthplace, Samos. The strength of his generic paternity claim rests on two works among his many: the Ικαρομένιππος (*Ikaromenippos*) and the Αληθής Ιστορία (*Alēthēs Historia*, the 'True History'), both written sometime between AD 160 and 180. The first of these is a dialogue between Menippos and a friend, in which the former relates how he attached an eagle's wing to his right arm and a vulture's wing to his left, and used them to fly up to the heavens (in the manner of the myth of Ikaros, or Icarus; hence the title). Menippos, frustrated by the contradictory squabblings of earthly philosophers, each of whom claims unique truth, decides to ask Zeus about the real state of affairs. He flies first to the Moon, where he gains a splendid vantage point from which to survey the Earth below him; then he flies past the Sun and up to Heaven itself to consult Zeus. The narrative, in other words, moves through science fiction to theological fiction.

The *Alēthês Historia* is even more famous, and may have been a satire on the more extravagant adventures recorded in Antonius Diogenes' *Ta huper Thulên apista*. The tale involves an eventful sea voyage westward across the Atlantic. In the first episode of this peripatetic narrative, the voyagers stop at an island where rivers of wine flow, and where 'Vine Women', with female upper parts but rooted vine trunks below, ensnare several of the narrator's crew. The second episode sees the voyagers' ship swept up into the sky by a massive whirlwind:

> While our vessel hung in the sky the wind caught her sails and propelled her onwards. For seven days and nights we sailed through the air, and on the eighth day we saw a large country in the sky, like an island, bright and circular and shining with light. (*Alēthēs Historia* 1: 110)

Landing on this 'sky island', which is of course the Moon, the voyagers are captured by soldiers riding three-headed flying 'horse-vultures' and taken to the lunar king, Endymion. Thereafter they become involved in a war between the peoples of the Moon and the peoples of the Sun. King Endymion deploys his horse-vultures and a troop of vegetable birds, and is reinforced by allies: 'thirty thousand Flea-archers and fifty thousand wind-runners'; King Phaethon, lord of the Sun, puts on the field his ant cavalry ('huge beasts with wings, resembling the ants we have but much larger') and sky-mosquitoes. The war is won by the Sun, and a peace treaty agreed. Lucian details some of the 'strange and wonderful things' about Moon life, and then takes his ship back through the sky and onto the ocean again. (It is significant that he proceeds no further into outer space; the Sun and Moon are assumed to be within earthly reach; to travel further out would run the risk of impiety.) When he splashes down, we are perhaps a quarter of the way into the *Alēthēs Historia*; very many strange adventures follow.

A proper account of these two texts needs to read them in the context of Lucian's 80 or so titles, many of which detail fantastical, parodic or extraordinary adventures, although only these two include journeys to the Moon. As the précis of the *Alēthēs Historia* suggests, the Moon interlude is only one episode in a larger peripatetic work. Returning to Earth, the travellers go on to be devoured by an enormous whale and live inside its belly for two years; they visit islands on which milk flows like water, see nations of men living on the surface of the ocean supported by feet made of cork instead of flesh, sojourn with the famous dead on the Isle of the Blessed, visit the Isle of Dreams and encounter mariners who sail in hollowed-out pumpkin shells, and others who sail on their backs, with their (presumably sizeable) penises straight up as a mast, from which a sail is hoisted, steering by 'holding their testicles in their hands' (*Alēthēs Historia*, 2: 45). This last detail in particular gives a sense of the tone of the whole: outrageous, inventive, bizarre and very funny. It is also a highly intertextual work, woven from quotations, allusions, pastiche and parody of a wide range of other works, from Homer through Greek philosophy, history and geographical works, to the fantastic romances popular in the first century. The ironic title indicates the ways in which the book explores the playful exuberance of lies and lying. If we want to determine to what extent we can call *Alēthēs Historia* 'science fiction', we need to have a sense of how 'falsehood' inflects the interaction between science and fiction. It perhaps overstates the case to assert, as John Griffiths (among many) does, that 'Lucian's *True Histories* does not qualify as science fiction' because it 'rather deliberately sets out to be ridiculous and implausible in order to make its point' (Griffiths, p. 33). But we can at least suggest that for Lucian a visit to the Moon is not a serious possibility precisely because the whole of the *Alēthēs Historia* refuses the discourse of 'seriousness'.

Many critics, however, do find it meaningful to talk of the *Alēthēs Historia* (to cite the title of Georgiadou and Larmour's detailed study) as 'Lucian's Science Fiction Novel':

Scientific knowledge in the fields of geography, astronomy, zoology and anthropology pervades the narrative, even if it is visible mainly through the lens of parody: the reader is constantly presented with information about islands, and rivers; the sun, moon and stars; plants, birds, fish and animals; and the customs and appearance of other beings. (Georgiadou and Larmour, pp. 45–6)

For other critics, the science-fictionality of Lucian's work resides less in its parodic discourse of science and more in the powerful cognitive estrangement it achieves. According to Fredericks, Lucian is 'like a modern SF writer' because he 'takes the science and other cognitive disciplines available to him and pictures alternative worlds which can dislocate the intellects of his readers in such a way as to make them aware of how many of their normal convictions about things were predicated upon cliché thinking and stereotyped response' (Fredericks, p. 54). But it is probably more accurate to describe Lucian as an allegorical, or mythic, writer. When Menippos leaves the Moon to fly to Heaven, the Moon herself addresses him as a personified female figure. She asks him to petition Zeus on her behalf to destroy the philosophers who 'pour dreadful abuse' on her: 'some of them say I am inhabited, some that I am suspended over the sea like a mirror, and others say whatever their imagination suggests to them. Recently they have even declared that my light is stolen and illegal, coming from the Sun.' What especially annoys the moon-goddess is that she herself sees these same philosophers committing 'shameful, appalling' deeds by night, 'committing adultery, or burglary, or anything else best suited to night-time ... yet although I see all this, I say nothing about it, I don't think it's proper to expose these nocturnal activities ... I gather my garment of cloud about me and veil my face, so that the ordinary people can't see these old men bringing shame upon themselves and on virtue' (*Ikaromenippos*, pp. 20–1). Lucian's sympathy is very plainly with the mythic, not the scientific, mode. The conception of the Moon as a goddess watching Earth from the sky and drawing cloud about her as a garment clearly owes nothing to second-century scientific understanding of the cosmos; in fact, such understanding (represented by the 'philosophers' that Lucian consistently attacks) is specifically repudiated. Similarly, in *Alēthēs Historia* the trip to the Moon finds that world presided over by the legendary Endymion (who in myth was a mortal lover of the moon-goddess). Lucian finds a harmony and imaginative strength in the religious, mythic discourses of the cosmos, where he tends only to ridicule the philosophical or 'scientific' discourses. As Georgiadou and Larmour put it:

In the lunar adventure of (*Alēthēs Historia*) and *Ikaromenippos*, the main subject is the disagreement among the various groups of philosophers ... in *Alēthēs Historia* the dispute is presented as a literal battle between fantastic forces, which, on at least one level, represent the outlandish notions and theories of the philosophers, or the arguing philosophers themselves. (Georgiadou and Larmour, p. 16)

It would be more accurate to see Lucian as anti-SF rather than proto-SF: but anti-SF nevertheless involves an engagement in the terms of SF.

Conclusion

'Ancient SF' is not a single-minded or 'pure' idiom: it mediates, on the one hand, scientific speculation and *voyages imaginaires*, and, on the other, religiously conceived fable.

There is a curious irony in the traditional belief that Lucian is the first science fiction author. In fact, he comes at the end rather than the beginning of a vigorous tradition of fantastic voyages into the sky and to the planets in the classical world. The many gaps in our record of Ancient literature mean that what was in all likelihood a more or less continuous genre of imaginative, speculative adventure stories seems partial and chronologically broken up. The missing texts between the ingenious fantasies of Attic comedy in the fifth century BC and the speculative scientific romances of the fifth century AD give a misleading sense of discontinuity; but the gap that ensues, when Hellenistic and Roman culture collapses into what historians still call 'the Dark Ages', is very real. For over 1,000 years SF fell into abeyance as a literary mode. Its disappearance was connected, very obviously, with the more general collapse of literary culture, and of literacy itself. But the delay in its re-emergence presents us with a more interesting problem. A varied and rich literary tradition reappeared early in medieval culture, but it was many hundreds of years after this that SF is written again. The reasons for this hiatus are discussed in the following chapter.

References

Antonius Diogenes, 'The Wonders Beyond Thule', in Photius, βιβλιοθηκη, ed. and transl. René Henry, 11 vols. (Paris: Société d'Édition 'Les Belles Lettres' 1960); 'Antoine Diogène, *Les merveilles incroyables d'au delà de Thule*', 2: 108b–111b

Aristophanes, *Comoediae*, ed. F. W. Hall and W. M. Geldart (Oxford: Clarendon Press 1900)

Cicero, Marcus Tullius, *On Friendship* and *The Dream of Scipio*, ed. J. G. F. Powell (Warminster: Aris and Phillips 1990); 'Somnium Scipionis' 136–46

Doody, Margaret Anne, *The True Story of the Novel* (New Brunswick, NJ: Rutgers University Press 1996)

Euripides, *Selected Fragmentary Plays*, ed. C. Collard, M. J. Cropp and K. H. Lee, 2 vols. (Warminster: Aris and Phillips 1995)

Farrington, Benjamin, *Greek Science* 2 vols (Harmondsworth: Penguin 1944)

Fredericks, S. C., 'Lucian's *True History* as SF', *Science Fiction Studies* 3 (1976), 49–60

Georgiadou, Aristoula and David H. J. Larmour, *Lucian's Science Fiction Novel* True Histories: *Interpretation and Commentary* (Leiden: Brill 1998)

Griffiths, John, *Three Tomorrows: American, British and Soviet Science Fiction* (London: Macmillan 1980)

Hammond, N. G. L. and H. H. Scullard, *The Oxford Classical Dictionary* (2nd edn., Oxford: Clarendon 1970)

Lucian, *Works*, transl. A. M. Harmon, 8 vols. (London and New York: Heinemann-Macmillan 'Loeb Classical Library' 1913–67); 'A True Story' 1: 247–357, 'Icaromennipus, or the Sky-Man' 2: 267–323

Jones, C. P., *Culture and Society in Lucian* (Cambridge, MA: Harvard University Press 1986)

Photius, βιβλιοθηκη, ed. and transl. René Henry, 4 vols. (Paris: Société d'Édition 'Les Belles Lettres' 1960); 'Antoine Diogène, *Les merveilles incroyables d'au delà de Thule*', 2: 108b–111b

Plutarch, *Moralia* transl. Harold Cherniss and William C. Helmbold, 15 vols. (London and New York: Heinemann-Macmillan 'Loeb Classical Library' 1913–69); 'On the Sign of Socrates' 7: 361–509, 'Concerning the Face which Appears in the Orb of the Moon' 12: 1–223

Russell, D. A., *Plutarch* (London: Duckworth 1973)

Swanson, R. A., 'The True, the False, and the Truly False: Lucian's Philosophical Science Fiction', *Science Fiction Studies* 3 (1976), 228–39

Interlude: AD 400–1600

There is one question any historian of science fiction must answer. We have seen the vigour of the genre in classical literature, and we shall see the flourishing of science-fictional titles which emerge in the early seventeenth century – but in between there was a 1,100-year hiatus in the history of the form. The question is: Why? What happened to SF from approximately AD 400 through to the beginning of the seventeenth century?

There was no shortage of fantastical tales and romances over these dozen centuries: many fantastic voyages were written, quests and adventures that are limited to a contemporary, earthly arena. Some authors even detailed voyages away from the Earth that inhabited a wholly religious, theological and supernatural idiom. Of the latter the most prominent example is the epic *Divina Commedia* ('Divine Comedy') by the Italian poet Dante Alighieri (1265–1321), most likely written between 1307 and the poet's death. In the first portion of this devoutly Catholic poem, *Inferno*, the narrator travels through Hell, conceived geographically as a world within the (hollow) Earth; the second, *Purgatorio*, traces a journey up an improbably elevated mountain in the Antipodes; and the third, *Paradiso*, out through the solar system. But every line of this massively influential poem is inflected by a profound spiritual and theological purpose. In a profound sense Dante's universe is not a material one, but is instead sacramental, connected at every point to the grace of God.

There is more than a merely pigeonholing instinct at work in excluding such religious works from a history of science fiction. Take, for example, the 'Katherine Group', a number of Middle-English prose works composed around 1200 and detailing the lives of various saintly women for devotional and religious purposes. Alexandra Barratt observes of one of these prose legends, *Katherine of Alexandria*, that it is 'a lively story with plenty of fairy-tale and romance elements':

> a magnificent minster appears and disappears; doors open of their own accord; an aged hermit overgrown with hair mysteriously materializes in Katherine's study; a phantasm takes her place in Alexandria while she attends a mystical nuptial mass in the desert; her martyred body is transported by angels to Mount Sinai. (Barratt, p. 234)

These extraordinary occurrences might be thought of as Suvinian nova, and therefore as approaching the 'cognitive estrangement' of SF; and, indeed, equivalents can be found in canonical SF (*Star Trek*'s self-opening doors and mysterious materialisation device, the film *Metropolis*'s 'phantasmic' copy of Maria, and so on). But in a vital sense these marvellous events are not 'nova' in the Suvinian sense: they are very particularly *miracles*, and as such non-material.

There is, of course, an extensive tradition of secular romance during the period 400–1600, but even in this non-miraculous idiom religion shaped imaginative literature. As Margaret Anne Doody puts it, 'the great change in religion in the West, as Christianity took hold and commanded the ethical centre of the lives of millions of men and women, certainly had a great and even permanent effect upon literature, including narrative fiction' (Doody, p. 181). She goes on to quote the twelfth-century French poet Jean Bodel de Arras, who declared that

> Ne sont que iii matières à nul homme attendant
> De France, et de Bretaigne et de Rome la grant.

(There are only three 'matters' that any man can attend to/The matters of France, of Britain and of Rome.)

By 'matter' Bodel meant a body of stories and myths that provided proper topics for medieval romance. The matter of France concerned the stories of the Emperor Charlemagne, the great Knight Roland and associated figures; the matter of Britain was the legend of King Arthur and his knights. Both 'matters' inflected romances as heavily chivalric, and although they contained many magical and supernatural elements they had very little to do with SF. The 'matter of Rome' covered anything from myths of Troy through to Roman Caesars, and tended to be treated in medieval texts (such as Benoît de Sainte-Maure's *Roman de Troie*, 1160) in like chivalric mode to Arthurian and Charlemagnian tales. In addition to these more elevated poetical works, European medieval culture produced a large number of shorter comic tales (*fabliaux*) and many stories that were of oral provenance and circulation, and which have accordingly survived only poorly to the present day (such as the stories of Robin Hood). All this must in turn be placed in a cultural context in which the majority of books produced (pre-printing) were devotional and practical. When an anonymous landowner from Asia Minor died in 1059 and bequeathed his library to a monastery, the contents give us some sense of the representative spread of topics on which books in this period were written: 57 Bibles, liturgies or patristic works; five saints-lives and seventeen secular works, including a book on the interpretation of dreams, an edition of Aesop's fables and one Greek novel.

In other words, literature from the end of the Dark Ages to the Renaissance consisted of explicitly religious texts or of religiously inflected chivalric romance. It is also, as Doody notes, more often than not in the form of poetry rather than prose: 'writers of the Middle Ages, from the tenth to the fourteenth century, tend to produce fiction primarily in verse.' Scholars have agreed no single explanation for this, although 'we may surmise that prose narrative largely disappears' in this period 'because city life is essential to the production of novels' (Doody, p. 183).

Of course, literature did not stop being religious in 1600, and in many ways it continues being religious in the twenty-first century, in SF as much as anything. But a vital shift of conceptual emphasis opened the possibility of a science, or more properly a technology fiction, only in the seventeenth century. Medieval romance was concerned with human action, and was therefore limited to the Earth. Beyond the universe, in Ladina Bezzola Lambert's expressive phrase, was a 'finite geocentric cosmos ... largely an amalgam of (neo)Platonic philosophy, Aristotelian cosmology and Christian theology. Its most conspicuous characteristic was the distinction between a terrestrial realm opposed to a metaphysical-transcendental one in the heavens.' The celestial realm 'was believed to consist of so superior and pure an essence as radically to defy any comparison with earthly matter' (Lambert, p. 2). As with the Greek novel, translation into the heavens was translation into a divine rather than a material realm; but unlike the Greeks medieval interplanetary voyages partook of a monotheistic unity, allied to a totalitarian religious authority, that denied the imaginative possibilities that SF requires. This culture can produce a Dante, but not an Asimov.

An example is the trip to the Moon (sometimes cited as an originary point for SF) to be found in the epic-romance *Orlando Furioso* ('Mad Roland', 1532) by the Italian poet Ludovico Ariosto (1474–1535). This poem relates the adventures of the great knight Roland, subject of Charlemagne (and therefore part of the matter of France), who loses his wits and runs naked through the forest destroying everything he encounters. Since Charlemagne's kingdom is under threat from the Saracens it is imperative that Roland be cured. To effect this, the English knight Astolfo mounts on the back of a Hippogriff, a fabulous winged creature, and flies first to the terrestrial paradise on top of a high mountain, where he meets John the Evangelist. From there St John accompanies Astolfo as they fly to the Moon – because it is to the Moon that all the things lost on Earth, such as Roland's wits, make their way. To be precise, things lost on Earth reappear on the Moon in metaphoric form: poetry rehearsed to flatter great lords appears as burst crickets; deeds of charity that came too late appear as spilt soup; the beauty lost by ladies as they age appears as multitudinous snares covered with bird-lime, and so on. With St John as interpreter, Astolfo is made aware of the true significance of all the variegated lumber he encounters on the Moon. Lost wits such as Roland's take the form of bottles filled with a delicious liquor bearing the name of their former owners. Astolfo retrieves Roland's and returns to Earth.

When Astolfo first approaches the Moon, it seems to him 'come un acciar che non ha macchia alcuna' ('like polished steel without any blemish', Ariosto, *Orlando Furioso*, canto xxxiv, stanza 70), but as he flies closer he notices that it has a landscape like Earth's, with 'altro fiumi, altri laghi, altre campagne ... c'han le cittadi, hanno I castelli suoi' ('other rivers, other lakes, other fields ... they have their own cities, their own castles', Ariosto, *Orlando Furioso*, canto xxxiv, stanza 72). As Lambert points out, this double apprehension reflects the 'two diametrically opposed, though equally current, symbolic readings of the Moon', as a symbol of Purity associated with the Virgin Mary, and as a symbol of corruption and inconstancy (Lambert, p. 24n.). Indeed, this symbolic ambiguity is reflected in precisely the status of the Moon: as the dividing line between the corruptible, material 'sublunary' world and the pure, eternal, divine celestial realm above. Astolfo does not genuinely travel to another planet: he journeys (with divine sanction, and under the chaperonage of Saint John) to the limit of the material realm. Despite the similarities with the lunar voyages of the seventeenth century, Ariosto does not begin a new form of 'science fiction' in the *Orlando Furioso*. Rather, he takes material Man as far as he is permitted within the theological constraints of the pre-Copernican cosmos. SF doesn't happen in that place.

References

Ariosto, Ludovico, *Orlando Furioso* (1532), ed. Cesare Segre (Milan: Mondadori 1976)

Barratt, Alexandra, 'St Katherine of Alexandria: The Late Middle English Prose Legend in Southwell Minster MS 7', *Notes and Queries* 240 (June 1995), 234

Doody, Margaret Anne, *The True Story of the Novel* (New Brunswick, NJ: Rutgers University Press 1996)

Lambert, Ladina Bezzola, *Imagining the Unimaginable: the Poetics of Early Modern Astronomy* (Amsterdam and New York: Rodopi 2002)

3
Seventeenth-Century Science Fiction

Science fiction was reborn in one year, 1600, the year that the Catholic Inquisition burned Giordano Bruno the Nolan at the stake for arguing in favour of the notion that the universe was infinite and contained innumerable worlds. Bruno's was a fundamentally science-fictional conception.

In his discussion of this period (which is not concerned with SF) Howard Margolis has argued that 1600 is the key turning point in the development of modern science. By way of evidence he itemises nine 'fundamental scientific discoveries' made around that year (including the laws of planetary motion, the magnetism of the Earth, and the distinction between magnetism and electricity). According to Margolis, if one compares the notable scientific discoveries of the previous fourteen centuries one discovers 'nothing at all'. The title of Margolis's book summarises his view of the whole discourse of modern science: *It Started with Copernicus*.

It is to state the obvious to say that modern SF would be impossible without Copernicus. Nevertheless I want to suggest that Bruno is the more appropriate symbolic starting point for a history of science fiction. The eminent physicist Alfred North Whitehead (1861–1947) acknowledges the crucial importance of Copernicus, but identifies 'the origins of modern science' with Bruno: 'his death in the year 1600 ushered in the first century of modern science' even though 'the cause for which he suffered was not that of science, but that of free imaginative speculation' (in Tauber, p. 53). 'Free imaginative speculation' is the same thing I refer to, in this history, as 'science fiction'.

Nikolaj Kopernik or Koppernigk (1473–1543, usually referred to by his Latinised name, Nicolaus Copernicus) was a churchman and astronomer from Ermland, in what is now Poland. His name is particularly associated with a revolution in cosmology that was, by undoing the scientific authority of the Catholic Church, to have the most profound effect on the development of thought and necessarily therefore on science fiction. Those alive in the medieval world who cared about such things believed in a model of the cosmos usually called Ptolemaic, after the Alexandrian astronomer Claudius Ptolemaeus (AD ?100–?150). Ptolemaeus' Μαθηματικὴ Σύνταξις (*Mathēmatikē Suntaxis*, 'Mathematical Syntaxis'; *c*. AD 145) described a solar system centred on a stationary earth, such that the Sun, Moon, five planets and a sphere of fixed stars revolve diurnally about our world. The medieval world believed this model to be consonant with the (few) accounts of the cosmos found in the Bible, and augmented Ptolomaeus' scientific text with the more 'spiritualised' geocentric model elaborated in Cicero's *Somnium Scipionis* (51 BC). This is the cosmos through which Dante takes his reader in the *La Divina Commedia*.

Kopernik's own observations suggested to him that this account was wrong, and that the Earth and other planets rotated about the Sun. A devout Catholic, he was cautious about publishing his revolutionary theory under his own name. Accordingly, a young follower named Georg Joachim von Lauchen, known by his Latin name 'Rheticus', issued his first-hand account of Kopernik's theories as *De revolutionibus orbium coelestium* ('On the Revolution of the Celestial Orbs') in 1543. Kopernik's theories spread only slowly, hampered by the hostility of the Church, the small print-run of the book and the inertia of the scholastic traditions of the learned. Nevertheless, by the end of the sixteenth century most scholars, whether they accepted or rejected it, knew about the theory.[1] These new theories were so deeply implicated in new theories of religion, that the Renaissance (associated with the former) and the Reformation (associated with the latter) are aspects of the same underlying cultural logic.[2]

In other words, the Copernican re-evaluation is a crucial and shaping event in the development (or re-development) of science fiction. Before Kopernik's new map of the heavens any fantastic voyage beyond the Earth necessarily took place in a realm understood to be divine rather than material, and therefore within a theological context. The difficulty in analysing Dante as a writer of science fiction, for instance, is that, although his protagonists leave the Earth and travel through the solar system, this Ptolemaic cosmos is, in effect, merely an extension of the Earth into the divine element. Dante's Heaven is actually humanity's earthly environment seen from a spiritual point of view (much the same is true of his Hell). The entire architecture is comprehensible only in a theological idiom. After Kopernik the cosmos not only expands tremendously in scale and scope, it becomes necessarily materialised. A cleavage opens up between the cosmological accounts of science and religion.

Giordano Bruno (1548–1600) was a Neapolitan speculative thinker who, inspired by the Copernican model of the cosmos, taught across Europe, particularly in the Protestant countries of the north. Returning to Venice in 1591 he was arrested by the Inquisition, interrogated and finally executed in 1600. It is true to say that 'magic' and hermetics played as important a part in Bruno's philosophy as 'science' (as we understand it today), but nevertheless his imaginative speculations concerning the nature of the universe are in essence SF. His 'On the Infinite Universe and Worlds' (1584) imagines the cosmos as infinite, containing an infinite plurality of worlds (see below p. 49), each of which can be likened to an organism, all of which contribute to the totality of the cosmos. Reading Bruno is likely to remind the contemporary SF fan of Olaf Stapledon's *Star Maker* (1937); both writers advance very similar sense-of-wonder versions of the cosmos, material yet semi-mystic (although Stapledon is likely to have received Bruno's influence at second- or third-hand, through German idealist philosophy of the nineteenth century). Such a vision was an imaginative expansion and inhabitation of the newly opened-up Copernican cosmos, and since it excluded God (Bruno believed the total 'Cosmos-Soul' of the universe was a different entity from God) and challenged the teaching of the Church, it is perhaps not surprising that Bruno was executed as a heretic.

Throughout the sixteenth and seventeenth centuries Catholic persecution of the Copernican version of the cosmos was energetic and prolonged. In 1616 the Italian astronomer Galileo Galilei (1564–1642) was forbidden by the Church from endorsing the Copernican account of the cosmos. Galileo nevertheless published a scientific work arguing in Copernicus' favour in 1632, and the Inquisition condemned him in 1633. Galileo recanted his 'heresy', denying that the Earth moved round the Sun and affirming that the Earth was the stationary centre of the cosmos.

At the beginning Protestantism was also alarmed by the new thinking. As early as 1549 the Protestant thinker Philip Melanchthon challenged what he saw as the dangerous implications of Copernicus' discovery. Grant McColley summarises:

> The most vital argument to Melanchthon is his last, wherein he states that there is but one Son of God, our Lord Jesus Christ, who was sent into this world, was dead, and was resurrected. He did not appear in other worlds, nor was He dead and resurrected there. Nor is it to be thought that if there are many worlds, something not to be imagined, that Christ was often dead and resurrected. Nor should it be considered that in any other world, without the sacrifice of the Son of Man, men could be brought to eternal life. As Melanchthon reasons, to accept a plurality of worlds is to deny or to make a travesty of the Atonement. (McColley, pp. 412–13)

None the less, the Protestant countries of Europe (Britain, Holland, parts of Germany) were not as repressive in their hostility towards the scientific and – importantly for our purposes here – imaginative exploration of the new cosmology.

This is not to say that seventeenth-century thinkers in the Protestant tradition enjoyed absolute intellectual freedom. Far from it. Indeed, a striking feature of the science fiction written in this century is the anxious insistence with which it continually interrogates the question of the primacy of Christ's sacrifice. Francis Godwin (1562–1633) was the Anglican Bishop of Hereford. His posthumously published space journey adventure *The Man in the Moone: or, A Discourse of a Voyage Thither by Domingo Gonsales, the Speedy Messenger* (1638) illustrates the interpenetration of religious and scientific discourses. The first action of Godwin's protagonist on arriving on the Moon and seeing its inhabitants is to call out 'Jesus Maria'. This results in the Lunarians '[falling] all down upon their knees, at which I not a little rejoiced' (Godwin, *Man in the Moone*, p. 96). This rejoicing reflects the confirmation of the uniqueness of Christ's incarnation. In similar vein, Wilkins' *Discovery of a World in the Moone* (1638), for instance, postulates lunar inhabitants, but goes on to worry whether such beings 'are the seed of Adam, whether they are in a blessed estate, or else what means there may be for their salvation'. Wilkins quotes Thomas Campanella to the effect that Lunarians must be 'liable to the same misery [of original sin] with us, out of which, perhaps, they were delivered by the same means as we, the death of Christ' (Wilkins, pp. 186–92). When the protagonist of Cyrano's comical voyage to the Moon, *L'Autre Monde ou les Etates et Empires dans la lune* (1657) arrives on the other world his first encounter is with the biblical figure of Elijah who tells him: 'cette terre-ci est la lune que vous voyez de votre globe; et ce lieu-ci où vous marchez est le paradis, mais c'est le paradis terrestre' ('This land is indeed the same moon that you can see from your own globe, and this place in which you are walking is Paradise, but it is the Earthly Paradise', Cyrano de Bergerac, p. 44). We discover that the Garden of Eden was removed from the Earth to the Moon by God after the expulsion of Adam and Eve, 'reflecting a popular belief that paradise must have been located on the Moon because the floodwaters [of Noah's flood] would then not have been able to reach the just inhabiting it' (Harth, p. 13). Wilkins makes a similar case in *Discovery of a World in the Moone*, describing the Moon as a 'celestiall earth, answerable, as I conceive, to the paradise of the Schoolemen ... this place was not overflowed by the flood, since there were no sinners there which might draw the curse upon it' (Wilkins, pp. 203–5). Cyrano becomes jocular with the pious-minded Elijah, interjecting his own bawdy interpretations of the biblical stories the prophet is retelling (for instance, suggesting that the serpent that tempted Eve was a form of penis). Elijah expels him from 'le paradis terrestre' in disgust, and he goes on to stranger adventures among the other inhabitants of the Moon. These surreal elements are the more famous parts of Cyrano's lunar voyage; the lengthy Elijah episode is much less widely known and the reader encountering it today is perhaps struck by its oddness. One twentieth-century critic, Remy de Gourmont, went so far as to excise the whole section from his edition of *L'Autre Monde*, calling it 'une longue digression incompréhensible', expressing 'une théologie bizarre' (De Gourmont, pp. 154–5n.). But to suggest so is to misunderstand the cultural context in which the book was produced. Cyrano's book is indeed comic, grotesque, a burlesque rather than a serious theological intervention; nevertheless, the 'bizarre theology' of the section, even in so self-consciously ludicrous a book as this, suggests the ways in which the buried anxieties of early SF are religious in form. Jean de la Bruyère may have insisted that 'la lune est habitée' in 1688, but he immediately added, 'si nous sommes convaincus l'un et l'autre que des hommes habitant la lune, examinons alors s'ils sont chrétiens' ('if we're all convinced that men live in the Moon, let us consider whether they are Christians', La Bruyère, p. 474).

Moreover, this anxiety – that the new cosmology undermines traditional Christian revelation – has never entirely left SF. In various ways, as this study will demonstrate, the tension between 'humanist' or 'Protestant' perspectives (which veer towards materialism and the unmediated individual exploration of the cosmos) and 'sacramental' or 'Catholic' perspectives (which stress a spiritual, transcendental, divinely mediated and fundamentally magical universe) intimately shapes the development of the genre.

The Copernican cosmos and 'sense of wonder'

The new cosmology, and in particular the dangerous notion of an infinity of worlds associated with a number of late sixteenth- and seventeenth-century thinkers, did two things absolutely crucial to the development of science fiction as we understand the form. First, it created an imaginative space in which humanity might encounter radically different beings – aliens, the Other, the material embodiments of the alterity that drives the mode. Dante traverses the entire cosmos and finds only other human beings, or angels who very much resemble human beings, because the whole cosmos has been (he believed) created by God and populated with creatures made in His image. But once the Ptolemaic egg had been cracked open, and once an infinity of possibilities presented themselves to the imaginative explorer, radical alterity can enter.

The second thing that the new cosmology reveals is the much grander scale of the universe, which in turn permits and indeed requires a corresponding aesthetic of the sublime. This attachment to the sublime – or, to invoke the common phrase from twentieth-century SF, the 'sense of wonder' – provoked by gigantic scale, enormous devices or very long stretches of time may not define the whole genre of SF, but remains integral to many fans' appreciation of the form. Arguably the sublime is core to interplanetary or interstellar SF, and central to most aspects of the genre. More than this, it depends fundamentally on an infinite, Copernican cosmos. The Englishman Thomas Digges (?1532–?1595) printed an account of 'the infinity of the universe' in 1576 (cautiously hiding his work by tucking it in as an appendix to an edition of a much less inflammatory book by his father called *A Prognostication Everlasting*). William Empson notes the 'appalling splendour' of Digges' vision, and quotes him at length:

> Thys baull of the earth wherein we move, to the common sorte seemeth greate, and yet in respect of the Moones Orbe is very small, but compared with *Orbis magnus* (the Great Orb) wherein it is carried, it scarcely retayneth any sensible proportion, so merveilously is that Orbe of Annuall motion greater than this little darcke starre wherein we live. But that Orbus magnus beinge ... but as a poynct in respect of the immensity of that immouveable heaven, we may easily consider what little portion of God's frame, our Elementare corruptible worlde is, but never sufficiently be bale to admire the immensity of the Rest. ('Thomas Digges his infinite universe', quoted in Empson, 1: 219)

Artists of the sublime, or of 'sense of wonder', dilate on the insignificance of the 'little dark star' on which we live when compared to the immensity of the universe; Douglas Adams' twentieth-century conception of the 'total perspective vortex', in which a machine compels individual minds to understand exactly how small they are in comparison to everything, thereby destroying them, is a comic version of this same understanding: comic because the reality underpinning the notion is indeed so unsettling, so appalling, that we prefer not to contemplate it.

It is undeniable, of course, that this 'sense of wonder' is connected to discursive traditions of religious awe and a sense of 'the sublime' conceived of as divine even as it expresses the new cosmos that science opens in space and time. A hundred years after Digges, Thomas Burnet (1635–1715), a doctor of divinity and one-time candidate for Archbishop of Canterbury, published *Telluris Theoria Sacra* ('Sacred Theory of the Earth', 1681), translating it into English as *The Theory of the Earth* in 1684. Popular well into the eighteenth century, this book elaborates a theologico-geology based loosely on Genesis. Burnet believed that God had created the world as a perfectly smooth spheroid; all the depressions and elevations, trenches and mountains had been carved out by Adam's sin and the subsequent deluge. What makes the work more than just one more example of batty biblical exegesis is the sheer verve of Burnet's imaginative entry into the fantastic narrative he is advocating. There is unmistakable sense-of-wonder in his vivid descriptions of global catastrophe:

> The pressure of a great mass of water falling into the abyss ... would carry it up to a great height in the air, and to the top of any thing that lay in its way, any eminency, high fragment, or new mountain. And then rolling back again, it would sweep down with whatsoever it rush'd upon, woods, buildings, living creatures, and carry them all headlong into the great gulf. Sometimes a mass of water would be quite struck off and separate from the rest, and tossed through the air like a flying river ... (quoted in Sutherland, p. 390)

What is worth stressing is the way in which the material correlative for this sublimity, the actual universe in which we exist, which is in turn amenable to physical exploration, simply does not exist for pre-seventeenth-century writers and artists (A. J. Meadows points out that Burnet was attacked by his contemporaries 'still unaccustomed, or in some cases actively opposed, to any attempt to explain the miraculous in terms of the natural', Meadows, p. 123). But it is precisely this conceptual arena in which modern SF germinates.

Seventeenth-century science-fictional prose romances

A fuller understanding of the rebirth of SF needs to be set in the literary context in which it occurred. The typical sixteenth- and seventeenth-century prose fiction (using the term 'typical' here with caution) was in the mode of what more traditional critics tend to call 'classic Romance'. A great many books were published that told stories of courtly love and of knights and noblemen adventuring, questing and falling in love. It is a misapprehension to describe such texts as 'realist', or even 'realistic', although of course many of them did bear deictic or satirical relationships to the world in which their readers lived. But most of the conventions that appear over and again in these tales were borrowed wholesale from the Ancient Greek novel (and many Ancient novels were translated into English for the first time during this period). The plot devices include children separated from their families after being kidnapped by pirates, growing up and returning to their native lands ignorant of their parentage; one lover apparently dying so that the other lover falls into despair, only for this apparent death to be revealed as a ruse or improbable mistake later in the story.

Some sense of the relative popularity of science-fictional tales in England during this period is indicated by Paul Salzman's 1985 study *English Prose Fiction 1558–1700: a Critical History*. Salzman 'lists all known extant works of fiction published' between his two dates – 582 titles in all: 105 Elizabethan fictions and 477 seventeenth-century works. These Salzman breaks down into various genres and sub-genres. Twenty-one of these are arguably science fiction (all published in the seventeenth century; Salzman lists most of them under 'Imaginary Voyage/Utopia/Satire'). This figure compares with 24 picaresque novels, 26 'popular chivalric romances' and 15 Oriental tales. In other words, science fiction was as popular a genre as most others during this period, although the large variety of different genres and modes of novelistic writing at this time means that only a small proportion of all novels produced were SF.[3] These are the beginnings of a genre that would grow enormously in size and influence over the following two centuries.

This flourishing of the novel in the seventeenth century is, of course, something shaped by the continuing religious reformation. As Margaret Anne Doody somewhat colloquially puts it: 'Protestantism set the cat among the pigeons in the matter of reading' (Doody, p. 233). If this is true of reading more generally conceived, it is true *a fortiori* of the speculative cosmological and epistemological fictions of SF. To put it briefly: SF is the genre that mediates the discourses of 'science' (or 'fact') and 'magic' (or, subsequently, 'imagination', 'fiction'); and it comes into generic being at precisely the historical moment when competing cosmic discourses were in the process of separating themselves into rationalist Protestant and ritualist-magical Catholic religious idioms. I should add that by saying this I am following Norman Davies' understanding of the Reformation (which some historians would challenge) as a deliberate attempt to 'take the magic out of religion'; and that although 'the Protestant onslaught on magic enjoyed only partial success', nevertheless 'Protestant Christianity was supposedly magic-free' (Davies, p. 405). By attacking 'magic', Protestantism was challenging beliefs very deeply rooted in human culture and in many human psyches, as well as a great deal of anxiety, cultural and personal; and 'SF' in its broadest sense can be understood as a textual strategy to mediate this dialectic cleavage. SF texts from the early seventeenth century mark this disjunction most clearly, and none more evidently than Kepler's *Somnium* (written *c.* 1600, published 1634), which has a good claim to be the first work of modern SF, and which deserves to be treated in greater detail.

Kepler's *Somnium*

The *Somnium*, although little read today, is an absolutely key text in the development of science fiction. It articulates precisely the dialectic between 'Protestant' rational science and 'Catholic' magical/demonic imaginative expansiveness that shapes the emerging genre.

Johann Kepler (1571–1630) was a Protestant German astronomer who established three important laws of planetary motion. But in addition to his scientific studies he wrote one work of science fiction, *Somnium, sive Astronomia Lunaris* ('A Dream, or Lunar Astronomy'). The exact chronology of composition is uncertain; he may have written an initial draft in the 1590s, revising it over subsequent decades without publishing it. The book was in press when Kepler died in 1630, although it was not until 1634 that one of his sons issued a complete edition.

Somnium is a short work. A brief narrative of a journey to the Moon and the conditions that obtain there is followed by a series of detailed scientific and explanatory notes. It utilises a number of embedded narratives. The outer frame is related by an unnamed narrator who tells us that one night, having observed the stars and Moon, he fell asleep and dreamed a certain dream in which he was reading another book: the life story of 'Duracotus', a man born in Iceland to a witch-mother called Fiolxhilda. After spending time in Europe working with the famous Danish astronomer Tycho Brahe (1546–1601; in his youth Kepler had himself been Brahe's assistant), Duracotus returns to Iceland to learn from his mother about demons (she calls them 'sapientissimi spiritus', 'most wise spirits') who travel between the Earth and 'Levania' (the Moon). On occasion these demons even carry human beings between the two worlds. She calls down one such spirit, and the two humans cover their heads with a sheet (as the magical pact requires) while this 'daemon ex Levania' talks about the true nature of the Moon. Almost all the remainder of this first part of *Somnium* is given over to this demonic speech.

This second embedded narrative is more expository than the first. The demon reveals how his kind travel to the Moon during a lunar or solar eclipse, when the cone of shadow touches both Earth and Moon and serves as their path. The demons can carry human beings with them, we learn, although the journey is 'most hard' ('durissima') for mortals, since the space between the worlds is characterised by extreme cold and great difficulty breathing. Demons remedy this first problem by creating heat from within their own bodies, and the second by pressing 'humid sponges' ('spongiis humectis') to their passengers' mouths. There follows an account of the natural history of the Moon. From a Moon-based point of view, the dominant object in the sky is the Earth, or 'Volva', as the Levanians call it. The Moon's month-long revolution on its own axis and its monthly orbit of the Earth means, of course, that one lunar hemisphere is always facing the Earth and one facing away. The former hemisphere of Levania is called by its inhabitants Subvolva, or 'Under-Earth', and the latter Privolva or 'deprived-of-Earth'. Kepler correctly intuits that the consequent changes of lunar temperature are extreme, from the great cold of the fortnight-long Levanian night to the great heat of the fortnight-long day – so hot is the lunar day, indeed, that the inhabitants of that world retire into deep caves and caverns to escape it. In Privolva, the lack of the Earth makes a great difference, and life is described in nightmarish terms:

They live an unfixed life, without permanent habitation. They roam in great crowds over the whole globe during one of their days, some on legs which are longer than are our camels', others flying through the air, others still in boats follow the fleeing water. (Kepler, *Somnium*, p. 46)

Most of the Privolvans are divers ('urinatores'), who can seek relief at the bottom of the waters. Their skin is thick and porous, and under the fierce sunlight may become scorched on its outermost layers; this layer is shed like a husk when evening comes. Mostly their forms are snake-like ('natura uiperina in uniuersum praevalet'). Some of these creatures loll in the midday sun; others stop breathing in the heat and only return to life in the evening. Life in the Subvolvan hemisphere is less frantic; some relief is accorded by the presence of Volva itself, and this portion of the world is mostly covered by rain clouds.[4] At this point the dreamer awakes, his head under cloth (like Duractos') – in reality, his bedclothes – and with a rainstorm (like the demon's Levanian description) rattling against the window. *Somnium* ends on this abrupt note.

This fantastic, grotesque little romance is striking enough (William Empson dismisses it as 'Bosch-like stuff', Empson, p. 239), but it is thrown into sharper relief by the substantial number of notes that follow. An appendix contains 223 notes, some of them very long, followed by a Selenographical Appendix, which in turn is followed by 34 numbered and 31 lettered notes. Occasionally, these notes explain Kepler's inspiration for specific details in the *Somnium*, but the majority contain detailed data on the moon, astronomy and general scientific matters.

One of the main problems facing the interpreter of *Somnium* is to reconcile these two idioms: the one fanciful and grotesque, the other soberly scientific. But in context it is easier to see that this dynamic between the 'magical' (witchcraft and demons and their attendant visions) and the 'scientific' precisely embodies the problematic out of which SF develops. The effectiveness of *Somnium* derives from its interplay of fantastical-grotesque imaginative play and solid scientific grounding. This linking of apparent opposites, this (to use the rhetorical term) oxymoron, is precisely the aesthetic principle behind *Somnium*. Kepler worked this principle into his text at several levels, not least the formal. Note 206 calculates that the Moon is one quarter the size of the Earth ('paulo maior diameter lunaris parte Quarta terrestris'). This dimension is then formally embodied in the text itself, the first part of the *Somnium* being 3,800 words and the notes over 15,000 words (which is to say, the text of *Somnium* is a quarter the size of the notes to *Somnium*). In other words, the larger, weightier scientific portion of the whole functions as the grounded, Earth-like element; and the smaller, more fanciful portion functions as flightier, more lunatic transport.

The witch Fiolxhilda, summoning the demon from Levania, was very much an up-to-date touch for the early 1600s. Kepler himself lived close to the heart, geographically and chronologically, of the infamous European witch craze, and at one point had to defend his own mother from the charge of being a witch. Olwen Hufton records that 'between 1560 and 1660 ... about 100,000 witches were condemned, of whom about 30,000 were from Germany with a particular emphasis on small states with a troubled religious history' (Hufton, p. 340). Although we often think of this fascination with demonic magic that characterised this period as anti-scientific and irrational, this was not the case. Stuart Clark has demonstrated at length that, instead of being opposed discourses, 'magic' and 'science' were viewed by most thinkers in the period as complementary and even aspects of the same truth. Questions of the capabilities of demons and the power of witchcraft 'were essentially scientific in character, and ... here demonology became thoroughly embroiled with some of the pressing concerns of late Renaissance natural philosophy' (Clark, p. 315). Kepler himself, we recall, worked as both an astronomer and astrologer, although he regarded this latter occult art as astronomy's sometimes stupid daughter (Thorndike, p. 17), both had their part in understanding the heavens.

Writers on demonological matters all agreed that it was possible for demons to transport people physically from place to place, a practice known as 'transvecting'. The fifteenth-century thinker Marsilio Ficino (1433–1499) wrote at length on demonic abilities in his *De vita coelitus comparanda* (1489), taking the term 'daemon' to mean both 'guardian angel/good demon' and 'devilish or bad demon'. Demons, he argued, 'are primarily planetary, though there are super-celestial and elemental ones. They have souls and aetheric or aerial bodies, according to their status (the super-celestial ones have no bodies) ... there are bad demons, of a low status and with aerial bodies who trouble men's spirits and imagination' (Walker, p. 47). Cardano's mid-sixteenth-century *De rerum varietate* 'restricted demonic activity to the aerial regions' (Clark, p. 237). The higher planets (in the Ptolemaic model) are pure, unchanging and partake of the divine, but the Moon and the areas below it are changeable and corruptible, and it is in this arena that the bad demons operate. Kepler seems to have restricted his demons to the aerial regions, and to have limited their intercourse with humanity, perhaps following Cardano, who 'spoke rather disparagingly of spirits having few significant dealings with men and women' (Clark, p. 237). In general, his perspective precisely negotiates the space between the 'magical' and strictly scientific, separating out a properly scientific fiction from a broader, theological-magical Fantasy.

Interplanetary travel

The relationship between the SF narrative of interplanetary travel and the chivalric-romance narrative of exotic terrestrial travel is especially close. Both Dante, in his *Paradiso* (c. 1320), and Ludovico Ariosto, in his *Orlando Furioso* (1532), imagined characters travelling to the Moon. Ariosto is sometimes called the 'father of science fiction' for this reason, although Dante rarely is; but in both cases there is a flaw in the ascription. For Dante and Ariosto the heavens are not material in the manner of the sublunary world. In their shared Ptolemaic view of the solar system, the heavens are unchanging, perfect, divine: a supernatural arena. Giambattista Marino (1569–1625), once 'the leading poet of Italy and the outstanding figure on the European literary scene' (Mirollo, p. vii) (although now largely forgotten), also wrote one science fiction work: his baroque, twenty-canto epic *L'Adone* (1622). This retelling of the love between the divine Venus and the handsome mortal Adonis includes an interplanetary interlude: in cantos 10 and

11 Adonis is carried to the heavens to travel through the solar system, flying past the Moon to visit Mercury and Venus. In one sense this journey to the planets is fantastical and mythological rather than science-fictional; carried up by a god, Marino's Adonis moves from the Earth to the Moon, to Mercury and to Venus as if moving up through the Ptolemaic spheres. But in an important sense this poem moves beyond the pre-Copernican fancies of Ariosto (to whom Marino's poem is heavily indebted). *L'Adone* is not an antiquary's resuscitation of old myth; it is consistently engaged in the most up-to-date science of Marino's day. So, for example, one of the heroes of the epic is Galileo, whose newly invented telescope is praised as revealing the truth of the heavens; '[te] Galileo ... potrai senza che vel nulla ne chiuda / Novello Endimion, mirarla ignuda', 'you Galileo will be able to gaze at the nakedness of Endymion [i.e. the Moon] with no veil intervening' (*Adone*, canto 10, stanza 43; Marino, p. 529). As he approaches the Moon, Adonis asks Mercury whether the heavens are composed of the same matter as the Earth, or whether their temper is radically otherwise, 'incorrottibili' (incorruptible) as he has heard. This for our purposes is the crucial question. At a similar moment in Dante's *Paradiso* this same question is asked, and Dante replies with the theologically orthodox explanation that heaven is composed of an incorruptible and perfect material, purer and fundamentally different from the corruptible matter of the Earth. But Marino's poem reverses Dante's explanation with a materialist rather than a spiritualist account of space:

> La material del ciel, seben sublima
> sovra l'altre il suo grado in eminenza,
> non pero dala vostra altra si stima:
> nulla tra gl'individui ha differenza.
> (canto 10, stanza 18;
> Marino, p. 523)

(The material of heaven, though sublimely / sovereign over others in its degree in its eminence / is nevertheless not other than yours [*i.e. not different from Earth*] / it is not qualitatively different.)

This is a point of the greatest significance. It means that Adonis's explorations of the moon, Mercury and Venus take place in a material rather than a spiritual dimension. It turns out that the Moon, instead of being a religious stepping-stone on a themed symbolic journey towards God, is a world just like Earth, with 'altri mari, altri fiumi ed altri fonti / citta, regni, province e piani e monti', 'other seas, other rivers, other springs / cities, kingdoms, provinces, and plains and mountains' (canto 10, stanza 40; the echo of Ariosto in fact emphasises the radical difference between the two poems). It is this stress on the *material actuality* of the Moon as another inhabited world that makes the allegorical journey of Marino's Adonis science fiction rather than merely mythological fancy. It may have been its Galileo-like intellectual daring that led to the poem being censored by the Catholic Church (who placed it on the *Index Librorum Prohibitorum* in 1624).

Marino illustrates one of the two main strategies for moving protagonists extraterrestrially. Marjorie Hope Nicholson's 1948 study *Voyages to the Moon* contrasts 'supernatural voyages' on the one hand, with voyages that, to one degree of plausibility or another, utilise *material* forms, on the other. The former are rarer: Nicholson describes only Kepler's *Somnium* and an English derivation of that work, *Conclave Ignatii* (1611, the 'Conclave of Ignatius') by the metaphysical poet John Donne (1572–1631) (she might also, but does not, mention Marino). Nicholson comments that 'the journey of Milton's Satan' flying through the cosmos in Book 2 of *Paradise Lost* (1667) 'is the last truly "supernatural" cosmic voyage written in England' (Nicholson, p. 56).

But of the latter, material methodology Nicholson adduces a great many examples: flight into space 'by the help of fowls' such as Godwin's *Man in the Moone* (1638); flight to the Moon aided by mechanically constructed wings or other flying machinery: Wilkins' *Discovery of a World in the Moone* (1638) looked forward to lunar colonies 'as soon as the art of flying is found out' (Wilkins, p. 208). The minor poet Francis Harding's collection of Latin verses *In Artem Volandi* ('On the Art of Flying', 1692) imagines the future invention of flying machines as a boon: the rich will leave the Earth for the other planets leaving their estates to the poor, and a new British aerial navy will establish peace on the Moon. Cyrano's various ingenious devices for interplanetary travel move the business of flight from fantastical into scientific and more importantly *technical* idioms. The anonymous prose work *Selenographia* (1690) is subtitled *The Lunarian, or Newes from the World in the Moon to the Lunaticks of This World*, and concerns a flight from the Earth to the Moon by means of a large kite.

An even more peculiar method of interplanetary travel occurs in Margaret Cavendish's *The Blazing World* (1666). In this eccentric tale a merchant elopes with a beautiful young woman from 'a foreign country', but his ship is blown by a tempest to the North Pole where he and his male crew die of cold. The woman, surviving, passes Antonius Diogenes-like into 'the blazing world' of the title; our North Pole having 'another Pole of another world ... joined close to it'. The cosmological rationale of these two linked planets is only obscurely related:

> It is impossible to round this world's globe from Pole to Pole, so as we do from East to West, because the Poles of the other world, joining to the Poles of this, do not allow any further passage to surround the world that way, but if anyone arrives to either of these Poles he is either forced to return or to enter into another world; and lest you should scruple at it, and think if it were thus those that live at the Poles would either see two suns at one time, or else they would never want [*i.e. lack*] the sun's light for six months together, as it is commonly believed, you must know that each of these worlds having its own sun to enlighten it, they move each one in their peculiar circles, which motion is so just and exact that neither can hinder or obstruct the other. (Cavendish, *Blazing World*, p. 254)

Attempts further to elucidate only thicken the cloud of unknowing ('for they do not exceed their tropics, and although they should meet, yet we in this world cannot so well perceive them by reason of the brightness of our sun, which, being nearer to us, obstructs the splendour of the suns of the other worlds, they being too far off').

The unnamed lady meets the Empress of this blazing world, is shown the marvels of the new planet, discusses science and metaphysics, and meets various alien humans, from worm-men and fly-men to spirit-creatures, although Cavendish is disarmingly vague about many details ('in what shapes of forms', she says at one point, 'I cannot exactly tell'; Cavendish, p. 291). Learning that living people can embody the souls of others such as 'Galileo, Gassendus, Descartes, Helmont, Hobbes, H. More, etc.' (Cavendish, p. 306), the Empress takes into herself the soul of the author, 'the Duchess of Newcastle', Margaret Cavendish herself, later leading an army from her world to conquer the Earth, after which she establishes a sort of terrestrial utopia. The sheer oddity of this text has provoked ridicule in some, although recent critics have seen merit in it. Geraldine Wagner, for instance, argues that 'it is a testament not to the text's incoherence but to its breadth and complexity that, like its author, it cannot be reduced to singleness. Consider first how it is at once a utopian experiment, a romance adventure, an unconventional autography, and a philosophical/scientific exposition' (Wagner, p. 5). Clearly Cavendish is articulating a sense of the dialectic of the sexes through her fable of two separate worlds, and it might be argued that the sheer weirdness of their cosmological relations is expressive of a metaphorical dislocation of cultural profundity. More, she is the first identifiable female SF author, the creator of a bold SF vision.

Savinien de Cyrano de Bergerac

Cyrano de Bergerac (1619–1655) probably wrote his two tales of cosmic travelling in the 1630s, although they were not published until after his death. The first to appear was *Histoire comique par Monsieur de Cyrano Bergerac, contenant les états et empires de la lune* ('Cyrano de Bergerac's Comical History, Containing the States and Empires of the Moon', 1657). In it Cyrano decides to travel to the Moon to discover for himself whether it is inhabited. He tries two inventions to achieve this journey; first, he flies by attaching to himself many bottles of dew ('de fioles pleines de rosée', Cyrano, *Voyage dans la lune*, p. 32) which are drawn up by the sun's heat, drawing Cyrano up with them; but instead of flying to the Moon he travels only as far as Canada. Here he constructs a sort of rocket, powered by 'fusées volantes' ('sky rockets', Cyrano, *Voyage dans la lune*, p. 40) that propels him into outer space. Landing on the Moon he meets the Old Testament prophet Elijah, who tells him that this is the garden of Eden (Elijah, we discover, had previously travelled to the Moon in a magnetised chariot). Travelling across the Moon Cyrano encounters its aboriginal inhabitants – gigantic beast-men, not unlike large horses. He converses with a more humanoid alien born on the Sun but now resident on the Moon. One social class of Lunarians uses music instead of language; food is inhaled rather than ingested; cities are mobile. Kept as a pet in a sort of zoo by the Lunarians, Cyrano meets Domingo Gonsales (the fictional hero of

Francis Godwin's *Man in the Moone*). He is then put on trial to decide whether his contention that the Earth is inhabited is actionable or not, although he is later released. After a lengthy meal, and discussion with a moon-being concerning religions and cosmology, Cyrano decides that his interlocutor is possessed by the devil, and clings on to him as he flies up into the air. He is thereby transported back to Earth, landing in Italy.

A second fantasy was published in fragmentary form: *Fragment d'histoire comique par Monsieur de Cyrano Bergerac, contenant les états et empires du soleil* ('Cyrano de Bergerac's Comical History, Containing the States and Empires of the Sun', 1687). In this work, Cyrano's account of his time on the Moon has landed him in trouble with conservative European authorities: he escapes imprisonment, and manages to return to outer space by means of a third flying device: a craft powered by mirrors that deflect the sun's rays to provide levitation. Flying through space, Cyrano eventually lands on the Sun, which he finds to be populated by a utopian society of intelligent birds, a 'Parlement des oiseaux'. These creatures put him on trial for the crime of being a human being and sentence him to death, although the intercession of a parrot (whom Cyrano had once released from a cage) reprieves him.

Cyrano's adventures blend fantastical humanist speculation on the material nature of existence with lengthy, to modern sensibilities, rather tedious disquisitions on theological questions. The point, for seventeenth-century SF, is that as an emergent mode it negotiates precisely these two quantities: the 'scientific' and the religious, the imaginative possibilities of the first dialectically engaged with the human anxieties about the erosion of the latter.

More specifically, Cyrano returns repeatedly to the question of the plurality of worlds, the issue over which Bruno probably lost his life, and which continued to be a live issue throughout the century. Before he even leaves the Earth, Cyrano's protagonist is having earnest discussions with a French military officer as to whether 'les étoiles fixes sont autant de soleils ... [ou] que le monde serait infini', their conclusion being that 'comme Dieu a pu faire l'âme immortelle, il a pu faire le monde infini' ('just as God made the soul immortal, so He made the universe infinite', Cyrano, *Voyage dans la lune*, pp. 37–8). Lunarians also seek to demonstrate to Cyrano's protagonist that there are an infinite number of worlds within an infinite universe ('Il me reste à prouver qu'il y a des mondes infinis dans un monde infini', one of them announces, before elaborating a lengthy argument by microcosmic analogy, Cyrano, *Voyage dans la lune*, p. 92). This particular notion goes to the heart not only of Cyrano's SF tales, but of the development of the genre itself.

On the plurality of worlds

I regard it as axiomatic that science fiction depends on this notion of a plurality of worlds, and might even be defined as that genre that enters imaginatively into precisely this 'otherness' (of worlds, of inhabitants and, by extension, of times). But investigation into the plurality of worlds was, to begin with, limited by orthodox religious restrictions.[5] Nor were these restrictions limited to Catholic cultures. The Protestant Lambert Daneau (1530–1596), in *The Wonderfull Woorkmanship of the World* (1578), rejects the very idea that there could be 'another world like unto ours, and other heavens, and another Sun, and a Moon, and all other things in them as in ours' precisely because nobody can determine 'what is their state, order, condition, fall, constancy, Saviour, and Jesus' or say 'what likewise is their life ever-lasting, and from whence cometh the salvation of this second or third world' (quoted in Empson, 1: 201). Unsupported by scriptural authority, the very notion of other inhabited worlds flirts with heresy, which lends the topic a dangerous flavour for more than 100 years. The speculative cosmology of the *Zodiacus Vitae* ('Living Zodiac'), originally published in Italy in 1537, caused its author, Palingenius (whose full name was Pietro Angelo Manzoli, and whose date of birth and death are unknown), to be classified as a heretic of the highest class in the Papal Index. There is a palpable anxiety about the shrinkage of human importance in this (anonymous) English translation of Palingenius's *Zodiacus Vitae* from 1580:

> But some have thought þt every starre a worlde we may well call,
> The earth they count a darkned starre, whereas the least of all.
>
> (quoted in McColley, p. 402)

Significantly, even so tentative expression of such sentiment was a dangerous business.

In other words, the fascination with the plurality of worlds running through seventeenth-century writing has at its roots theological anxiety rather than purely scientific curiosity (even if we could, as we cannot in this period, separate out 'religion' and 'science'). Since arguing in favour of an inhabited terrestrial Moon was the first step on the intellectual road to asserting the plurality of inhabited worlds, even so modest a science-fictional device is treated cautiously: neither Kepler nor Cyrano, for instance, published their works during their lifetimes. Charles Sorel (*c.* 1600–1676) buries his speculations in the middle of his dense comic picaresque *La vraie histoire comique de Francion* (1626), whose protagonist declares late in the novel, 'vous sçavez que quelques sages ont tenu qu'il y avoit plusieurs mondes ... moi, je crois qu'il y en a un dans la lune' (Sorel, *Vraie histoire*, p. 425).

Later in the century some of the theological sting had gone out of the debate, and a freer set of speculations began to be published. The English philosopher James Howell (1596–1666) could assert confidently as early as 1647 that 'Every Star in Heaven ... is coloniz'd and replenish'd with Astrean Inhabitants' (Howell, p. 425). More influential were works such as Pierre Borel's *Discours nouveau prou-vant la pluralité des mondes* ('New Discourse Proving the Plurality of Worlds', 1657) and Bernard le Bovyer de Fontenelle's very popular *Entretiens sur la pluralité des mondes* ('Dialogue on the Plurality of Worlds', 1686). This latter work takes the form of a witty conversation between the author and a noblewoman, divided between five evenings' conversation. The second and third evenings' discussion explores the notion that 'la lune est une terre habitée' ('the Moon is a habited land'), with 'autres fleuves, d'autres lacs, d'autres montagnes, d'autres villes, d'autres forêts, et ce qui m'auroit bien surpris aussi, des nimphes qui chassoient dans ces forêts' ('other rivers, other lakes, other mountains, other towns, other forests and it wouldn't surprise me other nymphs hunting through these forests', Fontenelle, p. 87). Like Kepler's Levanians, these Lunarians 'incommodez par l'ardeur perpetuelle du soleil' must make their homes underground (Fontenelle, p. 135). The fourth evening considers the 'particularitez des mondes de Venus, Mercure, Mars, Jupiter et Saturne' (p. 95) and the fifth and final evening suggests 'que les etoiles fixes sont autant de soleils, dont chacun éclaire un monde' ('that the fixed stars are also suns, and therefore each illuminates a world'; p. 129); although the narrator later retracts his belief in the inhabitation of the Moon.

The work apprehends something of the sheer size of the Copernican cosmos: Fontenelle's interlocutor declares herself flabbergasted ('mon imagination est accablée') by the thought of 'la multitude infinie des habitans de toutes ces planetes' ('the infinite multitude of inhabitants on all these worlds', p. 150). More importantly, and despite occasional pious interjections to the effect of praise, 'à Dieu … qui fixât les gons dans les places qui leur sont naturellement convenables!' (p. 184), the work is remarkably secular in its speculations. The Moon and other planets are compared to Australia and the Americas, populated worlds ready to be discovered by humanity. The whole book is scientifically explained; indeed, 'Fontenelle's promotion of Copernican ideas led to its being placed on the Papal index' (Hawley, 1: 29). This is not to say that these new 'materialist' cosmic speculations entirely supplanted the more traditional theological orthodoxies. An influential example of the latter is found in the peculiarly cloying piety of the Jesuit priest Athanasius Kircher (1602–1680), who in 1660 published a supernatural tour of the cosmos: *Iter exstaticum coeleste, quo mundi opificium, id est, coelestis expansi siderumque tam errantium* ('A journey in the form of a trance to the heavens, or how the universe is made; which is to say, the idea that the heavens and constellations extend to such a large degree is mistaken'; an earlier Roman edition of the book, 1656, was called *Itinerarium exstaticum …*). Kircher tells of a fantastic voyage through the solar system in order to dramatise his own religious-cosmological theories. To the accompaniment of the music of the spheres the two protagonists, Theodidactus (whose name means 'taught by God') and Cosmiel (an angel), travel from planet to planet, conversing with the intelligent life forms they encounter on the way. Kircher rejected the Copernican cosmology, preferring the model proposed by Tycho Brahe (who believed that the Sun orbits the Earth and is in turn orbited by the planets and the fixed stars). This model allowed Kircher to maintain the geocentric orthodoxy demanded by his faith while also exploring the solar system as it was being revealed by contemporary science. Indeed, the whole voyage takes place within a strictly Catholic Christian framework; the title-page illustration shows, to the right, the worlds in their Brahean orbits; but also represents, beyond the cosmic horizon, the mysterious Hebrew Name of God in a roiling mass of cloud.

In contrast to the religious-mystical Kircher, the Dutch polymath Christaan Huygens (1629–1695) composed his account of the solar system *Cosmotheoros* (posthumously published in 1698) according to the principle of rigorously scientific falsifiability: 'I must acknowledge that what I here intend to treat of is

Frontispiece of Kircher's *Iter exstaticum coeleste* (1660)

not of that Nature as to admit of a certain knowledge; I cannot pretend anything as positively true (for how is it possible), but only to advance a probable Guess, the Truth of which everyone is at his own liberty to examine' (Huygens, vol. 1, pp. 9–10). He deduces an inhabited solar system from theological grounds (God could not have made all these planets to no purpose; *ergo* they must be inhabited by rational creatures). Huygens' own scientific accomplishments included mathematical advances in the field of calculus, the discovery of Saturn's ring-system, works on clocks and lenses and much else.

Marjorie Hope Nicholson finds the book replete with Huygens' belief that 'Justice, Honesty, Kindness and Gratitude' must be omnipresent in God's cosmic creation, making him, in effect, a utopian writer: 'We journey on to find everywhere "solemn troops and sweet societies" of highly civilised men who ... bend their minds to scientific invention.' Nicholson thinks that this form of *voyage extraordinaire* lacks vigour: 'the supernatural voyage had lost the vitality it once had had in the hands of Lucian and Kepler. Kepler had lived in an age of superstition when witches were more common than were scientists. His descendants of the eighteenth century lived in the cool clear light of reason' (Nicholson, p. 62). But this is a simplification. It is true that the *Cosmotheoros* can strike the modern reader as dull, but Huygens bases his 'conjecturae' on a rational-scientific principle that, although things in an infinite universe may be infinitely different from things observable on Earth, just because this is possible does not make it *necessarily* so ('neque idcirco necesse esse'). Huygens' conjectures proceed from a principle of uniformity in Nature that tempers the more extravagant posturing of much early SF with a coolly understated scientific restraint that has its own imaginative power.

By stressing the materialist nature of the cosmos, Huygens details a space into which mankind can spread; and indeed, among much else, he contemplates future plans to colonise the various moons of the solar system. Seventy years earlier, Charles Sorel's fictional hero Fancion had similarly seen the inhabited Moon in terms of a threat to the Earth, wondering if 'il y aura là un prince comme Alexandre qui voudra venir compter ce monde-ci. Il fera provision d'engins pour y descendre ...' ('there will be a prince like Alexander the Great up there, planning to come down and subdue this world of ours. He'll need to provide engines for descending to our world ...', Sorel, *Vraie histoire*, p. 425).

This same Charles Sorel (*c.* 1600–1676) was a wonderfully inventive writer, much less well known than Cyrano but of particular significance for the development of SF. Sorel's novel *Gazettes et Nouvelles ordinaires de divers pays lointains* ('Gazettes and News from Various Faraway Countries', 1632) includes among its various wonders an artificially constructed, metallic woman fluent in all the world's languages: the creature with the best claim to be the first humanoid robot in literature. Sorel followed up on this idea with *Recueils de pièces en prose* ('Collections of Prose Pieces', 1644 and 1658), in which a locksmith invents a number of metal beings to act as servants. An earlier work, the playful anti-pastoral *Le Berger extravagant* ('The Extravagant Shepherd', 1627), includes among its wonders a race of men with skin as transparent as grease paper ('une peau transparente comme du papier huilé', p. 382) through which bones and organs can be seen.

It is the sheer imaginative inventiveness of Sorel that delights most. In *Le Courier véritable* ('The True Courier', 1632), an Australian-set utopian tale (for which sub-genre see the following section) we discover an ingenious mechanism for long-distance communication: sponges that soak up sound and voices, and can be carried great distances before squeezing the sound out again ('certaines éponges qui retiennent le son et la voix articulée, comme les nôtres font les liqueurs'). Other fanciful ideas for long-distance communication in Sorel's work include view screen-like magic mirrors ('miroirs magiques') in *Le Berger extravagant*.

Terrestrial utopias

Many seventeenth-century interplanetary romances might more usefully be considered as utopias; and this sub-class of imaginative fictions requires consideration in a separate section. Fictions of ideal societies take their name, of course, from Sir Thomas More (1477–1535), whose *Utopia* was published in Latin in 1516. This slim book relates a voyage to an imaginary island called 'Utopia' (the name a three-way pun: suggesting 'ou-topos' or 'nowhere', 'eu-topos' or 'good-place' and 'u-topos' or 'u-shaped land', which describes the shape of More's imaginary island). Society in Utopia is organised along improved lines: goods are held in common, education is widespread, the population are well ordered, productive and happy, and so on.

It is, as I discussed in the Preface, one of the assertions of the present history that 'utopia' – a valid and often fascinating literary mode – evolves separately from the *voyage extraordinaire* of science fiction more generally conceived. Nevertheless they are an important component of the ongoing development of SF precisely because they mediate a general cultural fascination with otherness in material terms: lands that might actually be reached by a voyager, strange but *material* new forms of human life and society and so on. The sub-genre, in other words, runs parallel to the extraterrestrial *voyage extraordinaire* of SF.

There are also, of course, crucial differences in the two modes of writing. Utopias take their germ from a static, Catholic work rather than from the mobile Protestant *voyage extraordinaire* of Kepler's *Somnium*. There *is* a voyage to More's *Utopia* (1516), but the bulk of the book is remarkably rooted and stay-at-home. The Utopians themselves are not travellers or explorers (they must obtain a licence even to visit another of the Utopian cities), and in Tony Davies' words, 'what is interesting' about the book 'is how un-futuristic it all seems, its peaceful, equitable community combining monastic simplicity of life with the imagined tranquillity of a long-forgotten golden age' (Davies, p. 73).

As Peter Ackroyd points out, the Utopians, in addition to their manifold improvements in social practice, also 'encourage euthanasia, condone divorce and harbour a multiplicity of religious beliefs – all of which actions were considered dreadful by More himself and by Catholic Europe'. Ackroyd is surely right to argue that these 'errors' are a function of the fact that More's Utopians 'throughout their history of 1760 years have been denied the truths of divine law ... [Utopia] may be no ideal commonwealth after all, but a model of natural law and natural reason taken to their unnatural extreme' (Ackroyd, pp. 168–9). At all points in the text this imaginary land is connected deictically to More's own country, and More's devoutly held Catholic beliefs. The dimensions of this land are, as Ackroyd points out, exactly the same as England, with the same number of city-states as England has shires. The main city of Utopia, Amaurotum, has the same expanse as the city of London, with a main tidal river like the Thames, a grand stone bridge like London Bridge and many other smaller points of identification. In short, 'it is London redrawn by visionary imagination' (Ackroyd, p. 167). But the didacticism of that visionary imagination is informed at all points by a traditional Catholicism: Utopia is a highly restrictive, authoritarian model of society.

A very extensive tradition of utopian writing grew from More's seed; but the extent to which this sub-genre cross-fertilised the developing tradition of SF is hard to assess. Certainly in so far as utopias allowed thinkers to exceed the imaginative impedimenta imposed by the Catholic Church, they provided conceptual possibilities that were akin to science fiction proper. The Italian churchman Tommaso Campanella (1568–1639) wrote his More-like utopia, *La Citta del Sole* ('The City of the Sun'), in 1602, and published it in Latin as *Civitas Solis* (1623). It describes an ideal city, built of seven concentric circular walls, ruled by a benign philosopher-king, called Hoh, who is also chief priest. As in More's *Utopia* property is held in common, although Campanella's vision is more technologically inventive: land-carts driven by great sails and self-propelling ships and flying machines are mentioned in passing, and elements of science and culture are written on the walls for public edification. Indeed, their utopian harmony is so advanced that none among the population suffers from catarrh or farting. Like More's, Campanella's Latin title contains a pun (it is more obvious in the original Italian title): his city is both 'Solis', 'of the sun', and 'Solus', 'of the self': its circular embedded design makes it a perfect figure for the emerging bourgeois subject, centred on a Reason-ego and in perfect harmony with itself.

Francis Bacon, later Baron Verulam (1561–1626), is often credited with advancing the cause of what today we would call 'science'. His fragmentary utopia, *New Atlantis: A Work Unfinished* (1627), has inaccurately been called 'the first science-fiction novel' (Carey, p. 63). A truncated work (in one modern edition it runs to little over 30 pages), *New Atlantis* does contain a certain amount of speculative science, including glancing allusion to submarines and automata. But it lacks narrative, and indeed its second part is reduced to a mere list by the manuscript's incompletion at Bacon's death ('The prolongation of life. The restitution of youth in some degree. The retardation of age. The curing of diseases counted incurable', and so on through 33 items).

In this same tradition of society-as-novum, there was a number of seventeenth-century *voyages extraordinaires* that, like More's or Bacon's, did not leave the Earth. Joshua Barnes' *Gerania* (1675) describes an imaginary kingdom of miniature humans 'on the utmost Borders of India', which some critics have seen as a precursor to Jonathan Swift's Lilliputians. The unfortunately named Richard Head (fl. 1660–80) composed a number of fables of imaginary lands, including *The Floating Island* (1673) and the supernatural *voyage extraordinaire O-Brazile: or The Inchanted Island* (1675). *The Isle of Pines* (1668) by Henry Neville (1620–1694) relates how an English mariner, George Pine, discovers and populates a new land near the coast of Australia, and produces 12,000 descendants in less than a century. This imaginary kingdom is distinguished by a remarkable sexual explicitness: one modern critic lists the erotic episodes ('polygamy, voyeurism, cross-class intercourse … miscegenation and orgiastic sexual indulgence') and notes that ' "pines" is an anagram of "penis" ' (Bruce, pp. xxxvii–viii). It is debatable how useful it is to describe this 'pornotopia' as science fiction; but Neville's fantasy of sexual fulfilment as social-*novum* was no one-off.

The French author Gabriel de Foigny (*c.* 1650–1692) spent some time in his youth as a Franciscan monk, but later fled to Switzerland and converted to Protestantism. His *La Terre Australe connu, c'est-à-dire les Aventures de Jacques Sadeur dans la Découverte et le Voyage de la Terre Australe* (1676) imagines Australia to be populated by a race of hermaphrodites.[6] These red-skinned creatures ('d'une couleur qui tire plus sur le rouge que sur le vermeil', Foigny, p. 83) live peaceful and utopian lives: naked, sexless, rational and harmonious, 'les Australiens sont exempts de toutes passions' (Foigny, p. 140). The hero, Sadeur, is by chance a hermaphrodite himself. Shipwrecked on the Australian coast he is happily accepted by the Australians. It transpires that the inhabitants of this utopia cannot tolerate imperfection of any kind. Foigny's red-skinned Australians throttle at birth any of their children who are born with only one sex, and they prosecute a vigorous war of annihilation against conventionally gendered natives ('Les Fondins', or 'half-men'). Sadeur falls in love with a Fondin woman, and eventually leaves this supposedly ideal land.

This strange utopia reflects a series of authorial anxieties about the sexual and social constraints of late seventeenth-century Europe. The text itself was bowdlerised in later publication. James Burns notes that

> In chapter five, which was almost entirely cut from the 1692 edition, Sadeur discusses such issues as public nudity as a means of ridding people of dangerous preoccupations, male tyranny over women, and sexual equality. Despite his noble pretences in this chapter and throughout the work, it is clear that Foigny, who was run out of several French towns for bigamy and other sexual crimes, used the landscape and society of his Austral world to indulge imaginatively in less wholesome practices. We glimpse the sexual appetite that got the author in trouble when Sadeur discusses his participation in a battle against the Fundians and his attempt to rape two female prisoners of war. (Burns, p. 5)

But although modern interest in Foigny's work has, for understandable reasons, tended to dwell on the sexual features, the importance of the book for the development of SF is rather different. In his own day Foigny's hermaphroditic Australians caused offence less because of erotic unconventionality, and more because he represented them as born without original sin. 'Foigny was persecuted', notes Geoffroy Atkinson, 'because the Australians of his novel did not sin in Adam' (Atkinson, II: 17). In other words, Foigny's book represents a theological otherness, an alien land outside the conceptual schema of the Church.

Foigny's work was part of a vigorous tradition of works that located utopias and satirical dystopias on the opposite side of the globe. One of the earliest is Joseph Hall's *Mundus alter et idem sive Terra Australis ante hac semper incognita lustrata* ('A World Other and the Same, or the Land of Australia until now unknown', 1605) which has his hero Mercurius Britannicus ('A British Messenger') journey to Australia on his appropriately named ship *Phantasia*. Mercurius visits five Australian territories: Crapula, populated exclusively by gluttons; Yvronia, full of drunkards; Viraginia, dominated by women; Moronia, peopled by fools; and Lavernia, a land of brigands. Hall's tone is largely satirical, with Australia functioning as a straightforwardly figurative contemporary London, although Hall's delight in the strangeness of his imagined land gives the work a more engaging imaginative flavour than might otherwise be the case. It was certainly popular, and Hall, Bishop of Norwich and an energetic Protestant evangelist, was known as the 'English Seneca'.

A number of other Australian utopias enjoyed success: Thomas Killigrew's English *Miscellanea Aurea: The Fortunate Shipwreck, or a Description of New Athens in Terra Australis incognita* (1720) narrates a shipwreck's discovery of 'New Athens' in Australia, in which descendants of ancient Greeks and native Australians live in harmony. Nicolas Edme Restif de la Bretonne (1734–1806) published an account of a voyage to Australia by flying machine, *La Decouverte Australe par une homme-volant* ('The Discovery of Australia by a Flying Man', 1781; discussed in the following chapter). By the eighteenth century, however, the number of actual *terrae incognitae* on Earth was rapidly diminishing. Utopian writers, and writers interested in wholly new imaginative voyages of exploration, were forced to turn their imaginations in other directions.

One alternative, intermittently utilised by fantastic authors, is that new worlds might be discoverable *beneath* the surface of the Earth rather than upon it. The German picaresque novel *Simplicissimus* (1668; a considerably expanded version was published 1671) by Hans Jacob Christoffel von Grimmelshausen (1621–1676) follows its hero widely across the surface of the globe in largely realist-satirical mode, but also includes various more fantastical interludes. One of these latter is an SF-utopian episode in which Simplicissimus visits the happy society of 'den Sylphis in das Centrum terrae' ('the sylphs who inhabit the middle of the Earth', Grimmelshausen, p. 427). But this episode is short and rather at odds with the grimly comic verisimilitude of the protagonists' misadventures across war-torn Europe which constitutes the bulk of the novel. Another strategy was to envisage utopia in a spiritual realm. Jean de La Pierre's *Le Grand Empire de l'un et l'autre monde, divisé en trois royaumes: le royaume des aveugles, des borgnes et des clair-voyants* (1630) deals, as its title suggests, with a 'Great Empire' divided into three kingdoms: of the blind, the one-eyed and the 'clear-seeing'.

Robert Applebaum argues that Joseph Hall's satirical Australian dystopia in *Mundus alter et idem* in fact marks a limiting fiction for terrestrial imaginary voyages. He identifies what he called 'Hall's most incisive yet subtle argument ... that what was once believed to have been *out there*, in the world beyond European experience, structuring our hopes and desires, our ontological sense of who and where we are, was never anything but an absurd, wishful projection of what we already are *in here*.'

> What we have discovered thanks to the new geography, in other words, isn't an object world of enchanted or useful phenomena, a land of riches ready to be appropriated and exploited, but rather ourselves, our disenchanted selves, confronting the vanity of our worldly wishes. Viraginia, Crapulia, Lavernia, Moronia – that's what's out there, and these sinful countries are clearly nothing but projections of our worst appetites and failings ... Terra Australis is in effect England itself, cut off from itself, spatially inverted, distorted, and magnified, but nevertheless its *self, idem*' (Applebaum, p. 11)

This takes the Morean *Utopia*, with its clear and manifold points of application to More's own England, to a logical extreme. Only off-world, we might think, can the conceptual space of possibility, and of radical otherness, be preserved. There is, we might say, a dead end-ness in the very project of utopia.

Future tales

But if 'utopia' becomes increasingly untenable in *this* world at *this* time, this fact provides an impetus for writers who chose not to relocate their utopias to another

world, instead to imagine how things might be better in another *time*. One of
the most important literary inventions of the seventeenth century is the tale of the
future. It would not be until the twentieth century that 'the future' would become
the default setting, as it were, for the SF tale; and most science fiction in the
eighteenth and nineteenth centuries was located in versions of its present day.
The few exceptions are interesting primarily as precursors to the twentieth century's 'cult of the future', but the earliest examples of 'futurological' fiction do
illustrate the ways in which SF in general was separating itself from religious discourse. The two earliest examples of futurist SF are Francis Cheynell's (1608–1665)
six-page political tract *Aulicus: His Dream of the King's Second Coming to London*
(1644) and Jacques Guttin's *Epigone, histoire du siècle futur* (1659). The first is an
anti-royalist political squib ('aulicus' means 'princely' or 'belonging to a prince's
court'); the second a more considered attempt to portray a future society. The
irruption of future-imaginings into literature is clearly of the utmost significance.
Paul Alkon's *Origins of Futuristic Fiction* contains a compelling reading of Guttin's
novel, to which my own account is deeply indebted.

In a sense it misrepresents this novel to see it as 'futuristic'. On the contrary, in
fact, it rehearses romance and epic tropes in a fully nostalgic manner. The frame
narrative, describes how Epigone (whose name means 'posterity' or 'after-born')
and his friends, caught in an enormous storm at sea, are shipwrecked on the
apparently African coast of 'Agnotie' (Guttin's own glossary defines this as 'terre
inconnue'). They are taken inland to a mighty city, where by virtue of a crystal
'translation' artefact they are able to communicate with the strange natives, to
whose monarch Epigone relates his various adventures. All this, despite being
explicitly located in the future age, draws heavily on Vergil's *Aeneid*. The adventures themselves, including an interlude in a female-ruled Amazonian kingdom
devoted to sensual pleasure, echo Homer's *Odyssey*. The reversals, adventures, love-
story elements, sword fights, escapes and all the usual baggage of conventional
'romance' dilute the notional futurity. As Alkon notes, 'Guttin's "future century"' is
not unequivocally a future at all' (Alkon, *Origins*, p. 37). It would be another 100
years before SF systematically imagined the future as an *other*.

It would be wrong, of course, to say with Paul Alkon that, prior to these two
works, 'writers from antiquity to the Renaissance never tried future settings'
(Alkon, *Science Fiction*, p. 21). There were many books that imagined the future;
the point is that they all, taking their inspiration from biblical prophecy, located
their future-visions in an exclusively *religious* idiom. Indeed, Alkon knows this
very well; in his *Origins of Futuristic Fiction* he clarifies that before Guttin's romantic
adventure, 'the future was reserved as a topic for prophets, astrologers, and practitioners of deliberative rhetoric' (Alkon, *Origins*, p. 3). This is the significance of
Epigone: not that it convincingly dramatised a possible future world, but that it
invoked 'futurity' in a secular, non-religious idiom at all.

It is because of the opening up of a secular idiom for cosmological speculation
in the seventeenth century that a new sub-genre of specifically secular futurology
can come into being, itself a form of conceptual *voyage imaginaire*, temporal rather
than spatial. Just as the seventeenth century was the time that provided some

writers with a newly materialist physical universe to explore, so the same period enabled the imagining of futures not determined by the Revelation of St John. Alkon also notes that 'Guttin's achievement as the first to endow any kind of fictional future with a significant aesthetic role is all the more striking for its isolation. No one followed his lead' (Alkon, *Origins*, p. 45). The next tale set in a consistently imagined future was not published until 1733.

Developments of science

It is worth stressing again the importance of the development of the discourses of 'science' through the seventeenth century to the growth of the form of SF. Earlier in this chapter I laid a certain emphasis on the particular religious component of the cultural logic of the period: the separation of a theological-magical 'Catholic' aesthetic from an imaginative-expansive 'Protestant' one, such that SF begins to emerge under the second ideological umbrella. Its rapid development across the seventeenth century was linked to a particular emerging discourse of 'science'. Kopernik's revolution, though apparently limited to cosmological models, in fact gave an (ultimately) unstoppable impetus to new sciences across the range of human enquiry. The shorthand for the distinction is between the traditional 'scholastic' thinkers, who believed that 'science' consisted of the accurate interpretation of traditional authorities, and the newer 'humanist' thinkers, who wanted to expand science in original directions.

Many 'scientists' (or, more properly, natural philosophers) of the sixteenth and seventeenth centuries worked within an inherited scholastic tradition that was at root Neoplatonic. It depended, as Desmond Clarke puts it, on a belief in 'the certainty of necessity of genuine knowledge claims, and on their universality', on 'the claim that our knowledge of physical nature depends ultimately on the reliability of our everyday observations and judgements', and on 'the very widely used distinction' between *matter* and *form*, a distinction that goes back to Platonic and Aristotelian models (Clarke, pp. 259–60). 'Forms' were thought of as ideal essences, 'matter' as contingent appearance. So (for example) insects come in a profuse variety of shapes, sizes, colours and behaviour whilst still all being recognisable as insects; according to the Platonic model, the common essence that links all insects would be thought of as the *form*, with the variable non-essential aspects of actual examples of insect life in the world as the *matter*. A great deal of pre-Enlightenment science involved definition and pronouncement upon the 'form' of things.

It was René Descartes (1596–1650) who effectively challenged this scholastic tradition, redefining 'science' in the process. He argued that talking about form was 'both redundant and pseudo-explanatory' (Clarke, p. 266). To say, for instance, that a magnet attracts metal because it possesses 'magnetic form' does not go beyond tautology, although such an explanation can give the investigator a false sense of having solved the question of magnetism. Descartes' writings encouraged a shift from the speculative to the practical in science, from essentialist thinking about 'forms' to thinking about the mechanisms that operated in the physical world, although he 'continued to accept the scholastic assumptions that we should construct our metaphysics first' (Clarke, p. 281) before proceeding to do science. As he put it in his *Principia Philosophiae* (1644):

> The whole of philosophy (today we would say 'knowledge') is like a tree. The roots are metaphysics (we would say 'philosophy'), the trunk is physics ('general material science'), and the branches emerging from the trunk are all the other sciences, which may be reduced to three principal ones, namely medicine, mechanics and morals ('biology', 'physics' and 'ethics'). (Cottingham et al., 1: 187)

The more science itself became an empirical, experimental discourse, and therefore the less place the speculative impulse had in the practice of science, the more important science fiction became.

Descartes exercised an immediate and direct influence on science fiction. The French Jesuit Gabriel Daniel (1649–1728) published *Voyage du monde de Descartes* in 1690 with the specific intention of critiquing Cartesian ideas; the book was quickly translated into English by Thomas Taylor in 1692 as *Voyage to the World of Cartesius*. Daniel makes reference to Lucian's fantastical adventures as a precursor, but he uses this device not for free imaginative play, but rather to challenge 'l'hypothèse de [Descartes'] tourbillons, qui est cependant le fondement de tout ce qu'il enseigne touchant le mouvement des planetes, [et] le flux et le reflux de la mer' ('his hypothesis of vortexes, which is the foundation of all his teaching about the movements of the planets and the tides', Daniel, p. 8).

Most of the volume is a dry discussion of the particulars of Cartesian philosophy and the ways they confirm or deny religious orthodoxy. The protagonists eventually travel to a 'third' heaven postulated by Descartes, although how they do so is not described: 'Je ne vous dirai rien du détail de ce voiage. J'espère dans quelques jours vous le faire à vous-même' ('I won't tell you any details about this voyage. I hope in a few days you'll undertake it yourself', Daniel, p. 56).

They walk for a long time in the 'grands déserts de l'autre monde', which represent (we are told) the chaos of Cartesian thought, and have lengthy conversations about Cartesian science, and finally convince him that his theoretical 'tourbillons' are not the true explanation for cosmic activity. In place of a materialist explanation of the cosmos, Daniel retreats to a Catholic doctrinal one: for example, the physical translation of narrator to this distant world is, he insists, akin to the translation of divine substance in place of bread at the Catholic sacrament (p. 70). The point here is that 'Descartes came under criticism' by many thinkers, precisely 'because his theory of vortices was thought to support a merely mechanical conception of physical law, and to leave no scope for divine will to operate' (Sutherland, p. 377). Even at the end of the seventeenth century the religious anxieties of Copernicus' revolution still haunted scientific, and science-fictional, exploration. They would continue to do so as the genre developed, although in increasingly buried forms.

References

Ackroyd, Peter, *The Life and Times of Thomas More* (London: Chatto 1998)

Alkon, Paul K., *Origins of Futuristic Fiction* (Athens: University of Georgia Press 1987)

Alkon, Paul K., *Science Fiction before 1900: Imagination Discovers Technology* (1994; London: Routledge 2002)

Appelbaum, Robert, 'Anti-geography', *Early Modern Literary Studies* 4.2 (September, 1998): 12. 1–17

Atkinson, Geoffroy, *The Extraordinary Voyage in French Literature* 2 vols (Paris 1922)

Bruce, Susan (ed.), *Three Early Modern Utopias: Thomas More, Utopia; Francis Bacon, New Atlantis; Henry Neville, the Isle of Pines* (Oxford: Oxford University Press 1999)

Burns, James R., 'Review of *Writing the New World: Imaginary Voyages and Utopias of the Great Southern Land* and *The Southern Land, Known*', *Early Modern Literary Studies* 2.2 (1996): 11. 1–7

Carey, John (ed.), *The Faber Book of Utopias* (London: Faber 1999)

Cavendish, Margaret, *The Description of a New World, Called the Blazing World* ('The Blazing World') (1666); in Paul Salzman (ed.), *An Anthology of Seventeenth-Century Fiction* (Oxford: Oxford University Press 1991), pp. 249–348

Cervantes, Miguel de, *Don Quixote* (first part published 1604, second part 1614; transl. J. M. Cohen, Harmondsworth: Penguin 1950)

Claeys, Gregory (ed.), *Modern British Utopias 1700–1850* 8 vols (London: Pickering and Chatto 1997)

Clark, Stuart, *Thinking with Demons: the Idea of Witchcraft in Early Modern Europe* (Oxford: Oxford University Press 1997)

Clarke, Desmond M., 'Descartes' Philosophy of Science and the Scientific Revolution', in John Cottingham (ed.), *The Cambridge Companion to Descartes* (Cambridge: Cambridge University Press 1992), pp. 258–85

Cottingham, John, R. Stoothoff and D. Murdoch (eds), *The Philosophical Writings of Descartes* 2 vols (Cambridge: Cambridge University Press 1985)

Cyrano de Bergerac, Savinien de, *L'Autre Monde ou les Etats et Empires de la lune* (*Voyage dans la Lune*) (1657; ed. Maurice Laugaa, Paris: Garnier-Flammarion 1970)

Cyrano de Bergerac, Savinien de, *Les oeuvres libertines de Cyrano de Bergerac* (*Oeuvres*) (1921; ed. Frédéric Lachèvre: Geneva: Slatkine Reprints 1968)

Daniel, P. Gabriel, *Voyage du Monde de Descartes* (1690) (Document électronique: http://gallica.bnf.fr/scripts/ConsultationTout.exe?E=0&O=N088113)

Davies, Norman, *Europe. A History* (Oxford: Oxford University Press 1996)

Davies, Tony, *Humanism* (London: Routledge 1997)

De Gourmont, Remy (ed.), *Cyrano de Bergerac, L'Autre Monde, et Physique, ou Science des Choses Naturelles* (Paris: Mercure de France 1926)

Ducos, Michèle, *Johann Kepler: Le Songe, ou Astronomie Lunaire* (Nancy: Presses Universitaires de Nancy, 1984)

Empson, William, *Essays on Renaissance Literature*, ed. John Haffenden, 2 vols (Cambridge: Cambridge University Press 1993)

Ferns, Chris, *Narrating Utopia: Ideology, Gender, Form in Utopian Literature* (Liverpool: Liverpool University Press 1999)

Foigny, Gabriel de, *La Terre Australe connu, c'est-à-dire les Aventures de Jacques Sadeur dans la Découverte et le Voyage de la Terre Australe* (1676; ed. Pierre Ronzeaud; Société des Textes Français Modernes 1990)

Foigny, Gabriel de, *The Southern Land, Known*, transl. and ed. David Fausett (Syracuse, NY: Syracuse University Press, 1993)

Fontenelle, Bernard de, *Entretiens sur la pluralité des mondes* (1686) (Document électronique: http://gallica.bnf.fr/scripts/ConsultationTout.exe?E=0&O=N088383)

Godwin, Francis, *The Man in the Moone: or, A Discourse of a Voyage Thither by Domingo Gonsales, the Speedy Messenger* (1638; ed. John Anthony Butler, 'Publications of the Barnaby Riche Society No. 3', Ottawa, Canada: Dovehouse Editions 1995)

Grimmelshausen, Hans Jacob Christoffel von, *Simplicissimus* (1668/71; München: Winkler-Verlag 1967)

Harth, Erica, *Cyrano de Bergerac and the Polemics of Modernity* (New York and London: Columbia University Press 1970)

Hawley, Judith (gen. ed.), *Literature and Science, 1660–1834* 8 vols (London: Pickering and Chatto 2003)

Howell, James, *Epistolae Ho-elianae, or familiar letters, domestic and foreign* vol. III (1712)

Hufton, Olwen, *The Prospect Before Her: a History of Women in Western Europe. Volume One 1500–1800* (London: HarperCollins 1995)

Huygens, Christaan, *The Celestial World's discover'd* (*Cosmotheoros de wereldbeschouwer*) (London, 2nd edn. 1722)

Kepler, Johannes (*Somnium*) *Mathematici Olim Imperatorii Somnium. Seu Opus Posthumum De Astronomia Lunari*, ed. M. Ludovico Kepler (1634).

La Bruyère, Jean de, *Oeuvres Complètes*, ed. Julien Benda (Paris: Gallimard, Bibliothèque de la Pléiade 1962)

Lambert, Ladina Bezzola, *Imagining the Unimaginable: the Poetics of Early Modern Astronomy* (Amsterdam and New York: Rodopi 2002)

Levi, Anthony, *Renaissance and Reformation: the Intellectual Genesis* (New Haven, CT: Yale University Press 2002)

Margolis, Howard, *It Started with Copernicus* (New York: McGraw-Hill 2002)

Marino, Giovan Battista, *L'Adone* (1622); vol. 2 of Giovanni Pozzi (ed.), *Tutte le Opere di Giovan Battista Marino* 5 vols (Milan: Arnoldo Mondadori 1976)

McColley, Grant, 'The Seventeenth-century Doctrine of a Plurality of Worlds', *Annals of Science* 1:4 (1936), 409–20

Meadows, A. J., *The High Firmament: A Survey of Astronomy in English Literature* (Leicester: Leicester University Press 1969)

Mirollo, James V., *The Poet of the Marvellous: Giambattista Marino* (New York: Columbia University Press 1963)

More, Thomas, *De Optimo Reipublicae Statu Deque Nova Insula Utopia*, ed. Edward Surtz and J. H. Hexter, transl. G. C. Richards, *The Yale Edition of the Complete Works of St. Thomas More* (New Haven, CT: Yale University Press 1965), vol. 4

More, Thomas, *Utopia*, ed. George M. Logan and Robert Adams, transl. Robert Adams (Cambridge: Cambridge University Press 1989)

Nicholson, Marjorie Hope, *Voyages to the Moon* (1948; New York: Macmillan 1960)

Olin, John C. (ed.), *Interpreting Thomas More's Utopia* (New York: Fordham University Press 1989)

Partington, Angela (ed.), *The Oxford Dictionary of Quotations* (4th edn., Oxford: Oxford University Press 1992)

Rosen, Edward (ed. and transl.), *Kepler's Somnium or Posthumous Work on Lunar Astronomy* (Madison, WI: University of Wisconsin Press 1967)

Salzman, Paul, *English Prose Fiction 1558–1700: a Critical History* (Oxford: Clarendon Press 1985)

Slawinski, Maurice, 'The Poet's Senses: G. B. Marino's Epic Poem *L'Adone* and the New Science', *Comparative Criticism: an Annual Journal* 13 (1991), 51–81

Sorel, Charles, *La vraie histoire comique de Francion* (1626; ed. Emile Colombey. Paris: Garnier 1909)

Sorel, Charles, *Le Berger extravagant: où parmi des fantaisies amoureuses on void les impertinences des romans & de poésie. Remarques sur les XIV livres du Berger extravagant, où les plus extraordinaires choses qui s'y voyent sont appuyées de diverses authoritez, et où l'on treuve des recueils de tout ce qu'il y a de remarquable dans les romans ...* (Toussainct de Bray, 1627)

Sutherland, James, *English Literature of the Late Seventeenth Century* (Oxford: Clarendon Press 1969)

Tauber, Alfred I., *Science and the Quest for Reality* (New York: New York University Press 1997)

Thorndike, Lynn, *A History of Magic and Experimental Science: Volume VII, The Seventeenth Century* (New York: Columbia University Press 1958)

Wagner, Geraldine, 'Romancing Multiplicity: Female Subjectivity and the Body Divisible in Margaret Cavendish's *Blazing World*', *Early Modern Literary Studies* 9.1 (May, 2003): 1. 1–59

Walker, D. P., *Spiritual and Demonic Magic from Ficino to Campanella* (London: Warburg Institute 1958)

Wilkins, John, *The Discovery of a World in the Moone. Or, A Discourse Tending To Prove that 'tis probable there may be another habitable World in that Planet* (1638; introd. Barbara Shapiro, Delmar NY: Scholar's Facsimiles and Reprints 1973)

4
Eighteenth-Century Science Fiction

The Enlightenment

With the advent of the eighteenth century we come to that period known as the 'Enlightenment'. Pigeonholing a flux as complexly variegated as history is clearly a dubious business, but a certain sketching-in of definition for the term Enlightenment is unavoidable because it has such an important role to play in the development of science fiction. By 'Enlightenment' we adduce an eighteenth-century philosophical consensus that agreed (mostly) on the primacy of reason and the importance of experimental and evidential science, and challenged the older religious myths and superstitions. Diderot's *Encyclopédie* (1751–65), a prestigious 17-volume collection of all that was then known, stands as one icon of the age; and the French *philosophes* ('philosophers') who contributed to it include some of the most famous names of the period, names such as d'Alembert, Rousseau and Voltaire. It is no coincidence, therefore, that the majority of significant SF produced in this century, particularly in its latter decades, was French, and that it tends to embody precisely this rationalist and questioning ethic.

The *Encyclopédie* represents a belief in the virtue of complete knowledge, especially scientific, material knowledge. In the words of Isaiah Berlin, 'the eighteenth century was perhaps the last period in the history of Western Europe when human omniscience was thought to be an attainable goal' and when almost all thinkers, despite their disagreements, agreed that 'the truth was one single, harmonious body of knowledge', the acquisition of which would solve humanity's problems (Berlin, p. 14).

Jonathan Rée expresses a very salutary suspicion of the habit of dividing history into neat cultural periods ('the Middle Ages, Renaissance, Romanticism, Modernism and so on'). But even Rée concedes that various different forms of 'period thinking' all acknowledge that a key epistemic shift happened in the eighteenth century.

There might be dozens of alternative histories of the present age, but they all intersected at some point in the 18th century known as the Age of Reason, or more vividly the Enlightenment (or *les siècle des lumières, die Aufklärung* or *l'illuminismo*). Essentially the Enlightenment was taken to be Europe's concerted effort to cleanse itself of the last residues of barbarism and medieval superstition and replace them with liberalism, science and secular philosophy. (Rée, p. 21)

Rée's portrait is a deliberately simplified account of a complex period, but it pinpoints the extent to which the cultural logic broadly associated with the Enlightenment included an openness to rational, logical frames of thought that facilitated speculative and 'scientific' fictions, leading to works that modern readers of SF are more likely to recognise as consonant with the protocols of the genre as they understand it.

One crucial reason for this was the fact that (to quote A. J. Meadows) 'the early eighteenth century saw an almost universal acceptance of a belief in a plurality of worlds' (Meadows, p. 126). Meadows' 'universal' means universal among scientists; nevertheless this is a vital staging-post in the development of the genre. As I have argued, it is easy to underestimate the significance of this theocracy-eroding belief. Theologically speaking, it can lead believers away from the Trinity and towards Deism. Culturally, it opens the cosmos as a material space, available to the imaginative explorations and colonisations we call SF.

In a sense it perhaps gives the wrong impression of the period to stress the scientific rationalism implicit in the term Enlightenment. In SF, as in culture in general, the tension between materialist and theological cosmic narratives is still acute, and the problematic of Catholic/Protestant perspectives on other worlds which had given birth to SF is still a live issue. It is striking, for instance, that as late as 1704 the philosopher John Locke (1632–1704) felt it worthwhile asserting that 'astronomers no longer doubt of the motion of the planets about the sun', as if this weren't already common enough knowledge to be taken for granted. And most scientists and astronomers of the period worked with the specific aim of reconciling material and theological versions of the universe. William Derham (1657–1735) was an astronomer and Fellow of the Royal Society who worked for many years cataloguing important observations of sunspots and the eclipses of Jupiter's moons. Yet the thesis of his most famous book is clear from its title: *Astro-Theology, or a Demonstration of the Being and attributes of God from a survey of the heavens* (1714). Like Derham, scientists of the age largely advocated what would come to be called natural theology. Edward Young (1684–1765), one of the early century's most celebrated poets, put it succinctly in his *Night Thoughts* (1742): 'An undevout Astronomer is Mad' (ix: 771). The point, which needs stressing, is that Enlightenment debates about the proper balance between materialist and theological ways of understanding the cosmos almost never involved 'secularism' as we understand the term. Almost all writers in this and the previous century believed it was a question of understanding the proper balance between God and His world, not of establishing the material universe as a God-free zone. But there probably were ways in which the Protestant north of Europe was more hospitable to scientific advance than the Catholic south.[1]

Newton and science-poetry

To a much greater degree than is the case in the seventeenth century, the eighteenth century sees close cross-pollinations between science and literature, particularly poetry. Isaac Newton (1642–1726) was perhaps the greatest scientist in the western tradition. Amongst his landmark scientific works are 'De Motu Corporum' ('On the Motion of Bodies', 1685) which codified what are now known as Newton's laws of motion and gravitation; the *Principia Mathematica* ('Principles of Mathematics', 1687), *Opticks* (1704) and *Arithmetica Universalis* ('Universal Arithmetic', 1707). Newton was part of the first great wave of modern scientific development, a period of rapid advancement in knowledge which included the work on atmospheric pressure of Robert Boyle (1627–1691), the natural and microscopic science of Robert Hooke (1635–1702), whose *Micrographia*, published in 1665, contained many drawings of microscopic creatures and objects; the infinitesimal calculus of Gottfried Leibniz (1646–1716), and the philosophy of John Locke (1632–1704). Neal Stephenson's *Baroque Trilogy* (2003–4) extravagantly dramatises the bustling vigour of Restoration and early seventeenth-century science precisely as a precursor discourse to SF.

Newton's eminence as a scientist, and the novelty of the world-view he elaborated, gave inspiration to a crowded school of scientific poets who stand in the margins of the development of SF. James Thomson (1700–1748), after Alexander Pope the leading poet of his age, wrote a poetry that was, in the words of Bonamy Dobrée, in 'thraldom to Newtonian physics' (Dobrée, p. 482). Thomson's early panegyric 'A Poem Sacred to the Memory of Sir Isaac Newton' (1727), published a year after the scientist's death, was the first of a great many Newtonian versefictions. Richard Glover (1712–1785) a merchant and MP, wrote 'A Poem on Newton' (1728) in which he asserted that 'Newton demands the Muse ... [he] shall raise her to the Heliconian height, / Where on its lofty top enthron'd, her head / Shall mingle with the stars'.[2] This anticipation of sublime possibilities in linking science and literature was intermittently justified only in the decades that followed. Often Newtonian science-poetry strayed into bathos as it attempted to recreate imaginatively scientific knowledge. This from *The Universe* (1752) by Moses Browne (1704–1787):

> Convenient Form, that round his *central Sun*
> The circling Planet might his period run;
> That purging *Tides* might unresisting flow,
> And seasons change, and genial Breezes blow[.]

An authorial footnote explains that the seasons and breezes depend on the deviation of the Earth's axis from the angular plane. This sort of deadening apprehension of contemporary science also informs the poetry of John Reynolds (1667–1727), who was deeply struck that planets remain in 'this elliptic Race, / Nor gallop out into the Fields of Neighb'ring Space' (*A View of Death, represented in a Philosophical Poem*, 1725). David Mallet's (1705–1765) *The Excursion* (1728) is a *voyage extraordinaire* through the solar system on the back of Pegasus – a device that points up the awkward symbiosis in the poem between poetical convention and Newtonian scientific imagining. Nevertheless the latter portions of this work, as the point of view moves past planets and approaches the stars, do achieve a degree of SF sense of wonder:

> Where unknown suns to unknown systems rise,
> Whose numbers who shall tell? Stupendous host!
> In flaming millions thro the vacant hung,
> Sun beyond sun, and world to world unseen,
> Measureless distance unconceiv'd by thought!
> Awful in their order; each the central fire
> Of his surrounding stars, whose whirling speed,
> Solemn and silent, thro the pathless void,
> Nor change, nor error know.
>
> (Dobrée, p. 506)

This delight in the changeless, error-free operation of Newtonian laws marks these Protestant poetics from the animistic apprehension of the cosmos found in earlier epochs. Mallet, Reynolds and Browne were all Deists, and although all include hymns to the Supreme Being in their poetry, there is no doubting that the sublime vastness of the universe they attempt poetically to evoke is a material, not a spiritual, environment.

But Newton, in addition to being the icon of the rationalist New Science, was also fascinated throughout his life by alchemy, and was always a fiercely devout man. The same SF dialectic between reason and magic filters through into much of the Newtonian poetry of the age. Samuel Edwards' (n.d.) poem *The Copernican System* (1728) attempts to blend Newtonian physics with astrological mysticism. In the same year John Theophilus Desaguliers (1683–1744) made plain in his title the hierarchical political agenda of his epic *The Newtonian System of the World, the best Model for Government. An Allegorical Poem* (1728). Bowden's 'A Poem Sacred to the Memory of Sir Isaac Newton' (1735) imagines the now deceased scientist's soul travelling through the solar system, stopping to observe wonders on the way.

> Mark where he halts on Saturn, tipt with snow,
> And pleas'd surveys his theory below;
> Sees the five moons alternate round him shine,
> Rise by his laws, and by his laws decline,
> Then thro' the void takes his immortal race,
> Amidst the vast infinity of space.
>
> (Meadows, p. 117)

This mystical-religious trope of souls touring the *material* solar system ('Saturn tipt with snow') was to become exceedingly popular in the later eighteenth and nineteenth centuries.

Science-poetry continued to be written throughout the eighteenth and nineteenth centuries; and although much of it must be considered science-fictional it is hard to deny that it constituted a dead-end in terms of the development of the genre. Why this should be is not immediately obvious; it may have to do with the great shift in poetics that occurred with 'Romanticism' at the end of the century; a Europe-wide reconfiguration of aesthetic value that superseded the Enlightenment poetic virtues of clarity, rationality and structured mimesis with new virtues of passion, intensity, the image and subjectivised poesis. Prose, and above all prose SF, is seen by many (though for reasons that are at root arbitrary) as a preferable medium for rationality and 'science'. It is certainly true most influential and important works of eighteenth-century SF were in prose. We shall from time to time in the pages that follow deal with significant works of SF poetry, but in a diminuendo.

Swift's *Travels*

The two key SF texts of the period were both written before a third of the century had passed: *Travels into Several Remote Nations of the World* (commonly known as *Gulliver's Travels*, 1726) by Jonathan Swift (1667–1774) and Voltaire's *Micromégas* (written 1730, published 1750). What these two works articulate is something crucial about SF itself. The many other *voyages extraordinaires* published during the century (see pp. 74–81 below), although often interesting, do little more than add marvels to a traditional narrative framework. Swift and Voltaire, however, rewrite the rules of imaginative speculation, freeing it from both the choking literalism of 'science poetry' and the deadening constraints of conventional religious thought.

Swift's *Travels* remains one of the most famous novels of the eighteenth century. Lemuel Gulliver, afflicted with a mania for travel, leaves England and sails the world. Shipwrecked on the island of Lilliput, he discovers a kingdom of people 'not six inches high' (Swift, *Travels*, pp. 55–6). He takes the Lilliputian side in their war with the equally diminutive Blefuscuns. He puts out a fire in the royal quarters by pissing on it, and in doing so, despite saving the palace and many lives, falls into disfavour at the court; the king decides his punishment must be to put out his own eyes and to avoid this fate he leaves the island, eventually returning to England. In the second volume, Gulliver travels again, this time to the land of the Brobdingnags where everything, including the inhabitants, is twelve times as large as conventional humanity. In the third volume, Gulliver sets out on a new voyage and encounters a number of new islands, such as Laputa over which a magnetically powered flying island called Balibali hovers. The fourth and final volume sees Gulliver encountering a utopian race of sapient horses, the Houyhnhnms.

Critics today are divided over whether it is proper to call Swift's *Travels* science fiction. For Brian Aldiss, the work 'does not count as science fiction, being satirical and/or moral in intention rather than speculative' (Aldiss, p. 81) – a strange reason to exclude it, we might think, since there is a great deal that is speculative about the book, and satire and speculation are by no means mutually exclusive, particularly in SF. Kingsley Amis thought that there was a difficulty in calling Swift's novel 'science fiction' in that 'there is no science (or technology) as such in the first two parts'. He suggests ways in which this 'problem' might be remedied:

> The Lilliputians perhaps represented as the fruits of an experiment in genetic microsurgery, the Brobdingnagians as those of mutation – though when one comes to think of it the births of the first-generation of Brobdingnagian babies to normal-sized mothers would raise acute difficulties. (Amis, pp. 12–13)

Amis's attitude reflects a widespread critical misunderstanding that takes the book to be 'unscientific' ('there is no science (or technology) as such ...'), or even *anti*-scientific. The latter reading is common enough, and is advanced by critics who contrast between the absurdity of life aboard the flying island of Laputa (in part 3), where the inhabitants are devoted to 'natural philosophy' on the one hand, and on the other the purity of life amongst the Houyhnhnms (the equine utopians of part 4), who are so removed from 'science' that they have not even discovered metallurgy. But it needs to be stressed that not only is Swift's great novel *inherently* science-fictional, all four parts are deeply steeped in science, and to such a degree that it becomes hard to avoid reading the book as being *about* science, or more particularly about the relationship between science and representation. Which latter phrase might function, we might think, as a short-hand definition of science fiction itself.

Such a reading, I should add, is at odds with most critical analyses of the book. The most common interpretation of Swift's *Travels into Several Remote Nations of the World* sees it as being 'about difficulties of identity and problems of judgment' (Erskine-Hill, p. 3). It is almost a commonplace among critics that Gulliver's wide-ranging travels in the external world are actually internal explorations of the individual psyche and eighteenth-century codes of subjectivity. In Terry Eagleton's opinion, the world explored by Gulliver is actually Gulliver himself, and his travels reveal himself to himself as 'an area traversed and devastated by intolerable contradiction' (Eagleton, p. 58). This reading of *Travels into Several Remote Nations of the World* as an ideological text has a great deal to recommend it and seems even common sense. The Lilliputians, for instance, function as a means of ridiculing the minute triviality of western court politics, with wars fought over which end of the egg to break when eating it, and political office won by those able to jump the highest.

But this is not the whole picture. Gulliver has nothing but praise for most of Lilliput's affairs: he admires the way that 'they look upon fraud as a greater crime than theft', the way the law not only punishes delinquency but actively rewards virtue (anyone who obeys all the laws for 'seventy-three moons' can claim certain privileges and is paid money out of the public purse), and the way the children are bred up 'in principles of honour, justice, courage, modesty, clemency, religion, and love of their country' (Swift, *Travels*, pp. 94–7). In other words, Swift's portrait of Lilliput engages *at one and the same time* in satirical mockery and quasi-utopian celebration. The same is true of the Brobdingnagians, a people from whom Gulliver receives wisdom and insight (such that, at the end of the book, he adjudges them 'the least corrupted' of the Yahoo-humans he has encountered, Swift, *Travels*, p. 41), although in their land he also suffers from court intrigue, is an unwilling participant in lechery and lives as a prisoner and freak to be exhibited.

Despite Amis's assertion that 'there is no science' in the first two parts of the novel, Swift returns to a number of scientific and technical discourses. The Lilliputians, for instance,

> are most excellent mathematicians, and arrived to a great perfection in mechanics by the countenance and encouragement of the Emperor, who is a renowned patron of learning. This prince hath several machines fixed on wheels for the carriage of trees and other great weights. He often buildeth his largest men-of-war, whereof some are nine-foot long, in the woods where the timber grows, and has them carried on these engines three or four hundred yards to the sea. (Swift, *Travels*, p. 61)

The first two parts of Swift's novel not only represent mathematics, they *embody* it, and engage the reader in a continual process of multiplication and division that works through this particular scientific method. As soon as the six-inch archer appears on Gulliver's chest, which is to say as soon as we readers understand that Lilliput has the dimensions of a one-twelfth scale model of a conventional-sized kingdom, and that Brobdingnag a twelve-times scale, then maths becomes core to our appreciation of the narrative.

Despite being excellent mathematicians and very capable machinists, the Lilliputians have not encountered clockwork and react with bemusement on discovering Gulliver's pocket-watch. The Brobdignagians do possess clockwork, which technology is apparently raised to 'very great perfection' (Swift, *Travels*, p. 142); but they lack military ordnance, and Gulliver tries, but fails, to interest the king in metal cannon fired by gunpowder. In other words, these various sciences, embodied by Swift in his fiction, relate to one science in particular: naval navigation. The Lilliputians, for instance, are represented as either expert or innocent of a range of sciences, all of which specifically relate to the business of navigation. Similarly, the analogies Gulliver uses to describe Brobdingnagians are mostly drawn from the world of shipping – unsurprisingly, perhaps, given that an early eighteenth-century writer looking for correlatives to great size would naturally look to nature, architecture and shipping, and that only the last of these categories combines great size with mobility.[3] The Brobdingnagians possess excellent maps, in addition to possessing excellence in mathematics and 'all mechanical arts' (Swift, *Travels*, p. 176), but they lack even the basic knowledge of the sort of armament a British ship of the line would carry as a matter of course. This encourages us to read *Travels into Several Remote Nations of the World* along the lines its title suggests: not (as the vulgarisation of that same title, 'Gulliver's Travels', prompts us) as being about Gulliver himself, whether as bourgeois subjectivity or cipher for Swift; but as about *travelling* into *remote* parts of the world. The 'science' in this eighteenth-century science fiction is the science of ocean navigation that enabled

Swift's contemporaries to travel to places practically further away and less well known than was effectively the case for those 1960s Americans who travelled to the Moon.

The third part of *Travels*, 'A Voyage to Laputa', is more obviously concerned with science. Gulliver is taken aboard a floating island and discovers it populated with a culture of scientists so entirely caught up in speculative astronomy that they have lost touch with reality. They can only converse when servants, 'flappers', knock their mouths with little bladders, and only pay attention to what is said to them when these same flappers knock their ears. Their scientific studies have created in them enormous disquietude:

> Their apprehensions arise from several changes they dread in the celestial bodies. For instance; that the earth by the continual approaches of the sun towards it, must in course of time be absorbed or swallowed up. That the face of the sun will by degrees be encrusted with its own effluvia, and give no more light to the world. That the earth very narrowly escaped a brush from the tail of the last comet, which would have infallibly reduced it to ashes. (Swift, *Travels*, p. 206)

Meanwhile their houses are poorly built and their wives have sex with strangers. As a satire on a particular caste of intellect, externalising the tendency of thinkers to 'have their heads in the clouds' by imagining a whole city that is literally up in the sky, this is very effective. But Swift goes to some pains to rationalise his fantasy island. Its mechanism for flight is a 'loadstone', 'in shape resembling a weaver's shuttle ... in length six yards, and in the thickest part at least three yards over'. The island of Balnibarbi over which Laputa flies is made of a particular stone which this loadstone forcibly attracts or repulses, depending on its orientation – a reasonable extrapolation from the then new science of magnetism. This more technically described SF *novum* relates, none the less, to the underlying ideological factor of the novel. It functions as an imaginative translation of the ocean-going ship, the 'wooden world' that carries an entire society with it as it goes. Translated into allegorical terms, the flying island is a fantastic extrapolation of the British fleet, an expensive aspect of the same standing army of which the Tory Swift disapproved.

Swift's text acknowledges the relative power balance between a nation that travels abroad and a nation that stays at home, in ignorance of others. Gulliver's narrative concludes with a plaintive appeal that 'those countries which I have described do not appear to have any desire of being conquered, and enslaved, murdered or driven out by colonies' (Swift, *Travels*, p. 344), but his rationale for wanting his discoveries left inviolate is unconvincing, deliberately so we may assume. For instance, he claims that the countries do not abound in gold, although the Brobdingnagians for one use myriad gold coins 'each piece being about the bigness of eight hundred moidores' (Swift, *Travels*, p. 140). More to the point, it requires a sort of blindness of imperial ideological logic to argue, as Swift ironically has Gulliver do, that 'the Lilliputians, I think, are hardly worth the charge of a fleet and army to reduce them', or to imply that the Brobdingnagians and Houyhnhnms would be too formidable as opponents, despite having already established that neither people possess the destructive technologies of artillery or explosives (Swift, *Travels*, pp. 342–3). We read past the end of the novel, following its own logic and applying our sense of the ideological conditions of early eighteenth-century Europe, and we see colonisation, exploitation, expropriation, slavery and death visited upon Lilliputians, Brobdingnagians and Houyhnhnms alike. Such is, of course, one of the key facets of the scientific rationalism of Swift's age.

Voltaire's aliens

Voltaire, the pseudonym of François-Marie Arouet (1694–1778), is for many the epitome of Enlightenment genius. Liberal where Swift was Tory, Voltaire rejected the notion of 'Providence' as validating social principle and argued instead for concepts such as humanity and morality. In part he expressed his political philosophy through written commentaries, histories and polemics, but he is best known today for a series of *Contes Philosophiques*, 'philosophical tales', which ridiculed and interrogated contemporary attitudes, the first of which was *Zadig* (1747), an Oriental tale. The most famous of these *contes* today is *Candide, ou l'optimisme* ('Candide, or Optimism', 1759), a book which makes serious as well as comic points from the disjunction between the protagonist's optimistic world-view (he is candid by nature as well as name) and the harder realities of life in eighteenth-century Europe.

Certainly Voltaire's *Micromégas* ('Littlebig', written 1730, published 1750) is a philosophical tale that displays a number of powerful SF tropes. Micromégas, the protagonist, is a gigantic alien, 8 leagues (5 km) tall, from the star Sirius. He travels through the galaxy, befriending a native of the planet Saturn who is only 1,000 fathoms high, a mere pigmy beside the protagonist. Together these two travel to Earth where they encounter a shipload of philosophers who are returning from an exploratory voyage to the Arctic Circle. Micromégas lifts the ship out of the ocean and examines the people aboard. Once the two extraterrestrials have got over their amazement that such *insectes invisibles* could possess intelligence and soul, they enter into conversation with them. The aliens question the philosophers on matters of physics – the distance from the Earth to the Moon, the weight of the Earth's atmosphere – and are impressed at humanity's knowledge of these matters. But when Micromégas extends the questions to inner matters ('dites-moi ce que c'est que votre âme, et comment vous formez vos idées', 'tell me about the nature of your soul, and how you form your ideas', *Micromégas*, p. 111) the philosophers reveal a multitude of conflicting theories, 'de Descartes … de Malebranche … de Leibnitz … de Locke'. Further questioning reveals man's essential ignorance. 'Mais qu'entends-tu par espirit?' asks Micromégas of the Cartesian ('What do you understand by *spirit?*'); the human replies, 'Que me demandez-vous là … je n'en ai point d'idée' ('Why are you asking me that? I've no idea', *Micromégas*, p. 111). Another thinker promises that everything to do with souls is to be found in the *Summa* of Thomas Aquinas, assuring the two extraterrestrial giants that they themselves, their worlds and stars, were all made 'uniquement pour l'homme' ('solely for mankind's benefit'). Micromégas and the Saturnian 'se laissèrent aller l'un sur l'autre en étouffant de ce rire inextinguible' ('fell over one another laughing with such inexhaustible laughter', *Micromégas*, pp. 112–13) at this. Before departing, Micromégas gives mankind a book of philosophy which he promises contains all the truth about things; this book is donated to the Paris Academy, but when opened its pages are revealed to be all blank. Voltaire's Lockean sympathies are evident throughout the work, not least in this gift of a *tabula rasa*, a blank book, at the tale's end.

Critics have frequently noted how Voltaire's fable re-uses the trope of giants and midgets from Swift's *Travels*. What is not so often remarked is how he inverts the dominant seventeenth-century SF premise: instead of travellers from the Earth encountering aliens and quizzing them about their Christian orthodoxy, he imagines aliens coming to Earth: the first such story. The earthlings act according to SF type and try to convince the aliens of the universal applicability of the Christian revelation; but Micromégas erases such certainties.

The work is science fiction not only in its premise of alien visitors, but in its ubiquitous connection to and fictionalisation of the scientific discourses of the day. As Roger Pearson observes:

> The celestial journeys of Micromégas to Saturn and then of Micromégas and the Saturnian to Earth are based on the very latest cosmology. As well as Newton's *Principia*, this included the work of Christiaan Huygens – especially his *Systema saturnium* (1659) but also his *Cosmotheoros* (1698) – as well as the work of Kircher, Keill and Wolff. There is nothing intrinsically fantastical about these journeys, for Micromégas has a sure knowledge of 'les lois de la gravitation et toutes les forces attractives et répulsives', and he is so well organized that he never has to stop for a comet. (Pearson, p. 59)

Voltaire's premise allows him to extrapolate the Copernican cosmological revolution into human affairs. Just as the Earth is no longer the physical centre of the cosmos, so mankind cannot be considered the philosophical or theological focus of the universe. Micromégas marvels at the physical insignificance of our planet's microscopic inhabitants, and his perspective is large enough to reveal the absurdities of human endeavour, which in turn focuses Voltaire's satire. So, we see the ridiculousness in the fact that, as one of the philosophers explains, 'il y a cent mille fous de notre espèce, couverts de chapeaux, qui tuent cent mille autres animaux couverts d'un turban' (at that very moment '100,000 idiots of our kind who wear hats, are killing 100,000 fellow creatures who wear turbans', *Micromégas*, p. 110) for the sake of a tiny portion of the, to Micromégas, insignificant globe.

The key to the text can be found in its animadversion towards the very notion of *comparison*. Micromégas and the Saturnian, before arriving on Earth, discuss Nature:

– Oui, dit le Saturnien; la nature est comme un parterre dont les fleurs ...
– Ah! dit l'autre, laissez là votre parterre.

– Elle est, reprit le secrétaire, comme une assemblée de blondes et de brunes, dont les parures ...

– Eh! qu'ai-je à faire de vos brunes? dit l'autre.

– Elle est donc comme une galerie de peintures dont les traits ...

– Eh non! dit le voyageur; encore une fois, la nature est comme la nature. Pourquoi lui chercher des comparaisons?

– Pour vous plaire, répondit le secrétaire.

– Je ne veux point qu'on me plaise, répondit le voyageur; je veux qu'on m'instruise.

('Yes,' said the Saturnian. 'Nature is like a flower bed in which the flowers ...' 'Ah!' said the other. 'Let it alone with your flower bed!' 'She is like,' the Secretary continued, 'a gathering of blondes and brunettes, whose dresses ...' 'Eh! What have I to do with your brunettes?' said the other. 'She is, then, like a gallery of paintings in which the individual traits ...' 'No!' said the voyager. 'I'll say it again: nature is like nature. Why do you look about for comparisons?' 'To please you,' replied the Secretary. 'I don't want to be pleased,' said the voyager. 'I want to be instructed.') (*Micromégas*, pp. 98–9)

This hostility to the very idea of simile is integral to Micromégas's approach to the universe; as with the misguided human philosophers at the end of the tale we understand that it is a mistake to try and translate the universe into metaphorical 'other' terms. This in turn provides an explanation for the various enormities within the narrative. Voltaire is not using shifts of scale for metaphorical purposes; he is highlighting the *actual* enormity of the cosmos, the sheer hugeness of the universe that eighteenth-century astronomy was beginning to reveal. Micromégas is as tall as he is, and the Saturnian as tall as he is, because these –Voltaire is saying – are the scales of the universe in which we live. *Micromégas* is not, in this sense, a metaphorical text. It is, precisely, an 'instructive' one.

Eighteenth-century *voyages extraordinaires*

Both Swift's *Travels* and Voltaire's *Micromégas* are examples of the *voyage extraordinaire*; the one of a western bourgeois subject travelling out into wondrous places, the other of wondrous aliens travelling to visit us. Between them, driven by the approximate ideological and religious similarities of their Swift–Voltairean, Tory–Protestant/Catholic–liberal (or 'magic–materialist') determinants, these two works establish the axes of future science fictional textual production.

The emergent western culture of quasi-imperial exploration and expansion determines both works, of course; and there is nothing surprising in finding many fantastical extrapolations of the principle of voyages of discovery and even conquest. Swedish writer Olof von Dalin's (1708–1763) story of extraterrestrials visiting Earth, 'Saga om Erik hin Götske' ('The tale of Erik and the Goths', 1734) might have had more impact had it been written in either Latin or a major European language. Eliza Haywood (1693–1756) published *The Invisible Spy* (1755) under the pseudonym 'Exploralibus'. This shapeless but entertaining novel is based on two particular pseudo-technological nova: a 'belt' that renders its wearer invisible, and a Dictaphone-style 'wonderful Tablet' which records 'every word that is spoken in as distinct a manner as if engraved' (Haywood, p. 5). These enable Exploralibus to spy on whomsoever he fancies and pass on to us, his readers, his accounts. The book, an accumulation of stories, anecdotes and bits and pieces, purports to be the result of his spying. Although these technical facilitators are treated partly in the tradition of the magical-fantastic (the narrator inherits them from 'a certain venerable person ... descended from the ancient Magi of the Chaldeans', Haywood, p. 3) they are none the less actual material objects subject to material problems – the book ends when the narrator loses the ability to delete what is written on his magic tablet, so that it fills up and he is forced to publish. In this lies its significance: its secularisation of a trope that had previously been the province of a magical Romance tradition. Haywood has taken the device from French writer Alain-René Lesage's (1668–1747) *Le Diable boiteux* (1707; frequently reprinted in English in the eighteenth century as *The Devil upon Two Sticks*). In Lesage's novel the protagonist chances to free a crippled devil from a glass jar; the demon proceeds to lift the house-roofs of sleeping Madrid to reveal the secret histories of the people beneath. By replacing this wholly magical premise with a quasi-technical one, Haywood shows how half a century had changed the inflection of the fantastic adventure.

An earlier Haywood novel is *The Adventures of Eovaai, Princess of Ijaveo* (1736; reprinted as *The Unfortunate Princess* in 1741). The beautiful heroine of this utopian novel, set in a land that is temperate despite being located near the South Pole, is snatched into the sky by a wicked magician who flies away with her on a hybrid monster 'part Fowl, part Fish'. Although the book is rather stifled by its anti-Walpole satirical intention, it does illustrate one persistent concern of eighteenth-century SF: *flight*, either in the form of alien creatures with the innate ability to fly or else humans aided by machine.

Flying men are the inventive *novum* of the novel *The Life and Adventures of Peter Wilkins* (1750) by Robert Paltock (1697–1767). The bones of the story in Paltock's novel are conveyed in the lengthy subtitle to be found on the title-page, as concerning the protagonist (Wilkins, a Cornish mariner) and 'his shipwreck near the South Pole; his wonderful Passage thro' a subterraneous Cavern into a kind of new World; his there meeting a Gawry or flying Woman, whose life he preserved, and afterwards married her; his extraordinary Conveyance to the Country of Glums and Gawrys, men and women that fly'. These 'Glums and Gawrys' are humanoid terrestrial aliens, whose extensive wings (illustrated with four beautiful woodcuts in the first edition) are worn as a sort of clothing when they walk on the ground, to be unfolded – leaving them naked – when they wish to fly through the air. Carried through the air by a group of these flying humanoids, Wilkins finds out about their country and customs, and intervenes on behalf of his wife's nation in a civil war, ensuring victory and reconciling the two nations. After his wife's death, certain that his children will be well provided for, Wilkins yields to his homesickness and returns home. Clearly influenced by Swift's *Travels*, although with an ingenuousness of tone and invention quite at odds to Swift's sharp genius, *Peter Wilkins* is a striking little *voyage extraordinaire*, with well-realised alien creatures and a compelling knack for narrative verisimilitude. It is a tale of radical otherness that is nevertheless rationalised cognitively. Indeed, the tale is more expressive of the dialectic at the core of the developing genre than many eighteenth-century works.

For the most obvious aspect of Paltock's tale is that it functions precisely as a material rationalisation of fables of angels. The eighteenth-century craze for narratives of angelic encounter, and the exact nature of angelic being, was just coming to the boil. Peter Wilkins' Christian name might remind the reader of the first Catholic Pope; and although it was also a common enough Protestant name in the period, there does seem to be something significant in Paltock's nomenclature. Peter ('rock') begins his adventure with a precipitous shipwreck on 'a Rock of extraordinary Height' (Paltock, p. 62), referred to many times subsequently in the narrative as 'the Rock'. It is through this externalisation of his Christian name that Wilkins passes, sucked by a cataract, into the land of the Gawrys beyond. Wilkins' first encounter with Youwarkee, the Gawry he will later marry, is preceded by a dream of his English wife dead and transformed into an angel. Immediately on waking Youwarkee falls from the sky, as in Adam's dream. Her beauty, purity and nakedness (she 'had [no] other Covering than what they were born with', namely her wings, Paltock, p. 117) align her with angelic representation. The Gawry live more or less Edenic lives, eating the fruit of trees (when Wilkins brings them fish and birds to eat they take them to be a strange form of fruit) and inhabiting cities laid out on a grid-like pattern that resembles, in a diagram included in the text, nothing so much as a Christian cross (Paltock, p. 315). The Gawry take Wilkins to be a saviour in accordance with an ancient prophecy, uttered by an eminent Ragam, or priest, who had reformed the Gawry religion, overturning the Papistical-sounding 'Country-Worship of the great Image' only to be thwarted by 'the rest of the Ragams opposing him'. He declares on his deathbed that because the people have 'rejected the Alteration in [their] Religion' they will endure generations of civil war between West and East which only the coming saviour, Wilkins, will be able to end (Paltock, p. 243). But the point of this novel is not to present a satirical allegory of Catholicism, or Christianity, but precisely to refigure supernatural topoi in secular and material terms. Symbolically *Peter Wilkins* explores questions of sin, of otherness.

Wilkins comes in like a missionary, banishes Gawry superstition and, as he puts it, 'settles Religion' (Paltock, p. 276); but he does this in a remarkably non-doctrinal way, informing the Gawrys only that there is a 'Supreme Being, Maker of Heaven and Earth, of us and all things' (Paltock, pp. 280–1) (although he does later translate the Bible into their native tongue). Instead of theology he brings a wide range of technological skills and devices into this new world, and works on a resolutely material level, for example abolishing slavery, leading his people to victory in their civil war, and so on. Paul Baines has shown how thoroughly the novel inhabits a technological idiom: Wilkins' narrative is a detailed account of tools, mechanics, domestication of animals, rationalisation and the individual mastery of his environment; but his various devices (from pennywhistles to guns and cannons) are taken by the innocent Gawry to be forms of magic. According to Baines it is the novel's 'juxtaposition of technological power and imaginative fantasy ... its dramatic encounter between mechanics and superstition' (Baines, p. 21) that is most distinctive about *Peter Wilkins*. In the larger context of the development of SF we can see that it is indeed precisely the balance between sober-minded Protestant technological materialism and the symbolic level of transcendent spiritual associ- ation that the novel mediates; the textual conflict between religious and technical discourses expresses that still potent cultural anxiety that determined the origins of the genre to begin with. Another interesting 'flying man' fantasy, Restif de la Bretonne's *La découverte australe par un homme volant, ou le Dédale français* ('The discovery of Australia by a flying man, or the French Dedalus', 1781) is discussed below.

Subterranean adventures and interplanetaries

In 1741, the Norwegian-Danish writer Ludvig Holberg (1684–1754; at this time, Norway and Denmark were politically connected countries; Holberg was born in Norway but educated and lived in Denmark) published his hollow earth fantasy *Nikolai Klimi iter subterraneum* ('Nikolai Klim's Journey Beneath the Earth') in Latin; the work was translated into most European languages and remained popular well into the nineteenth century. Klim, a penniless ex-student, narrates how he fell into a chasm through the Earth's surface and into an interior cosmos, lit by a sun at the Earth's centre around which various worlds orbit. Klim himself falls into orbit around one of the planets of this internal system. A biscuit from his pocket goes into orbit about him, turning him, as he himself puts it, into a planetary body in his own right: an eloquent emblem of the materialising shift in emphasis of the *voyage extraordinaire*. In Greene's *Planetomachia* ('Battle of the Planets', 1589) the planets are supernaturally figured as individuals; a century and a half later Holberg quite properly reverses this, making a person into a planet, wittily pointing to the truth that in a Newtonian cosmos any material body might so become.

Landing on the world of 'Nazar', Klim discovers a quasi-utopian society populated by walking, talking, intelligent trees. Brian Aldiss does not think much of Holberg's arboreal aliens ('trees', he opines, 'do not effectively serve didactic purposes. Their bark is better than their bite', Aldiss, p. 79), but the potent imagi- native strangeness of Klim's subterranean voyage, its rather frenetic succession of satirical alien societies and details, lends the world a distinctive and impressive flavour. To take only the sentient trees: what is especially interesting about them is the way Holberg inhabits and materialises a religious-mythic convention of the pre-SF literary imagination: the forest of the suicides from Dante's *Inferno*, or the magical transformations of humans into trees scattered through Ovid's *Metamorphosēs*.

Works like *Nikolai Klimi iter subterraneum* (which is to say, *voyages extraordinaires* into worlds within the Earth) enjoyed considerable popularity throughout the eighteenth century, following Athanasius Kircher's dull work of speculative science *Mundus Subterraneus* in 1678. The anonymous French work, *Relation d'un voyage du pole arctique au pole antarctique par le centre du monde* ('Account of a Voyage from the Arctic to the Antarctic Pole', 1722) is a lively and interesting adventure narrative in this mode. An anonymous English text, *A Voyage to the World in the Centre of the Earth* (1755), locates a utopian society of animal-respecting vegetarians at the heart of the globe. Charles de Fieux, chevalier de Mouhy combined the popular genres of Oriental tale and subterranean adventure with an Egyptian hero in *Lamekis, ou les voyages extraordinaires d'un Égyptien dans la terre intérieure* (1787).

One of the most interesting of these subterranean stories was by the famous sexual adventurer Giacomo Casanova de Seingalt (1725–1798, more usually known as Casanova), who towards the end of his life published a lengthy subterranean imaginary voyage: *Icosameron, ou Histoire d'Édouard et d'Élizabeth Qui Passèrent Quatre-Vingt Un Ans chez les Mégamicres Habitans Aborigènes du Protocosme dans l'Intérieur de Notre Globe* ('Icosameron: or the History of Edward and Elizabeth who spent 81 years with the "Mégamicres", original inhabitants of the protocosmos in the interior of our world', 5 vols, 1788). Casanova's 'Mégamicres' (or 'Biglittles') owe an obvious debt to Voltaire's Micromégas; but the conception of these interior aliens engages directly with the sorts of theological anxieties that are more common in seventeenth-century SF. The idyllic life of the Mégamicres is revealed as a function of the fact that their world pre-dates, and was not contaminated by, Adam's original sin; but at the same time, and so as not to trespass on conventional theology, they are revealed as lacking souls. Their name derives from their 'little' stature but 'great' spirits.

Adventures within the solar system were also common. The German author Eberhard Christian Kindermann's (n.d.) novel *Die geschwinde Reise auf dem Luft-Schiff nach der Oberen Welt, welche jüngsthin fünf Personen angestellt* ('The Rapid Journey by Airship to the Upper World, Recently Taken by Five People', 1744) takes its five protagonists to Mars by balloon. Didacticism does not overpower Le Chevalier de Béthune's *Relation du Monde de Mercure* ('An Account of the Planet Mercury', 1750); a work of early SF unusual in not using the description of an imaginary Mercurian society as a vehicle for political satire or utopian fantasy or satire. Béthune's aliens are diminutive winged creatures who are ruled by mysterious but benevolent beings who dwell in the sun. Béthune insisted, in a preface, that the book was merely 'une fable, dans laquelle on a essayé de joindre à des idées amusantes par leur nouveauté, quelques observations utiles' ('a fable in which I have tried to link ideas amusing because of their novelty with a number of useful observations'); but the charm and power of the book belie this understatement.

The Swedish theographer Emmanuel Swedenborg (1688–1772) followed in the tediously mystical-religious footsteps of Kircher's *Iter exstaticum* (1656) with his allegedly visionary *De Telluribis* ('Of Earths', 1758; translated into English by John Clowes in 1787 as *Concerning the Earths in Our Solar System, which are Called Planets, and Concerning the Earths in the Starry Heavens; Together with an Account of their Inhabitants*). This work does involve imaginative voyages to a number of different planets, but the whole is suffused with a dogmatic if wishy-washy magical mysticism; the planets and aliens encountered are not differentiated in any but a spiritual sense, and the whole works directly towards justifying the religious cult Swedenborg was establishing. That the work has little value does not mean it had no influence; quite the contrary, some very major literary talents were inspired by Swedenborg's vision, most notably the Romantic poet William Blake (1757–1827). Blake as a young man was probably a straightforward Swedenborgian, although his greatest poetry was written in angry reaction against Swedenborg's influence. *The Marriage of Heaven and Hell* (1790) is a prose-poetry parody of Swedenborg's writings which surpasses the original so stratospherically as to render it irrelevant. It is also, in part, a subterranean fantasy in which the protagonist descends into a huge underground realm inhabited by some very alarming creatures through an entrance behind the altar of a church. His earlier *An Island in the Moon* (written 1774–75, though not published until long after Blake's death) belongs to the sub-genre of satirical Moon visits (see the following section of this chapter), and expresses Blake's profound hostility for the systematising tendencies of western scientists.

The Frenchwoman Marie-Anne de Roumier's *Les Voyages de Milord Ceton dans les sept Planettes* ('The Voyages of Lord Ceton in the Seven Planets', 1765) is more astrological than scientific: an English lord and his sister flee the court of Charles I and, carried up by an angel called Zachiel, tour the solar system, where they discover Mars to be a planet of war, Venus a planet of love, and so on. But although this veers towards the pre-scientific (not unlike Greene's *Planetomachia*, which perhaps inspired it) it is nevertheless part of a broader 'scientific' idiom. This is true to an even greater degree of French scientist Louis-Guillaume de La Folie's *Le Philosophe sans prétention ou l'homme rare* ('The Philosopher without Pretension, or the Rare Man', 1775). This strange but rather wonderful work concerns a visitor from the planet Mercury called Ormisais who flies to earth in an electrically powered sky-chariot, which he breaks by crash-landing it on Earth. Aided by an Earthling named Nadir, Ormisais searches for materials to mend his spacecraft. The Russian Vassily Lyovshin's *Noveishete puteshestviye* ('The Newest Voyage', 1784) is another materialist fantasy of interplanetary exploration, rationalised according to the latest discourses of eighteenth-century science.

The eighteenth-century Moon

As a location unmistakeably other and yet close enough to be reached by man, the Moon is the prime SF site. But the plausibility of the Moon as an actual site of habitation or exploration was eroded by eighteenth-century advances in astronomy. At the beginning of the century, in 1701, Nehemiah Grew (1614–1712) could take it wholly for granted that the Moon 'is another Terraqueous Orb, having its Atmosphere, Winds, Seas, and Tides; and herewithal a suitable tho' perhaps a different Furniture of Animals, Plants and Mines' (Grew, p. 10). By the end of the century, however, more accurate observations, especially of stars viewed on the Moon's rim, revealed that the satellite lacked an atmosphere, and therefore must be bereft of 'winds, seas and tides'. Samuel Taylor Coleridge (1772–1834) jotted in his notebook in 1794:

> Moon at present uninhabited owing to its little or no atmosphere but may in
> Time – an atheistic Romance might be formed – a Theistic one too. – Mem!
> (Coleridge, *Notebooks*, p. 1)

Though he never wrote this particular romance, Coleridge's balance of 'atheistic'-
materialist and 'Theistic'-spiritual perspectives on this tale of future lunar colonisation is very much to the point. In the 1830s Coleridge's view of possible alien life
in the solar system was similarly negative: where most commentators were assured
that some if not all planets were occupied, Coleridge complained, 'Must *all* possible Planets be lousy? None exempt from the Morbus pediculus ("louse disease")
of our verminous man-becrawled Earth?' (Coleridge, *Marginalia*, II: 887). This is a
remark that carries a flavour of the way a previously spiritual realm had been
contaminated by the material observations of the scientists.

Somewhere between this ideal inhabitable Moon and the airless desert is to be
found in English poet Thomas Gray's (1716–1771) Latin poem 'Luna habitabilis'
('The Inhabitable Moon', 1737), which populates the satellite with bizarre aliens
worthy of Kepler. The poem looks forward to the times when Imperial Britain will
turn the Moon into a colony, conquering the terrifying and apparently cyborg
aboriginals: 'acies ferro, turmasque biformes, monstraque feta armis, et non
imitabile fulmen' ('the army of iron, regiments of two-formed [*literally: 'consisting
of two parts of different kinds'*] monsters, great beasts full of armed men, and their
inimitable lightning'). Gray styles this war as a necessary pre-emptive strike
against a 'Lunae in orbe tyrannus, se dominum vocat' ('a tyrant in the Moon's
orb who says he is *our* master'); and there is an unambiguous triumphalism in his
portrait of the first British colonists ('primosque colonos') crossing space in
'classem volantem' ('flying boats') to settle the pacified new world (Gray, p. 301).

But by far the most common appearances of the Moon in eighteenth-century literature are as a location of comico-satirical ludicrousness. A work such as Aphra
Behn's (1640–89) *The Emperor of the Moon: A Farce As it is Acted by Their Majesties
Servants, at the Queens Theatre* (1687) makes no attempt to rationalise the science
of its satire. *The Consolidator, or Memoirs of Sundry Transactions from the World of the
Moon* (1705) by Daniel Defoe (1660–1731) is an uncharacteristic offering from this
brilliantly entertaining writer – uncharacteristic in the sense that it is so constipated with satiric and contemporary-allegorical reference as to approach a state of
absolute unreadability.

More readable, in a knockabout sense, is the quasi-Rabelaisian *Cacklogallinia* (1727)
by 'Captain Samuel Brunt'. Murtagh McDermot (the name is a pseudonym)
published *A Trip to the Moon ... containing Some observations, made by him during his
Stay in that Planet, upon the Manners of the Inhabitants* in 1728. According to
'Pythagorolunister' (John Kirkby, 1710–1780), in his *A Journey to the World in the
Moon* (1740), 'no material Engines, nor any possible Inventions can ever convey our
Bodies to the World in the Moon' and therefore it is only possible to 'visit these
Regions by *spiritual Analogy*' (in Claeys, 2: 5). *A Trip to the Moon: Containing an Account
of the Island of Noibla* (1764) by 'Sir Humphrey Lunatic' (also, obviously, a pseudonym)

is another work of this type. *The Life and Astonishing Transactions of John Daniel* (1751) by Ralph Morris (about whom little is known; the name is probably a pseudonym) also involves a journey to the Moon. William Thomson's *The Man in the Moon; or, Travels into the Lunar Regions by the Man of the People* (2 vols 1783) sees 'the Man in the Moon' descending to snatch up English radical politician Charles Fox (sarcastically described as 'the man of the people') on a tour of a satirical-allegorical moon.

Much of the point of the lunar location of these satires is its unarguable distance from Britain, a function that was also served by setting satires in far-flung portions of the globe. Tobias Smollett's (1721–1771) peculiar and scatological allegory *The History and Adventures of an Atom* (1769) uses Japan in this role. The title character is a sentient atom that, lodging itself in the pineal gland of one Nathaniel Peacock, is able to communicate to him its adventures. It was originally part of the arse of the Japanese prime minister Fika-kaka (whence it passed through a duck and a sailor into Peacock) and it relates the hectic, slapstick events of the Japanese court, invoking a series of teenagesque comic names ('Sti-phi-rum-poo', 'Nin-kom-poo-po' and so on) by way of re-presenting English political life of 1757–67.

These sorts of lunar fantasy were also popular on the continent. Cornelie Wouters, Baronne de Wasse's *Le Char Volant; ou, Voyage dans la lune* ('The Flying Chariot or Trip to the Moon', 1783 is one). Another lunar fantasy was the work of Polish churchman Michal Dymitr Krajewski (?1735–1801). *Wojciech Zdarzyński, życie i przypadki swoje opisujący* ('The Life and Adventures of Wojciech Zdarzyński', 1785) takes its titular protagonist, Cyrano-like, to the Moon to explore the inhabitants and civilisations thereon.

Two slightly more interesting examples can be isolated from this mass of satirical-fantastical *jeux-d'esprits*. The anonymous *A Journey Lately Performed Through the Air in an Aerostatic Globe* (1784) has its narrator fly in his balloon, improbably enough, all the way to Uranus (or to 'the lately discovered planet Georgium Sidus' as it was then known) – the astronomers, he asserts, who declared it large and distant are wrong, for it is small and near by. Slight as this utopian satire is, it may be the first interplanetary voyage by balloon: a sub-genre that was to demonstrate surprising longevity. The protagonist of the anonymous *A Voyage to the Moon, Strongly Recommended to All Lovers of Real Freedom* (1793) also travels into space by means of 'that curious machine an Air-balloon' (Claeys, 4: 281), flying to the Moon, here called 'Barsilia', which he finds populated by serpentine aliens. These hard-working creatures are, in a parody of Tory England, oppressed by the 'Great Snake', who taxes six-sevenths of all their earnings, using most of the money to keep a pointless 'great wheel' turning in the capital city. The satire is of the sledgehammer sort (' "As to truth and justice", I replied, "they seem to be treated here as imaginary or contemptible things" ', Claeys, 4: 1997), but the detail is deftly handled and the whole thing achieves a surprising degree of freedom from the merely polemical.

Science fiction and Gothic fiction

Although SF was being written throughout Europe during the eighteenth century, two nations emerged as the key producers: Protestant England and Catholic France. This in turn reflects cultural and historical contexts: unlike pre-unification

Italy and Germany, England and France enjoyed a textually nutritive culture; unlike the smaller nations of Europe both countries were engaged at the violent nascence of technological and imperial augmentation. They were also two of the countries with the most pronounced history of Catholic/Protestant religious reformation, and therefore cultures determined by the dialectic at the core of SF.

But a graph of SF textual production throughout this century would demonstrate one curious feature: towards the end of the century the number of SF texts being written in English dropped away, a circumstance not reflected in France. This fact is made more remarkable when we consider that it was exactly this period in England that saw the great craze for Gothic fiction; and most historians of SF link the birth of the genre with Gothic writing. Brian Aldiss, in an influential argument, goes so far as to define SF as an offshoot of Gothic: he begins his critical history with 'the dream world of the Gothic novel, from which science fiction springs' and defines the genre as 'characteristically cast in the Gothic or post-Gothic mode' (Aldiss, p. 25). Given the wide currency of this argument, we may wonder if it is correct.

Gothic fiction is a popular category of academic pedagogy and research: a usefully delimited subgenre of fantastic literature of which many works remain very readable today (and which are therefore popular with students and with the framers of contemporary university syllabuses). Typically, a Gothic novel includes mysterious and sinister goings-on, usually involving supernatural agency such as ghosts or devils, although sometimes these events are explained away in rational terms. Many Gothic novels are located in distant, wild places, castles or monasteries in inaccessible portions of central Europe, where innocent young women are terrified, men have commerce with the Devil and there is much to do with graveyards, ruins and madness, all flavoured by a distinctive atmosphere of eroticised suspense, shock and terror.

There was an enormous vogue for such novels in Britain in the latter decades of the century. The start of this subgenre is usually taken to be 1764, with the publication of the hysterically overheated haunted house novella *The Castle of Otranto* by Horace Walpole (1717–1794). Writers such as Ann Radcliffe (1764–1823), Matthew Lewis (1775–1818, whose novel *The Monk*, 1796, is a particularly extreme tale of devilish imposture and priestly corruption) and Charles Maturin (1782–1824) enjoyed tremendous success. By the time Maturin's soul-sold-to-the-devil yarn *Melmoth the Wanderer* (1820) was published, the craze for this kind of writing had waned; nevertheless the influence of Gothic writing continued unmistakeably in other genres, in works as diverse as Coleridge's poetry, Emily Brontë's *Wuthering Heights* (1847), Bram Stoker's vampire story *Dracula* (1897) and latterday horror fiction and films.

But by the same token, there *is* something inimical to SF in Gothic writing. I say this with no desire to revisit the tedious rule-games engaged in by so many critics seeking to define science fiction (see chapter 1), but rather to restate the basic thesis of this present Critical History of SF. Gothic is another version of the magical Romance, although one with a distinctively dark flavour and widespread popularity. On those few occasions where apparently supernatural events are revealed not to be supernatural (the most famous instance of this is Ann Radcliffe's *The Mysteries of Udolpho*, 1794) the novel in question folds back into bourgeois love-story (the events of *Udolpho* are revealed as impedimenta designed to make the course of true love between Radcliffe's heroine and hero a more interesting read). In general Gothic fiction is irrational and magical in a pseudo-Catholic sense (or specifically, inspired by English Protestant paranoia about the supposedly monstrous aspects of Continental Catholicism). SF, as it had been developing for 200 years, was modulated by discourses of rationality and Protestant Deism. The closest any of the hundreds of Gothic novels published between 1764 and 1820 come to science fiction are the little known *The Balloon, or Aerostatic Spy* (1786, author unknown), which stays within the logic of ballooning established by the Montgolfiers; and the even more obscure *The Invisible Man, or Duncam Castle* (1800, author unknown) which recycles the 'belt of invisibility' from Eliza Haywood's *Invisible Spy* (1755; see above). Otherwise Gothic fiction was a non-SF idiom.[4]

But the reader who is prepared to accede to my assertion that SF does not begin with Gothic may nevertheless be curious why SF diminishes in England in the last decades of the eighteenth century whilst flourishing in France. In fact, Gothic is part of this answer. The shaping cultural context of European life was shaken up radically by the French Revolution of 1789; rapid social change presented itself as a pressing reality, and artists reacted, as might be expected, by attempting to mediate these new social circumstances via their art. French and Continental writers, most of them caught up in excitement about the new political developments, wrote (amongst other things) forward-looking rationalist and fundamentally revolutionary science fiction. Many English writers, working in a climate generally hostile to these developments, retreated into a Gothic rehearsal of fears and terrors. Putting it so baldly inevitably distorts a more complex actual set of textual problematics, but goes some way, I think, towards explaining why almost none of the remaining writers mentioned in this current chapter are English.

The French Revolution: pre-revolutionary and revolutionary science fiction

Robert Darnton, in his persuasive literary history *The Forbidden Best-Sellers of Pre-Revolutionary France* (1996), has shown the extent to which the repressive instincts of the French *ancien régime* were thwarted by an underground network of writers, booksellers and readers, circulating clandestine works that challenged the prevailing ideological orthodoxies. Many of these were science fiction, often utopian works which implied criticism of the imperfections of contemporary life by contrasting it with an alternative and often future ideal.

Indeed, romances set in the future formed the backbone of late eighteenth-century Continental SF. After a stalled beginning in the seventeenth century (Guttin's *Epigone*, 1659, was almost wholly a one-off), future fiction became increasingly popular in the early eighteenth century. The Irish churchman Samuel Madden (1687–1765) published *Memoirs of the Twentieth-century, or Original Letters of State under George VI* in 1733. This is a partly satirical work, but makes dull reading, although it is interesting for its Protestant author's hostility to the Pope, and also for its ambiguous mediation of SF and Fantasy. On the one hand, Madden is very interested in the advances of science; on the other, as in Kircher's *Iter Exstaticum* (which Madden mentions), it is an angel that brings the documents from the future that provide the subject of the *Memoirs*. The anonymous English cod-history *The Reign of George VI: 1900–1925* (1763) details a wish-fulfilment English victory over France led by the hero-monarch of the work's title; although the supposedly future world differs in very few particulars from the eighteenth century in which it was composed. Another anonymous work, *Private Letters from an American to his Friends in America* (1769), imagines a world only a few decades beyond its date of composition, but dramatises some major changes: a largely depopulated Britain ruled by America (rather than the other way around). More serious literary considerations of 'time' were also being written. Eighteenth-century Russia's greatest poet Gavriil Derzhavin (1743–1816) published his poem 'Na smert knyazya Meshcherskogo' ('On the Death of Prince Meshchersky') in 1779. In this ('his finest poetic work', according to Ilya Serman) Derzhavin contemplates 'the relationship between time and eternity' and 'the irreversibility of time's flow' (Moser, p. 88).

But it took the French Revolution, and its pre-revolutionary ideological climate, to turn future fictions into a major mode of late eighteenth-century SF. By far the most influential practitioner of this kind of writing was Louis Sébastien Mercier (1740–1814), political radical, writer and journalist who was active during the French Revolution. His political sympathies were evident a decade before this in his future-fantasy *L'An deux mille quatre cent quarante: Rêve s'il en fut jamais* ('The Year 2440: a Dream if Ever One Was', 1771). This popular work (Darnton calls it 'the supreme bestseller' of clandestine pre-revolutionary French novels; it went through at least 25 editions, Darnton, p. 115) uses the device of a dream to move into its future world. Its narrator, dissatisfied with contemporary life, falls asleep and wakes in the future Paris of the title date to find a utopian city run on rationalist and republican lines. Citizens cohabit peacefully. The Catholic Church has been abolished (replaced by a universal, rational Deism), as has slavery and colonialism. School education is based not on outmoded subjects like Greek and Latin but rather on algebra and physics. Everything is immeasurably improved from the *ancien régime* France of the late eighteenth century.

L'An 2440 circulated widely despite being considered a dangerously incendiary work by the pre-revolutionary French authorities; indeed, this contemporary political *cachet* goes a long way to explain its popularity, for it is a very dull read when judged by other criteria. Robert Darnton, quite accurately, describes its 'heavy and bombastic' moralising, 'always straining for sentimental effect, never betraying the slightest sense of humor' (Darnton, p. 115). But it is a significant text none the less; as Paul Alkon perceptively notes, its date is well chosen: 'what Mercier conspicuously avoids is some future date, such as the year 2000 or 2666, with possible millenarian or other religious significance' (Alkon, p. 122). The very specificity and the unremarkability of a year such as '2440' set an interesting precedent for subsequent SF future-visions. The replacing of Catholic-magical sensibilities not with atheism (John Carey notes that there is 'not a single atheist in the whole kingdom' and that 'if one were found he would be despised as a stupid wretch', Carey, p. 159) but with a Deism more familiar from writers in the High Protestant tradition, points to what would become the mainstream of nineteenth- and twentieth-century SF. And *L'An 2440* provided undeniable imaginative liberation to a great many writers.

Mercier's influence is unmistakable in the work of German writer Johann Albrecht (1752–1816). His *Dreyerley Wirkungen: Eine Geschichte aus der Planetenwelt* ('Three-way Effects: A Story from the Planet-world'; eight volumes appeared between 1789 and 1792) catches the French revolutionary mood of its time, criti- cising King Friedrich Wilhelm II via a romance ostensibly set on the planet 'Hidalschin'. Albrecht's democratic political sympathies also inform *Urani: Königin von Sardanopalien in Planneten Sirius* ('Urani: Queen of Sardanopolis on the Planet Sirius', 1790), in which SF reference is used as a way of coding subversive political sympathies. The title character of *Urani* is a straightforward version of Marie Antoinette transferred to the monarchy of a distant planet. The Swiss writer Heinrich Zschokke's (1771–1848) three-volume *Die schwarzen Brüder* ('The Black Brotherhood', 1791–95) also expresses the radical-democratic opinions of its author; the third volume being set in a 24th-century dystopia. Another Mercier- influenced text is the future-set play *Anno 7603* (1781) by the Norwegian John Hermann Wessel (1742–1785), which wittily interrogates assumptions about gender roles.

The prolific French writer Nicolas-Edme Restif de la Bretonne (1734–1806; some- times known as 'Rétif de la Bretonne') also used SF as a means of exploring radical reappraisals of conventional social mores. Indeed he was so much a person of his time that he 'took upon himself nothing less than the total reform of society' (Poster, p. 4). Perhaps his most famous book is *La Découverte australe par un homme volant, ou le Dédale français* ('The Discovery of Australia by a Flying Man, or the French Dedalus', 1781). The young protagonist, Victorin, invents a mechanical flying device that comprises cape-like wings and a head-worn umbrella-device. Utilising this he carries his girlfriend away to 'Mont-Inaccessible', an otherwise unreachable alpine summit, where they live happily together and start a family. From there they fly to Australia where Victorin establishes a colony, marries his son to a giant native from the island of Patagonie, and encounters many beast- hybrids (monkey-men, bear-men, dog-men and so on through sheep-, goat- and bird-men). The novel concludes with a description of the land of Mégapatagonie, an Antipodean anti-Europe whose capital 'Sirap', occupying the exactly opposite point on the globe to Paris, has natives (as we might guess from the city's name) who talk French backwards, and wear shoes on their heads and hats on their feet.

Restif had something of a reputation in his own day for licentiousness, but this reflected a belief in 'free love' rather than any Sadean excess (indeed, Restif 'hated Sade bitterly' and wrote an antidote to Sade's novel of sexual violence and rape *Justine* replacing that book's cruelty with 'eroticism'; Porter, pp. 385–6). Indeed, part of the point of *La Découverte australe* is to articulate Restif's idiosyncratic cos- mology (he conceptualised the solar system in terms of 'the copulation of the sun and the planets. All matter was alive and emitting seminal fluid; the sun gave forth light which fertilized the planets ... there was a God in Restif's universe who resembled the Deist's God in that he was the remote force who gave momentum to all matter. But in Restif's strange vision this God was composed of pure passion ... he was seminal fluid', Poster, p. 38). The beast-men are, in part, a cod-Rabelaisian satire on contemporary European life; but they also represent an inchoate but none the less interesting effort on Restif's part at a sort of evolutionary theory, a description of 'l'origine de l'homme et des animaux' that sees the different species as originally linked.

Restif's other science fiction works belong to a similarly radical climate of moral and political reappraisal. His late novel *Les Posthumes* ('The Posthumous Ones', 1802) is subtitled 'Lettres du tombeau', 'Letters from the Tomb', and is a compendium of various tales purportedly contained in letters written by the Président de Fontlhète, who, in the knowledge that he must die within one year, writes to his wife Hortense in order, gently and over time, to break the news to her. His strategy is to encourage her to believe in life after death, and thereby lessen the shock of his own demise. Volume One largely concerns the story of the lovers Yfflasie and Clarendon who die in an earthquake at the moment of consummating their marriage, and whose souls then float about learning the secrets of reincarnation. Volume Two is more interesting for our purposes: Fontlhète has invented mechanical wings and flies about the globe. On his travels he meets the Duc de Multipliandre, a man several thousand years old who is able to insert his own consciousness into other people by displacing their souls. Using Fontlhète's wings and his own psychic ability, Multipliandre becomes lord of the earth and establishes a utopia (based, of course, on Restif's own utopian writing). He then travels around the solar system, meeting various alien humanoids, to whose inmost thoughts he has access through his ability to displace their souls. Multipliandre discovers that comets are living creatures and planets merely the corpses ('ne sont que des cadavers flottants dans le fluide solaire') of dead comets. The closer he approaches the Sun, including to an unknown planet inside the orbit of Mercury called Argus, the more wisdom he discovers, from the inhabitants of Argus Multipliandre discovers the true nature of the cosmos ('la vie est le produit de la copulation ineffable de Dieu', 'life is the product of the ineffable copulation of God').

References

Aldiss, Brian, with David Wingrove, *Trillion Year Spree: the History of Science Fiction* (London: Gollancz 1986)

Alkon, Paul K., *Origins of Futuristic Fiction* (Athens: University of Georgia Press 1987)

Amis, Kingsley (ed.), *The Golden Age of Science Fiction* (Harmondsworth: Penguin 1981)

Backscheider, Paula R. (ed.), *Selected Fiction and Drama of Eliza Haywood* (Oxford: Oxford University Press 1999)

Baines, Paul, ' "Able Mechanick": *The Life and Adventures of Peter Wilkins* and the Eighteenth-Century Fantastic Voyage', in David Seed (ed.), *Anticipations: Essays on Early Science Fiction and its Precursors* (Liverpool: Liverpool University Press 1995), pp. 1–25

Berlin, Isaiah, *The Age of Enlightenment* (1956; Oxford: Oxford University Press 1979)

Carey, John (ed.), *The Faber Book of Utopias* (London: Faber 1999)

Claeys, Gregory (ed.) *Modern British Utopias 1700–1850* 8 vols (London: Pickering and Chatto 1997)

Clute, John and Peter Nicholls, *Encyclopedia of Science Fiction* (2nd edn., London: Orbit 1993)

Coleridge, Samuel Taylor, *Marginalia* vols. 1–2, ed. George Whalley (Princeton, NJ: Princeton University Press 1980–4); vols. 3–5, ed. H. J. Jackson and George Whalley (Princeton, NJ: Princeton University Press 1992–2000)

Coleridge, Samuel Taylor, *Notebooks: A Selection*, ed. Seamus Perry (Oxford: Oxford University Press 2002)

Darnton, Robert, *The Forbidden Best-Sellers of Pre-Revolutionary France* (London: HarperCollins 1996)

Dobrée, Bonamy, *English Literature in the Early Eighteenth-Century 1700–1740* (Oxford: Clarendon Press 1959)

Eagleton, Terry, 'Ecriture and Eighteenth-Century Fiction', in *Literature, Society and the Sociology of Literature*, ed. Francis Barker et al. (Colchester: University of Essex 1980), pp. 55–8

Erskine-Hill, Howard, *Jonathan Swift: Gulliver's Travels* (Cambridge: Cambridge University Press 1993)

Gray, Thomas, 'Luna habitabilis' (1737), in Thomas Gray and Oliver Goldsmith, *Poems*, ed. Roger Lonsdale (London: Longman 1969)

Grew, Nehemiah, *Cosmologia Sacra: or. A Discourse of the universe as it is the creature and kingdom of God* (London 1701)

Hawley, Judith (gen. ed.), *Literature and Science, 1660–1834* 8 vols (London: Pickering and Chatto 2003)

Haywood, Eliza, *The Invisible Spy* (1755; *Novelists Magazine* 1788)

Meadows, A. J., *The High Firmament: A Survey of Astronomy in English Literature* (Leicester: Leicester University Press 1969)

Moser, Charles A., *The Cambridge History of Russian Literature* (revised edition, Cambridge: Cambridge University Press 1992)

Nicholson, Marjorie Hope, *Science and Imagination* (Ithaca, NY: Cornell University Press 1956)

Paltock, Robert, *The Life and Adventures of Peter Wilkins* (1750; ed. Christopher Bentley, intro. James Grantham Turner; Oxford: Oxford University Press 1990)

Pearson, Roger, *The Fables of Reason: a Study of Voltaire's 'Contes Philosophiques'* (Oxford: Clarendon Press 1993)

Porter, Charles, *Restif's Novels, or an Autobiography in Search of an Author* (New Haven, CT: Yale University Press 1967)

Poster, Mark, *The Utopian Thought of Restif de la Bretonne* (New York: New York University Press 1971)

Rée, Jonathan, 'The Brothers Koerbagh', *London Review of Books* 24 (2002), 2: 21–4

Ross, Angus and David Woolley (eds), *Jonathan Swift* (Oxford: Oxford University Press 1984)

Swift, Jonathan, *Travels into Several Remote Nations of the World (Gulliver's Travels)*, ed. Peter Dixon and John Chalker, intro. Michael Foot (Harmondsworth: Penguin 1967)

Voltaire, *Romans et Contes*, ed. Henri Bénac (Paris: Editions Garnier 1960); *Micromégas*, pp. 96–113

Voltaire, *Candide and Other Stories*, transl. with an introduction and notes by Roger Pearson (Oxford: Oxford University Press 1990)

Whicher, George Frisbie, 'Mrs Eliza Haywood' (PhD dissertation; Columbia University 1915); published online at www.gutenberg.net/1/0/8/8/10889/10889.txt (EBook #10889)

5
Early Nineteenth-Century Science Fiction

Most SF written during the earliest years of the nineteenth century continued to excavate the premises determined by the late eighteenth-century revolutionary turmoil, especially with regard to future-fantasies. But the period as a whole came to be dominated in SF terms by two Anglophone writers, who had the greatest impact on the continuing development of the genre and will be discussed in some detail in this chapter. They are Mary Shelley and Edgar Allan Poe.

But taking the whole busy century in overview, we can see a number of broad fascinations working their way through the SF writing: an increased interest in the mystical and theological component of interplanetary or interstellar Romances; reflections in imaginative literary form of nineteenth-century advances in science, technology and industry; in some cases a direct mapping of Imperialist or political concerns into SF or utopian fantasy; and above all a much greater emphasis on the *future* as the arena for science-fictional storytelling.

Visions of the future and 'last man' fictions

For many the distinctive bias of SF is towards a representation of the future; and according to the eminent critic Darko Suvin, the 'central watershed' of the development of SF as a specifically futuristic fiction can be located 'around 1800, when space loses its monopoly upon the location of estrangement and the alternative horizons shift from space to time' (Suvin, 1979, p. 89). As we have seen, romances of the future came into vogue with Mercier's *L'An 2440* in 1771; but this became a particularly pervasive trope in the Romantic literature of the early nineteenth century. It is not clear why the late eighteenth to early nineteenth century marks such a watershed (Suvin himself notes that 'this turning, that cuts decisively across all other national, political and formal traditions in culture, has not so far been adequately explained'; Suvin, p. 73). It may have something to do with increasing levels of knowledge about the Earth itself following extensive European exploration, the need, in other words, to locate topoi for more radical alterity than the too well-known globe permitted.[1] The important consideration is not only that SF became a predominantly futuristic fiction, but that, early in the century, so many of these visions of the future looked to the very end of time, and to the figure of the last living human.

One particularly influential example of this sort of writing was *Le dernier homme* ('The Last Man', 1805) by Jean-Baptiste François Xavier Cousin de Grainville (1746–1805). Ridiculed on original publication so savagely that Grainville is reputed to have drowned himself in the Somme in misery, this book later came to be regarded in its own country as a great epic in prose. Indeed, to bring out its epic qualities it was rewritten in verse by Auguste-François Creuzé de Lesser in 1831 as *Le dernier homme, poeme imité de Grainville* (it was later rewritten in verse by Elise Gagne, in 1859, as *Omégar ou le dernier homme*). The idea was not original to Grainville: the concept of a single 'last man' looking around him on the end of the world goes back at least as far as La Bruyère's *Caractères* (1688). But it was Grainville's work that had the greatest impact in its native country. In fact Grainville's posthumous grip on early century French literary imaginations probably had much to do with its intoxicating intercalations of religion and secularism. Alkon's lengthy chapter discussing the novel (to which I am largely indebted) is entitled: 'the Secularization of Apocalypse'.

Grainville had been a priest, but had been defrocked during the French Revolution and married. His book is balanced between visions of religious apocalypse and a rational, materialist view of the end of the world. Some of the narrative is explicitly biblical (Adam watching a procession of damned souls into Hell, for instance); but the key characters, Omegarus and Syderia, the last human couple (fertile but childless) live, as Alkon points out, 'in a world without any trace of Christianity. Remnants of an advanced technology' (such as airships and the ruins of advanced cities) 'provide appealing glimpses of what human civilization might achieve if its reversion to barbarism could be turned around' (Alkon, *Origins*, pp. 165–6). The complicated plot hinges on the question of whether Omegarus and Syderia will become a new Adam and Eve, repopulating the world; but the story mediates explicitly religious and explicitly materialist idioms in a way that makes it impossible to separate them. The novel ends with a general apocalypse, one partly familiar from the biblical *Revelation of Saint John*, but also described in ways incompatible with scripture: this end of the world is, rather nihilistically, the end full stop.

Gloomy visions of specifically secular future apocalypse were something of a staple of English Romantic poetry. Charlotte Smith's poem 'Beachy Head' (published posthumously in 1806) imagines a Britain fallen into ruins, an 'almost postapocalypse world in which the remaining human inhabitants are forced to return to pre-civilised existence and inhabit the wrecked shells of once impressive buildings' (Bradshaw, p. 6). Anna Barbauld (1743–1824) put her radical political sympathies into the very many heroic couplets of *Eighteen Hundred and Eleven* (1812), a poem which imagines the end of Britain's 'Midas dream' of prosperity, and dramatises wealthy American tourists ('from the Blue Mountains, or Ontario's lake', l. 130) sightseeing in ruined London, 'the fractured arch, the ruined tower, / Those limbs disjointed of gigantic power', ll. 153–4). A much darker and more complete evocation of apocalypse was 'Darkness' (1817), a broodingly powerful poem by George Gordon, Lord Byron (1788–1824), in which the sun is extinguished and the inhabitants of earth live out the brief remainder of their days in desperation and despair:

> I had a dream that was not all a dream.
> The bright sun was extinguished, and the stars
> Did wander darkling in the eternal space,
> Rayless, and pathless, and the icy earth
> Swung blind and blackening through the moonless air.
>
> (Byron, p. 40)

Byron's acquaintance Thomas Campbell (1777–1844) wrote a Grainville-inspired poem entitled 'The Last Man' (1812) set 'ten thousand thousand years' in the future, although its solitary protagonist piously and un-Byronically declares that even the virtual extinction of mankind cannot 'shake his trust in God' (Campbell, pp. 232–4). A more wincing contrast to the power of Byron's gloriously pessimistic work can hardly be imagined. Mary Shelley's novel *The Last Man* (1826) (discussed below) and Thomas Hood's (1799–1845) rather goofy poem 'The Last Man' (1829) are later examples of the same sub-genre.

Not all future-fantasies were limited to this apocalyptic 'last man' idiom. The eighteenth-century Enlightenment impulse to imagine rational possible future worlds continued to inspire. More balanced, neither overly pessimistic nor deadened by piety, was the detailed future world portrayed by the German writer Julius von Voss (1768–1832) in *Ini: ein Roman aus dem ein und zwanzigsten Jahrhundert* ('Ini: a Novel from the Twenty-First Century', 1810). Félix Bodin (n.d.) published *Le Roman de l'avenir* ('The Novel of the Future') in 1834. For a long time an almost wholly obscure work, this novel has recently been resurrected by Paul Alkon's compelling reading of it in *Origins of Futuristic Fiction* (1987) as providing, for the first time in the genre, a 'poetics for futuristic fiction'. Any analysis of this work will inevitably be influenced by Alkon's critique, although it is possible for a reader to find the book duller and less inspiring than he does. Pitched somewhere between 'un burlesquement sérieuse et sérieusement burlesque', *Le Roman de l'avenir* is as much a disquisition on whether imagining the future is best done in an optimistic or pessimistic frame of mind as it is a novel. Top-heavy with elaborately elongated dedication, preface and introduction, the novel's episodic narrative breaks off before it concludes; yet, as Alkon points out, 'no writer of futuristic fiction before Bodin actually created a novel whose elements all worked coherently to elicit a sense of the marvellous within a plausible framework of realistic setting and action' (Alkon, p. 246). This did indeed become a governing aesthetic for the long tradition of subsequent future-fiction.

Another strategy for mediating the strangeness of possible futures was comedy. Emile Souvestre's (1806–54) *Le Monde tel sera qu'il sera* ('The World as it Will Be', 1846) takes a more dystopian perspective, its future Tahiti blighted by industrialism and mechanisation, although its vision is leavened with a well-handled humour. An English dystopia from a few years later, the anonymously authored *The Last Peer* (1851) is set in a grimly over-industrialised twentieth-century Britain, although it does not treat its topic with this saving wit.

Extraordinary voyages and automata

The traditions of eighteenth-century *voyages extraordinaires* continued in the early nineteenth century. The American fabulist Washington Irving (1783–1859) ironically re-imagined the European colonisation of America in terms of an invasion of the Earth by Lunarians in *A History of New York from the Beginning of the World to the End of the Dutch Dynasty by Diedrich Knickerbocker* (1809). The men of the Moon 'riding on hyppogriffs (as in *Orlando Furioso*, 1534), defended with impenetrable armour, armed with concentrated sunbeams, and provided with vast engines, to hurl enormous moonstones' (Franklin, p. 252) overrun the Earth, treating its inhabitants as hopeless primitives. Irving's rather charming Lunarians have the distinction, as far as I am aware, of being the first *green*-skinned aliens in literature (they also have their heads under their arms, tails, many legs and only one eye). The English poet Percy Bysshe Shelley (1792–1822, husband of the more famous Mary) includes a journey round the solar system in his youthful allegorical poem *Queen Mab* (1813). More interesting than the rather conventionally elevated poetry of this work ('Earth's distant orb appeared / The smallest light that twinkles in the heaven; ... / Innumerable systems rolled, / And countless spheres') is the extensive prose annotation Shelley appended. Note 2, for instance, uses the vastness of the cosmos and the plurality of worlds as an argument in favour of Shelley's own atheism: 'the plurality of worlds, – the indefinite immensity of the universe is a most awful subject of contemplation. He who rightly feels its mystery and grandeur, is no longer in danger of seduction from the falsehoods of religious systems, or of deifying the principle of the universe' (Percy Shelley, p. 296). Shelley takes the theologically destabilising Brunonian ideas of the plurality of worlds to a more rigorous materialist conclusion than most early practitioners of SF, although it should be added that, despite professing atheism, Shelley in fact believed in a universal 'principle of Necessity' that approximates more closely to Deism.

English utopian writing during this period was thin on the ground; the cultural reaction against Napoleon's continental ambitions (until the Battle of Waterloo in 1815) creating a climate in which only radicals were prepared to follow the French rationalist ideals. The four fat volumes of James Henry Lawrence's *The Empire of the Nairs; or the Rights of Women, A Utopian Romance* (1811) allow what could have been an interesting premise – a society which grants rights to women – to die slowly of fatty degeneration of the imagination: the longer the book goes on, the more boring it becomes. Indeed, though utopian novels continued to be published throughout the century, it was not until the 1880s and 1890s that they enjoyed widespread popularity.

The Dutch writer Willem Bilderdijk (1756–1831) was predominantly a religiously inclined poet, but his novel *Kort verhaal van eene aanmerklijke luchtreis en nieuwe planeetontdekking* ('Short Account of a Remarkable Aerial Voyage and Discovery of a New Planet', 1813) took the seventeenth- and eighteenth-century traditions of *voyages extraordinaires* in a new direction. Bilderdijk's balloonist protagonist flies up to a planetoid that orbits the Earth within the atmosphere. This world is described in plausible terms (flora and fauna but no indigenous

civilisation), and the book, though brief, has considerable charm. But its publication in Dutch, and the lack of translation into other European languages (until the
late twentieth century), seems to have severely limited its possible impact on the
development of the genre.

Much more influential in the subsequent development of the genre was *Der
Sandmann* ('The Sandman', 1816) by Ernst Theodor Amadeus Hoffmann
(1776–1822), the German writer and musician. This spooky story relates how a
supersensitive and poetic young man, Nathanael, falls in love with the beautiful
Olimpia; but he cannot see, until the end, what is obvious to everybody else in the
story: that Olimpia is an automaton, constructed by the sinister 'Doktor
Coppelius'. This simple-seeming story has had an astonishingly long life, in its
original form, and also in interpretations such as Léo Delibe's (1836–1891) ballet
Coppélia (1870) and, more famously, Jacques Offenbach's (1819–1880) popular
opera *Tales of Hoffmann* (1881). It remains a marvellously potent little piece.
Hoffmann's story was also the subject of one of Sigmund Freud's most important
pieces of literary criticism, *Die Unheimlich* ('The Uncanny', 1919).

Actual automata (relatively simple clockwork devices) were popular in the
eighteenth and nineteenth centuries; and often appear in literature as a means of
commenting comically or satirically upon the regimentation of human society.
An early piece of Dickens' journalism, for instance, imagines the ludicrous
possibilities of automated policemen.[2]

But what gives Hoffmann's robot its enormous purchase on the European
imagination is precisely its refusal of the comic possibilities of its premise. As Freud

'Automaton Police Office'; George Cruikshank's illustration to Charles Dickens' *Mudfog
Papers* (1838)

notes, Hoffmann's robot is 'uncanny', unsettlingly neither human nor non-human; a borderline creation of technology that forces the reader to reappraise her own relationship to notions of 'humanity' and 'nature'. In this respect it anticipated, by only two years, an even more influential SF tale (there is no evidence of direct influence) in which an 'uncanny' technological creation unsettles our assumptions about the identity of the human. This book has often, and persuasively, been called the origin-point of science fiction: *Frankenstein, or The Modern Prometheus* (1818) by Mary Shelley (1797–1851).

Mary Shelley's *Frankenstein* (1818)

This study, as is evident, does not concur with the belief – so commonly stated by critics as almost to approach dogma – that 'Science fiction starts with Mary Shelley's *Frankenstein*' (Alkon, *Science Fiction*, p. 1). Nevertheless, it cannot be denied that this tale has a good claim to the title of 'most influential nineteenth-century novel', and its presence in subsequent SF cannot be denied. Chris Baldick's acclaimed study *In Frankenstein's Shadow* (1987) traces its presence in myriad other texts, through Thomas Carlyle, Charles Dickens, Karl Marx and into the twentieth century, via cinema, cartoons, comics and many other examples of cultural dissemination. Indeed, so famous are these often distorted versions of the central mythos of Shelley's tale that it can be something of a shock for a reader to return to the original text. The narrator, Robert Walton, is a restless English gentleman who takes passage on a ship bound for polar realms in order to 'accomplish some great purpose' for which his soul yearns, although he is not specific about what (Shelley, p. 15). In the Arctic he meets the scientist Victor Frankenstein, who is near death, and tells his own story: how 'deeply smitten with the thirst for knowledge' and after long researches in both alchemical and scientific texts, he resolved upon 'the creation of a being like myself' (Shelley, pp. 6, 52). Frankenstein is deliberately uncommunicative about how he accomplished this: 'I dabbled', he says, 'among the unhallowed damps of the grave, or tortured the living animal to animate the lifeless clay … I collected bones from charnel houses' (Shelley, p. 53), although whether as raw material or merely as a model to copy is not made explicit. Nor does he explain how he was able to 'infuse a spark of being into the lifeless thing' (Shelley, p. 56). Some critics have seen this vagueness as a flaw in the novel. Perhaps the 'spark' is a metaphorical turn of phrase; there is no mention in the 1818 novel of electricity, although a second edition of the novel in 1831 included a preface in which Shelley suggests ways in which this conceptual gap might be filled in ('perhaps a corpse would be re-animated; galvanism had given token of such things: perhaps the component parts of a creature might be manufactured', Shelley, p. 8). This oblique hint was taken up by later adaptors of the novel, the most famous of which, James Whale's 1931 film *Frankenstein*, revisions Shelley's dour creation scene with a climactic electrical storm in which a monster, not 'manufactured' but rather assembled from dead body parts, is brought to shuddering life.[3]

This is an important point, because Shelley's studied ambiguity of creation makes it hard for some critics to read the novel in the context of its contemporary discourses of science or technology. Frankenstein refuses to tell Walton the secret of his creation: 'I will not lead you on, unguarded and ardent as I then was, to your destruction and infallible misery. Learn from me ... [how miserable is] he who aspires to become greater than his nature will allow' (Shelley, p. 52). The emphasis of the novel, in other words, is on the hubris of Frankenstein himself, and his ambition to usurp the status of God:

> A new species would bless me as its creator and source; many happy and excellent natures would owe their being to me. No father could claim the gratitude of his child so completely as I should deserve theirs. (Shelley, pp. 52–3)

The automaton (constructed from 'lifeless matter', not bodies 'devoted ... to corruption', Shelley, p. 53) appals Frankenstein on its coming to life. In the first of a number of implausible developments, he flees his laboratory and apparently forgets that he has even created the creature in the first place. The monster, whose own tale is the second narrative embedded within Frankenstein's (in turn embedded within Walton's), wakes as an intelligent but blank being, a Lockean *tabula rasa*. Lacking a teacher – the role Frankenstein irresponsibly abandoned – he soaks up experience from his direct environment. But, hideous in appearance, he encounters only hostility and persecution and flees to the wilderness. In a splendidly implausible development he learns to speak *and read* purely by eavesdropping on a rural family through a hole in the wall of their house. His nature now formed by the reactions people have had to him, and by textual identification with the three books he has read (*Paradise Lost*, *The Sorrows of Young Werther* and Plutarch's *Lives*), he has adopted a melodramatically antisocial and self-consciously tragic persona. Potentially good, he becomes malign, violent and even murderous. Tracking down his creator he demands that Frankenstein make him a mate. Fearing for the lives of his friends and fiancée, Frankenstein at first agrees; but having constructed a female monster he has a change of heart, fearing that this monstrous bride would result in 'children, and a race of devils would be propagated upon the earth' (Shelley, p. 160). He tears the inanimate body of the female monster to pieces with his own hands. In revenge the monster murders Frankenstein's bride on their wedding night, and in turn Frankenstein pursues him across Europe and into the Arctic wastes. The novel ends with Frankenstein's death and Walton's own encounter with the monster, who carries away the corpse of his maker, declaring 'I shall die ... polluted by crimes, and torn by the bitterest remorse, where can I find rest but in death?' (Shelley, p. 214).

Despite moments of rather adolescent over-writing *Frankenstein* remains a very powerful piece of fiction. As Brian Aldiss points out, despite her youth, Shelley achieved something very remarkable in this novel: the creation of a new archetypal figure (Aldiss, p. 145). Indeed, later versions of the book (as 'film icon, breakfast cereal, figure of speech', Clery, p. 126) have effectively superseded the original text. One crucial difference between the original and its cultural dissemination, as Emma Clery points out, is the very loquaciousness of the monster: the eloquent narrator of the central third of the book, yet reduced to grunts and incoherence in later cinematic versions. In one crucial sense, the point of this text is precisely *to give a voice* to the monstrous outsider.

In place of the nebulous 'science' of this science fictional tale, Franco Moretti has advanced a reading of the tale as 'ideology-fiction' – something not as far removed from science fiction as might be thought (ideology is a technology, after all, in the Heideggerean sense of an enframing of the world as 'ready-to-hand'). Moretti's persuasive reading sees the monster as an emblem of the increasingly ubiquitous 'alienated' or 'monstrous' figure from the industrial proletariat. The reason the novel has been so successful, Moretti implies, is that it struck a chord in early nineteenth-century culture that continues to resonate: dercinated, estranged from his or her labour, isolated and characterised as 'monstrous' by the ruling classes, the industrial worker bore much of the brunt of nineteenth-century economic downturns. Moretti suggests that 'Between Frankenstein and the monster there is an ambivalent, dialectical relationship, the same as that which, according to Marx, connects capital with wage-labour' (Moretti, p. 83). What cannot be denied, as Baldick's study shows, is how common an icon Frankenstein's monster became as a visual shorthand for the threatening working-class mob, or for Fenian Irishmen, or any kind of extra-bourgeois group. According to this reading, what is 'monstrous' about the monster (a creature, like the proletariat, with enormous potential for good that is thwarted and turned to destructive ends) is precisely its aggregate nature, its brute strength, the fact that it stands outside the discourses of polite society.

Shelley's other SF novel, *The Last Man* (1826), is a much less engaging work. The lengthy and frankly dull first part concerns a twenty-first century England after the abolition of the monarchy, centring on the political career of Raymond (modelled on Lord Byron) who aspires to become president. Things become more interesting with Raymond's death from the plague in Constantinople, which disease depopulates the entire globe, leaving the titular protagonist and narrator, Verney, based in Rome, the last human on the planet.

But, as we have seen, this sort of secular apocalypse was a common trope in late eighteenth- and early nineteenth-century SF. In fact, Shelley's *The Last Man* fails precisely where *Frankenstein* succeeds so well; it cannot generate an archetypal resonance from its premise. But to say so is not to detract from the enormous impact Shelley has had upon the development of SF with *Frankenstein*.

Science fiction of the 1820s and the 1830s

Shelley's influence, though it has proved enormous, was not immediate. Through the 1820s and 1830s most SF rolled along familiar grooves of subterranean adventure, future fantasy and journeys off-world. The whiff of the tedious-eccentric attaches to many of them. Thomas Erskine's (1750–1823) *Armata* (1817) treats of a Margaret Cavendish-like sister planet to the Earth. A ship *en route* from New York to China is blown by storms to the South Pole, and thence into the world of Armata, connected to the Earth by a bizarre and apparently fluid pathway. The narrator at first believes that this new world proves that the Earth has 'a ring like

Saturn, which, by reason of our atmosphere, could not be seen' and 'which was accessible only by a channel so narrow and so guarded by surrounding rocks and whirlpools, that even the vagrancy of modern navigators had never before fallen in with it' (Claeys, 6: 6). Later, he decides that, rather than being a ring, this other world must be a connected globe, 'the earth and its counterpart ... like the chain of double-headed shot, both of which might revolve around the sun together, and the moon around both (Claeys, 6: 7). (Rather implausibly, though, the traveller notes that 'the heavens above presented new stars' than any that could be seen 'from either of our hemispheres', Claeys, 6: 13). Armata is a country on this world, a sort of anti-England; at war with Capetia (as Claeys points out, 'France, after Hugh Capet (938–996), the founder of the Capetian dynasty which reigned in France until the revolution', Claeys, p. 28n.). The narrator is given a tour of this improved version of England and is impressed by what he sees.

In one sense *Armata* is a work in the grand tradition of science fiction. The narrator, having described the society and culture of the Armatans, notes with some satisfaction: 'I found they had a Revelation as we have – simple, eloquent, bearing throughout the stamp of divine truth, communicating, like our own, a fallen condition and a mediatorial redemption' (6: 141). Close enough to the earth to be in effect an extension of it, Armata need not trouble the Christian faith of its author by introducing the idea of a plurality of worlds.

John Trotter's *Travels in Phrenologasto* (1829) takes its hero to the Moon by balloon, where he finds the Lunarians to be not aliens but humans, descended from the Ancient Egyptians. The answer to the mystery lies in Trotter's enthusiasm for phrenology. It turns out that 'so great was the progress which our ancestors made in craniology ... that at length they brought the development of all faculties of the mind to the very highest perfection' (7: 167), invented a balloon and colonised the Moon – a utopia founded on 'the important truth ... that the basis of all knowledge is virtually situated in the shape of the skull' (7: 168). Phrenology leads to a perfect utopian state, because, as Trotter says, 'as everybody in the state is legally obliged to go with his head uncovered, this very happily precludes all deception, and enables the world to know at once a knave from an honest man' (7: 218).

Adam Seaborn (presumably a pseudonym, although the identity of the author has not been uncovered) published his *Symzonia: a Voyage of Discovery* in 1820. The title alludes to the theories of the American John Symmes (1780–1829), who issued a brief manifesto of his belief that the Earth is hollow in 1818, and who may have been the author of the fiction. 'Seaborn' imaginatively explores this idea – a notion with a long history in science fiction (as we have seen, at least back to Kircher in the seventeenth century and Holberg and Casanova in the eighteenth) – relating his own voyage to the North Pole, which he describes as a concavity. Travelling through the polar hole brings Seaborn to an ideal land of 'gently rolling hills within an easy sloping shore, covered with verdure, chequered with groves of trees and shrubbery, studded with numerous white buildings ... here was nothing wanting to a perfect landscape' (Seaborn, ch. 7). The natives salute him by pulling on their noses and give him a tour of their utopian environment by flying machine. The whole interior world is illuminated by the same sun that lights the exterior world, its rays refracted through the polar holes in accordance with Symmes' theories.

Symmes was regarded as a crank by many (the phrase 'Symmes' hole' became a slang term in the 1820s and 1830s, meaning something 'quackish' or 'fake'). Nevertheless, the US Congress thought seriously enough of his theories to vote funds for an expedition to the South Pole to uncover his supposed hole (the Wilkes Exploring Expedition of 1838–42 failed to locate it). *Symzonia* is more a curiosity than a satisfying novel, too largely given over to dry lectures on geography and social structures. But hollow earth adventures continued to be a popular mode of writing: *The Fountain of Arethusa* (1848) by Robert Eyres Landor (1781–1869) tells of a journey into a utopia located in the Earth's centre, reached by underground river below Derbyshire. This realm has its own sun and is modelled on an idealised ancient Greek and Roman model.

A novel with rather more advocates as a significant work of nineteenth-century SF is *The Mummy! A Tale of the Twenty-Second Century* (1827) by the British author Jane Loudon (*née* Webb, 1807–1858). A spirit gives the narrator a scroll which is 'the Chronicle of a future age', urging her to turn it into a novel. The tale that follows includes some retrospectively (from a twenty-first century perspective) interesting speculations about future society, but is dominated by a wild and rather crazy Gothic plot. In the twenty-second century, Britain is again a Catholic country; a tunnel links England and Ireland; technological advance has resulted in dirigible transportation and electrically operated lawyers, although Loudon's tone is largely satirical, stressing the incompetence of the latter and the pollution of future modes of transport. In Egypt two characters penetrate the Pyramid of Cheops and reanimate the mummy therein with electricity ('by [the] horrid and uncertain glare, Edric saw the mummy stretch out its withered hand, as though to seize him'; quoted in Alkon, p. 238). Cheops escapes to England where he becomes involved in a plot to control the choice of the next Queen of England. By the novel's end it transpires that the mummy is actually a force for good, reanimated not by electricity but by 'God's will'. More explicitly than Mary Shelley's ur-text, Loudon's novel dramatises the dialectic between technology and religion that continues to determine the development of the genre.

The Russian author Prince Vladimir Odoevsky (1804–1869) is often compared by critics with Hoffmann, who certainly seems to have been the inspiration behind the story 'Skazka o mertvom tele, neizvestno komu prinadlezhashchem' ('Fairy Tale about the Danger of Young Ladies Walking along Nevsky Prospekt', 1833), in which a beautiful girl is kidnapped by a shop-owner, turned into a mannequin and sold to a young man. By the time the latter realises that she is alive, it is too late. Odoevsky's later '4338 i-god' ('The Year 4338', 1840) is a liberal technological fantasy of the relatively distant future, although it was never completed.

Another future-fantasy was the British writer R. F. Williams' (n.d.) *Eureka: a Prophesy of the Future* (1837), which looks forward to a time when Africa is the dominant world-power and Britain a forgotten backwater. Alfred Tennyson (1809–1892) provided one compact little future-narrative encompassing aerial trade, aerial war and a World State, in his poem 'Locksley Hall' (1841):

> For I dipt into the future, far as human eye could see,
> Saw the Vision of the world, and all the wonder that would be;
>
> Saw the heavens fill with commerce, argosies of magic sails,
> Pilots in the purple twilight dropping down with costly bales;
>
> Heard the heavens fill with shouting and there rain'd a ghastly dew
> From the nations' airy navies grappling in the central blue;
> ....
> Till the war-drum throbb'd no longer, and the battle-flags were furled
> In the Parliament of man, the Federation of the world.
>
> (Tennyson, pp. 106–7)

The very compression of these images gives them a compact power missing from the often prolix prose future-imaginings of Tennyson's contemporaries, such as the 5,000-years-in-the-future fantasy *The Air Battle. A Vision of the Future* (1859), by Herrmann Lang (an unidentified pseudonym), where Tennyson's 'airy navies' and their battles are dramatised in detail – although in this novel they are flown not by Britain but by the empires of Brazilia, Madeira and Sahara. Britain has been surpassed as an imperial power.

In the comical 'Christmas Story' by Charles Rowcroft (?1800–1856), in *The Triumph of Woman* (1848), a meteor crashes into a German astronomer's house. From it comes a humanoid alien who learns German instantly by pressing a finger to the astronomer's head and telepathically absorbing it. He also possesses the ability to turn metal into gold; but he has left his planet because there are no women there (' "But how do you keep up your population?" asked the apothecary. "By magnetism", replied the planetarian', Claeys, 8: 291). The book shifts gear into a love story, as the visitor, Zarah, falls in love with the astronomer's daughter, and we find ourselves in the realms of Victorian domestic fiction.

On the other side of the Atlantic the journalist Richard Adams Locke (1800–1871) concocted a rather lumpen hoax, 'Great Astronomical Discoveries Lately Made by Sir John Herschel ... at the Cape of Good Hope' for the New York *Sun* in 1835. On its publication many people believed it to be real, although it is hard in retrospect to see why they were taken in. The account details an implausibly effective reflecting telescope through which Herschel was reported to have observed lunar vegetation, varieties of cow and sheep, and several species of lunarian alien, some of them winged. Locke's *jeu d'esprit* is mentioned now only because it annoyed one of the great SF fabulists of the age, who attacked it in print, and wrote his own lunar fantasy to surpass it: Edgar Allan Poe.

Edgar Allan Poe

The American writer and journalist Edgar Allan Poe (1809–1849) has exercised a particularly powerful influence on subsequent generations of American readers and writers. He wrote in a rather rank 'high' style, of a sort that people who don't know any better sometimes describe as 'fine writing'. He is prone to be prolix, his frequent attempts at humour are heavy and not amusing; his characterisation is thin; and yet the potency and brilliance of his imagination, especially in his SF tales, can root his writing in a reader's mind in a way that better writers often fail to do. Most of his stories are 'Gothic', tales of haunting and communication beyond the grave, or uncanny events and bizarre murder mysteries; but a significant portion (perhaps a fifth) of his output was SF, and this includes some of his very best work.

Just as some critics describe Mary Shelley as the starting point of modern SF, so Edgar Allan Poe has his enthusiasts as the originator of the genre. Thomas Disch says it straightforwardly: 'Poe is the source' (Disch, p. 34). Brian Aldiss, who thinks Mary Shelley is 'the source', nevertheless traces the notion that Poe is 'the father of Science Fiction' all the way back to an anonymous review in 1905. 'The notion has rattled about ever since', Aldiss comments, a little sourly, 'like the living dead'. Aldiss thinks that 'Poe's best stories are not science fiction, nor his science fiction stories his best … far from being the Father of Science Fiction, this genius bodged it when he confronted its themes directly' (Aldiss, pp. 58–9, 63). He detects a fundamental inarticulacy in his work which is more artistically debilitating than the immaturity and schlock that other critics have sometimes identified as fatal flaws. Disch admits the immaturity and schlock, arguing that SF as a whole has often traded in the same currencies, and asserting, surely correctly, that for all his over-the-top faults there is *something* about Poe, some powerfully imaginative ability to reach the reader. That instinct, and that rightness, find their science-fictional expression above all in a dialectical balance between 'science' and 'magic', between rationalism and mystic fantasy, that is (as this study has been arguing) the precise determinant of SF as an historical mode of writing. An early long poem, 'Al Aaraaf' (1829; the title is, Poe's own note makes clear, the name of a star 'discovered by Tycho Brahe which appeared suddenly in the heavens' and then 'as suddenly disappeared') is almost seventeenth-century in its assumption of a galaxy barred to mankind to save alien life from the sin of Adam. God addresses the spirit of this new star, instructing it to

> 'Divulge the secrets of thy embassy
> To the proud orbs that twinkle – and so be
> To ev'ry heart a barrier and a ban
> Lest the stars totter in the guilt of man!'
> (Poe, p. 43)

A similarly archaic religious sensibility flavours the otherwise 'hard' SF narratives of other tales. 'The Conversation of Eiros and Charmion' (1839) concerns the destruction of the world by a comet, not by simple collision (although this is what the globe's inhabitants fear) but because its near passage infuses so much oxygen into the Earth's atmosphere that apocalyptic fires become inevitable. It is a compelling little disaster story, but Poe frames it with the conversation of the figures mentioned in the tale's title: two disembodied spirits dwelling in 'Aidenn' after the end of the world, recalling the disaster as a theological object lesson ('let us bow down, Charmion, before the excessive majesty of the great God!', Poe, p. 363). A similar tale, 'The Colloquy of Monos and Una' (1841), efficiently encompasses future humanity's abuse of its world: 'huge smoking cities arose, innumerable. The fair face of Nature was deformed' (Poe, p. 451). The narrator recalls her life, and then her death and rebirth, into the 'Life Eternal'. In 'The Facts in the Case of M. Valdemar' (1845) a sick man is mesmerised on the point of death, turning him into a sort of undead zombie.

This reaches a climax in Poe's last major work, *Eureka: a Prose Poem* (1849), subtitled 'An Essay on the Material and Spiritual Universe'. This lengthy prose disquisition attempts to marry contemporary scientific and astronomical knowledge with a quasi-Idealist religious sensibility in which the whole cosmos is governed by a divine 'Oneness'. *Eureka* recounts Laplace's theories on the origin of the solar system and Mädler's theory that the Milky Way has at its core a massive object that is attracting all the stars. It also posits a version of an originary Big Bang, and insists that everything will tend towards a Big Crunch, not because Poe has any observational data to this effect (the work begins with an elaborate dismissal of empirical science in favour of 'instinctive' or 'intuitive' reasoning), but because he believes the cosmos, or God, to be a unity that operates according to dialectical principles of 'attraction' and 'repulsion'. Some recent critics have attempted to resuscitate *Eureka* as a legitimate work of philosophical cosmology, but it is not: it is metaphysically muddled, and for every lucky guess (for instance, that the cosmos originated in a big bang) there are a dozen errors that are culpable even by the standards of mid-nineteenth-century science (that electricity is a 'principle of repulsion', that the planet Venus is 'wildly self-luminous', that the evolution of new species on Earth was provoked by the successive creation of planets out of the solar nebula, and even the old Keplerian chestnut that a harmonious mathematical relation governs the respective distances of the planets from the sun – Poe believes this to be the inverse square law, which he thinks a ubiquitous mathematic truth of the universe).[4] But, in a sense, this does not matter, for Poe puts all his scientific reading and astronomical speculation into the same mill to grind out one conclusion: the Big Crunch at the end of time will unite matter and spirit, and turn everybody into God:

> When the bright stars become blended ... [Man] ceasing imperceptibly to feel himself Man, will at length attain that awfully triumphant epoch when he shall recognize his existence as that of Jehovah. In the meantime bear in mind that all is Life – Life – Life within Life – the less within the greater, and all within the *Spirit Divine*. (Poe, pp. 1358–9)

There are respectable cosmological scientists today (I am thinking of Frank Tipler's work of speculative science, *The Physics of Immortality: Modern Cosmology, God and the Resurrection of the Dead*, 1995) who concur with Poe's windy mysticism here, but it represents a Fantasy rather than an SF vision.

The best of Poe's SF tales are those in which this misty-eyed religiosity is abandoned in favour either of a Gothic sublimity or a satirical levity, either of which more effectively leavens Poe's science-based premises than conventional Theism. 'Mellonta Tauta' (1849; the title is Sophoclean Greek for 'those things that are to be') is set in the year 2848 and uses the form of a balloon journey across the Atlantic to comment, satirically, on the US of a thousand years before. This tale uses draft material that was later incorporated into the piously straight-faced *Eureka*, but works much more effectively, with its glimpses of future technology and a future world – its emperor, its ideology that 'no such thing as an individual is supposed to exist', and its affluent inhabitants watching the cosmos through powerful telescopes:

> [I] watched with much interest the putting up of a huge impost on a couple of lintels in the new temple of Daphnis in the moon. It was amusing to think that creatures so diminutive as the lunarians and bearing such little resemblance to humanity, yet evinced a mechanical ingenuity so much superior to our own. (Poe, p. 882)

Poe's most famous literary achievement, his short novel *The Narrative of Arthur Gordon Pym* (1838), moves from a conventional sea-adventure tale (though an extremely bloodthirsty adventure, written in an unusually gnashing manner) into something far stranger at the end, where the narrator discovers bizarre lands and cultures towards the South Pole. Indeed this novel shares with the earlier short story 'MS Found in a Bottle' (1833) a hidden 'hollow earth' theme. In the latter, a brilliantly effective Poe tale, a mariner is swept aboard a massive and technologically advanced ship crewed exclusively by very old individuals who ignore him and pursue their course relentlessly through stormy seas to the South Pole. Arthur Gordon Pym likewise details a voyage to the far south. In both cases the incidental details resemble *Symzonia* (which it is reasonable to assume Poe read), and in both cases the stories end abruptly, hinting that both protagonists are headed for an interior world.

The key piece of Poe science fiction, however, is the superb 'The Unparalleled Adventure of One Hans Pfaall' (1835), a quasi-scientific narrative relating a journey to the Moon by balloon. Here in effect Poe rewrites Kepler's *Somnium* (Poe's *Eureka* praises Kepler, 'that divine old man', as a greater genius than Newton: Poe, p. 1270) substituting *fancy* for the darker witchery-magic of Kepler's original, but otherwise maintaining a close interrelation between the 'science' and the 'fiction' throughout. The text is divided into three parts. A four-page opening section describes the 'high state of philosophical excitement' of the Dutch city of Rotterdam, occasioned by the appearance of a balloon made of dirty newspapers piloted by a very strange figure: 'two feet in height' with a huge nose and no ears (Poe, 'Hans Pfaall', p. 953). This seeming lunar alien brings the balloon to within 100 feet of the ground, drops 'a huge letter sealed with red sealing wax' over the side, and then ascends rapidly. The second portion of the story, constituting the bulk of the narrative, gives us the contents of this letter, Hans Pfaall's own account of his outlandish adventures. He introduces himself as a Rotterdam bellows-mender with a family to support, who had fallen on hard times. In debt, he hatched a scheme to build a balloon and (inspired by 'a small pamphlet treatise on Speculative Astronomy ... by Professor Encke of Berlin') to pilot it to the Moon. The canopy is not filled with air, but with something far lighter, on the nature of which Pfaall is a little reticent: 'I can only venture to say here, that it is *a constituent of azote* so long considered irreducible, and that its density is about 37.4 times *less than that of hydrogen*' (Poe, 'Hans Pfaall', p. 958). This spacecraft is launched on 1 April. The next 30 pages are given over to a detailed account of his voyage to the Moon, flying through the attenuated but not vacuum interplanetary atmosphere. On the nineteenth day of his voyage he finds himself plummeting towards the Moon's surface and is forced to throw overboard all his ballast including, eventually, the car itself:

> And thus, clinging with both hands to the net-work, I had barely time to observe that the whole country as far as the eye could reach, was thickly interspersed with diminutive habitations, ere I tumbled headlong into the very heart of a fantastical looking city, and into the middle of a vast crowd of ugly little people. (Poe, 'Hans Pfaall', p. 993)

But here Pfaall breaks off his narrative, with the promise of more interesting revelations to come if the burghers of Rotterdam are prepared to give him 'a pardon for the crime of which I have been guilty in the death of my creditors upon my departure' (Poe, 'Hans Pfaall', p. 995). In a page-long coda Poe relates the 'astonishment and admiration' of the people of Rotterdam, and then immediately undercuts the veracity of the narration by itemising certain salient facts: that 'an odd little dwarf and bottle conjurer, both of whose ears, for some misdemeanour, have been cut off close to the head, has been missing for several days from the neighbouring city of Bruges', that 'the newspapers which were stuck all over the little balloon were newspapers of Holland and therefore could not have been made in the moon', and that Pfaall himself, 'the drunken villain', has been seen drinking 'in a tippling house in the suburbs' with the 'three very idle gentlemen styled his creditors' (Poe, 'Hans Pfaall', p. 996).

Poe's appetite for hoaxes is one aspect of his genius for which critics today have little sympathy, and the heavy-handed 'hoax' ending has perhaps done more to sink 'Hans Pfaall' in current critical estimation than anything else. Criticism nowadays has little purchase on the hoax as literary form; once the critic has distanced herself from the anxiety of 'being taken in', and once the acknowledgement has been made that hoaxes are supposed to be *funny*, there is little more to say apart from laboriously explaining the joke – and a joke explained ceases to be funny.

Harold Beaver plots out the various fooleries in the text, notes that Pfaall lifts off on April Fool's Day, that his balloon is shaped like a 'fool's cap', and that the *burgermeisters* all have ridiculous names: Professor Rubadub, Mynheer Superbus Von Underduk, and so on. Pfaall actually flies to the Moon by utilising the principle of levity, and metaphorical 'levity' is what Poe aims for with his puns and jokes. His hoax inverts normal expectations and turns the logical world upside down. 'Invert "phaal" ', Beaver notes (referring to one of Poe's variants of the name 'Pfaall'), 'what sound do you hear but *"laugh"*?' (Beaver, p. 339).

But Poe is doing much more with his 'hoaxing' textual strategies than simply having a laugh. He uses some of the conventions of the April Fool's joke, playing them off against the codes of scientific investigation, precisely in order to explore the dialectic relationship between playfulness and 'scientific' seriousness. This same dialectic is also the aesthetic underpinning of science fiction: the interplay between the imaginative and the scientific.

Poe's skill is in the balancing of the two elements. We are given a narrative of journeying to the Moon in a balloon: it is clearly more 'plausible' that this is a hoax, a fantastical story, than that Pfaall *actually* travelled to the Moon in such a manner: in the real world, such journeying is impossible. But when Poe divides his narrative between the real world of Holland in the 1830s on the one hand, and the balloon journey through space to the Moon on the other, it is the real world that is rendered in a fantastical manner, and the implausible balloon journey that is treated with pseudo-scientific precision. Rotterdam, a real city, is populated with people named 'Rubadub' and 'Underduk', with earless dwarfs and balloons made out of newspaper. When Poe details the actual, he adopts an archly satirical tone: Pfaall complains that it is the general march of the time towards 'liberty, and long speeches, and radicalism, and all that sort of thing' that bankrupted him as a bellows-mender: 'if a fire wanted fanning, it could readily be fanned with a newspaper' (Poe, 'Hans Pfaall', p. 955). This, together with the 'balloon manufac-tured entirely of dirty newspaper' in which the dwarf/moonling descends at the tale's opening, points to some self-referential satire on behalf of Poe, a newspaper-man himself. Newspapers, the implication runs, fill people's heads with ridiculous notions like 'liberty' and 'radicalism', things of which the conservative Poe broadly disapproved. At the same time as paying him and giving him an outlet for his imag-ination, papers like the *Southern Literary Messenger* and the New York *Sun* floated their gullible readers metaphorically off solid earth into airy, lunatic speculation.

Poe's real stroke of genius was to reserve the ludicrous and satirical purely for the 'real' world; the implausible balloon journey itself is recorded with a scrupulous stylistic exactitude that lifts the whole out of the historically-specific quagmire of 1830s satire into a mind-expanding realm of SF. There is a greater imaginative gravity and appeal, a greater literary punch, in this aspect of Poe's conception than in the puffed-up ridiculousness of the 'real' world described. He gives us numerous pseudo-scientific observations, performs experiments upon the birds and cats he has brought with him, and leavens his account with various precise-looking num-bers. On 4 April he declares the balloon to have reached '7,254 miles above the surface of the sea', an impressive height but a tiny fraction of the '231,920 miles' or '59.9643 of the earth's equatorial *radii*' he has calculated as the distance he has to go. When Mary Shelley writes the fantastic elements of *Frankenstein*, she adopts a strained, elevated Gothic sublime tone of voice; when Poe writes about the fantastic in this story he does so in as matter-of-fact a manner as he can.

This, in the final analysis, is Poe's most significant contribution to the genre. It is in his conviction, expressed in *Eureka*, that the *intuitive imagination* (rather than deductive or inductive reasoning) should be the motor of advancements in science. In other words, his philosophy was Feyerabendian *avant la lettre*, and it was when he granted his own astonishing imagination the freest reign, when he yoked it least to conventional religious piety, that he wrote his very best SF; and when he blazed a path for subsequent writers in the genre.

References

Aldiss, Brian, with David Wingrove, *Trillion Year Spree: the History of Science Fiction* (London: Gollancz 1986)

Alkon, Paul K., *Origins of Futuristic Fiction* (Athens, GA: University of Georgia Press 1987)

Alkon, Paul K., *Science Fiction before 1900: Imagination Discovers Technology* (1994; London: Routledge 2002)

Baldick, Chris, *In Frankenstein's Shadow: Myth, Monstrosity and Nineteenth-Century Writing* (Oxford: Oxford University Press 1987)

Barley, Tony, 'Prediction, Programme and Fantasy in Jack London's *The Iron Heel*', in David Seed (ed.), *Anticipations: Essays on Early Science Fiction and its Precursors* (Liverpool: Liverpool University Press 1995), pp. 153–71

Beaver, Harold (ed.), *The Science Fiction of Edgar Allan Poe* (Harmondsworth: Penguin 1976)

Bradshaw, Penny, 'Dystopian Futures: Time-Travel and Millenarian Visions in the Poetry of Anna Barbauld and Charlotte Smith', *Romanticism on the Net*, 21 (February 2001); http://users.ox.ac.uk/~scat0385/21bradshaw.html

Byron, George Gordon, 'Darkness', in *Complete Poetical Works*, ed. J. J. McGann, 5 vols (Oxford: Clarendon 1980–86), IV: 40–3

Campbell, Thomas, *Complete Poetical Works of Thomas Campbell*, ed. Walter Jerrold (London: Oxford University Press 1906)

Carey, John (ed.), *The Faber Book of Utopias* (London: Faber 1999)

Claeys, Gregory (ed.), *Modern British Utopias 1700–1850* 8 vols (London: Pickering and Chatto 1997)

Clery, E. J., *Women's Gothic from Clara Reeve to Mary Shelley* (Tavistock: Northcote House/British Council 2000)

Clute, John and Peter Nicholls, *Encyclopedia of Science Fiction* (2nd edn., London: Orbit 1993)

Davy, Humphrey, *Consolations in Travel; or, The Last Days of a Philosopher* (1830)

Disch, Thomas, *The Dreams Our Stuff is Made of: How Science Fiction Conquered the World* (New York: Simon and Schuster 1998)

Franklin, H. Bruce (ed.), *Future Perfect: American Science Fiction of the Nineteenth-Century: An Anthology* (rev. and expanded edn., New Brunswick, NJ: Rutgers University Press 1995)

Grainville, Jean-Baptiste François Xavier Cousin de, *The Last Man* (1805; trans. I. F. Clarke and M. Clarke, 'Wesleyan Early Classics of Science Fiction Series', ed. Arthur B. Evans; Middletown, CT: Wesleyan University Press 2002)

Hood, Thomas, *Complete Poetical Works of Thomas Hood*, ed. J. Logie Robertson (London: Oxford University Press 1907)

James, Edward, 'Science Fiction by Gaslight: An Introduction to English-Language Science Fiction of the Nineteenth-Century', in David Seed (ed.), *Anticipations: Essays on Early Science Fiction and its Precursors* (Liverpool: Liverpool University Press 1995), pp. 26–45

Moretti, Franco, *Signs Taken for Wonders* (London: Verso 1983)

Neff, D. S., 'The "Paradise of the Mothersons": *Frankenstein* and *The Empire of the Nairs*', *Journal of German and English Philology*, 95.2 (1996), 204–22

Philmus, Robert, 'Science Fiction: from its Beginnings to 1870', in Neil Barron (ed.), *Anatomy of Wonder: Science Fiction* (New York: R. R. Bowker 1976), pp. 3–32

Poe, Edgar Allan, *Poetry and Tales*, ed. Patrick F. Quinn (New York: Library of America 1984)

Shelley, Mary, *Frankenstein, or The Modern Prometheus* (1818; ed. Maurice Hindle, Harmondsworth: Penguin 1992)

Shelley, Percy, *The Complete Poetical Works of Percy Bysshe Shelley*, 2 vols, ed. Neville Rogers (Oxford: Clarendon Press 1972)

Suvin, Darko, *Metamorphoses of Science Fiction: on the Poetics and History of a Literary Genre* (New Haven, CT: Yale University Press 1979)

Suvin, Darko, *Victorian Science Fiction in the UK: The Discourses of Knowledge and of Power* (Boston: G. K. Hall 1983)

Tennyson, Alfred, *The Oxford Authors: Tennyson*, ed. Adam Roberts (Oxford: Oxford University Press 2000)

6
Science Fiction 1850–1900

The 1850s

Increasingly, as the nineteenth century progressed, advances in science were changing the way human beings thought about their position in the cosmos, with profound implications for the development of SF. The geologist Charles Lyell (1797–1875) challenged the Bible-inspired notion that the Earth was less than 6,000 years old in *Principles of Geology* (1830–33), introducing the idea of 'deep time' to a wide audience. A few years later, Charles Darwin (1809–1882), perhaps the most famous scientist between Isaac Newton and Albert Einstein, published his world-changing *On the Origin of Species by means of Natural Selection* (1859), which confirmed the vertiginously long time-scale in its portrayal of the valueless proliferation and evolution of life. William Beddoes' (1803–1849) posthumously published *Death's Jest Book, or the Fool's Tragedy* (1850) gives a sense of the way this new understanding of time in its aspect of 'the Sublime' inflected the far-future imagination. In place of the personalised 'last man' narratives of secular apocalypse from earlier in the century, Beddoes tolls a more impersonal and impressive knell:

> Tis nearly passed, for I begin to hear
> Strange but sweet sounds, and the loud rocky dashing
> Of waves, where time into Eternity
> Falls over ruined worlds.
> (Beddoes, *Death's Jest Book*, IV.iii.107–10)

The same sense of decay would increasingly inform much late century future fiction.

The German scientist Rudolph Clausius (1822–88) coined the term *entropie* in his 1865 description of the tendency of the amount of energy unavailable for conversion into work to increase. Popularised by John Clarke Maxwell (1831–1879) and others, entropy entered popular consciousness as a belief that order inevitably disintegrates and chaotic order inevitably increases, until the universe perishes in heat-death. This isn't quite what Clausius was saying, but it is in this form that a fascination with 'degeneration' enters late century SF as a corollary to the 'evolution' proposed by Darwin. (The French medical scientist Bénédict Auguste Morel (1809–1873) was the first to elaborate a comprehensive physical and social theory of 'degeneration'.) These various theories were not only science; they were, and are, powerfully coloured with philosophical and ethical aspects. Darwinism may only describe, in a properly passionless manner, the likeliest hypothesis for the development of different species; but it was received very passionately indeed as advocating either a deplorable erosion of religious values or, conversely, a creditably progressive narrative of ongoing materialist improvement. Similarly, 'degeneration', though in essence a purely 'scientific' premise, was never discussed in a morally neutral manner.

It would not be correct to suggest that all the SF in the latter half of the nineteenth century could be located on a notional scale somewhere between 'positivist optimism' or 'degeneration pessimism'. But these two ways of responding to the changes in the present, and the possible directions the future might take, do determine much of the fictional speculation created over this period.

Many people were hopeful about future possibilities. The American poet Walt Whitman (1819–1892) stood 'Facing West from California's Shores' (as the title of his poem puts it) in 1860, pondering that the 'circle' of global exploration was 'almost circled'. Eleven years later, in 'Passage to India' (1871), he looked beyond the exploration of the Earth to new frontiers, the 'sun and moon and all you stars! Sirius and Jupiter!', although he did not make the imaginative leap to wonder what such new 'passages' might involve. Andrew Blair's (?1830–1885) *Annals of the Twenty-Ninth Century, or The Autobiography of the Tenth President of the World-Republic* (1874) looks forward to a future of a global state and interplanetary travel. Brian Stableford says, with some justice, that this work founders 'under [its] own ponderous weight' (in James and Mendlesohn, p. 23); but in its optimism and ebullient expansiveness it is very much representative of its time. Edward Maitland's (1824–1897) *By and By, An Historical Romance of the Future* (1873) is a utopia set in a future Africa. Its emphasis is much more on the mystical and psychic (and vegetarian) dimensions of the ideal future life, and the book should be considered in the context of the many mystical SF tales being written in the nineteenth century (for which see below).

Others, particularly later in the century, found a mournful cadence in the imaginative possibilities of SF. The great Hungarian poet Mihály Vörösmarty (1800–1855) published a number of pessimistic cosmic epics towards the end of his life, including *Az emberek* (1848), in which the whole of human history is revealed to be circularly tragic, and *Előszó* (1850), which projects the 'tragedy' of Hungarian history onto a cosmic scale.

The question of whether SF manifests itself as an 'optimistic' or 'pessimistic' idiom depends largely on the broader cultural attitudes to science and technology. Writers persuaded by the Enlightenment project were liable to write positive extrapolations in which society and human life were improved by progress in these areas. But of course many writers were more sanguine about such projects, perhaps part of the anti-Enlightenment reaction which in part inspired the late century craze for psychic phenomena. This dialectic works itself through into a significant cultural binary in the early twentieth century (see Chapters 8 and 9 below), but is also evident in the work written in the mid- and late century.

The Irish-born American writer Fitz-James O'Brien (1828–1862) wrote a number of SF short stories which were as significant for their impact on the then nascent form of the short story as for their fame in generic SF (the term 'short story' was not coined until 1884). Inventive in premise and brilliant in execution, what most strikes the reader of O'Brien's short works is the distinctive atmosphere or mood he creates: a rather mournful, unsettling and marvellous tone. 'The Diamond Lens', published in the *Atlantic Monthly* in 1858, took the principle of Swift's diminutive Lilliputians to a new degree. A scientist, using a microscope fitted with the lens of the title, observes a whole universe within a drop of water, and falls in love with a beautiful though infinitesimal humanoid he christens 'Animula' ('the perfect roundness of her limbs formed suave and enchanting curves. It was like listening to the most spiritual symphony of Beethoven', O'Brien, p. 302). When her water drop evaporates, she shrivels and dies; the scientist's grief drives him mad. Part of O'Brien's technique in this as in his other SF stories is to establish what H. Bruce Franklin calls 'a surface realism' (Clute and Nicholls, p. 884) against which the poetic strangeness of O'Brien's themes can be more powerfully contrasted; a strategy that becomes so common in twentieth-century SF as to become generically normative. Much of 'The Diamond Lens' is concerned with the technical challenges of building the instrument, but the pith of the tale is its mystical inflection of the theme of lovers doomed to separated existences. O'Brien's narrator laments of Animula in her minuscule water droplet that 'the planet Neptune was not more distant from me than she' (O'Brien, p. 302). But he was only able to build the telescope in the first place because of the help he received from the famous dead microscopist Leeuwenhoek, via the spirit medium Madame Vulpes. This blend of scientific-technical and mystic idioms is not only characteristic of nineteenth-century SF, it is also a penetrating gloss on the theme of the story itself, because although it presents itself as a love story, 'The Diamond Lens' is actually about death, the veil of which would (many believed in the 1850s) soon be pierced by science.

In Charlemagne Ischir Defontenay's (1814–1856) *Star ou psi de Cassiopée: Histoire merveilleuse de l'un des mondes de l'espace* ('Star, or Psi of Cassiopeia: The Marvellous History of One of the Worlds of Space', 1854) a case is discovered in the Himalayas which contains a wealth of information about the alien inhabitants of the planet 'Psi', which orbits three different-coloured suns in the constellation of Cassiopeia. Defontenay includes details of the alien physiognomy and society as well as samples from their literature, and indeed tends to overwhelm his readers with his inventiveness, making *Star* a rather exhausting book.

Anti-gravity

Poe's solution in *Hans Pfaall* (1835) to the problem of how to lift a human crew from Earth's gravitational field (by balloon) was implausible even in the 1830s.

Other SF approaches to the problem faced certain scientific difficulties. Despite the fact that Jules Verne famously propelled his astronauts into space by firing their craft from a gigantic cannon in 1865, in general it was believed that too sudden an acceleration would kill passengers (in fact the danger was largely overestimated: in the early years of the century some people believed that steam trains, travelling too rapidly, could prove fatal). Accordingly, although the principles of ballistics were well understood, few SF authors before Verne (or since) ever proposed cannon-launched or rocket ascents. This left the problem of escaping the gravitation pull of the Earth; one way round this was the utilisation of anti-gravity devices.

The first such device can be credited to 'Joseph Atterley', the pseudonym of the American author George Tucker (1775–1861), whose *A Voyage to the Moon* (1827) is accomplished in a craft coated with an anti-gravitational metal. (Poe plagiarised several pages from this otherwise rather undistinguished novel.) J. L. Riddell (n.d.) was the author of a little-known lunar romance, *Orrin Lindsay's Plan of Aerial Navigation* (1847), in which the titular scientist-protagonist uses an anti-gravity field created by 'magnetics' to pilot a spacecraft to the Moon. A more widely read novel in the same idiom was written by 'Chrysostom Trueman', a pseudonym that critics have yet to penetrate, although Darko Suvin makes a good case for James Hinton (1822–1875). Trueman's *The History of a Voyage to the Moon, with an Account of the Adventurers' Subsequent Discoveries* (1864) features an early use of an anti-gravity device to propel its protagonists to the Moon: an ore called 'repellante' mined from the Colorado mountains by Trueman's two protagonists, Stephen Howard and Carl Geister. They fly to the Moon in a repellante-powered spaceship, where they crash-land. Here they discover a utopian lunar society of four-foot-tall humanoids called 'Notol'. After a year living among these people and exploring the Moon, they inscribe their adventures on metal tablets and fire these back to Earth by dropping them into an exploding lunar volcano. The book speculates that the Lunarians are actually reincarnated human souls, something, as we have seen, true of the seventeenth- and eighteenth-century traditions of material-mystical lunar journeys.

The English author Percy Greg (1836–1889) picked up on the concept, which he christened 'apergy', for his *Across the Zodiac: the Story of a Wrecked Record* (1880). A veteran of the American Civil War is shipwrecked on a strange island in 1867 where he sees a UFO crash ('it had a very perceptible disc ... I came upon fragments of shining pale yellow metal ... [and a] remarkably hard impenetrable cement', Greg, p. 9). From this wreckage he extracts a Latin manuscript which tells of its anonymous narrator's invention of 'apergy' and his use of it in an 1820 expedition to Mars in a spaceship called the *Astronaut* (the first recorded use of this term). Greg correctly anticipates the weightlessness of space travel, and fills the earlier chapters of his book with carefully recorded scientific data, including detailed linguistic tables and declensions of the humanoid Martians' language. The tale sags in the middle, with lengthy accounts of the society and Utilitarian morals of the Martian inhabitants, and his rather listless love for a Martian maid, Eveena; but it ends excitingly, with political intrigue and attempted assassination. Eveena sacrifices her life for the narrator and he leaves Mars in grief.

Greg promised a second volume of these adventures, but it was not forthcoming.

Several subsequent interplanetary romances also used the concept: the American writer John Jacob Astor (1864–1912) adopted both the concept and the word 'apergy' as the motile principle for his *A Journey to Other Worlds* (1894). Frank Stockton's (1834–1902) 'A Tale of Negative Gravity' (1884) is a more playful variation of the same principle, in which an American inventor creates an anti-gravity backpack which he uses to go hiking, mountain-climbing and lifting heavy objects, before putting it away for fear of destroying his and his wife's happiness with the fame that would ensue if the invention were made public. The long-lived alien protagonist of *Willmoth the Wanderer, or The Man from Saturn* (1890, by Charles Curtis Dail, 1851–1902) flies about the solar system by smearing himself with an anti-gravity ointment. He arrives on a prehistoric Earth and creates Homo sapiens by selective breeding. The most famous dramatisation of the concept is H. G. Wells' *The First Men in the Moon* (1901), where the anti-gravity substance is called 'Cavorite' after its fictional inventor; this novel is discussed in the chapter following.

These anti-gravitational spacecraft function as a purely material attempt to escape the gravitation pull of the Earth; but they do more than that. The very popularity of this conceit indicates the way anti-gravity functions as an objective correlative to the very imaginative freedom that makes SF distinctive. Although nineteenth-century authors employed a wide range of devices to get into space, from balloons to rockets, the two predominant assumptions were anti-gravity and 'will' (this latter, in which characters either 'will' themselves into space, or travel as disembodied spirits, has a long pedigree in SF; see the section on 'mystical SF' below). The former involves a materialist rationale, the latter partakes of spiritual or mystical discourses, but they both function exactly as externalisations of the imaginative liberty of SF.

Contemporaries of Verne

It was during the 1860s that one of the most famous of all science fiction writers, Jules Verne (1828–1905), started publishing his *voyages extraordinaires* (the first of which, *Cinq semaines en ballon*, appeared in 1863). Verne is discussed in greater detail in Chapter 7, but without jumping forward, it is enough to note that the increasing popularity of the extraordinary voyage throughout the 1860s and 1870s drew on antecedents that long pre-dated Verne, however much it received a huge boost from his immense popularity. Indeed, some sense of the sheer variety of extraordinary voyages and future-fictions published in this period makes Verne's imagination seem rather small-scale.

Achille Eyraud's *Voyage à Vénus* (1865) is praised by historians of rocket science as the first fictionalisation of a reaction propulsion system (in this case, the projection of water to propel the spacecraft). What is interesting about this work is that, although it was much less popular than Verne's *De la terre à la lune* ('From the Earth to the Moon') of the same year, its rocketry premise was so much more plausible than Verne's astronauts' journey inside a shell fired from a giant gun.

There was a marked optimistic tone to many mid- to late century fictions. The Hungarian Mór Jókai (1825–1904) was a writer whose 'popularity was enormous' despite the fact that 'serious critics held strong reservations about the aesthetic qualities of his works from the very beginning' (Pynsent and Kanikova, p. 166). Among his many novels were several intriguing SF tales, especially *A jövö század regénye* ('A Novel of the Next Century', 1872), a positivist future-history in which new technologies powered by (and indeed built out of) a new substance called 'ichor' lead, via war, to world peace and prosperity, and thence to the colonisation of the solar system.

This sort of optimistic extrapolation into the future was very common in the later nineteenth century. One corollary of such future speculation was the creation of speculations about how the past might have developed differently, a now popular sub-genre of SF known as 'alternate history'. Charles Renouvier (1815–1903) wrote one of the earliest examples of this sort of book: *Uchronie (l'Utopie dans l'histoire), esquisse historique apocryphe du développement de la civilisation européenne tel qu'il n'a pas été, tel qu'il aurait pu être* ('Uchronia (Utopia in History); an Apocryphal Historical Sketch of the Development of European Civilisation not as it Was but as it Might Have Been', 1874) which traces the path history might have taken had the Roman Empire not collapsed after the death of the emperor Marcus Aurelius.

A more conventional utopia is James Davis Ellis's (?1847–1935) *Pyrna, a Commune, or Under the Ice* (1875), which concerns an ideal society located beneath the surface of the Earth, in this case underneath a glacier. The use of the politically charged term 'commune' in the title gives some sense of the ideological complexion of this form of utopia (the radical socialist-reformist Communards had taken over Paris in 1870, before being ruthlessly suppressed by French and Prussian troops in 1871). Ellis's sternly rational society, organised on eugenicist grounds, will strike many modern readers as more or less repellent. *Etymonia* (1875), by the same author, is another utopian story, this time set on an island.

A different tone is set by the English author Anthony Trollope (1815–1882). Trollope, enormously prolific though he was, published only one science fictional work, the curious *The Fixed Period* (1882). This short novel is set on an imaginary British colony, Britannula, located somewhere in the vicinity of Australia. Having gained independence from its colonial master, the unhabitants of Britannula (its name suggests both 'Little Britain' and 'Britain annulling') intend to establish compulsory euthanasia for all citizens when they reach the age of 66 to relieve the country of the burden of supporting a useless elderly population, and supposedly to relieve the elderly from the burden of living through such a disagreeable period of their existence. The satire here is not as Swiftian, or as obvious, as some commentators suggest. The Britannulan president, Neverbend, in addition to instituting 'the fixed period', has established universal progressive education, abolished capital punishment and in other respects has turned his country into a very attractive environment. In the novel Britain re-colonises Britannula, ostensibly to prevent the 'fixed term' practice. The debate, although it is one to which SF will often return (for instance, in the 1976 movie *Logan's Run*), does less to engage us as readers than the occasional glimpses of Trollope's future-world: 'the glittering spires of Gladstonopolis in the distance' (Trollope, p. 170) or the rules of future-cricket, in which bowling is done by machine.

Another difficult to classify work, charming and appealing but lacking depth (in a direct sense), was conceived by Edwin Abbott (1839–1926), an English clergyman. His *Flatland: a Romance of Many Dimensions* (1884) remains popular to this day. Essentially a witty and didactic fable designed to elucidate certain mathematic and topological premises, the book possesses a certain charm that makes it more than its rather dry premise. The narrator, A. Square, is a two-dimensional being living in a two-dimensional cosmos, which is described in some detail. In a dream, he visits a one-dimensional world ('lineland') and is in turn visited by a sphere from a three-dimensional world (spaceland). Despite some attractive satirical observations about the limitations of Victorian society, and gender relations in particular, this work never quite manages to flesh out its two-dimensional conceit (if that is not too contradictory a statement). A deft sequel by Ian Stewart, a professor of mathematics at Warwick University, *Flatterland* (2001) is subtitled 'Like Flatland, but More So' and takes its protagonist (the great-great-granddaughter of A. Square) through a much more comprehensive set of mathematically determined worlds.

Mystical science fiction

As we have seen, most seventeenth- and eighteenth-century SF braided together the discourses of materialism and spiritualism. In the nineteenth century the interrelation of these idioms was assumed by many to have been established on a scientific basis (some people continue to believe this today). Organisations such as the Centre for Psychical Research attempted to place so-called 'supernatural' or 'psychic' phenomena on a scientific footing.

Humphry Davy (1778–1829) was a professor of chemistry at the Royal Institution who did much to advance human knowledge of chemistry and physics, and invented the miner's safety-lamp which bears his name. But he was also a committed believer in psychic phenomena. His SF fantasia *Consolations in Travel; or, The Last Days of a Philosopher* (1830) concerns the meeting between the protagonist, Philalethes, and a disembodied spirit he calls his Genius. This Genius relates the true nature of the cosmos, in which material atoms exist alongside 'spiritual' atoms ('the quantity, or the number of spiritual essences, like the quantity, or the number of atoms in the material world, are always the same; but their arrangements ... are infinitely diversified' (Davy, pp. 41–2; see also Stableford, in Flammarion, pp. xx–xxii)), from which are constructed spiritual beings who can travel about the universe unconstrained by physical laws. The physical universe strives continually upwards to the preferable spiritual existence; after death humans move up through a number of increasingly elevated spiritual incarnations as extraterrestrials on other planets – a description of sextuple-winged and many-tentacled aliens flying through the atmosphere of Saturn is particularly impressive. This model of a spiritual cosmos interpenetrating the material as a desired destination was common to spiritualist discourses of the nineteenth century; and the idea that extraterrestrials might in fact be beings of this nature was fairly common in the century's SF. Poe's *Eureka* (1849, discussed above) is part of this tradition.

The major figure of nineteenth-century mystical SF is the French astronomer Camille Flammarion (1842–1925). He worked as a young man for the Paris Observatoire, and his first published work drew on astronomical science for some sober speculation about the likelihood of life on the other planets of the solar system which he published under the Fontanellean title *La Pluralité des mondes habités* ('The Plurality of Inhabited Worlds', 1862). The success of this work (it went through thirteen editions in the next 30 years) encouraged him to pursue a career as a writer; but significantly his next work was a piece of psychical speculation about the after-life, *Les inhabitants de l'autre monde: révélations d'outre tombe* ('The Inhabitants of the Other World: Revelations from beyond the Tomb', 1862) based on 'revelations' via a spirit-medium Mlle Huet. Brian Stableford thinks that Flammarion may have seen this book 'as a companion piece to its predecessor' (Flammarion, *Lumen*, p. x), and may have been discouraged enough by its reception to separate his mystical interests from his more conventional science (his *Des Forces naturelles inconnues* ('Of Natural but Unknown Forces', 1865) appeared under the pseudonym 'Hermès'). But the mystic and the scientific are so closely interwoven in almost all Flammarion's science fiction that it is difficult to believe that such a separation was ever attempted. In fact, Flammarion's own style of often very powerful, and certainly influential, mystical SF was not only characteristic of his age (respectable interest in the paranormal and the psychic remained strong throughout the latter half of the century) but also reflects what this study has argued is the fundamental and determining dialectic of the genre.

Les Mondes imaginaires et les mondes réels ('Worlds Imaginary and Real', 1864) was non-fiction, including sober astronomical speculation about the solar system with a comprehensive compendium of previous fictional imaginary voyages around the solar system, this latter having an understandable bias towards the French literary tradition. Flammarion continued making astronomical observations, and took balloon journeys to experiment on the atmosphere (he published the results as *L'Atmosphère* in 1871). His general introduction to the business of astronomy *Astronomie populaire* (1875) was immediately successful, and (frequently updated) is still in print. He moved to an estate south of Paris at Juvisy-sur-Orge where he constructed a large telescope, and where he also conducted frequent séances. Both these forms of research informed his fiction.

Lumen (collected in *Récits de l'infini*, 1872) dramatised much of the same material as *Les Mondes imaginaires et les mondes réels*. The book contains five conversations between a mortal man and a disembodied spirit, the Lumen of the book's title, who is free to travel throughout the cosmos, the latter illuminating the former on many features of alien life as well as on the spiritual truths of reincarnation and the benign God who providentially governs everything. Two other SF stories from *Récits de l'infini* express this dual nature of Flammarion's interest. 'Histoire d'une comète' tours the solar system from the point of view of a comet; whereas 'Dans l'infini' is about a mysterious communication from the spirit world.

The point where these two sides of Flammarion's interest coincide is in an affection for the sublime. *Lumen* is a powerfully affecting read, suggesting the sheer scale of the universe with, in its last two 'récits', a parade of brilliantly inventive tableaux concerning the forms alien life might take. The effect of reading it is one of a sense of the enormous scale and variety. It is a novel that broadens the reader's mind.

It also uses its material premises to adumbrate a specifically ethical and religious point. The major theme of *Lumen* (the word means 'light' in Latin) is the way the finite speed of light determines the cosmos: if light from Earth takes 65 years to reach a certain distant star, then an observer on that star (assuming they could command improbably efficient powers of magnification) would see life on Earth as it was 65 years ago: in 1854 they would see the French Revolution of 1789. Flammarion explores the various implications of this: if one travelled at the speed of light, the scene would appear frozen; if one travelled at slightly less than the speed of light, the scene would unfurl in slow motion; and so on. Flammarion's conceit is that this circumstance renders all the cosmos always present and apprehensible: nothing is lost; good or evil deeds from any historical period are instantly accessible. We discover, for instance, that Napoleon Bonaparte (having been the cause of the deaths of five million individuals who would have lived on average 30 more years had they not been killed) is responsible for 185 million years of human life: and that his spirit will be prevented from developing for this period of time.

Stella (1877) and *Uranie* (1889) are both concerned with reincarnation. *Stella* flirts with sentimentality in its tale of lovers separated by death but then reunited. *Uranie* is a broken-backed accumulation of various SF mystic bits and pieces, although some of the descriptions of strange life-forms on other worlds have the power of *Lumen*. More effective, but still saturated with religious and mystical discourse, is *La Fin du monde* (1893–94). In this novel a comet narrowly avoids crashing into twenty-fifth-century Earth, and the anticipation and passage of the disaster is related with all Flammarion's considerable astronomical skill, combined with an effective description of the social panic attending the event. Flammarion, as is the case in most of his work, mixes in a great deal of factual information relating to previous historical disasters, often to good effect. But after expected disaster is averted, Flammarion, in a bold narrative move, fast-forwards his tale. By the hundredth century human beings have evolved past resemblance to monkeys with enormously refined sensory capacities and extensive psychic capacities. Europe is abandoned and has fallen to ruin. Flammarion then coolly steps several million years further on, to the cooling of the Sun and the slow freezing of the world. The last surviving humans are a young boy and girl, Omegar and Eva; but there seems no possibility of a repopulation, the Earth is doomed. In fact, Flammarion is recycling the (in France) very popular pair of last human lovers from Grainville's *Dernier homme* (1805; see above p. 89), although he gives that story a positive and rather bizarre twist: Eva's dead mother revives to inform the two that Jupiter is where humanity, now cleansed and purified, lives in spiritual form. Omegar and Eva, ready to die, are instead carried off by the spirit of Cheops, King of Egypt, to the spirit utopia of Jupiter. Rather abruptly, Flammarion concludes his story by relating the eventual death of the whole solar system and the cosmos itself, to make way for a new infinite universe. Flammarion exhibits the increasingly intimate relation between materialist and mystic impulses in late century SF; based on his meticulous astronomical-scientific knowledge his fictions enjoyed much of their wide popularity because of the spiritualist flavour of their narratives.

Something similar is true of the hugely popular (in their own day) potboilers of the British writer Marie Corelli (1855–1924). Her first novel, *The Romance of Two Worlds* (1886), reads like an especially prolix throwback to the mystical voyages of Kircher and Swedenborg. The book details a trip round the solar system undertaken by the narrator-heroine in the company of Azùl, an angel, taking in perfected societies of spiritual life on Saturn, Venus and Jupiter, the voyage achieved via a mystical variant of electricity. Its sequel, *Ardath: the Story of a Dead Self* (1889), has an even more occult emphasis, and includes time-travel back to 5000 BC. Corelli's idiosyncratic 'Gospel of Electricity' won many fans, although it was ridiculed in its day and meets with more robust ridicule nowadays.[1]

But the later years of the nineteenth century saw a remarkable increase in interest in psychical and supernatural affairs, with special interest in telepathy, 'spirit rapping' (communicating with ghosts at seances by means of tapping on the tabletop), ghostly apparition and reincarnation: a set of beliefs often dignified by scientific pseudo-explanation and often endorsed by respectable scientists. Taken in the broader context of the development of SF, this blurring of the boundary between the mystic and the material can be seen as a determining dialectic for the genre itself. It ought not, therefore, surprise us that so much late nineteenth-century SF straddled precisely this boundary.

Realmah (1868) by Arthur Helps (1813–1875) includes among its various stories a sort of externalisation of heaven, a planet which receives the souls of the dead. Mortimer Collins' (1827–1876) *Transmigration* (1873), a fantasy of multiple incarnations which includes a sojourn on a utopian Mars, is conceived rather like ancient Athens. The American writer John Astor's *A Journey in Other Worlds, A Romance of the Future* (1894; see above p. 110) is set in AD 2000 and details a tour round the solar system in a spacecraft powered by anti-gravity. The book begins with an impressive array of material details of the future world: from projects on the largest scale (the Earth is purged of its axial tilt to ensure a pan-global temperate and regular climate) to the smallest (ocean-going ships carry metallic windmills deployed when in port to store wind energy, thereby providing 'a great part of the energy required to run them at sea', Astor, p. 43). There is also a eugenicist agenda not unfamiliar from the discourses of psychical and theosophist believers (the non-white races, we are told, 'show a constant tendency to die out' and 'the places left vacant are gradually filled by the more progressive Anglo-Saxons', Astor, p. 74). But the tone changes in the latter portions; the space travellers encounter a number of spirits, including the soul of a dead American bishop, who instruct them in standard nineteenth-century mystical terms about the nature of the after-life floating through the ether of space. ('Though many of us can already visit the remote regions of space as spirits, none can as yet see God; but we know that as the sight we are to receive with our new bodies sharpens, the pure in heart will see Him, though He is still as invisible to the eyes of the most developed here as the ether of space is to yours', Astor, pp. 385–6.) The book ends with a sermon in which its SF speculations are reconciled to scripture: it is difficult to imagine a more fell dramatic device.

The prolific English novelist Edward Bulwer-Lytton (1803–1873) came late in his career to mystical-scientific romances. His *A Strange Story* (1862) rationalises spiritualism and the posthumous survival of the soul in pseudo-scientific terms. *The Coming Race* (1871) takes its traveller, in now traditional SF mode, into a subterranean world inhabited by beings superior to Homo sapiens in a number of ways. These creatures utilise a nebulously defined but enormously powerful energy, 'vril', something between electricity and spiritual force, to power their lives. The novel ends with the protagonist escaping to the surface and attempting to warn the complacent surface dwellers that the underworlders are soon to emerge and take over the world, an alarming prospect. The book was a hit, and 'vril' became a term with cultural currency (the makers of a concentrated beef paste cashed in on its vogue by calling their product 'bovine vril', or 'bovril'). But what is most interesting about the book's success was precisely the way it mediates science and mysticism.

Another mystical writer of some interest to the development of the genre, because of his influence on H. G. Wells (discussed in the next chapter), is the English mathematician Charles Howard Hinton (1853–1907). Hinton yearned to codify his own religious and spiritual views by apprehending them mathematically. His collection *Scientific Romances* (1884–5) gathers together essays and other short pieces, many of them of a scientific-mystical sort; for instance, the mathematical algorithms needed for the precise quantification of sin and virtues such as might affect a soul's judgement after death. One theme of Hinton's was that time might be conceptualised as a fourth dimension, something that Wells very possibly read as he was imagining his own *Time Machine*: for if time is a dimension like length, breadth and depth, then perhaps people might travel along that dimension.

Future war and invasion fantasies: militaristic extrapolation

Historians of SF identify a significant strand of late century British (and to a lesser extent European) SF as beginning with Chesney's *The Battle of Dorking* (1871), a near-future fable in which a fearsomely efficient German army invades Britain and defeats the enthusiastic but poorly organised and armed British reserve troops. Lieutenant-Colonel George Tomkyns Chesney (1830–1895; knighted in 1890 and promoted to General in 1892) had served with the Bengal Engineers in British India. Invalided home, he concerned himself with what he saw as British military unpreparedness, and made a number of suggestions for military reorganisation to the War Office. *The Battle of Dorking: Reminiscences of a Volunteer*, a short story published as a pamphlet in 1871, was part of this campaign. Its intrinsic interest is small: it reads today as a rather bland first-person narrative blended with an annoying and hectoring Tory militarism: 'a little firmness and self-denial, or political courage', laments its narrator, 'might have averted the disaster', the cause of which he ascribes to the fact that 'the lower classes, uneducated, untrained in the use of political rights' had usurped the powers of 'the class which had used to rule ... and which had brought the nation with honour unsullied through former struggles'. But there is no denying the tremendous popularity it enjoyed, and the chord of British Imperial anxiety it touched. *Blackwood's Magazine*, where the story was first published, reprinted six times to meet demand; issued as a pamphlet it sold 110,000 in two months. Gladstone, the prime minister, attacked it in the House of Commons as alarmist. It was translated into most European languages, and other authors rushed to write plagiarisms or counter-blasts to Chesney's slim tale.

It was by no means the first tale of future war and invasion – we have already noted R. F. Williams' *Eureka: a Prophesy of the Future* (1837), and we might add *The Invasion of England (A Possible Tale of Future Times)* (1870) by Alfred Bate Richards (1820–1876). But neither of these works achieved the enormous success that Chesney enjoyed. The songwriters Frank Green and Carl Bernstein wrote a popular music-hall song, 'The Battle of Dorking: A Dream of John Bull's' (1871), which rewrote the story in England's favour ('England invaded, what a strange idea! / She, the invincible, has nought to fear', quoted in Clarke, p. 37). *Punch* similarly insisted that 'JOHN BULL's not yet the brainless ass that *Blackwood's* prophet would make him' (quoted in Clarke, p. 76). The vigorous public debate about the desirability, and extent, of British rearmament was touched by this to the extent that it was often couched, seriously or light-heartedly, in science fictional terms. An example is this logically extrapolated supergun Dreadnought from *Punch*, 1875:

'The warship of the (Remote) Future', *Punch*, 1875

Chesney sought to capitalise on his success. A second future-war tale, *The New Ordeal* (1879), imagined developments of weaponry that would eventually make war obsolescent, but was much less popular. His novel *The Lesters* (1893) revisits Verne's *Les Cinq cents millions de la Bégum* (1879, see the next chapter), recasting it in more straightforwardly utopian mode: its hero chances upon a huge fortune which enables him to found an ideal new city which he names after himself, 'Lestertia', and which embodies all Chesney's right-wing crotchets.

In the words of I. F. Clarke, *The Battle of Dorking* 'was the beginning of a great flood of future war stories that continued right up to the summer of 1914' (Clarke, p. 15). There are too many books and stories like this to list here (probably more than 60 titles could be adduced), although it is worth noting that in the 1870s and early 1880s they tended to trade in an atmosphere of fear and paranoia (for example, Horace Lester's *The Taking of Dover*, 1888); whereas by the 1890s and 1900s such stories had often acquired a more triumphalist tone which, as Brian Stableford notes, 'helped to generate the great enthusiasm which Britons carried into the real war against Germany when it finally came' (in Clute and Nicholls, p. 1297). Louis Tracy (1863–1928), a popular journalist, wrote several such books. *The Final War* (1896) sees future-Britain and its Empire pitted (successfully) against the rest of the world. In *The Lost Provinces* (1898) an American who has wound up running France leads that country in a successful war against Germany. Other works had even greater contemporary impact, particularly Erskine Childers' (1870–1922) *The Riddle of the Sands* (1903) which treats, excitingly, of the discovery of secret German invasion plans; and William Le Quex (1864–1927), who mined this particular seam to great effect. Le Quex created an enormous public stir with anti-German fictions such as *The Great War in England in 1897* (1894) and especially *The Invasion of 1910: With a Full Account of the Siege of London* (1906), which were both serialised in the right-wing popular newspaper *Daily Mail*. John Sutherland puts it well when he notes that, 'pseudo-documentary in style', *The Invasion of 1910* 'charted the German advance over those parts of England where the newspaper's readership was particularly strong'. Sutherland also mentions the fact Le Quex's political views 'as reflected in his many novels' became increasingly 'anti-Semitic and pro-Fascist' as time went on (Sutherland, p. 372). Which fact stands as a commentary, I think, on this whole sorry subgenre.

Albert Robida

Robida's (1848–1926) future-war tale *La Guerre au vingtième siècle* ('War in the Twentieth Century', 1887) is sometimes included among the previous class of militaristic fantasies, but unfairly so. Robida was a unique figure in nineteenth-century SF, who worked as both an illustrator and a writer, and who produced what Philippe Willems anachronistically but suggestively called 'hypermedia SF novels' (Robida, p. xiii), works in which inventive and witty text was supplemented by marvellous line drawings.

The first of his three future-France masterpieces was *Le vingtième siècle: roman d'une parisienne d'après-demain* ('The Twentieth Century: the Tale of a Parisian of the Day after Tomorrow', 1882). Indeed, in this work the illustrations rather overwhelm the sometimes thin narrative line. The protagonist Hélène, a beautiful young orphan, adopted into the wealthy Ponto family, returns to future-Paris from school on the coast. Robida uses her as an ingénue to whom all the features of Paris unusual to his 1880s readership can be explained; indeed, Hélène's ignorance of the features of future-French society is so complete it is hard to swallow; she is startled or baffled by the telecommunications network (even though this network is global) by which images are projected onto 'crystal screens' with sound accompaniments from around the world; she must have explained to her the workings of the centralised food factories (which pipe soup and food directly to householders; the Ponto house is even flooded with hot *potage* after a broken pipe), the gender-equalised academies of politics, law and literature, and so on. Later in the novel M. Ponto's son Philippe has to be rescued from Britain, which has become a fanatical Mormon colony in which polygamy is a legal requirement and bachelors are imprisoned as dangerous criminals. After this successful escapade Hélène marries Philippe, and they go on a world-wide honeymoon. Travelling through the Pacific, their submarine hits a mine left over from the world war of 1910, and the passengers are marooned on one of the many artificial islands that dot the seaways. There Philippe conceives a plan to build a sixth continent in the Pacific, and the novel closes with his plan being put into action.

We discover more or less incidentally that the Moon has been dragged closer to the Earth ('to a mere 675 kilometres' indeed), apparently for no other reason than to 'brighten our nights'; that Russia has been entirely blown up and flooded during a war; that Italy has been bought by commercial interests and turned into a giant holiday-land (and that Monaco, another nation-state amusement park, is so anxious to preserve its market share that it is prepared to go to war on its neighbour). But if the deliberately light-hearted vivacity of all this invention tends to distract the reader from too complete an emotional investment in the story as actual, the pictures draw the reader in. They are illustrations to pore over, somewhere between Phiz and Heath Robinson and dripping with lovely details; airships, cannons and futuristic machinery are rendered with all the curlicues and fripperies, all the wrought-iron ornamentation and frilly dresses of late century fashion.

The success of *Le vingtième siècle* encouraged Robida to extend his imaginative creation of twentieth-century life with *La Guerre au vingtième siècle* ('War in the Twentieth Century', 1887), which details a war in 1945 with an even greater proportion of illustration; and *La Vie électrique* ('The Electric Life', 1892), which praises the possibility of the force that Robida assumed would power most twentieth-century technology. But endearing as Robida's prose imaginings are, his greater anticipation was in the form of SF; he anticipates by nearly a century the major development SF would take in the late twentieth century, towards becoming a visual-verbal text instead of merely a written thing.

Late century utopias

Although utopias were written throughout the century, becoming if anything more popular in the latter decades, one particular utopian novel was more influential than any other: *Looking Backward 2000–1887* (1888) by Edward Bellamy

(1850–1898), the author whom H. Bruce Franklin has called 'the most influential American science-fiction writer of the nineteenth-century' (Franklin, p. 255). *Looking Backward* takes its protagonist Julian West to the year 2000 by means of a hypnotic trance. On awaking, West discovers a harmonious collectivised America founded on the principles of 'Nationalism' and 'the Religion of Solidarity' which has banished the poverty and misery associated with the individualist capitalism of the past. Everybody, male and female, is recruited into the 'industrial army' between the ages of 21 and 45, where they work at whichever jobs they are considered best fitted for. Those who do not pull their weight are sent to solitary confinement until they buck up their ideas. Everybody receives the same income, although since cash has been abolished payment takes the form of a credit card. Technological advances are also detailed, including a machine that pipes music directly into people's homes. The novel ends with West waking up in the America of 1887, horrified that his visit to the year 2000 was only a dream and depressed by the squalor and degradation of nineteenth-century life. But in a final twist, it turns out that this revisiting of 1887 was itself a dream; West is safe, actually living in Bellamy's ideal future world.

Bellamy's 'utopia', though not uninteresting, reads as just one more example of a very prolific nineteenth-century sub-genre. It is a little puzzling, therefore, to note just how great was the popularity it enjoyed (John Carey claims it had 'a greater impact than any other single utopia', Carey, p. 284).[2] Within a few years of its publication more than a million copies had been sold and it had been translated into all major world languages. Hundreds of clubs were formed to lobby for Bellamy's 'Nationalist' collectivist ideals, and a Nationalist political party was formed in America which enjoyed considerable popularity. Opponents, motivated by party political or broader ideological worries, hurried into print. According to Tony Barley, 'by 1900 over 60 Bellamy-inspired titles had been published in the United States', including many refutations by authors such as the American Ignatius Donnelley (1831–1901), whose *Caesar's Column, a Story of the Twentieth Century* (1890) took Bellamy's premises in a dystopian rather than utopian direction. 'Bellamy was also debated in "straight" political tracts such as W. J. Ghent's ironically titled *Our Benevolent Feudalism* (1901)' (Barley, p. 156). The most famous of the anti-Bellamists was the British poet, writer and designer William Morris (1834–1896), whose *News From Nowhere, or An Epoch of Rest* (1891) details an anti-industrial, rural-idyllic future England that resembles the Middle Ages rather than any high-tech extrapolation of industrial progress. Morris's beautiful book was designedly the antithesis of Bellamy's collectivist vision of society as a giant machine: its representation of perfect social living is gentle and compelling, with a very attractive emphasis on beauty. Still in print today, it is to many socialists a sort of sacred text, although it no longer has its bite as a counterblast to *Looking Backward*.

That Bellamy could have such an impact is explicable at least in part with reference to the idiom in which he wrote: the imaginative recreation of utopia in a popular genre breathes life into, and casts an attractive light over, an otherwise rather repellent social programme. As Chris Ferns notes, Bellamy's future US is 'explicitly based on a military model', workers and women are disenfranchised and there is a quasi-eugenicist fascination with the purity of the race. Moreover, there is 'a significant gap between what is described and what is dramatized. While the emphasis of Bellamy's description is heavily on *work* ... his portrayal of how utopian life is *lived* shows only the utopians at leisure' (Ferns, pp. 76–83). With something amounting to disingenuousness, Bellamy portrays all the benefits of his centralised society and none of the costs. Something similar can be said of Morris's more elegantly written but, alas, even less plausible blueprint for social change: the practical steps, and even the general logic of the change that lead to his cod-medieval idyll are simply absent.

Bellamy's success certainly breathed new life into the utopian mode (and therefore into certain forms of SF) in the 1890s and 1900s: scores upon scores of utopias were published, of varying merit and interest. One significant strand of this movement was a rise in female or proto-feminist utopias. Indeed, utopian fic-tionalising of the state of women actually pre-dated the broader cultural discourse (anxiously debated in British society in the 1890s and 1900s) about the so-called 'New Woman' – women who to one degree or another refused the traditional subservient roles offered by Victorian society. Theatrical burlesques such as Gilbert a Beckett's (1811–1856) *In the Clouds: A Glimpse of Utopia* (first performed at the Alexandra Theatre in 1873), located on a gynocratic 'Island of Flying Women', embody a necessarily frivolous approach to the topic.[3] But more serious-minded exercises in gender thought-experiments followed: *Mizora: a World of Women* (1880) by Mary E. Bradley Lane; the rather austere man-free future Ireland of *New Amazonia: A Foretaste of the Future* (1889) by Elizabeth Corbett; the long poem *Women Free* (1893) by Elizabeth Wolstenholme (1834–1918) and the most famous feminist fantasy of this period, *Herland* (1915) by Charlotte Perkins Gilman (1860–1934).

L'Eve future (1884): Edison's android

A very different perspective on women is to be found in *L'Eve future* ('Future Eve', 1884), an intriguing but misogynistic Symbolist novel by Mathias Villiers de l'Isle-Adam (1840–1889). Symbolism was a late century literary movement, particularly associated with a loose affiliation of French poets (amongst them Charles Baudelaire, Stéphane Mallarmé, Paul Verlaine and Arthur Rimbaud) who wrote in reaction to the dominant literary modes of Realism and Naturalism. Oblique and suggestive, Symbolist writing aims to create certain resonant moods in their works, and used a number of symbols to evoke an often mysterious distillation of the material and spiritual. Influenced, as were several Symbolists, by Poe, Villiers de l'Isle-Adam wrote prose that aspired to the condition of poetry, which fact can make him a tiresome read; but for a long time he was highly regarded by many (Edmund Wilson's study of Symbolism, *Axel's Castle* (1931) took its title from Villiers de l'Isle-Adam's pompous 1890 drama *Axel*).

Excited and repelled to a greater degree than other Symbolists by contemporary advances in technology, Villiers de l'Isle-Adam wrote various short pieces that satirised nineteenth-century *mécanisme*. 'L'Affichage céleste' ('Celestial Advertising', 1873) concerns a new machine that will turn the sky into a medium for advertising purposes. 'L'Appareil pour l'analyse chimique du dernier soupir' ('An Apparatus for Chemically Analysing the Last Breath', 1874) pokes fun at the emotional sterility of the bourgeoisie – a favourite butt of the aristocratic Villiers' ponderous humour – positing a machine that offspring can use to analyse scientifically the last breath of their dying parents, relieving them of the need to feel any sadness. 'La Machine à Gloire' ('The Glorification Machine', 1874) satirically imagines the creation of an automated audience-response machine for the theatre, to applaud or boo as required: 'vingt Andréides sortis des ateliers d'Edison ... automates electro-humains' ('twenty Androids from the workshops of Edison ... automated electro-humans', Villiers, p. 593).

This mention of the real American inventor Thomas Alva Edison (1847–1931) looks forward to the peculiar conceit of *L'Eve future* (1884), a novel in which Edison is the main character. The popular reputation of Edison's boundless capacity to invent had turned him, as Villiers notes, into a legend in his own time: 'l'enthousiasme ... en son pays et ailleurs, lui a conféré une sorte d'apanage mystérieux, ou tout comme, en maints esprits' ('the enthusiasm for Edison in his own country and overseas has given him a sort of special mystique, or something like it, in many minds', Villiers, p. 765). For several decades Edison was likely to appear as a character in SF novels that celebrated the inexhaustible power of his inventive mind. An example is *Edison's Conquest of Mars* (1898) by Garrett P. Serviss (1851–1929), which was written as a more upbeat American response to H. G. Wells' Martian invasion story *The War of the Worlds* (1898); Serviss's novel dramatises a super-competent Edison inventing weapons of war, as well as the anti-gravity device necessary to carry them to Mars, where the alien threat is neutralised, many Martians killed and the rest colonised. Compared to the instrumental militarism of many 'Edisonades' (as Edison invention-adventures are called), Villiers de l'Isle-Adam's Symbolist-mystical treatment seems positively pacific.[4]

L'Eve future opens with Edison in his New York mansion. His friend, the English Lord Eward, visits; driven to distraction by his love for the physically beautiful but spiritually shallow Alicia Clary, Eward has resolved to commit suicide. Edison vows to save his life by constructing a copy of Miss Clary perfect in every detail but with greater profundity of soul. This he does after what sometimes seems like endless descriptions of the precise techniques he will use. This simulacrum is in every way better than the original; indeed, one of Villiers' themes in the novel is that under the logic of modernity the copy *is* preferable to the prototype. Eward, after initial doubts, falls in love with the android woman named Hadaly, abjures suicide and leaves for Europe with his new bride. In a twist at the end, a shipwreck destroys Hadaly, and Eward is left 'inconsolable'.

A near-hysterical fear of the seductive power of women permeates the novel. Women *en masse* are liars; beautiful to men only because of the 'lie' of cosmetics; to which lie the 'lie' of an android woman seeming human is preferable. Alicia Clary and women like her are presented not only as shallow and vain, but as a positive threat to the health and even life of the male sex. According to Edison, 'en Europe et en Amérique, il est, chaque année, tant de milliers et tant de milliers d'hommes raisonnables qui, – abandonnant de véritables, d'admirables femmes ... se laissent ainsi assassiner' by similar seductresses ('every year in Europe and America thousands upon thousands of reasonable men, leaving their true and admirable wives, allow themselves to be assassinated like this', Villiers, p. 904). In such a world, infested with lethal *femmes fatales*, the creation of perfect, pure artificial women is a simple service to male humanity. The fantasy here is not so much sexist (although, clearly it *is* that) as an adolescent wish-fulfilment very unedifying in a man in his mid-forties, as Villiers was at the publication of *L'Eve future*. Hadaly is an anatomically accurate life-sized doll who can be programmed (by fiddling with 20 rings worn on her fingers and thumbs) to perform any action and assume any personality in perfect submissiveness. She/it, moreover, is programmed to kill any other men who might attempt to take sexual advantage of her: 'elle ne pardonne pas la plus légère offense; elle ne reconnaît que son élu' ('she doesn't forgive the slightest offence; she recognises only her designated man', Villiers, p. 860).

Villiers' contemptuous dismissal of, for instance, the very idea that a white man would ever choose a woman from another race (Villiers, p. 864), combined with his wearyingly offensive repetitions of various anti-female libels, make *L'Eve future* a very difficult book to like. Yet this is a novel that has enjoyed a considerable posthumous reputation, in large part because of the sheer single-mindedness with which Villiers pursues his themes of the artificiality of contemporary existence. Those passages of the book that work best are the ones in which he integrates his vision of the allure of technical artificiality into a complete aesthetic. Going to collect his android bride, Eward realises that he has come to see the cosmos itself as unreal:

L'horizon donnait la sensation d'un décor ... Du sud au nord-ouest se roulaient de monstrueux nuages pareils à des monceaux de ouate violette, bordés d'or. Le cieux paraissaient artificiels.

('The horizon gave the impression of being only an ornamental painting ... from the south to the north-west rolled prodigious clouds like so many heaps of violet-coloured wadding fringed with gold. The heavens themselves seemed artificial', Villiers, p. 976.)

In so far as it looks forward to a set of concerns that in the late twentieth century would come to be called 'postmodern', a *Blade Runner*ish precession of the simulacrum, *L'Eve future*, however tainted with racism and misogyny, is a significant work. It forms a very interesting pairing with *Frankenstein* (written, as it was, as many years before the end of the century as Shelley's novel was after the beginning). Shelley's monster, though an artificial creation, was in many senses more authentic, more 'real' (more passionate, more intelligent, more alive), than its creator. Villiers' artificial woman is less real, and her ontological artificiality infects not only 'womankind' (a function of Villiers' sexism) but, by degrees, the cosmos itself. The twentieth century's Sartrean fear of inauthenticity had already arrived.

Science fiction in the 1890s

To employ a cliché, SF boomed in the 1890s, influenced by the continuing success of Verne and Wells, not to mention the vogue for utopias created by Bellamy. Many hundreds of titles were published in this decade; indeed the explosion in interest in the genre feeds through into the twentieth century as an ever-increasing gradient on the graph marking the productivity of SF authors. Of the many writers in this decade only a few who enjoyed particular success, or had particular influence on the development of the genre, can be noted here.

In Louis Boussenard's (1847–1910) *Dix mille ans dans un bloc de glace* ('10,000 Years in a Block of Ice', 1890) the protagonist is frozen in the titular block for the specified length of time, and awakes to find a utopian world-state populated by the diminutive descendants of Chinese and African humanity. Boussenard was very popular in France, and also in the (then) extensively francophone Russia, but has remained almost unknown in Britain and America. The title character of his *Les Secrets de Monsieur Synthèse* (1888–89) is a 'synthetic' man (he has, for instance, dispensed with food and exists by swallowing ten pills and ten ampoules of fluid a day, which he prepares himself). An early example of the 'mad scientist' type, 'Mr Synthetic' hopes to influence human evolution in a 'synthètique' direction. His *Monsieur Rien* ('Mr Never', 1907) concerns the adventures of an invisible man in Tsarist Russia.

American SF was also burgeoning. Milton Worth Ramsey's (1848–1906) *Six Thousand Years Hence* (1891) and Addison Peale Russell's (1826–1912) aerial adventure *Sub-Coelum: a Sky-Built Human World* (1893) are both interesting titles. In Robert D. Braine's (1861–1943) *Messages from Mars by Aid of the Telescope Plant* (1892) a Pacific island is discovered that has been in contact with Martians. Other writers were beginning careers that would last well into the twentieth century. J. H. Rosny aîné ('aîné' means 'eldest son') was the pseudonym of the Belgian writer Joseph-Henri Boëx (1856–1940; he sometimes shared the pseudonym with his younger brother Julian). In *Le Cataclysme* ('The Cataclysm', 1896) an electromagnetic entity from outer space comes to France, disrupting the usual laws of nature as a result. He began as a disciple of Emile Zola, and in 1887, after becoming immersed in palaeontology, began a reimagining of a mythical human history from the origins of humanity to the time when man is dethroned by a new and superior species, the Ironmagnetics ('Ferromagnétaux') in the books *Les Xipéhuz* (1887), *Un Autre Monde* ('Another World', 1888), *La Mort de la Terre* ('The Death of the Earth', 1910), *La Guerre du feu* ('The War of Fire', 1911) and *Les Navigateurs de l'infini* ('Navigators of the Infinite', 1925). The space explorers of the Englishman John Munro's (1849–1930) *A Trip to Venus* (1897) discover on Venus an idyllic and aesthetic utopia (' "The good of it?" rejoined the Venusian; "It is beautiful and gives us pleasure" ', ix. 173). This otherwise unexceptional novel has some interest as the first, presciently, to feature a liquid-fuelled rocket as a spacecraft (Munro was Professor of Mechanical Engineering at Bristol). The great Russian scientist Konstantin Tsiolkovsky's (1857–1935) famous paper on the use of liquid-fuelled rockets to travel through space, 'The Probing of Space by

Means of Jet Devices', was not published until 1898. Few SF novels from this period, however, relied on the most up-to-date science. The cave-dwelling Moon-beings of the Hungarian István Makay's (1870–1935) *Repülögépen a Holdbar* ('By Airplane to the Moon', 1899) owe as much to Kepler's *Somnium* as to any modern speculation.

One of the best late century interplanetary romances is *Auf Zwei Planeten* ('On Two Planets', 1897) by the German philosopher and writer Kurd Lasswitz (1848–1910). Lasswitz is sometimes called 'der deutsche Jules Vernes', but in fact his science fiction has a very different flavour to the eminent Frenchman's. A balloon expedition to the North Pole chances on a settlement established by Martian explorers. The humans discover the Martians to be socially and techno-logically superior, and insist that their ethical code prevents them from exploiting others, even aliens such as the humans. But as the book goes on the Martians, increasingly convinced of their own superiority, start treating the humans with condescension and even contempt. It is revealed that the Martians hope to exploit Earth's natural resources. Events deteriorate to the point where a battle breaks out between the Martians and a British warship, which the Martians easily win. Hostility between the British Empire and other Earth nations gives the Martians the excuse to declare the Earth a Martian protectorate, a military occupation that begins mildly but soon escalates into an oppressive autocracy. Human resistance, in part utilising Martian technology, is effective and the book ends with a peace treaty.

Auf Zwei Planeten is a very effective combination of SF technological speculation (the Martians possess anti-gravity, long-distance communicators and a machine called the Retrospektiv with which they are able to look into the past) and simple but effective social-political observation into the unavoidably malign nature of a supposedly benign imperialism. The combination of those two qualities, indeed, might be taken as a thumbnail definition of 'Golden Age' 'Hard SF': futuristic tech and individualist ideology. Perhaps this is why a number of luminaries from the Golden Age, including Arthur C. Clarke, have expressed their admiration for this book. But good though it is, *Auf Zwei Planeten* is not in the same class as Wells' *The War of the Worlds*, which was published the following year (almost certainly with-out being influenced by Lasswitz's novel).

'Will'

Looking back over the nineteenth century we can see a particular cleavage begin-ning to open up between cultures that would continue to separate out more markedly in the twentieth century: on the one hand, technological and rational optimists, working often in a quasi- (or explicitly) militaristic frame of mind, who imagined bringing society as a whole under the logic of the machine to the immense betterment of everybody; and, on the other, a nebulous group, more or less suspicious of technology, drawn to the spurious consolations of pseudo-scientific mysticism. The correlatives for these two broad camps were, on the one hand, 'anti-gravity' (as a symbol of science's ability to break the bounds of earth), and, on the other, 'Will' (the notion, very common in SF from this period, that simply by willing it, the body, or perhaps the astral body, could travel anywhere it wanted). Posterity has shown neither strategy to be effective in any practical manner; but whilst science is by and large dismissive of anti-gravity (an object with anti-gravity would also, presumably, possess 'anti-inertia' and 'anti-momentum'), the doctrine of 'Will' has a perfectly respectable philosophical pedigree.

The German philosopher Arthur Schopenhauer (1788–1860) distinguished between the world as it *appears* to our senses (which he called '*vorstellung*' or 'representation') and the way the world *actually is*, behind the veil of appearance. This base reality of the universe, he argued, was nothing other than 'Will'. He believed not only that 'Will' was prior to our bodies and other thoughts in the human experience, but that it was the fundamental nature of the cosmos, manifested in the actions of animals, the magnetic action of the Earth's poles and even gravitation. Schopenhauer's influence is evident on another influential German philosopher, Friedrich Nietzsche (1844–1900), who took Schopenhauer's rather pessimistic 'Will' and reinterpreted it as 'Will-to-Power', an aggressively amoral and joyous principle. Nietzsche despised Christianity as a religion of slaves and losers, and looked forward to a Homo superior, the *Übermensch* ('Overman') who will live beyond good and evil, and embody the Will-to-Power in a direct way.[5] If some twentieth-century SF writers can be classified as Schopenhauerean (the outstanding example is the British writer Olaf Stapleton, who uses 'Will' to traverse space and time in *Star Maker*, 1937), many more can be thought of as Nietzschean, particularly if we take the term to refer to the rather caricature version of Nietzsche's philosophy that circulated in many quarters in the first half of the century. John Carter, the *Übermensch* hero of Edgar Rice Burroughs' *A Princess of Mars* (1912), travels to Mars at the beginning of that novel simply by 'willing' it. Nietzsche's 'will to power' did not originally mean 'power over others' (it could equally well be rendered 'creative power' or 'self-knowledge'); but it was taken by many, particularly those with fascistic sympathies, to mean exactly that: and John Carter's triumphalist adventures on Mars, or Barsoom, see him defeating a bewildering array of alien antagonists in battle, and boasting, though blood-stained and battered, 'I still live!' However thrilling this is to readers, particularly adolescent males, it is nevertheless informed by a quasi-fascist understanding of the Will-to-Power. A more sophisticated book in this respect is Alfred Bester's very highly regarded *The Stars My Destination* (1956). In Bester's imagined future (which we would call 'cyberpunk' if the novel did not precede that movement by several decades) individuals have learned to teleport themselves merely by thinking clearly about the destination, a process referred to as the 'jaunte'. What this is, of course, is a concrete externalisation of the Will, and it takes a peculiar hero, the beyond-good-and-evil Gully Foyle, to refine and develop this skill to mankind's eventual benefit.

In this respect, twentieth-century SF writers are following in a tradition laid down by nineteenth-century thinkers and writers. In their own way Flammarion's spirit protagonists who can 'will' themselves to any point in the cosmos are part of the same broad cultural discourse that also produced the philosophy of

Schopenhauer and Nietzsche. If will continues to be important to SF today (witness the 'force' of *Star Wars*, or *The Matrix*, or Superman flying by simply willing himself to fly), then it owes its longevity to roots that are closely twined around the development of SF as a genre.

References

Aldiss, Brian, with David Wingrove, *Trillion Year Spree: The History of Science Fiction* (London: Gollancz 1986)

Alkon, Paul K., *Origins of Futuristic Fiction* (Athens, GA: University of Georgia Press 1987)

Alkon, Paul K., *Science Fiction before 1900: Imagination Discovers Technology* (1994; London: Routledge 2002)

Astor, John Jacob, *A Journey in Other Worlds, A Romance of the Future* (1894; ed. Charles Keller, Electronic Text Center, University of Virginia Library: http://etext.lib.virginia.edu/toc/modeng/public/AstJour.html 1999)

Baldick, Chris, *In Frankenstein's Shadow: Myth, Monstrosity and Nineteenth-Century Writing* (Oxford: Oxford University Press 1987)

Barley, Tony, 'Prediction, Programme and Fantasy in Jack London's *The Iron Heel*', in David Seed (ed.), *Anticipations: Essays on Early Science Fiction and its Precursors* (Liverpool: Liverpool University Press 1995), pp. 153–71

Carey, John (ed.), *The Faber Book of Utopias* (London: Faber 1999)

Claeys, Gregory (ed.), *Modern British Utopias 1700–1850* 8 vols (London: Pickering and Chatto 1997)

Clark, Maudemarie, 'Nietzsche's Doctrines of the Will to Power', in John Richardson and Brian Leiter (eds), *Nietzsche* (Oxford: Oxford University Press 2001)

Clarke, Bruce, *Energy Forms: Allegory and Science in the Era of Classical Thermodynamics* (Ann Arbor, MI: University of Michigan Press 2001)

Clarke, I. F. (ed.), *The Tale of the Next Great War 1871–1914* (Liverpool: Liverpool University Press 1995)

Clute, John and Peter Nicholls, *Encyclopedia of Science Fiction* (2nd edn, London: Orbit 1993)

Davy, Humphry, *Consolations in Travel; or, The Last Days of a Philosopher* (1830)

Disch, Thomas, *The Dreams Our Stuff Is Made of: How Science Fiction Conquered the World* (New York: Simon and Schuster 1998)

Ferns, Chris, *Narrating Utopia: Ideology, Gender, Form in Utopian Literature* (Liverpool: Liverpool University Press 1999)

Flammarion, Camille, *Lumen* (1887; transl. and with an introd. by Brian Stableford, Middletown, CT: Wesleyan University Press 2002)

Franklin, H. Bruce (ed.), *Future Perfect: American Science Fiction of the Nineteenth-Century: An Anthology* (rev. and expanded edn, New Brunswick, NJ: Rutgers University Press 1995)

Grainville, Jean-Baptiste François Xavier Cousin de, *The Last Man* (1805; trans. I. F. Clarke and M. Clarke, 'Wesleyan Early Classics of Science Fiction Series', ed. Arthur B. Evans; Middletown, CT: Wesleyan University Press 2002)

Greg, Percy, *Across the Zodiac: The Story of a Wrecked Record* (1880; e-text at www.bookrags.com/ebooks/10165/1.html)

Hood, Thomas, *Complete Poetical Works of Thomas Hood*, ed. J. Logie Robertson (London: Oxford University Press 1907)

James, Edward, 'Science Fiction by Gaslight: An Introduction to English-Language Science Fiction of the Nineteenth-Century', in David Seed (ed.), *Anticipations: Essays on Early Science Fiction and its Precursors* (Liverpool: Liverpool University Press 1995), pp. 26–45

Lagarde, André and Laurent Michard, *XXe Siècle: Les grands auteurs français – Anthologie et histoire littéraire* (Bordas 1988, updated 1993)

Moser, Charles A., *The Cambridge History of Russian Literature* (revised edn, Cambridge: Cambridge University Press 1992)

O'Brien, Fitz-James, 'The Diamond Lens' (1858), in H. Bruce Franklin (ed.), *Future Perfect: American Science Fiction of the Nineteenth-Century: An Anthology* (rev. and expanded edn, New Brunswick, NJ: Rutgers University Press 1995), pp. 285–306

Philmus, Robert, 'Science Fiction: from its Beginnings to 1870', in Neil Barron (ed.), *Anatomy of Wonder: Science Fiction* (New York: R. R. Bowker 1976), pp. 3–32

Pynsent, Robert and S. I. Kanikova (eds), *The Everyman Companion to East European Literature* (London: Dent 1993)

Robida, Albert, *The Twentieth Century* (transl. with an introd. by Philippe Willems, 'Wesleyan Early Classics of Science Fiction Series', ed. Arthur B. Evans, Middletown, CT: Wesleyan University Press 2004)

Stableford, Brian, 'Science Fiction before the Genre', in Edward James and Farah Mendlesohn (eds), *The Cambridge Companion to Science Fiction* (Cambridge: Cambridge University Press 2003), pp. 15–31

Sutherland, John, *The Longman Companion to Victorian Fiction* (London: Longman 1988)

Suvin, Darko, *Metamorphoses of Science Fiction: On the Poetics and History of a Literary Genre* (New Haven, CT: Yale University Press 1979)

Suvin, Darko, *Victorian Science Fiction in the UK: The Discourses of Knowledge and of Power* (Boston: G. K. Hall 1983)

Trollope, Anthony, *The Fixed Period* (1882; ed. David Skilton, Oxford: Oxford University Press 1993)

Villiers de l'Isle-Adam, Mathias, *Oeuvres Complètes*, ed. Alan Raitt, Pierre-Georges Castex and Jean-Marie Bellefroid (Paris: Gallimard: Bibliothèque de la Pléïade 1986)

7
Jules Verne and H. G. Wells

The French writer Jules Verne and the Englishman H. G. Wells remain, arguably, the two most famous writers of science fiction. Their names are conventionally linked, as in this chapter, although they never met, and in fact come from different generations (Verne was 38 when Wells was born). But, for reasons to do with the period in which they wrote as much as the individual excellence of their writing, their key SF texts consolidated the increasing cultural dominance of SF as a form.

Not that there is consensus among critics about the nature of their achievement. Some critics deny that Verne wrote SF at all. Trevor Harris suggests that it is 'misleading' to call Verne a science fiction writer, and quotes Jean-Pierre Picot's denial of any SF aspect to his writing (Harris, p. 109). Andrew Martin, whose *The Mask of the Prophet: the Extraordinary Fictions of Jules Verne* (1990) remains one of the best full-length studies of the writer, is embarrassed enough on Verne's behalf to express the desire to rescue him from categorisation as a science fiction writer altogether, a category Martin considers (quoting Kurt Vonnegut) to be restrictive and even urine-stained.[1] Yet Verne is so completely a science fiction writer that the embarrassment must adhere to those whose preconceptions blind them to the fact.

These critics want to make a case for Verne as a 'serious' writer, although many have taken him as a mere entertainment; and their animadversion against the very thought of him as writing SF is part of that. It is unfortunately true that Verne suffers, even today, from the belief that he was a provider of mere pabulum, his books only a confection of generic adventure storytelling mixed in with 'interesting facts' of a largely geographical or geological nature. Many readers certainly have enjoyed his tales of adventure and incident in an untroubled way as 'mere entertainment', but that in itself does not mean that his books are exhausted by such a reading.

In fact, the enormous popularity that Verne achieved, first with his novels and later with many adaptations and film versions of certain key tales, consolidated a particular sort of technology fiction as being core to SF as a genre: a story not only premised on one or other technological artefact (or in some cases on the technical facility of certain individuals, the *techne* ranging from invention, construction, potholing or whatever), but a story that *enframes* the world in a certain way. Arguably, it is this enframing that lies behind Verne's enduring appeal: global, capable, mobile and yet grounded at all points in comforting bourgeois social and cultural certainties. As Sarah Capitanio points out, despite the ceaseless to-ing and fro-ing in his books there is a sort of *stasis* at the heart of Verne's imaginings: 'situations tend to be played out in isolation and the characters then return to the world "as we know it" which itself remains largely unchanged and unchallenged'. This is connected to the fact that

> the characters themselves undergo no fundamental evolution as a result of their extraordinary experiences ...
> At the end of the novels the speculative objects and marvellous machines are destroyed, the world returns to comforting order, and the reader ... is encouraged to accept an unproblematical status quo. (Capitanio, pp. 70–1)

The interesting thing about this aesthetic inertia is precisely the way it exists in a powerful dialectic with the principle of movement. The elements of almost all Verne narratives are well summarised by Thomas D. Clareson: 'a man of reason ([often] a scientist) ... journey[s] to some exotic destination, generally somewhere on earth; and [experiences] a series of largely disconnected adventures, most often involving the threat of pursuit and capture' (Clareson, p. 38). Travel, in both a literal and metaphorical sense, is the core appeal of Verne's works. It is no coincidence that the great age of exploration was, in fact, coming to an end in the closing years of the nineteenth century. The world had been mostly explored, and so Verne's fictions tapped into the substratum of human desire that there be mysterious places still to uncover. Another way of putting this would be to say that Verne created an imaginative space into which the exploring urge could move itself, an 'enframing' of the world as an unknown still to be unveiled. This space was grounded at all times in contemporary discourses of 'the possible' and 'the known': Verne almost never extrapolated or speculated, his imaginative realm was not escapist, and in fact it was continually being brought back to the world with which his readers were familiar. All this represented, of course, an ideologically determined trajectory; it is in the nature of 'technology fiction' to return to the status quo precisely because it enframes the world as a ready-to-hand resource, and by doing so re-inscribes the bourgeois perspective on the world as a resource.

Wells, though in some senses a less attractive character than Verne, provided SF with a different model; a series of more thoroughgoing Feyerabendian protocols that dramatised not *stasis* but radical change, and are in fact so deeply embedded in an ideology of change that we might call them 'revolutionary'. In contrast to Verne's static characters in motion, Wells's characters are more often than not complex individuals who more or less *passively* experience adventure, not venturing further from home than they have to, often complacent and rather unadventurous. He excelled in portraits of lower-middle-class conventional individuals; indeed, those of his characters who do not fit the procrustean bed of English parochial respectability are often characterised precisely as dangerous, even diabolical (the two most prominent examples are the Invisible Man and Dr Moreau).

The archetypal Wells story concerns the irruption of the extraordinary into the ordinary. In other words, for Wells, the dialectic worked as an inverted version of Verne's: character and action were frequently static, but the conceptual and imaginative elements were not only in constant motion but related to a fundamental belief in the primacy of change. For many (myself included) this makes Wells by far the more interesting writer, although to say so is not to underestimate either the impact, or the exceptional skill, of Verne's adventure tales.

Jules Verne

Verne was born in the French port of Nantes in 1828 into an affluent bourgeois family (his father was a lawyer). In 1839 he tried to run away from home, taking a position as a ship's boy on a vessel bound for India. He was tracked down by his father at Paimboeuf, down the coast from Nantes, in the face of whose displeasure he is supposed to have promised 'je ne voyagerai plus qu'en rêve' ('I will no longer travel except in my dreams'). In later life his continuing attachment to an adventure-loving Byronic romanticism confined itself to a refusal to follow in his father's professional footsteps, moving instead to Paris in 1848 (on a parental allowance) to devote himself to a writer's vocation. For several years he wrote prolifically, composing many plays, only a few of which were staged, together with a number of short stories and other pieces. Some of these are interesting in the light of Verne's later development (for instance, the disaster story 'Un voyage en ballon', which was published in the magazine *Le Musée des Familles* in 1851); but none of them was particularly successful. 1852 saw the publication of his first long prose work, *Martin Paz*, a historical narrative about the conflict between Spanish explorers and Peruvian Indians – another work which failed to ignite the torch of his literary reputation. Despite his initial lack of success Verne resisted pressure from his father to become articled as a solicitor, although in 1854 he did become a stockbroker in Paris, a career he pursued for eight years, and which enabled him to marry in 1856. But he continued writing: short stories, poetry and even librettos for a number of operettas, including an ape-themed opera *Monsieur de Chimpanzé* (1860). Most of the work from this period remains unpublished. He took two tours abroad – to Scotland in 1859 and Scandinavia in 1861 (a book based on the former of these journeys, *Voyage en Angleterre et en Écosse*, remained unpublished until 1989). His only child, Michel, was born in 1861.

This rather unremarkable career was transformed in 1862. Having been introduced to the Parisian publisher 'Hetzel' (Pierre Jules Stahl, 1814–1886), Verne gave him the manuscript of *Cinq semaines en ballon*, a ballooning-across-Africa adventure. Hetzel not only published it, but signed a contract with Verne to issue a series of books under the general title *voyages extraordinaires*. The contract required three volumes a year (some of Verne's novels were published in three volumes, some in two), although this was later reduced to two. Over the decades that followed this partnership became a world-wide publishing phenomenon.

Whilst the *voyages extraordinaires* were appearing in volume form, Hetzel also launched a magazine aimed at children and young adults, *Le Magasin illustré d'Éducation et de Récréation* in 1864, to which Verne was a frequent contributor. 'Education' combined with 'Recreation' sums up the ethos of Verne's science fiction: exciting adventures are always laced with extensive factual material, often copied by Verne from a range of scientific textbooks and sources directly into the narrative (a fashion that later SF would describe by the uneuphonious coinage 'infodumps'). The Verne–Hetzel text combines an encyclopaedic didacticism with wide-ranging adventures in a forcefully propelled narrative, usually structured around a journey that is driven by an external force (escape from pursuers, the urgent search for the solution to a mystery or some other specific goal; more rarely the motive is simple exploration). The books were also, as the magazine's name suggests, 'illustrés'; indeed in the re-issues under the collective title 'Collection Hetzel: les Voyages Extraordinaire' in the early twentieth century Verne's novels are lavishly and gorgeously illustrated, almost to the point of becoming (like Robida's work) comprehensible primarily as image-texts. In this formal feature they anticipate the later developments of SF as a mainstream phenomenon. Verne's enormous popularity in France soon spread to non-francophone nations in translations which, although they served to disseminate Verne's work, were sadly incompetent in many respects, cutting, bowdlerising and sometimes mistranslating the originals.[2]

However successful and prolific it became, the relationship with Hetzel did not begin smoothly. Verne's second submission to the publisher was the future-fantasy *Paris au XXe siècle* ('Paris in the Twentieth Century', written 1863); inventive and engaging if narratively underpowered, this version of a future France tends towards the dystopian; which is to say, it was part of what was a vigorous nineteenth-century tradition of future fictions. But Hetzel rejected the work outright (it was not published until 1994), apparently telling the young author 'vous avez enter-prisé une tâche impossible', 'you have taken on an impossible task' (Harris, p. 120). This seems to have turned Verne away from more freewheeling imaginative speculation, although it did not discourage him from writing science fiction as such.

Prevented from working in the idiom of future utopia/dystopia, Verne instead excavated an older but (by the 1860s) less popular mode of science fiction in *Voyage au centre de la terre* ('Journey to the Centre of the Earth', 1864). In this novel Professor Lidenbrock discovers directions to the centre of the Earth in the writings of a long dead Viking called Arne Saknussemm and leads a party consisting of himself, his nephew Axel and a guide, Hans, in Saknussemm's footsteps. Where previous subterranean fantasies tend to hurry their protagonists into the subterranean realm so that the various utopian or other wonders can thereby be elaborated, the reader is halfway through Verne's novel before she even gets a glimpse of the inside of the extinct Icelandic volcano through which ingress to the 'centre de la Terre' is made. More strikingly, the bulk of Verne's underground adventures are rendered with a powerfully claustrophobic sense of close, dark, dangerous spaces. Where Nils Klim (for instance) passed into a massive airy space lit by a bright central sun, Verne's protagonists struggle through a repetitive series of lightless caverns, tunnels and crevasses. At one point, separated from his fellows, the narrator Axel despairs: 'j'étais enterré vif, avec la perspective de mourir dans les tortures de la faim et de la soif ... perdu, dans la plus profonde obscurité!' ('I was buried alive, facing death by the tortures of hunger and thirst ... lost in the deepest darkness!' *Voyage au centre*, pp. 176, 183). The effectiveness of this depends partly on the sense of realism with which it negotiates its story, which is to say, the reader's tacit belief that underground exploration would actually be like this vision of extreme and hazardous potholing, rather than the instant access to fertile lands of *Symzonia*. But there is more to the novel than its approach to verisimilitude. The text as a whole is driven by mystery: set in motion by a cryptographic mystery (a set of runes which Lidenbrock, over several chapters, decodes to reveal the path to the underworld); and continued by more conventional narrative devices. (What will they find at the centre of the Earth? Did Saknussemm truly precede them? Will they survive?) The darkness they encounter, in other words, functions as an objective correlative of precisely this principle of 'the unknown'.

What they do find, eventually, is something which functions as the symbolic or emblematic *locus classicus* of 'mystery' in Verne's writing: a vast sea ('un océan veritable ... desert et d'un aspect effroyablement sauvage', *Voyage au centre*, p. 193). Once inside this enormous and illuminated space (light created, we are told, by 'les nappes électriques', 'electric layers', which shine with 'une remarquable intensité') the party constructs a raft out of trees they find growing beside this 'middle sea' (Axel calls the ocean 'cette Méditerranée', p. 203) and presses on with their journey. They see gigantic mushrooms growing on the shore, and witness dinosaurs swimming and battling in the water. Their journey underground has been, in effect, a journey back in time. Eventually the raft is swept into a shaft, elevated rapidly by exploding lava, and all three adventurers are cast alive onto the surface of the planet, on the summit of Stromboli in Italy. That a novel so filled with detailed technical and scientific information should resolve, ultimately, into a grand symbolist narrative (the mysterious sea at the heart of the world functioning eloquently as a multivalent signifier: the primal scene; the subconscious; the Sea of Faith) helped cement one of the prime textual strategies of SF. Particularly when written in the Vernean tradition, SF often blends a formal realism with an aesthetic symbolism.

Voyage au centre de la terre was followed by an SF adventure oriented in the opposite direction: not down but straight up. *De la terre à la lune* ('From the Earth to the Moon', 1865), Verne's third *voyage extraordinaire*, concerns a plan by veterans of the American Civil War, the members of the Baltimore Gun Club led by Impey Barbicane, to travel to the Moon in a craft launched from a gigantic cannon. In fact, the novel is wholly concerned with the conception, planning and construction of this spacecraft: it ends with the manned projectile being fired from Florida, but tantalisingly tells us nothing more than that 'un nouvel astre' is then visible in the sky. Five years later Verne published a sequel, *Autour de la lune* ('Around the Moon', 1870) which finally reveals what became of the explorers. The tendency of critics has often been to treat of these two books as if they were one; but they are not.

De la terre à la lune, although it contains a certain amount of scientific information about the Moon (and rather more information on the history and physics of ballistics and gunpowder), is not really concerned with extraterrestrial affairs. Rather, the emphasis is on the bullish confidence of the American protagonists and on their frankly belligerent and imperialist ambitions: one member of the Gun Club reflecting on the end of the American Civil War, laments 'et nulle guerre en perspective! ... et cela quand il y a tant à faire dans la science de l'artillerie!' ('no war in view! ... and this when there is so much to do in the science of artillery!' Verne, *De la terre*, p. 12). Barbicane announces the flight to the Moon not in the interests of science but of 'conquête', promising that the name of the Moon 'se joindra à ceux des trente-six États qui forment ce grand pays de l'Union' ('will join the 36 States that make up this great country of the Union'; *De la terre*, p. 22). All the busy preparation serves to establish the discourse of martial expansion and triumph, and it carries the whole of the US along with it. The emblem of this novel is the enormous gun, bigger and more powerful than any made before.[3] *Autour de la lune*, on the other hand, is actually concerned with a flight in space; its mood and tone, quite apart from the balance of its scientific content, are very different.

One historical fact, over which Verne of course had no control, has massively overdetermined the way we read *De la terre à la lune* today: the actual voyages to the Moon that took place in the 1960s and 1970s. Commentators rarely resist the temptation to read Verne's fictional account through the lens of the Apollo missions, which often provokes a tedious itinerary of points where Verne was 'right' and where 'wrong' (so: he was correct in thinking the moon-shot would be launched from Florida and that the spacecraft would be steered with rockets, but wrong that the launch mechanism would be a giant cannon). Indeed, so widespread is this impulse that it bears reiterating how inappropriate it is. Although Verne inhabits many of the idioms of 'scientific fact', his books are not factual, and certainly not prophetic; but they are *perceptive*. The comments about war, conquest and the gun industry speak volumes about American mentality.

Nevertheless, it is perhaps instructive to consider a few of the points raised by commentators. One critic, for instance, praises Verne for realising that a flight to the Moon would be 'a vast engineering enterprise' requiring 'the labour of thousands and the expenditure of millions' (rather than the 'solitary genius' building a spaceship 'in a backyard' imagined by many SF writers); but then marks Verne down for 'mak[ing] no allowance for air resistance', suggesting that the shell would be 'vaporised before it ever left the gun muzzle', and – a point noted by many denigrators of Verne's SF imagination in this book – that 'the acceleration would have spread the heroic voyagers over the base of the inside at the initial shock' (Hammerton, p. 104). The former objection presumably reflects Verne's over-optimistic appraisal of the tensile strength of aluminium; but the latter is more interesting. In a novel crammed with calculations about the dimensions of the craft, the acceleration required to lift it out of the Earth's gravity, the explosive material needed to provide this and so on, it is strange that none of the characters considers the effects on the travellers' bodies of accelerating so rapidly. This seems doubly strange when we recall the extensive nineteenth-century discourse on the debilitating effects of rapid acceleration on the human body; the fearful speculation, for instance, that so mild an acceleration as could be provided by steam engine railway travel might prove fatal to passengers. Perhaps they *are* killed by the launch at the end of the novel.

Of course, in the book's sequel, *Autour de la lune* (1870), the astronauts are revealed to be alive. But this does not necessarily contradict what I am saying. It is interesting that in *Autour* Verne does mention (for the first time) an elaborate system of 'tampons ... cousins d'eau ... cloisons brisantes' ('buffers ... water cushions ... collapsible partitions') specifically designed to dampen the effect of 'cette vitesse initiale d'onze mille mètres qui eût suffi à traverser Paris ou New York en une seconde' ('this initial velocity of 11,000 metres which was enough to traverse Paris or New York in a second', Verne, *Autour*, p. 20). Nevertheless we may wish to infer that, *until* the sequel is written (with Verne's life-saving afterthoughts) the astronauts have indeed perished in the launch. This certainly makes sense of the rather mournful tone of the final pages of *De la terre à la lune*, in which everybody except the constitutionally optimistic J. T. Maston ('[le] seul homme [qui] ne voulait pas admettre que la situation fût désespérée', Verne, *De la terre*, p. 243) despairs. *De la terre à la lune* ends with an explosion which Verne describes as massively more destructive than its creators anticipated (not only does it destroy a large stretch of Florida, its effects are felt 'à plus de trois cents milles des ravages américains', 'more than three hundred nautical miles from American shores'). To read the book in this light is to suggest, in effect, that it functions primarily as a satire on America's love of guns, a national belligerence that Verne saw as unshaken by years of bloody civil war, and which, in this book, although supposedly channelled into pure exploration, in fact leads to a colossal explosion.

My larger point is that there is a deal of evidence that Verne's conception of his books changed, sometimes radically, between publishing one book and later writing a sequel. In *Vingt mille lieues sous les mers* ('Twenty Thousand Leagues under the Seas', 1869–70) the mysterious figure who captains the submarine *Nautilus* under the Latin pseudonym *Nemo* ('No man') is a revolutionary Polish aristocrat whose family was slaughtered by the Russians: Hetzel insisted on Verne removing most of the direct allusions to this identity as too contentious (see Butcher, pp. 434–43). In the book's sequel, *L'Île mystérieuse* ('The Mysterious Island', 1874) Nemo is revealed to be an Indian prince, whose animus is against the British rather than the Russians, although many details of his age and personal history do not tally with the earlier book. In another work, *Voyage à travers l'impossible* ('Journey across the Impossible', co-authored with Adolphe d'Ennery, 1882), Nemo, the radical and friend of the oppressed, has become 'a reactionary and a bigot' expressing 'Colonel Blimp-like' views (Butcher, p. 443). This is as much as to say that Nemo cannot be taken as fixed or 'given'; as with any fictional character, he is a textual construction in flux, shaped by the local demands of particular stories and the cultural idiom in which Verne was writing at any given time.

To say this is not to criticise Verne; rather it is to suggest that the ongoing project of the *voyages extraordinaires* was one formally *in*, as well as *about*, movement: a textual accumulation determined (to quote Nemo's motto) *mobilis in mobile*, 'mobile in the mobile element'. Another way of saying this would be to foreground the almost obsessive fascination with moving on, with travelling continually to new destinations, at the heart of all Verne's novels which finds expression not only in subject-matter but in form. In other words, Verne's novels are not only about the restless urge to move on, to explore further, to seek, to strive, to find and not to yield: they also embody this thematic in their own form. There is an evident thematic tension here between the impulse to fix and define (the 'Russellian' scientific apprehension of nature) and the overwhelming urge to *move on* (in travel, often galvanised by pursuit). This restlessness so often takes a circular trajectory – *around* the world, not *to* the moon but *around* the moon and so on – precisely because Verne is articulating an ideological *stasis*.

To call the appetite for sheer travel that characterises many of Verne's earliest books 'tourism' is to underplay both the scope and the importance of the concept. The three English protagonists of his first publication *Cinq semaines en ballon* (1862) spend, as the book's title says, 'five weeks in a balloon', their explorations taking them across but not beyond Africa. But *Les Enfants du Capitaine Grant* ('Captain Grant's Children', 1868) is more comprehensive. The English nobleman Lord Glenarvan catches a shark from his yacht the *Duncan* and pulls a bottle from its belly, inside which he finds three fragmentary messages in broken English, German and French. Piecing these together takes the *Duncan* on a round-the-world voyage in search of the survivors of a shipwreck. But the main thrust of this lengthy novel (it runs to about 250,000 words) is identified in its subtitle, *voyage autour du monde*, a circumnavigation that is related with extensive geographical, zoological and cultural detail. The frontispiece to the 'edition Hetzel' foregrounds this global scope with Verne's name spread, as if he were a continent rather than an author, over a map of the whole earth; and beneath it the *Duncan*, the agent of travel, filling the sky with as much smoke as the erupting volcano on the left-hand side.

This global range is also embodied in one of Verne's more famous titles, *Le Tour du monde en quatre-vingts jours* ('The Eighty-day Tour of the World', 1872), in which yet another Englishman, Phileas Fogg, combines the balloon of *Cinq semaines* and the circumnavigating ambition of *Les Enfants de Capitaine Grant*, winning a wager by circling the globe in the titular 80 days.

It is not coincidental that all these world-spanning heroes are English (Verne describes Phileas Fogg as 'Anglais, à coup sûr', 'most assuredly English'): Verne is apprehending, in fictively extrapolative terms, the expansive ideology underpinning British Imperial expansion of the later nineteenth century. In the broadest sense we can see the new primacy of 'science' as a discourse mapping itself globally reflected in Verne's fictions. But this is not to say that Verne, respectable bourgeois though he was, approved of Imperialism. On the contrary, many of his books valorise the revolutions of peoples against Imperial rule: the Franco-Canadian independence movement of the 1830s in *Famille-sans-nom* (1889) and the struggle for Home Rule in Ireland in *P'tit bonhomme* (1893) are only two examples. Andrew Martin notes that 'Verne appears especially critical of the atrocities committed in the name of the British Empire', especially in *Les Enfants du Capitaine Grant* and *Mistress Branican* (1891), although he adds that many novels 'seem to tolerate or even welcome the ineluctable necessity of colonial domination' (Martin, p. 23).

But in Captain Nemo, Verne's most enduring archetype, we find (to quote Martin again) 'the most lyrical and committed of anti-imperialists, a champion of the oppressed' (Martin, p. 23). The novel which Nemo dominates, *Vingt mille lieues sous les mers* (1869–70), is probably the most celebrated of Verne's titles. It begins, typically, with a mystery: the disappearance of a number of ships, which is blamed on a sea monster. The oceanologist Dr Pierre Aronnax, his companion, Conseil, and a Canadian harpooner, Ned Land, are hired by the US government to join an expedition on board the *Abraham Lincoln* to solve the mystery. After months of searching the *Abraham Lincoln* finds its monster, which turns out to be an enormous, electrically-powered submarine, aboard which Aronnax, Conseil and Land are held prisoner by the eccentric but charismatic Captain 'Nemo'. The captain permits the trio to live aboard his submarine, the *Nautilus*, as his guests, but tells them that, to preserve his secret, he can never release them. From this point on in the three-volume adventure, the *Nautilus* travels around the globe, and Aronnax witnesses a wealth of marvellous undersea sights, both the biology of the sea and such locations as the ruins of Atlantis. Nemo reveals his sympathy for oppressed people, and sinks a destroyer from an unspecified Imperial power. Eventually, the three do escape, whilst the *Nautilus* is caught up in a terrible maelstrom off the coast of Norway. The book ends ambiguously: 'mais qu'est devenu le *Nautilus*? A-t-il résisté aux étreintes du maelström? Le capitaine Nemo, vit-il encore?' ('But what happened to the *Nautilus*? Did it survive the clutches of the maelstrom? Is Nemo still alive?' *Vingt mille*, p. 616).

There had been submarine adventures before Verne's; indeed, the genre can be traced back to Bishop John Wilkins, whom we have already encountered (one chapter in Wilkins' *Mathematicall Magick* (1638) is entitled 'Concerning the Possibility of Framing an Ark for Submarine Navigations'). Verne named the *Nautilus* after an actual submarine built by the Englishman Robert Fulton for Napoleon I, and Butcher lists three submarine-set fantasies published in France 1867–89 alone (*The Depths of the Sea, Submarine Adventures* and *The Submarine World*, Butcher, p. xiv). But Verne's story captured the public imagination in a way none of the predecessors had.[4] Nemo can travel anywhere he likes without ever leaving home (and an exceptionally well-appointed and luxurious home it is). Everything he needs is manufactured from the materials the sea provides. He is pure circulation and yet pure bourgeois stability (which is to say, although he is purportedly a prince, in fact he embodies a fundamentally bourgeois conception of wealth). He is the first wish-fulfilment hero of the bourgeois SF novel.

Vingt mille lieues is in addition a pleasantly didactic read. Verne assimilated an enormous amount of data about the sea, and reproduced his facts very palatably (the illustrations are also, mostly, instructive). This is another facet of the *stasis* of Vernean fictionalising; he is much more comfortable with the known than the unknown. The reader of *Autour de la lune* discovers how uncomfortable Verne was with unsubstantiated speculation. The narrative of this novel very deliberately blends careful extrapolation on the state of contemporary astronomical knowledge (often supported with footnotes) with more fanciful speculation, this latter always put into the mouths of characters, and almost always contradicted by interlocutors. Passing above the dark side of the moon, the voyagers notice 'étranges rayons' ('strange rays') for which they cannot account. Michel insists that they are seeing light reflected on a vast ossuary, a desert of bleached bones of a thousand generations of Lunarians ('cette plaine ne serait alors qu'un immense ossuaire sur lequel reposeraient les dépouilles mortelles de mille generations éteintes', Verne, *Autour*, pp. 171–2). His companions disagree, and as the matter cannot be decided the voyagers do not land, and the mysteries of the far side of the Moon are preserved.

The surprising thing is just how effective a literary device this bet-hedging is: rather than unveiling the mystery of the Moon, Verne manages to increase that mystery precisely by probing it. Is there life in the craters and valleys of the far side of the Moon? Even if it is barren now, has the Moon ever been inhabited by humanoids? Could the whiter patches on the surface be vast fields of bones? Is the far side littered with monumental architectural ruins, or is that merely the trick of the observers' pattern-loving minds? Without having to commit to one specific rationalist narrative, Verne manages to hint at powerfully suggestive narratives – deep time, environmental disaster, humanoid lunar inhabitants fleeing the dying globe and coming (perhaps) to the Earth – whilst at the same time providing the reader with a defamiliarising raft of statistics that in fact do not elucidate.

As in many Verne novels, dramatic tension is generated by a particular lack of scientific preparation. Having splashed down in the Pacific, the projectile is assumed to have sunk and a submarine expedition is mounted; but of course the projectile – an airtight container whose displacement is far greater than its weight – floats ('tous ces savants', Verne tells us, 'avaient oublié cette loi fondamentale'; 'all the experts had forgotten this fundamental law'; *Autour*, p. 318). What are we to make of this frankly culpable forgetfulness on behalf of the 'savants'? Time and again in Verne's books, dénouements depend on the belated realisation of scientific facts that, by all reason, the protagonists ought to have known: straight-forward calculations that have been wrongly worked through provide catastrophic climaxes in *Les Cinq cents millions de la Bégum* (1879) and *Sans dessus dessous* (1889). It is almost as if Verne wishes to dramatise a human command of the discourses of science and technology that, whilst allowing phenomenal achievements, includes gaping blindnesses: a combination of brilliant insights and idiotic misapprehensions and omissions. I would argue that, quite apart from the narrative payoff of enabling Verne to generate twist-endings, this aporia in fact articulates the fundamental inability to close the gap between Verne's realist-technological (and bourgeois) ambitions and the mystical-symbolic aesthetic that, sometimes unconsciously, acts as the motor to his greatest fictions.

By the mid-1870s Verne's fortunes were still being shaped by the success of *Vingt mille lieues sous les mers*. His sequel to that work, *L'Île mystérieuse* ('The Mysterious Island', 1874), only reveals itself as sequel in the third of its three lengthy volumes. Indeed, it is a double sequel; for it carries on the story of one character, Aytoun, who had previously been abandoned on a desert island at the end of *Les Enfants du Capitaine Grant*. It begins as a balloon adventure; five men escape the siege of Richmond during the American Civil War by hot air balloon, but are blown wildly off course, eventually coming down on the island of the book's title. Much of this very long work concerns the efforts undertaken by the group to survive on the island, with many digressions on wildlife, natural history, engineering and the strategies of self-sufficiency. That they do not die, despite hostile natural conditions and an attack by pirates, is in part due to aid they mysteriously receive at crucial moments. The source of this help is finally revealed to be Captain Nemo himself.

The effect of this revelation, near the end of this very long novel, creates an atmosphere of circularity: the unknown is revealed to be known all along. This touches once again on the subject of the *stasis* of Verne's aesthetic. In *Hector Servadac, voyages et aventures à travers le monde solaire* ('Hector Servadac, Journeys and Adventures around the Solar System', 1877) a meteor crashes into the Earth, carrying off a sizeable chunk of northern Africa, on which are not only the eponymous Servadac (a French military officer) and his servant, Ben Zoof, but various other European characters: a Russian count, the crew of his yacht, a group of Spaniards, a young Italian girl, a Jewish merchant and a French professor, who is the first to understand that they are travelling through space on an object, which patriotically he names 'Gallia'. This new mini-world (still possessed of atmosphere and, apparently, full gravity) passes into the outer reaches of the solar system, freezing as it does so: the humans survive by retreating into volcanically heated caves. Eventually, and after observing many interesting astronomical sites, the meteor returns to the inner solar system. The humans construct a balloon out of the sailcloth of a ship and leave their world hoping to transfer themselves to the atmosphere of Earth and so return home; and in a bizarrely dreamlike conclusion they do just that, finding the world exactly as they left it. In fact, Verne wanted to portray a world devastated by the catastrophe, but Hetzel demurred, pressing on his author the need not to leave too tragic a taste in their readership's mouth. The net result is a book which, after seemingly taking the reader on the most fantastic of fantastic voyages, ultimately reveals that they have been literally nowhere.

Les Cinq cents millions de la Bégum ('The Bégum's Five Hundred Million', 1879) returns to the icon of the enormous gun. 'Begum' is Hindustani for 'queen or lady of high rank', and Verne uses this notionally Orientalist device as the springboard for a schematic utopian/dystopian fantasy set on the other side of the world, in America. The two main figures of the story are the Frenchman Dr François Sarrasin and the German Professor Schultz, who by virtue of both being related to the Begum, prove to be the sole heirs to the titular fortune of 525 million francs (lawyers top-slice a cool 25 million francs for their services, rounding the sum down neatly). With his portion of the money Sarrasin builds an ideal community called 'France-ville' in the wilderness of the American northwest. Professor Schultz uses his money to construct his own city, Stahlstadt ('Steeltown'), a place of rigid mechanisation and ruthless adherence to social and technological order. The main output of Stahlstadt is weaponry, culminating in a gigantic cannon with which Schultz intends to destroy France-ville by firing an enormous shell filled with compressed carbonic acid that will simultaneously gas and freeze the French

settlers. The frontispiece to the 'Collection Hetzel' edition of the novel is dominated by this enormous piece of ordnance, the mad Professor Schultz standing by
the muzzle, whilst below the ideal rural utopia of France-ville and the dark satanic
weapon factories of Stahlstadt are located, schematically opposed, in the bottom
left- and right-hand corners, respectively.

This novel, published only a few years after France's humiliating defeat at the
hands of the Prussian army in the Franco-Prussian War of 1870, has been read by
several critics as a crude piece of wish-fulfilment (Andrew Martin suggests that
'Verne obliquely recapitulates and rewrites German defeat of France ... making
France the victor this time round and reinscribing the irrevocable facts of history
in a more congenial form', Martin, pp. 60–1). Certainly, the libel on 'German-ness'
inherent in the text seems egregious. But the book works much more vividly like a
commentary on utopian tradition than it does as an example of recent historical
denial. As political commentary the very binary of 'good city'/'bad city' seems
abstract and removed from reality, located as it is in a thinly realised American
location (we never learn what the US Federal authorities make of these two communities forging their respective armies, and Stahlstadt's weaponry of mass
destruction, on US soil). But as a utopian meta-text the book is full of penetrating
insight into the relationship between different versions of utopian idealism: the
pastoral and the authoritarian. That Verne emblematises the mediation between
these two cities in terms of a gigantic cannon constellates destruction, militarism,
a degree of optimism (after all, this is a cannon that could, in Verne's cosmos, carry
men to the Moon) and a familiar technological giganticism. In the event Schultz
fires his cannon, but he has – as is so often the case with Verne's scientists –
mistaken his calculations: with a muzzle velocity of 'dix mille metres à la second'
the shell not only overshoots France-ville, but overshoots the horizon, putting it
into orbit and providing the Earth with 'un second satellite' (Verne, *Cinq cents
mille*, p. 183). Schultz, enraged by this failure, plans a general assault on France-
ville, intending to destroy it utterly, to turn it into 'une Pompéi moderne' such
that it would be 'l'effroi et l'étonnement du monde entier' ('the object of terror
and astonishment of the whole world', Verne, *Cinq cents mille*, p. 230). But before
he can send this order he is himself killed, frozen by the explosion of one of his
poisonous shells. At the very end this icon of rapid mobility becomes the medium
for another representation of absolute stasis.

Verne's 1880 novel *Le Maison à vapeur* ('The Steam House', 1880) contains a
'steam elephant', a pachyderm-shaped mechanical contrivance in which an
engineer named Banks travels about northern India – a straight lift from the
American Pulp-author Edward S. Ellis (1840–1916), whose *The Steam Man of
the Prairies* (1868) concerns a motile steam-engine in the shape of a giant man in
the American Midwest (see below p. 174). The title of *L'Etoile du Sud* ('The Southern
Star', 1884) refers to a giant 243-carat diamond, apparently artificially constructed
by Victor Cyprien, a French engineer working in South Africa. But although
published as by Verne, this adventure love-story was in fact written by
Paschal Grousset, an ex-Communard, who wrote under the pseudonym André
Laurie. Verne either appropriated the manuscript whole or else lightly reworked

it; something also possibly true of *Les Cinq cents millions de la Bégum*, which was also Grousset's idea. Later in his career, Verne's son Michel passed off his own stories under his father's name, a rather different case of publishing misdirection, since by then Verne's name was an immensely valuable property as a brand in its own right.

Robur le conquérant ('Robur the Conqueror', 1886), together with its sequel *Maître du Monde* ('Master of the World', 1904), reworks the Nemo mythos by locating it in the sky rather than the water. The first book begins, like *Vingt mille lieues*, with unexplained events happening around the world, in this case trumpets in the air and strange lights, and flags planted in impossible-to-reach locations; and as in *Vingt mille lieues*, the solution to the mystery is a technological artefact rather than any supernatural or monstrous explanation. *Robur le conquérant* continues with a lengthy debate in the 'Weldon Institute' in Philadelphia between those who believe that the future of air travel lies with lighter-than-air craft ('aérostat') such as balloons, and Robur himself, who scorns ballooning and insists on heavier-than-air craft ('aéronef'). To prove his point Robur kidnaps members of the Institute and takes them aboard his propeller-powered flying craft, *Albatross*. This aircraft is seemingly impregnable (its fuselage is built from *paper* compressed and treated to make it impervious to attack), and Robur reveals that he intends to use his literal superiority to police the world. By way of illustration he attacks and slaughters native Africans because of the supposed barbarism of their customs. The sequel, as is often the case with Verne, reconfigures the original conception quite radically. Robur returns, but as an out-and-out criminal rather than a benevolent though misguided Nemo figure. His *Albatross* is replaced with a craft called *L'Epouvante* ('Terror'), and the book is essentially concerned with the attempts by the authorities to track down Robur.

Sans dessus dessous ('Anti-Topsy-Turvy', 1889)[5] returns us to the Baltimore Gun Club, which had provided the ordnance to launch the moonship in *De la terre à la lune* (1865). In this novel their plan is even more ambitious: to alter the very rotation of the globe. Under the guise of 'the North Polar Practical Association' the Gun Club has been buying up worthless land in the area of the North Pole. They plan to fire a shell from a very large cannon, such that the recoil jolts the Earth into a more regular rotation, ending the seasons and rendering the entire globe temperate (and thereby freeing up the land they have purchased for profitable exploitation). In other words, this is a novel that functions almost as a hyper-trophic version of the Vernean fascination with mobility-in-*stasis*. If the Gun Club is successful, then a pseudo-technical regularity and consistency will be applied on a global level. The world will still turn and life will continue, but in a more machine-tooled and controlled manner. In the event the Gun Club's calculations are wrong; the gun is fired but the blast is insufficient, by a large measure, to affect the rotation of the world. One form of *stasis* (the return, at novel's end, to the status quo) replaces a more ambitious dream of *stasis* (the world's rotation absolutely regularised).

L'Île à Helice ('Propeller Island', 1895) is based on another sizeable technological premise: a massive artificial island designed to travel about the Pacific island. But

Verne does not really have a story to tell worthy of this massive construction. The narrative, which drops a group of musicians into the middle of a schematic political disagreement between the left-side and right-side occupants of the island, fails to grip. *L'Invasion de la mer* ('The Invasion of the Sea', 1905) details the adventures of certain French engineers, planning to create a new sea in Saharan Africa, who are attacked by the Tuareg. A number of posthumously published works were almost certainly written by Verne's son Michel (1861–1925), among them *La Chasse au météore* ('The Hunt for the Meteor', 1908) and *Le Secret de Wilhelm Storitz* ('Wilhelm Storitz's Secret', 1910). The first of these is about a meteor of solid gold that is approaching the Earth, and the frantic hunt to reach it after it crashes into the polar regions – despite the fact that, as is pointed out several times in the novel, the introduction of so much gold would cause the collapse of all the financial systems of the globe (the meteor is even addressed directly: 'c'est l'universel appauvrissement, la ruine générale qui s'y précipiterait avec toi!' ('it's universal impoverishment and general ruin that will fall with you!', Verne, *La Chasse*, p. 229)). In the event the immensely hot meteor slides into the polar ocean and is lost forever, averting the economic catastrophe (although why a Vernean super-submarine, from some other portion of the mega-textual universe of the *voyages extraordinaires*, might not be employed to recover it is not made clear). In *Le Secret de Wilhelm Storitz*, Storitz's secret is invisibility, and he uses it to try to prevent the marriage of a woman who had spurned him; the influence of Wells' invisible man is unmistakeable and debilitating. The short story collection *Hier et demain* ('Yesterday and Tomorrow', 1910) was also issued under Verne's name, although many of the stories in it were partially or perhaps wholly written by Michel. Two stories are of particular interest: 'Au XXIXme Siècle: la journée d'un journaliste américain en 2889' ('In the 29th Century: One Day of an American Journalist in 2889') and another far-future tale, although a much more successful one, 'L'Eternel Adam' ('The Eternal Adam'; the story is sometimes known by the alternative title 'Edom') in which a future-human called Sofr-Aï-Sr looks back over the thousands of years of history that have led to the world-spanning empire in which he lives. He discovers an ancient manuscript in which a figure from our day relates the cataclysmic flooding of the world and attempts by the few survivors to begin civilisation anew. After reading this story Sofr-Aï-Sr is struck with 'l'intime conviction de l'éternel recommencement des choses' ('an intimate conviction of the eternal recurrence of things'). Nor is this citation of the Nietzschean philosophical concept pompous or out of place. As several critics have pointed out, this brief novella recapitulates the themes of discovery, catastrophe and circularity so central to Verne's entire corpus. *Mobilis in mobile* is revealed as the principle by which the whole world operates, across long time. In Selenick's well-chosen words, Verne's 'recurrent themes and tensions disclose a fear of the imagination's ambivalent potential for dislocating a stable world' (Selenick, p. 2). The enormous, inventive *fort-da* game of Verne's fiction spools out and reels in the imaginative possibilities of radical change and radical departure. A dominant strand of twentieth-century SF, either directly or indirectly influenced by him, continued to explore that same anxious ambivalence.[6]

H. G. Wells

If, on pain of some unpleasant punishment, I were pressed to name the greatest novelist to have worked in the science fictional idiom, I would name Herbert George Wells (1866–1946). He innovated some new premises for SF; more often he adapted older SF tropes; but whatever he touched became alive in a distinctly modern way with a compressed poetry and a profound, perhaps intuitive, understanding of the dialectic that determines the genre as a whole. In his best books he is more eloquent, thought-provoking and quietly astonishing than any other SF writer. It is hard to deny, in Patrick Parrinder's words, that he is 'the pivotal figure in the evolution of scientific romance into modern science fiction. His example has done as much to shape SF as any other single literary influence' (Parrinder, *Science Fiction*, p. 10).

One of the key factors that shaped Wells's life, and therefore his fiction, is sometimes hard to convey to those not born into the peculiar bindweed complexities of the English class system. Wells's family were neither poor or unrespectable; but nor were they, quite, middle-class. His father had been a professional cricketer (the participants of which game, sacred to some Englishmen then as now, used to be divided into amateur 'gentlemen' and professional 'players', the latter being the inferior category) and was, at the time of Wells's birth, a shopkeeper. His mother, always important in his life, was in domestic service at a large Kent stately home. Initially, Wells was himself apprenticed to a local draper, but, powerfully driven, he instead maximised his academic potential, becoming a pupil-teacher at Midhurst grammar school. From this position he was able to win a scholarship to the 'Normal School of Science' in London, an institution without university status but with the benefit of the pedagogy of the eminent biologist and Darwinist Thomas Huxley (1825–1895). Wells later claimed that the year he spent on Huxley's course was 'beyond all question the most educational year' of his life; 'at the end of that time I had acquired a fairly clear and complete and ordered view of the ostensibly real universe' (Wells, *Modern Utopia*, p. 226). While both Huxley's scientific humanism and energetic proselytising for the theory of evolution are very evident in Wells's writing, they mediate a deeper social animus. In a meritocracy an individual of Wells's gifts would have risen easily; but Britain in the 1880s and 1890s was no meritocracy. Wells's social mobility was hard won and left him with a sense of social struggle that chimed with his sense of Darwin's theories of the world as the location of strenuous striving, and of the potential of the discourse of 'science' usefully to supersede the discourses of 'class' and 'religion'. It added to the imaginative brilliance and narrative deftness of Wells's writing a depth and sophistication of social relevance largely lacking in the more comfortably bourgeois Verne.

Wells worked as a schoolteacher, first in Wales and later in north London, and married in 1891. He began writing scientific journalism, selling occasional articles to various journals. Some of these are speculative science in the fullest sense. 'The Man of the Year Million' (published in the *Pall Mall Gazette*, 1893) extrapolated Darwinism over the very long time-scale of its title, imagining the humans who might evolve. In 1894 he left his wife and the following year married a former student, Amy Catherine Robbins, a woman who seems to have brought order to his otherwise rather hectic life. Wells's serial infidelities did not destroy this relationship, although (without being prurient) they do point to his belief that conventional rules and conventional morality did not apply to him – something that also informs his novels. There are, we might say, both advantages and disadvantages in a life so lived, and in a literature created from this fundamental premise.

Wells's writing career was lengthy, and by the end of it he had become one of the most famous writers on the planet. Critics sometimes distinguish between his writings up to 1914, when he was writing mostly fiction, and his career from the First World War until his death when he wrote in an exhausting range of different sorts of modes. For our purposes, however, the turn evident in much of his (less well known) late writings towards the religious-mythical and theological fable can be thought of as merely making manifest a core dialectic present in his writing from the earliest: that between, on the one hand, a *scientific* and, on the other, a *mystical* perspective on the cosmos: which is to say, the formal and generic tension between 'realist fiction' (which Wells also wrote) and 'science fiction', two forms that exist in an unusually intimate interrelation in his work.

Indeed, despite the fact that Wells was a prolific author into the 1940s, it was the decade from 1895 to 1905 that saw the production of almost all his SF masterpieces. His first novel, *The Time Machine* (1895), has proved one of his most enduring. It was followed by books a list of whose titles proves to be a checklist of SF masterpieces: *The Island of Doctor Moreau* (1896), *The Invisible Man* (1897), *The War of the Worlds* (1898), *When the Sleeper Wakes* (1897–8), *Anticipations* (1901), *The First Men in the Moon* (1901), *The Food of the Gods and How it came to Earth* (1904) and *A Modern Utopia* (1905).

Wells published the first version of *The Time Machine* in the small-circulation *Science School Journal* as early as 1888, under the rather ugly title 'The Chronic Argonauts'. This narrative went through 'no fewer than seven different versions' before appearing in book form in 1895 (Hammond, p. 79). The 'time traveller' (we are not given his name) has invented a machine that enables him to move backwards or forwards in time. He travels to the year 802,701 and discovers that mankind has evolved (we might say 'devolved' if that didn't imply an inappropriate sense that Darwinian 'evolution' is synonymous with progress) into two separate races: the beautiful but mindless Eloi who live hedonistic lives above ground; and the savage, ugly Morlocks who live below the ground, and who (the tale reveals) come out at night to devour the Eloi. The traveller later travels even further into the future and sees further 'devolution', with mankind becoming first rabbit-like creatures (a scene cut from the 1895 volume) and finally – in a scene of marvellously desolate vision – crab-like monsters scuttling about a terminal beach under a dying sun.

Conventional critical responses to this novella have seen it as a meditation on the class structure of *fin-de-siècle* Britain, or alternately (or also) as a powerfully condensed attempt to think 'long time', and particularly the implications of the Darwinian theory Wells had acquired via Huxley. The Eloi, 'those pale decadent artistic people', living in a neo-hellenic communistic pastoral paradise above ground, 'carry a flavour of the aesthete from the eighteen-seventies' (Aldiss, p. 118). It can't be denied that the Morlocks are identified in the tale as the Darwinian extension of the industrial proletariat: 'even now, does not an East-end worker live in such artificial conditions as practically to be cut off from the natural surface of the earth?' (Wells, p. 52). That the cannibalistic Morlocks literally eating the imbecilic if beautiful Eloi is easily read as a savage satire, or reverse-satire, on the inherent violence of class in late nineteenth-century Britain; but the Swiftian allegory of Morlocks and Eloi is intensified by Wells's provision of a quasi-scientific explanation for the fantastic and extreme state of affairs.

But this is not the only way to read this brilliant tale. Thinking of Wells as a 'philosopher', a 'quasi-scientist' or a 'prophet', a point of view endorsed of course by Wells himself, can distract us from his extraordinary abilities as a writer; and I want to argue, with a claim to primacy, that this book – before it is 'about' class, Darwinism, degeneration or prophecy – is 'about' narrative and genre. As against critics who see the premise of the novella (a machine that can travel through time) as a means of examining evolution in action, or as a facilitator for class satire, we can read it as a Suvinian 'novum' for narrative itself, a piece of self-reflexive textuality.

We see a consistent shape in much of his short fiction: from an 'ordinary' contemporary environment a device, object or circumstance opens vistas to strange new worlds. In 'The Door in the Wall' (1896), the protagonist finds a mysterious green door that permits him to leave the grimy reality of nineteenth-century London and enter 'a world with a different quality, a warmer, more penetrating and mellower light with a faint, clear gladness in its air' (Wells, *Short Stories*, p. 148). There are many subsequent stories that employ the same device. In 'The Remarkable Case of Davidson's Eyes' (1897), a malfunctioning scientific experiment replaces the protagonist's ordinary vision with vision of the exact opposite point of the globe. In 'The Crystal Egg' (1899) the object of the story's title gives its possessor, a London junk-shop owner, unexpected visual access to a scene on Mars, including a Martian house and flying Martians. This latter story epitomises the way this sort of tale operates: Wells draws a clear distinction between the shabby lower-middle-class existence of the shopkeeper who owns the crystal egg (Mr Cave with his shrewish wife and ungrateful children), and the fantastic, exotic world opened up by the egg itself. This contrast is integral to the functioning of the tale: as Wells said later in *Experiment in Autobiography* (1934), with reference to *The Time Machine*, 'I had realized that the more impossible the story I had to tell, the more ordinary must be the setting' (quoted in Lawton, p. xxxiv). In 'The Crystal Egg' what the egg is, in fact, is *science fiction itself*. It is that thing that gives us fantastic, other-worldly visions. By setting seedy Seven Dials junk shop against exotic Martian palace, the story balances the genre of late century 'realist' fiction of the sort that Wells also wrote, but which is more strongly associated with writers such as George Gissing and Arnold Bennett, with the sparkling possibilities of SF itself.

This is the key not only to *The Time Machine* but to all the fiction of Wells's 'great decade'. Instead of reading the tale as an allegorical 'coding' of contemporary class circumstances, we can read it as a deliberate mediation of the generic representation of those circumstances (Realism) and the escape from such quotidian, everyday representation (the time machine itself, or 'science fiction'). It is of course possible to say, as critics have done, that the time machine is a mechanism by which the author can, for instance, represent Darwinian time, but this is to relate the device at second hand. The time machine is 'like' a clock, a car, a weapon and all the various things that critics have read into the tale built around it; but the time machine *is* a literary device. We can put this another way by remembering how the novella begins. The (unnamed) time traveller 'was expounding a recondite matter to us', explaining the notion of 'space time':

> Any real body must have extension in four directions: it must have Length, Breadth, Thickness, and – Duration. But through a natural infirmity of the flesh ... we incline to overlook this fact. There are really four dimensions, three which we call the three planes of Space and a fourth, Time. (Wells, p. 10)

This 'fourth dimension' is invisible to us only because 'our consciousness moves intermittently' along it throughout our lives. This invisibility is one of the key themes of the novel, and is underlined by the notion that the time traveller's machine grows literally 'invisible' when it travels. This is explained in quasi-scientific terms: 'we cannot see it ... any more than we can see the spoke of a wheel spinning, or a bullet flying through the air. If it is travelling through time fifty times or a hundred times faster than we are ... the impression it creates will be only one-fiftieth or one-hundredth of what it would make if it were not travelling in time' (Wells, p. 17). At the end of the story we are shown this principle in action, as the narrator glimpses the time machine 'ghostly, indistinct ... transparent' (p. 90) as the time traveller journeys off on a new voyage. In the logic of the tale that the time traveller himself tells, this 'invisibility' or 'ghostly indistinctness' becomes realised as the relation between the Eloi and Morlocks, and more materially by the Morlocks themselves.

The Island of Doctor Moreau (1896) is a compelling reworking of *Frankenstein* filtered through a more explicitly religious idiom. Wells's scientist, the vivisection-ist Moreau, has sequestered himself on a tropical island where he has been surgically reworking various animals, making their physiques more human and enhancing the capacity of their brains. These creations are 'monstrous' rather as was Mary Shelley's monster, a marvellous combination of hideousness and a weird beauty based on dogs, pumas, pigs and monkeys, 'they wore turbans too, and thereunder peered out their elfin faces at me, faces with protruding lower jaws and bright eyes' (Wells, *Island*, p. 27). They have developed a rudimentary religion, with Moreau himself as a combined God of Mercy and Pain ('*His* is the Hand that wounds', the chant: '*His* is the Hand that heals'). The novel's scientific Eden also includes a version of the biblical command not to eat from the 'Tree of the Knowledge of Good and Evil': Moreau has ordered his beast-men not to taste blood. This command is, of course, transgressed, and the creatures revert to their bestial origins. We may wish to read this as a fairly straightforward satire on organised religion. But it is the *qualia* of this imagined world, deftly and plangently evoked by Wells's prose, that stick in the imagination. As a novum the beast-men enable Wells to write fluently about the balance between civilisation and bestiality in humankind. The advantage in embodying the story in a science-fictional idiom is that the connection between bestiality and humanity speaks to the larger dialectic of the material and spiritual realms. Wells later cheerily called the book 'an exercise in youthful blasphemy ... theological grotesque' (quoted in Kemp, p. 211).

The War of the Worlds (1898) is so powerfully written a novel, so current in contemporary culture (Steven Spielberg's 2005 feature film adaptation is the latest in a very long series) that we forget how crowded a sub-genre of late nineteenth-century SF the paranoid 'future invasion of Britain' story was (see above pp. 116–118). As is usual for those stories, the narrative centres on an ordinary Englishman's life, and then dramatises the extraordinary irrupting into it. Wells's brilliance was to imagine not Germans but aliens invading: a giant metal cylinder crash-lands near Woking; tentacled Martians climb out of this cylinder to make war on humanity from towering mechanical tripods, laying waste to south-east England before eventually succumbing to earthly bacteria against which (we are told) they have no natural defence.

Wells's Martians are imagined according to the scientific orthodoxy of the time. Mars was believed to be a much older planet than Earth; Wells's Martians, planning to acquire Earth's real estate because their own world is dying, are hyper-evolved. Wells tells us that they 'were heads, merely heads. Entrails had they none' (Wells, *War*, p. 119). They do not digest their food, but ingest its blood directly into their own circulatory systems; and their life is governed by a superior and rather cruel rationalism. They are, at the same time, monstrous to earthly eyes:

> Those who have never seen a living Martian can scarcely imagine the strange horror of their appearance. The peculiar V-shaped mouth with its pointed upper lip, the absence of brow ridges, the absence of a chin beneath the wedge-like lower lip, the incessant quivering of this mouth, the Gorgon groups of tentacles. (Wells, *War*, p. 19)

This brilliantly visualised icon of monstrous and horrific alien-ness is what the other now forgotten invasion fantasies from this period lacked, and it is key to *The War of the Worlds*.

Throughout the novel, indeed, Wells displays extraordinary control and expressiveness in his writing. Few writers in any genre can match the desolate beauty he evokes in a London emptied by the Martian threat and overrun with the red weed they have brought across space. At this point in the book the last Martian is ceasing its weird cry and dying:

> Abruptly as I crossed the bridge, the sound of 'Ulla, ulla, ulla, ulla' ceased. It was, as it were, cut off. The silence came like a thunder-clap.
>
> The dusky houses about me stood faint and tall and dim; the trees towards the park were growing black. All about me the red weed clambered among the

ruins, writhing to get above me in the dimness. Night, the mother of fear and mystery, was coming upon me. But while that voice sounded the solitude, the desolation, had been endurable; by virtue of it, London had still seemed alive, and the sense of life about me had upheld me. Then suddenly a change, the passing of something – I knew not what – and then a stillness that could be felt. (Wells, *War*, p. 159)

The pseudo-Arabic cry 'Ulla Ulla', with its echoes of 'Allah', is an interestingly suggestive touch (it would surely be hard to make with a beak-shaped mouth). *The War of the Worlds*, like the other invasion-fantasy books of the 1880s and 1890s, captures a fundamentally xenophobic fear of foreignness. Are the Martians merely ciphers for 'foreigners'? Darko Suvin thinks so:

The Martians from *The War of the Worlds* are described in Goebbelsian terms of repugnantly slimy and horrible 'racial' otherness and given the sole function of bloodthirsty predators (a function that fuses genocidal fire-power – itself described as an echo of the treatment meted out by the imperialist powers to colonized peoples – with the bloodsucking vampirism of horror fantasies). (Suvin, p. 78)

Many critics have noted how Wells's novel symbolically distilled the concerns of his age. His Martians are of course Imperialists, using their superior technology to invade a nation (England) which had been accumulating its own Empire throughout the century largely due to a superior technological sophistication. In other words, the arrival of the Martians and their mechanised brutalities are the symbolic forms Wells chose to explore a deeper set of concerns about the violence of Empire-building, and about the anxieties of otherness and the encounter with otherness that Empire imposes on the Imperial peoples. Many critics have explored this rather obvious observation into the undergrowth of cultural materialist and ideological-critical complexity, but it is at root a simple point, needless of over-elaboration. In Brian Aldiss's words Wells's novel 'showed the Imperialist European powers of the day how it felt to be on the receiving end of an invasion armed with superior technology' (Aldiss and Wingrove, p. 71). It rather overlooks the power of this book to reduce it to a political message. What works so well in this absolutely gripping book is the minuteness of Wells's grasp of the detail of his imagined drama.

The Invisible Man (1897) is a simple and briefly realised story: the scientist Griffin, an antisocial and (by the story's end) practically psychotic man, invents a means of rendering himself invisible, but is unable to reverse the process. He arrives in an English village swathed with clothes and bandages to disguise his condition, and tries from his inn room to further his researches. But the natives are understandably suspicious of him, and become more so after a series of inexplicable thefts and other events. Unclothed, the invisible man reveals megalomaniac desires and is eventually hunted down and killed, whereupon he becomes visible again. It is a book, of course, 'about' seeing; but it is not a coincidence that the invisible man himself is a scientist. Invisibility, the novella suggests, removes a man from social interaction and therefore from social responsibility (something the book conceives of as essentially scopological), with malign consequences. We might, as some critics have done, want to place *The Invisible Man* in a particular nineteenth-century tradition of penology associated with the British thinker Jeremy Bentham (1748–1832), who planned an ideal prison, the 'panopticon', which placed all inmates under the eye of a centrally located warder. The French philosopher Michel Foucault (1926–1984) took Bentham's ideas as expressive of the larger cultural logic of the nineteenth century itself.[7] Authority depends, in involved but inescapable reasons, on seeing. Removed from that logic the invisible man is immediately a social threat.

This is a text, in other words, about the dangers of 'the invisible' (the mystic, the unknown) as opposed to the known, the technical, the machinic (which is why the apparatus Griffin used to render himself invisible neither appears in the book nor is described). It is also a fantasy on the formal conventions of 'narration' itself: the unseen omniscient narrator who can travel unnoticed by the characters in the novel and yet is privy to their most private actions and even thoughts. As with all of Wells's best books it folds political, cultural, formal and speculative into one perfectly controlled textual whole.

The Invisible Man is, arguably, the last of Wells's books to have retained widespread cultural currency today (which is to say, the last of the books likely to be recognised by ordinary people – the last to have been adapted into many other idioms and modes). Although he had many scores of books yet unwritten, among them many masterpieces, and although he was to become in his own lifetime enormously famous, posterity has been much more selective than were his contemporaries. In part this is because Wells himself began, at around the turn of the century, to conceive of his vocation as a writer in a different manner.

When the Sleeper Wakes (1897–8) is another parodically religious book, although this fact is well disguised. Graham, an ordinary nineteenth-century Englishman, falls into a coma and awakens in the future to discover that the mysterious actions of compound interest on his savings have made him Master of the World, with a group of twelve trustees in his name (and who are, of course, greatly incommoded by his revival, despite protestations of delight). The megalopolis that London has become, and the political machinations that threaten Graham's life, are the entertaining subject of the book: but as Peter Kemp notes, it may be the 'religious burlesque' of this fable about 'a miraculous resurrection' that is its heart ('Graham, Master of the World, finds his interests being administered by dubious disciples, the Twelve Trustees, and ... speech sometimes becomes a sort of tongue-in-cheek plainchant, "Verily it is the Sleeper" ', Kemp, p. 211).

The First Men in the Moon (1901) takes the familiar late nineteenth-century tropes of an anti-gravitational metal (see above pp. 108–110; Wells calls his 'Cavorite') and the lunar *voyage extraordinaire* as the framework for a readable but not especially striking adventure. Two Earthmen inadvertently shift themselves to the Moon, where they run around a great deal to escape from the insectoid Selenites who live inside the globe. *Anticipations* (1901), a work of serious extrapolation rather than of fiction, enjoyed considerable success in its day, and is still of great interest in its sober-headed attempt to think through how the future might actually be. But in one sense it represents a deleterious development in Wells's career. The reader working his or her way chronologically through Wells's output begins to notice a dichotomy, sometimes rather poorly disguised, between the specifically dramatic conception prominent in the earlier books, and the specifically generalising, panoramic ambition of the *Anticipations*-inspired approach. Indeed, a footnote on the first page of *Anticipations* identifies fictive detail and definiteness (the facilitator of some of Wells's finest effects) as a problem that, the implication is, Wells hopes to overcome:

> Fiction is necessarily concrete and definite; it permits of no open alternatives; its aim of illusion prevents a proper amplitude of demonstration ... the very form of fiction carries with it something of disavowal; indeed, very much of the Fiction of the Future pretty frankly abandons the prophetic altogether, and becomes polemical, cautionary, or idealistic; and a mere footnote and commentary to our present discontents. (Wells, *Anticipations*, pp. 1–2)

We are entitled to ask the question: How could a fictional genius like Wells get it so very wrong?

The shadow of this resolution spreads across much of Wells's subsequent fiction. *The Food of the Gods and How it Came to Earth* (1904) is an example. The premise of the book is an entertaining if one-dimensional fantasia based on Swift's Brobdingnagian episode. A new nutrient ('Herakleophorbia' or 'Boomfood') enormously increases the size of the creatures that eat it. Chickens become large enough to devour cats and even men; giant rats and beetles attack humanity. The book shifts in tone towards the end, leaving this very effective grand guignol for a rather dissipated philosophising about the potential for 'greatness' (literal and metaphorical) the food affords those humans bold enough to eat it: 'Great and little cannot understand one another', Wells concludes. 'But in every child born of man ... lurks some seed of greatness – waiting for the Food.' This artistically debilitating lurch from the particular to the general stands as an emblem for the larger development of his career: from the powerfully evocative particular details of the early science fiction to the generalised and rather abstract pompousness of the post-*Anticipations* 'broad canvas' attempts to trace giant patterns of futurity.

In the Days of the Comet (1906) tells (like Poe's 'The Conversation of Eiros and Charmion', 1839) of how the world passes through the tail of a comet; but where Poe's story treats this as the end of the world, Wells's 'green vapours' have so beneficial an effect on the population that they turn the Earth into a utopia. Like *The Food of the Gods* this book falls into two unequal parts. The first is a brilliantly claustrophobic evocation of lower-middle-class life; gripping and vivid, it delineates with a horrible sense of inevitability the descent of its thwarted narrator into sexual jealousy and murderous rage. But the latter portion of the book replaces this particularity with a belief in panorama as an aesthetic virtue in itself: this is what the fatal success of *Anticipations*, Wells' first panoramic work, did to his writerly vision. The reformed world of *Days of the Comet* is interesting but rather distant. It is not merely that this is a particularly unbelievable utopia (it does follow, at least, logically from its premise) but rather that it is insufficiently particularised. Something similar is true of *The War in the Air* (1908), a book which extrapolates the then nascent technologies of flight (Louis Bleriot's first flight across the English Channel did not take place until 1909). Bert Smallways, a Wellsian lower-middle-class 'little man' hero, is caught up in the events of global war fought in the skies. The book ends pessimistically: this devastating new form of war brings worldwide disaster ('everywhere there are ruins and unburied dead. and shrunken, yellow-faced survivors in a mortal apathy … it is a universal dissolution'); but it is hard to care for this slightly abstract apocalypse – an odd observation to make about a writer who could (in the *Time Machine*) capture an almost inexpressibly plangent sadness in a single man faced with a beach full of crab-like creatures.

The early 1900s was marking a point of transition in Wells's world-view, and therefore in the type and variety of work he published. In the words of Patrick Parrinder:

> The calculations of planetary cooling reflected in both *The Time Machine* and *The War of the Worlds* [lost] their sway over contemporary scientific opinion. In *The Interpretation of Radium* (1908) – the book which led Wells to envisage the possibility of atomic warfare – Frederick Soddy wrote that 'our outlook on the physical universe has been permanently altered. We are no longer inhabitants of a universe slowly dying from the physical exhaustion of its energy, but of a universe which has in the internal energy of its material components the means to rejuvenate itself perennially over immense periods of time.' Wells [switched] shortly before the First World War from entropic pessimism to a position much closer to Soddy's thermonuclear optimism. (Parrinder, *Shadows*, pp. 46–7)

The first manifestation of this change in outlook was, perhaps counter-intuitively, a gap of five years in which Wells published almost no science fiction at all. Instead he published a series of mainstream novels (which he himself thought his most important work; a view some critics endorse). Radioactive matter appears, as the noisome but valuable 'quap', in *Tono-Bungay* (1909), a story rooted in semi-autobiography in which the Wells-like George Ponderevo escapes from the confining class into which he is born, and together with his Dickensian uncle makes a fortune by means of the quack-remedy of the novel's title, afterwards losing it. Sending a boat to harvest the quap becomes a desperate strategy (which fails) to stave off disaster. Indeed, the idea of 'radioactivity' in the novel stresses its pathological propensities: the ship bringing the stuff to England decays and collapses during its voyage. Other novels anatomised and challenged the mores of Edwardian England. The representation of a sexually self-confident woman in *Ann Veronica* (1909) caused a scandal in its day. The comedy of *The History of Mr Polly* (1910) has great charm, although there is perhaps something condescending about the portrait of the hapless lower-middle-class hero. *The New Machiavelli*

(1911), a novel about a politician whose career is ruined by a love affair, is so precisely located in the London political scene of the early 1900s that the loss of that topicality is fatal. Like *Marriage* (1912), a novel about an unhappily married couple, these books all stay close to the contours of Wells's own life. But the outbreak of the First World War in 1914 seemed to galvanise Wells's interest in SF again. Although he continued writing a prolific stream of mainstream novels, he also produced his second great spread of SF works.

The World Set Free (1914) is another tale of future-war, looking back from the perspective of 1970 on the world war of 1956. Atomic weapons bring enormous devastation, but in contrast to *The War in the Air* ultimate catastrophe is averted by the intervention of an elite of far-seeing politicians, led by 'God's Englishman', King Egbert. As J. R. Hammond points out, the notion that only rule by an oligarchy of brilliant men, scientists and philosopher-kings could ensure mankind's future became an *idée* increasingly *fixe* for Wells in his later life: 'again and again in his writings we find this emphasis on a world renaissance brought to fruition by and through an elite; in the New Republicans of *Anticipations*, the Samurai of *A Modern Utopia*, the Open Conspirators of *The World of William Clissold* and the "Modern State" movement of *The World Set Free* and *The Shape of Things to Come*' (Hammond, p. 110). Wells was by no means alone in this belief in the early years of the twentieth century; but many of the people who believed this converted to fascism in Europe of the 1930s. It would not be right to call Wells a fascist, although many commentators find his sympathy for a quasi-Nietzschean supreme elite, along with his advocacy of eugenicist theories, repellent.

But, perhaps by instinct, Wells had lighted upon a genuinely twentieth-century scientific idiom. Just a few years earlier, the chemical rationale of global change in *In The Days of the Comet* is implausible even by the standards of 1906 (the comet reacts with 'the nitrogen in the air ... [which] in the twinkling of an eye was changed out of itself, and in an hour or so became a respirable gas, differing indeed from oxygen, but helping and sustaining in its action', Wells, *Days of the Comet*, p. 172). By 1914, with *The World Set Free*, this naïve 'chemical' understanding has been replaced by a more contemporary 'atomic' one.

Many of Wells's later books rehearse these authoritarian if anti-establishment views. In *Men Like Gods* (1923) travellers step through a 'kink in space' into an alternative world run on utopian lines. The dreamer who has *The Dream* (1924) is Sarnac, a citizen of the far future; and the dream he has is the life of a certain Harry Mortimer Smith born 1895, died 1920. In trying to relate the details of this immersive dreaming experience to his fellow futurians, Sarnac provides a distant perspective on contemporary life. Wells's wit and intellectual dexterity does not save the book from a certain preachiness. But he was still capable of reflecting wittily on his own soapbox certainties. In *The Autocracy of Mr. Parham* (1930) the ordinary man of the title believes himself to be possessed by the spirit of a warlord from the planet Mars ('he realised that an immense power of will had taken possession of him'). He declares himself Lord Paramount of England and wages a holy war. The main political figures of the day (including Churchill and Mussolini) are introduced under various fictional names.

Wells had much greater success with the rather po-faced *The Shape of Things to Come* (1933). Mankind descends into the valley of collective death after global war; anarchy prevails, warlords prey on humanity; but everything is saved by an elite of technocrats who reconstruct civilisation as a unitary world-state orchestrated on rational lines. The book is written as if it were a history (Wells had published his nippy, thought-provoking *A Short History of the World* in 1922), which gives the vistas of human suffering and social reconstruction a rather chilly, distant feel, although it has the virtue of implying a suitably historical time-scale. A film made from this book, *Things to Come* (dir. William Cameron Menzies, 1936), is highly regarded by some critics, although it is difficult to see exactly why. Menzies' movie concertinas the novel into a three-act melodrama, and portrays the coming of the technocrats in a much more offensively fascistic manner; descending from the skies in white aircraft to rescue suffering humanity.

The little-known late novella *The Croquet Player* (1936) revisits some of the themes of *The Island of Doctor Moreau*, expressing the gloom Wells felt during the rise of European fascism. The story is premised on the notion that *Homo neanderthalis* (characterised in this book as 'invincibly bestial, envious, malicious, greedy') is still present genetically within Homo sapiens, and indeed is on the point of breaking through in modern man, destroying civilisation with an epidemic of atavistic barbarism. The mood of the novel is wonderfully dark and unsettling, and it not only captures the sheer pessimism of Wells's last years, but also functions as a brilliant little gloss on the rise of totalitarianism in Europe. On the other hand, Wells was also writing spurious *Übermensch* fantasies such as *Star Begotten* (1937), which suggests that 'great men' owe their greatness to certain rays beamed to the Earth by vastly superior Martian life-forms. The individuals who are susceptible to these rays seek one another out and plan a cleansed and improved world order; but this conceit (which the novel appears to take quite seriously) seems an almost incredibly obtuse fantasy to write in the teeth of rising European fascism. By the time of his death another world war was in the process of gorging on the entrails of Europe, and Wells's pessimism increased: in his eightieth, and final, year he published *Mind at the End of its Tether* (1945), a book in which, as its title makes clear, he holds out little hope for humanity.

With a writer as multifariously brilliant (and who produced at a high level over so long a period) as Wells, it is very difficult to provide a concise summary and assessment. The fact of his importance to the development of SF is so often reiterated by critics as to wear itself smooth in the ear; to be taken for granted. And yet there is something unique about Wells's achievement: an imaginative capaciousness that, in his best books, is blended with a superb aesthetic precision.

An account of the career of a writer of Wells's productivity, who produced so many mark-worthy books and stories, runs the risk, as the second half of this chapter demonstrates, of degenerating into a rather breathless list of titles. But, then again, it *is* difficult to draw simple, elegant conclusions from Wells's glorious, inspiring sprawl. David Smith's 1986 study, *H. G. Wells: Desperately Mortal*, argues that Wells's education in 'real science' resulted in an SF rooted in scientific precision rather than 'pseudo-science'. There are many places in Wells's fiction where this is clearly the aesthetic rationale. But Roger Luckhurst, disagreeing with Smith, surely gets closer to the target by insisting on the way Wells exemplifies precisely 'generic impurity and hybridity':

> Much recent criticism frames Wells within the Gothic tradition, and he certainly exploits the cultural strain of antipathy to scientific for his farrago of Gothic monstrosity in Moreau. Elsewhere he might dismiss the ghost hunters and mesmerists of psychical research in the august pages of *Nature*, but he also used this 'pseudo-science' to generate fictions about clairvoyance, the projection of psychic doubles, crystal-gazing, soul swapping and astral travel. This is hardly the stuff of 'real science', but it thrives opportunistically in the fissures and lacunae of materialism. (Luckhurst, p. 41)

It is in attacking these dialectics, Gothic/scientific, magic/materialist, from so wide a variety of angles that Wells works through and through the determining dichotomy of science fiction itself, an atheist writer who incubates both Catholic supernaturalism and rationalist 'Protestant' materialism into unique and compelling fiction. Wells did not invent science fiction; but he did revivify its core dialectic with promiscuous energy, and with lasting impact.

References

Aldiss, Brian, with David Wingrove, *Trillion Year Spree: the History of Science Fiction* (London: Gollancz 1986)

Alkon, Paul K., *Science Fiction before 1900: Imagination Discovers Technology* (London: Routledge 1994)

Bergonzi, Bernard, *The Early H. G. Wells* (Manchester: Manchester University Press 1961)

Butcher, William (ed. and transl.), *Jules Verne: Twenty Thousand Leagues under the Sea* (Oxford: Oxford University Press 1998)

Capitanio, Sarah, ' "L'Ici-bas" and "l'Au-delà" ... but Not as they Knew it. Realism, Utopianism and Science Fiction in the Novels of Jules Verne', in Edmund Smyth (ed.), *Jules Verne: Narratives of Modernity* (Liverpool: Liverpool University Press 2000), pp. 60–77

Clareson, Thomas D., 'The Emergence of the Scientific Romance 1870–1926', in Neil Barron (ed.), *Anatomy of Wonder: Science Fiction* (New York: R. R. Bowker 1976), pp. 33–78

Clute, John, 'Jules (Gabriel) Verne', in John Clute and Peter Nicholls (eds), *Encyclopedia of Science Fiction* (2nd edn., London: Orbit 1993), pp. 1275–9

Disch, Thomas, *The Dreams Our Stuff is Made of: How Science Fiction Conquered the World* (New York: Simon and Schuster 1998)

Hammerton, M., 'Verne's Amazing Journeys', in David Seed (ed.), *Anticipations: Essays on Early Science Fiction and its Precursors* (Liverpool: Liverpool University Press 1995), pp. 98–110

Hammond, J. R., *An H. G. Wells Companion* (London: Macmillan 1979)

Harris, Trevor, 'Measurement and Mystery in Verne', in Edmund Smyth (ed.), *Jules Verne: Narratives of Modernity* (Liverpool: Liverpool University Press 2000), pp. 109–21

Kemp, Peter, *H. G. Wells and the Culminating Ape* (2nd edn., Basingstoke and London, Macmillan 1996)

Lawton, John (ed.), *H. G. Wells: The Time Machine* (London: Dent, 'The Everyman Library' 1995)

Luckhurst, Roger, *Science Fiction* (London: Polity 2005)

Martin, Andrew, *The Mask of the Prophet: the Extraordinary Fictions of Jules Verne* (Oxford: Clarendon 1990)

Parrinder, Patrick, *Science Fiction: its Criticism and Teaching* (London and New York: Methuen 1980)

Parrinder, Patrick, *Shadows of the Future: H. G. Wells, Science Fiction and Prophecy* (Liverpool: Liverpool University Press 1995)

Senelick, Laurence, 'Outer Space, Inner Rhythms: the Concurrences of Jules Verne and Jacques Offenbach', *Nineteenth Century Theatre and Film* 30/1 (Summer 2003), 1–10

Smyth, Edmund (ed.), *Jules Verne: Narratives of Modernity* (Liverpool: Liverpool University Press 2000)

Suvin, Darko, *Metamorphoses of Science Fiction: On the Poetics and History of a Literary Genre* (New Haven, CT: Yale University Press 1979)

Verne, Jules, *Voyage au centre de la terre* (1864; Paris: Livres de Poche 2000)

Verne, Jules, *De la terre à la lune* (1865; Paris: Livres de Poche 2001)

Verne, Jules, *Autour de la lune* (1869; Paris: Livres de Poche 2000)

Verne, Jules, *Vingt mille lieues sous les mers* (1869; Paris: Livres de Poche 2000)

Verne, Jules, *L'Île mystérieuse* (1874; Paris: Librarie Hachette 1919)

Verne, Jules, *Hector Servadac, voyages et aventures à travers le monde solaire* (1877; Paris: Librarie Hachette 1919)

Verne, Jules, *Les Cinq cents millions de la Bégum* (1879; Paris: Livres de Poche 2000)

Verne, Jules, *Le Maison à vapeur* (1880)

Verne, Jules, *L'Etoile du Sud* (1884)

Verne, Jules, *Robur-le-conquérant* (1886)

Verne, Jules, *Sans dessus dessous* (1889; Paris: Magnard Collège 2002)

Verne, Jules, *L'Île à Helice* (1895)

Verne, Jules, *Maître du monde* (1904)

Verne, Jules, *L'Invasion de la mer* (1905)

Verne, Jules, *La Chasse au météore* (written 1904–5; published 1908; ed. Olivier Dumas, Paris: Gallimard 2002)

Verne, Jules, *Le Secret de Wilhelm Storitz* (written 1904–5; published 1910)

Verne, Jules and Michel Verne, 'Au XXIXme siècle: la journée d'un journaliste américain en 2889' (first published 1891; collected in *Hier et demain* 1910)

Verne, Jules and Michel Verne, 'L'Eternel Adam', also known as 'Edom' (collected in *Hier et demain* 1910)

Wells, Herbert George, *Complete Short Stories* (London: Ernest Benn 1927)

Wells, Herbert George, *The War of the Worlds* (1898; London: Dent 'Everyman' 1993)

Wells, Herbert George, *Anticipations of the Reaction of Mechanical and Scientific Progress upon Human Life and Thought* (London: Chapman and Hall 1902)

Wells, Herbert George, *In the Days of the Comet* (1906; London: Hogarth Press 1985)

Winandy, André, 'The Twilight Zone: Imagination and Reality in Jules Verne's Strange Journeys', transl. Rita Winandy, *Yale French Studies* 43 (1969), 97–110

8
The Early Twentieth Century: High Modernist Science Fiction

The twentieth century is the period when SF begins to approach cultural dominance, because it was in this century that the gradient of the graph marking technological and cultural change against time went nearly vertical. SF in this period becomes a – perhaps the – key way in which writers and readers tried to come to terms with what those changes meant. The next two chapters must work with a critical binary that runs the risk of crudeness and simplification. I state it here straightforwardly to be transparent: the first half of the twentieth century sees the opening of a cleavage between 'high art' and popular culture, something which, if not wholly unprecedented, had never before been as ideologically charged or divisive. On the one hand, with that literary movement today taught in academies as 'Modernism', we have a group of often brilliant writers dedicated to an aesthetic programme of 'making it new' (Ezra Pound's phrase), experimentation, focusing more deliberately on form and style than plot and character, often working dense textures of quotation and allusion into their texts. On the other hand, in the aftermath of the spread of mass literacy in the late nineteenth century, the huge new audience for popular narratives was catered for by an equally large and often talented but nowadays less well-known group of writers.

Fredric Jameson, in *The Political Unconscious*, talks of the contested state of literary culture at the end of the nineteenth century from which emerged 'not modernism alone, but rather two distinct literary cultural structures, dialectically interrelated and necessarily presupposing each other', 'High Culture' or 'elite' Modernism on the one hand, and mass culture on the other (Jameson, p. 207). Maria DiBattista, in her edited collection of essays on exactly this dichotomy, *High and Low Moderns: Literature and Culture 1889–1939*, earmarks the former group as 'self-conscious formalists wrestling with newly perceived instabilities in language and meaning ... writers whose imputed moral as well as aesthetic "difficulty" removed or elevated them from the prevailing low and middlebrow culture of their day'. The 'low' Modernists, on the other hand, are taken as 'more accessible (i.e. popular as well as easily readable) [and] morally transparent' (DiBattista, pp. 3–4). As DiBattista goes on to argue, the reality was that no such clear-cut distinction in fact existed; the two movements were dialectically interrelated. Both 'elitist' and popular-cultural artists faced a similar cultural problematic. The broad difference is that the 'elitists', or High Modernists, reacted in general with hostility to increasing technological change, whereas popular cultural artists reacted, generally speaking, with excitement and exhilaration. The following two chapters will flesh out this statement.

Such fleshing out must begin, of course, by noting the various exceptions to the rule. Not all High Modernists were hostile to the machine. Indeed, one subset, the 'Futurists', positively embraced the disarrangement of bourgeois conventions that the machines, and especially machines of *speed* such as motor cars and aeroplanes, brought with them. Italian poet Filippo Marinetti (1876–1944) launched the Futurist movement with a manifesto in February 1909:

Un'automobile da corsa ... un'automobile ruggente, che sembra correre sulla mitraglia, è più bella della Vittoria di Samotracia. Noi vogliamo inneggiare all'uomo che tiene il volante, la cui asta ideale attraversa la Terra, lanciata a corsa, essa pure, sul circuito della sua orbita

('A racing car ... a roaring car that seems to ride on bullets is more beautiful than the "Victory of Samothrace". We will write hymns for the man at the wheel, who hurls his spirit's lance from the Earth, along the curve of its orbit'; Marinetti, *Il manifesto del futurismo*, pp. 4, 5)

This adolescent excitement at the thought of rapid – and especially *world-spanning* – machines is present in the poetry, visual art and other work of the various Futurists. It provides an uncomfortable context for the pulp, populist idiom that also valorised the technological – uncomfortable because Futurism was at heart a life-denying and fascistic movement. Marinetti, in declaring 'war against Italian *pastism*', pledged Futurist artists to an aesthetic of total war: 'Dynamic and aggressive', he wrote in 1914, 'Futurism is now being fully realized in the great world war which it – alone – foresaw and glorified ... *the present war is the most beautiful Futurist poem which has so far been seen*' (quoted in Griffin, p. 26). The ideological consonance between Futurism and fascism was such that Marinetti and his followers enthusiastically endorsed Mussolini's regime, praising him as a specifically Futurist superman. By comparison the larger group of machine-hating *avant-garde* Modernists seem a much more appealing crew.

Futurists were, speaking broadly, not very characteristic of 'High' or 'Literary' Modernism. In terms of the SF dialectic I have been arguing for in this present study, the anti-machinists, and High Modernists, tend towards a mythopoeic vision that very often shades into mysticism or religion: to forms of art deeply embedded in 'tradition' (social traditions, literary antecedents, quotation, allusion and intertextuality) even as it self-consciously strives for newness. Works such as Marcel Proust's *A la Recherche du Temps Perdu* (1913–27), James Joyce's *Ulysses* (1918–22), Wyndham Lewis's *Apes of God* (1930), T. S. Eliot's *The Waste Land* (1922), Ezra Pound's *Cantos* (the first appeared in 1917, the last in 1970), Robert Musil's *Der Mann ohne Eigenschaften* ('The Man without Qualities', 1930–32) all shared experimental stylistic or formal ambition; and the SF written by 'High

Modernist' writers aimed to fracture and reassemble writerly practice in order to apprehend a similarly mythic-transcendent consciousness. Pulp SF, on the other hand, mediated its 'theological' aspects via a technological sublime that was much more obviously materialist, although a great deal of pulp works its way to the reader via concepts such as a Schopenhauerean–Nietzschean fetishisation of Will. Indeed, it is possible to reverse the binary usually implied by 'High Art' versus 'Popular Culture' in this period. On the one hand, J. R. R. Tolkien 'made it new' as radically as any other High Modernist (although his 'newness' was in genre and mode rather than prose style or form) while at the same time returning to a deliberately old-fashioned, religiously informed 'magic' Fantasy. On the other hand, the great achievement of the Pulp Era – the Pulp magazines themselves – fractured reading practice upon the brightly coloured coal-tar dyes of their illustrations and the restless forward-fidgeting ethos of their stories: narrative, excitement, ingenuity and a consistent *outwardness* that chimed with their readers' expectations but which also anticipated the godless fragmentation of world culture that was eventually nicknamed postmodernism.

Anti-machinists

Hostility towards technology took a number of forms and various degrees of intensity amongst the High Modernists: for some the danger was the increasing mechanisation of the social arena, with the belief that individual humans would come to be treated as mere cogs in a machine. For others the whole drift of contemporary society towards technology involved a deplorable loss of primitive 'naturalness' and contact with the organic, non-technological and spiritual. Valentine Cunningham identifies Aldous Huxley's *Brave New World* (1932) as 'the key dystopian fiction' of the 1930s, arguing that it crystallises a 'widespread Western disquiet over the triumph of machine-age materialism' (Cunningham, p. 399). Cunningham traces this influential (and we should add, 'mainstream' and 'elite') 'machine bashing' from D. H. Lawrence and T. S. Eliot's polemical opinions in the 1910s and 1920s, through

> the same line as Q. D. Leavis (neither Naomi Mitchison in *We Have Been Warned* nor Annabel William-Ellis in *To Tell the Truth*, rasped Mrs Leavis ... were capable of questioning 'the machine' as 'an absolute value' or entertaining 'any doubts about machine tending as the good life') ... It was natural enough for Evelyn Waugh to join in this machine-bashing, and Robert Byron ('conditioned reflexes, Ford lorries, and abortion clinics') and Osbert Sitwell ('Magnetogorsk, the Nuremberg Stadium and the Great West Road') and J. R. R. Tolkien ... George Orwell sid[ed] in *Wigan Pier* with *Brave New World* ('probably expresses what a majority of thinking people feel about machine civilisation'). (Cunningham, p. 399)

Cunningham adds Charles Williams, Wyndham Lewis, Peter Fleming and Julian Symons to these 'machine-bashers'. Linked to this hostility was a belief that the denial of the organic, natural world was part of what Stephen Spender in 1937 called a cultural 'death-wish', characterised by 'sentimental masochism'. Malcolm Muggeridge, writing in the last year of the decade, diagnosed a 'longing for death' and 'a reservoir of death-longing, ready to be tapped' (quoted in Baker, p. 58). Huxley based his character Mark Rampion, in *Point Counter Point* (1928), on his friend D. H. Lawrence; and Rampion expresses some typically Lawrentian views: 'the love of death's in the air', he complains, of his children's fascination for 'motor cars, trains, aeroplanes, radios':

> I try to persuade them to like something else. But they won't have it. Machinery's the only thing for them. They're infected with the love of death. It's as though the young were absolutely determined to bring the world to an end – mechanize it first into madness, then into sheer murder. (Huxley, *Point Counter Point*, p. 320)

Rampion's opinion of the mass of the working population is that 'they live as idiots and machines all the time, at work and in their leisure. Like idiots and machines, but imagining they're living like civilized humans' (Huxley, *Point Counter Point*, p. 305).

This view of the drift of the world is found often in the writing of this portion of the self-elected 'elite' of Modernist art. It does not overstate the case to say that for such writers 'science fiction' is an exclusively dystopian mode: the machines and technologies so often celebrated and even fetishised in Pulp SF of the period are presented as pernicious and dehumanising in works such as Karel Čapek's *R.U.R* (1921), Yevgeny Zamiatin's *We* (1920; translated into English 1924), Fritz Lang's *Metropolis* (1927), Aldous Huxley's *Brave New World* (1932), René Barjavel's *Ravage* (1943), Hermann Kasack's *Die Stadt hinter dem Strom* (1946), Robert Graves' *Seven Days in New Crete* (1949), George Orwell's *Nineteen Eighty-four* (1949) and Ernst Jünger's *Gläserne Bienen* (1957). In each of these cases an imagined future society has been ordered along technological or scientific lines to the detriment of the individual quality of life of that society's citizens.

We need to note immediately that it distorts *Brave New World* somewhat to place it in the company of unambiguously miserable or restrictive dystopias. Such categorisation tends to miss the subtleties of Huxley's approach. The mistake many critics make with this novel is to read purely for the content, and to see the book as an indictment of technological society, consumer society or mass-culture in general. John Carey thinks that the novel is 'intent upon establishing the superiority of "high" culture, and the baseness of the leisure pursuits preferred by the masses' (Carey, pp. 86–7). In fact, *Brave New World* is a commentary on the logics of utopian fiction and an engagement in the meta-textual questions of genre that include the dynamic between 'science fiction' and 'realist fiction'.

Brave New World is set in 'A.F. 637', which is to say 637 years after 'Ford' (Henry Ford, 1863–1947, the American industrialist whose motor car 'Model T' was the first to be assembled by mass-production methods). Huxley imagines an entire society based on the principles of specialist engineering, uniformity and a Taylorised communal ideology. The World State's two billion citizens are 'hatched' in commercial hatcheries rather than born (indeed, they consider biological parturition a disgusting notion); whilst still foetuses they are conditioned with a set of virtues, including passive obedience, material consumption and sexual promiscuity, and graded before birth into castes, with 'alphas' at the apex of society occupying professional jobs; betas occupying middle-ranking positions; and the inferior gammas, deltas and epsilons allotted menial work. All are given state handouts of a drug called 'soma' which is 'euphoric, narcotic, pleasantly hallucinant' (*Brave New World*, p. 48). Community solidarity is promoted through a range of ideological tactics, from hypnopedic indoctrination to Community Sings and Solidarity Services. Citizens lead long, disease-free, happy lives and are all materially productive (and more importantly for this capitalist ideal, materially consumptive). They have a variety of pastimes; sex, scent organs, tactile 'feelies' that have superseded the visual 'movies' of the twentieth century. The novel begins with a visit to a hatchery, and continues with the ordinary but obscurely dissatisfied alpha-plus male Bernard Marx (his unhappiness is explained in terms of a minor malfunction in his antenatal treatment). Marx travels to a reservation in New Mexico where small communities of 'Savages' live a more primitive existence; there he discovers a young man, the son of a Brave New World woman who had been mistakenly abandoned in the reservation decades before. Marx brings this Savage back to 'civilisation', where his naïf perspective allows Huxley to draw out the various aspects of his imaginative creation. In his *Brave New World* religion and art have been eliminated because they tend to destabilise the communal harmony. The Savage, who has schooled himself on heady draughts of Shakespeare, falls powerfully and destructively in love with the vacuous Lenina – inappropriate behaviour for a society in which 'family, monogamy and romance' have been eliminated. The intensity of John Savage's emotions leads him to despair. Increasingly alienated from the bland hedonism of the civilised world, he secludes himself in a tower; whips himself in self-disgust; and eventually hangs himself.

One climax in the book's construction is a lengthy discussion between the Savage and Mustapha Mond, the Resident Controller for Western Europe and one of the Ten World Controllers, in which the relative merits of poetically ideal misery and scientifically practical happiness are debated.

Huxley has drawn a world in which human happiness, and the stability of that happiness, is the defining quality. Utopian thinkers had hitherto always assumed one of two principles as the index of utopian success: either the increased efficiency of the (for example) militaristic-mechanistic aspects of society, or – more usually – the Utilitarian criterion of the maximisation of happiness for the greatest number of people.[1] Huxley was uninterested in the military utopia, but his innovation was wittily and profoundly to interrogate the utopian associations of happiness *per se*. This is a constant theme in his writing. Before Huxley it was generally taken for granted that increasing 'happiness' was a good thing. Huxley questioned that assumption. His thwarted poet character, Francis Chelifer, from 1925's *Those Barren Leaves* wonders

whether the ideal of happiness towards which we are striving may not turn out to totally unrealisable, or, if realizable, utterly repellent to humanity? Do people want to be happy? If there were a real prospect of achieving a permanent and unvarying happiness, wouldn't they shrink in horror from the boring consummation? (Huxley, *Those Barren Leaves*, p. 86)

This is the blueprint for *Brave New World*, a society of permanent and unvarying happiness. 'Happiness', observes Mustapha Mond, 'is a hard master – particularly other people's happiness' (Huxley, *Brave New World*, p. 207). In Huxley's hands it becomes a harder master even than Orwell's Big Brother, because it is, as ideology, so much more thoroughly internalised into the individual.

George Orwell thought *Brave New World* 'a brilliant caricature of the present', but insisted that the book 'casts no light on the future. No society of that kind would last more than a couple of generations'. The reason for this instability of dictatorship, Orwell thought, was the fact that the ruling caste in Huxley's world lacked 'a strict morality, a quasi-religious belief in itself, a mystique' (quoted in Baker, p. 13). What is especially interesting about Orwell's opinion is its fundamental misreading of Huxley's text – because the whole point of *Brave New World* is precisely its studied lack of 'mystique', of 'quasi-religious belief'. The banal hedonism of his imagined world acquires its horrible flavour from the absence, and indeed the ruthless suppression, of one particular 'numinous' quality. This is not religion as such, although conventional religion is one of its manifestations. Towards the end of the book the Savage argues with Mustapha Mond, the latter arguing that 'God is incompatible with machinery and scientific medicine and universal happiness', and explaining how the drug soma has replaced the religious impulse, it is 'Christianity without tears' (*Brave New World*, pp. 214, 217). God is an 'inconvenience' for this particular civilisation, and the Savage's insistence on desiring inconvenience for its own sake is one feature of a more thoroughgoing masochism that also sees him whipping himself and eventually hanging himself. 'I don't want comfort,' he tells Mond. 'I want God. I want poetry. I want real danger, I want freedom, I want goodness. I want sin.' Mond suggests that he is 'claiming the right to be unhappy', and when the Savage agrees, Mond points out that such a right also includes 'the right to grow old and ugly and impotent; the right to have syphilis and cancer; the right to have too little to eat; the right to be lousy; the right to live in constant apprehension of what may happen tomorrow; the right to catch typhoid; the right to be tortured by unspeakable pains of every kind' (*Brave New World*, p. 219). It is difficult to avoid conceding that he has a point.

The satire here is fully anti-Freudian; Freud's definition of mental health as the ability to work and love is caricatured in the typical citizen of Mond's world, 'a happy, hard-working, goods-consuming citizen' with unlimited access to sexual love, and as such 'perfect' (*Brave New World*, p. 215). The other obvious correlative to Huxley's satire is Bolshevik Russia; in the 1950s Huxley observed that 'the old-fashioned *1984*-style dictatorship of Stalin has begun to give way to a more up-to-date form of tyranny', and 'the Soviet system combines elements of *1984* with elements that are prophetic of what went on among the higher castes in *Brave New World*' (*Brave New World Revisited*, pp. 4–5). Again, in contrast to Orwell's more heavy-handed attack on State Communism the genius of Huxley's approach is to follow the logic of a Stalinist ideology through to its own stated conclusions. More straightforward ideological opponents would be content to stress the patently unpleasant consequences of this political practice. Huxley was certainly opposed to Bolshevism as he understood it, and his opposition reads as a gloss upon *Brave New World*:

> To the Bolshevik idealist, Utopia is indistinguishable from one of Mr Henry Ford's factories ... Into the Christian Kingdom of Heaven men may only enter if they have become like little children. The condition of entry into the Bolshevik Earthly Paradise is that they shall have become like machines. (Huxley, *Music at Night*, p. 152)

Nevertheless, Huxley considered this belief 'extravagantly romantic', and dubbed Bolshevism 'the new Romanticism'. But although he inserts some banal Middle England caveats (to the effect that 'men cannot live without a certain modicum of privacy and personal liberty') Huxley does understand the visceral appeal of 'romanticism'. In other words, the greatest achievement of *Brave New World* is not portraying dystopia; it is portraying dystopia *as utopia*. Teaching the text to undergraduates for many years, I have often been struck by how few of them would consider a long, disease-free life devoted to leisure and promiscuous sex anything other than a wished-for consummation.

It is possible to ignore, as many critics have, the sheer *reasonableness* of Mond's lengthy justification of his system. It is also possible – just – to sympathise with the Savage's adolescent, muddle-headed and masochistic Shakespearean idealism. But it is simply wrong to argue (as John Carey does) that 'Huxley is committed to an idea of the human spirit which requires the existence of pain and hardship' because 'by surmounting these, the spirit proves itself' and without them 'life becomes soft and ignoble'. Carey goes on to connect *Brave New World* with Nietzsche's *Beyond Good and Evil*, suggesting that both works express a 'notion of the human spirit as combative and aspiring ... suffering is necessary for the plant, man, to grow up vigorously' (Carey, p. 8). But Huxley has no illusions about pain; and it is not *suffering* that the hedonistic, herd-like, contented, cowardly citizens of his Brave New World lack. It is something else altogether.

The element missing from the picture is described in a 1931 essay called 'Meditation on the Moon' in which Huxley defines a 'god':

> How shall we define a god? Expressed in psychological terms (which are primary – there is no getting behind them) a god is something that gives us the peculiar kind of feeling which Professor Otto has called 'numinous' [from the Latin *numen*, a supernatural being]. Numinous feelings are the original god-stuff from which the theory-making mind extracts the individualized gods of the pantheon. (Huxley, *Music at Night*, pp. 60–1)

For Huxley this 'numinous' feeling is a core aspect of the healthy psyche. It does not relate to the actual existence or non-existence of a divine being, but rather to the psychological make-up of the human animal. The unrelenting happiness of *Brave New World* is dystopic not because it excludes suffering, but because it excludes this 'numinous' element. In the novel Huxley balances the ridiculousness of the Savage's self-flagellation with the beauty of his repeated quotations from Shakespeare – the numinous element of the poetry obscuring the fundamental implausibility of an autodidact Mexican peasant from the 27th century becoming so expertly versed in the work of an English poet and dramatist from the sixteenth. In poetry this Huxleyan 'numinous' quality is called 'beauty', but in his own life Huxley went on to argue that, practically speaking, a pharmochemical form of the numinous was required if the masses were to be catered for. His own experiences with mescaline and LSD in the 1950s persuaded him that social solutions lay in this direction. His famous accounts of his drug-taking, and the numinous sense of wonder they opened for him in quotidian experience, are *The Doors of Perception* (1954) and *Heaven and Hell* (1956). Unlike the drug 'soma' from *Brave New World*, Huxley found LSD to be fundamentally *sacramental*; he describes observing both the outer world and his own inner landscape under the influence of the drug and finding them both 'self evidently infinite and holy' (Huxley, *Doors*, p. 38). He suggests that the drug should be introduced to the general public in preference to the commonly used stupefacients like alcohol and tobacco. Utilising Wells's metaphor of 'the door in the wall' he warns:

> The problems raised by alcohol and tobacco cannot, it goes without saying, be solved by prohibition. The universal and ever-present urge to self-transcendence is not to be abolished by slamming the currently popular Doors in the Wall. The only reasonable policy is to open other, better doors ... what is needed is a new drug which will relieve and console our suffering species without doing more harm in the long run than it does in the short ... (Huxley, *Doors*, p. 53)

In all, it is this underlying philosophy that renders *Brave New World* one of the inescapably great SF novels of the twentieth century: because it takes 'SF' as a genre to one sort of logical anti-mystical conclusion, a world – happy, healthy, bland – purged of religious magic (in the strong sense of that word). In other words, *Brave New World* in fact exactly elaborates the dialectic that has been shaping SF since the 1600s.

Mystical and religious science fiction

Also expressive of this 'deep' SF dialectic are the various hybrid novels of the period that cross SF tropes with expressly religious or spiritual ambitions. In part, these sorts of books reflect the continuance of the subgenre of 'mystical science fiction' from the later nineteenth century (see above pp. 112–116). An unbroken chain of such books can be traced through the century.

The Austrian artist Alfred Kubin (1877–1959) published only one novel, which he also illustrated: *Die andere Seite* ('The Other Side', 1909), an apocalyptic extravaganza in which a journey to the fantastic city 'Perle' mirrors the protagonist's journey into his own soul. People compare the often oppressive absurdity of mood conjured by Kubin in this book with that of fellow Austrian Franz Kafka (1883–1924). Kafka is the more famous of the two although his apprehension of alienated modern life was starker and less forgiving. In *Die Verwandlung* ('The Metamorphosis', 1915) Gregor Samsa wakes to find himself inexplicably mutated into the form of a gigantic insectoid (the German *'Ungeziefer'* does not mean 'insect' only, but has a wider semantic field: vermin, bug, grub). The reactions of Samsa's family to this situation (initially disturbed, later indifferent and neglectful) constitute the meat of the tale. More obviously SF is *Der Prozeß* ('The Trial', 1925). Joseph K is arrested, and then elaborately and somewhat mysteriously persecuted by shadowy and possibly governmental agents. He is never told the crime of which he is accused. That this is a novel that was, effectively and after Kafka's death, translated from its idiom of dystopia into that of realism by a number of totalitarian regimes in the twentieth century does not make it any less SF, and if anything only enhances its potent materialist-fantastic flavour. In its dramatisation of the absurdist arbitrariness of affairs in its bleakly rendered alternative Viennese society, as much as its powerful insistence on the radical passiveness and hopelessness of individual human life, it remains perhaps the most relentlessly pessimistic vision of what the Enlightenment tradition had made of ordinary life at the start of the century.

The penultimate chapter of *Der Prozeß* takes Joseph K to the cathedral where a priest offers not the hope of salvation but an obliquely depressing fable. Faith in Kafka is something painfully noticeable by its absence. The contrast with, for instance, the much less sophisticated British writer Guy Thorne (1874–1923), whose novel *Made in his Image* (1906) imagines a futuristic dystopia only as a prelude to a fable of man redeemed by the love of Christ, tends only to show how much more penetrating Kafka's vision was.

An idiosyncratic planetary romance with strongly mystic-religious overtones is *A Voyage to Arcturus* (1920), by the eccentric Scottish writer David Lindsay (1878–1945). This novel remains relatively little known, although it has disciples who praise it extravagantly. It tells the story of Maskull, a 1920s man transported to the planet Tormance, a world in the constellation of Sirius. There he awakens with his body changed into a humanoid but alien form and embarks on a series of intensely imagined, rather disorienting and bewildering adventures. *Pilgrim's Progress*-like (although Lindsay, like Tolkien, repudiated allegory) he advances towards an understanding of the spiritual nature of the cosmos: two alien beings, Muspel and Crystalman, are revealed as Manichean spirits in contention: Muspel is the source of spiritual light which is broken into material fragments by Crystalman – Lindsay dramatises these fragments (which is to say, the material world, and more particularly *us*) as maggot-like 'green corpuscles' striving 'towards Muspel' but 'too feeble and miniature to make any headway' and 'danced about against their will' by Crystalman, in the process of which they 'were suffering excruciating shame and degradation' (Lindsay, *Voyage*, p. 296). The book ends with a transformed Maskull recognising that only through pain can salvation from Crystalman be achieved.

A Voyage to Arcturus has a rather Marmitey, love-or-loathe taste about it. Some readers find its involuted complexities frustrating and self-indulgent. For others they are insightful and eloquent revelations about the true nature of life. What makes the book characteristic of High Modernism is more than just the serious-minded elevation with which it is composed. More importantly all its baroque adventures, weird characters and bizarreness tend towards the elaboration of Lindsay's private mythology, an ethos and a *Weltanschauung* that valorises the mortification of the flesh. It is easy to say (as I am tempted to) that Lindsay's repeatedly expressed disgust at sexual desire and his fetish for 'purity' ('sparks of living, fiery spirit hopelessly imprisoned in a ghastly mush of soft pleasure ...' *Voyage*, p. 298), together with his ultimate belief in life as pain (which, he says, though intense and unremitting must be endured and even celebrated as our only validation), are straightforwardly repellent. Others may find these crotchets less blots and more profound insights into the underlying spiritual logic of the cosmos. Certainly, this is a novel unlike any other: a work of genuinely startling if unconventional mythopoeic grandeur.

Some of this grandeur is apparent in the classical music of the period. The Russian-born composer Alexander Scriabin (1872–1915) moved from inspiration by the Nietzschean doctrine of the Will-to-Power into more mystical beliefs, matching atonal and experimental musical languages to quasi-religious vision. He died before completing *Mysterium*, a piece of such ambition that he literally believed its performance (in, he hoped, the Himalayas) would herald the end of the world itself: an ambition slightly less comical if seen as part of a longer-standing chiliastic or science fictional apocalyptic tradition. A mystical and quasi-religious drift is also apparent in the astrologically inspired orchestral fantasia *The Planets* (first publicly performed in 1920) by the British composer Gustav Holst (1874–1934). This suite begins with music that characterises the planets as corre-lating to human concerns, as its essentially astrological trope might suggest: the blasting, machine-gun-stuttering rhythm of 'Mars, the Bringer of War', and the human-scale sensuality of 'Venus, the Bringer of Peace'; but as the planets repre-sented recede further from Earth a distinct other-worldly chill enters the music. 'Jupiter, the Bringer of Jollity' is jolly enough, but 'Saturn, the Bringer of Old Age' expresses not so much human dotage as the estranging age of the galaxy itself. 'Uranus, The Magician' and 'Neptune, The Mystic', with their chilly beauties, provide a distanciated strangeness rare even in good SF. That it is essentially religious in inspiration is hard to deny.

Unambiguously religious is the writing of C. S. Lewis (1898–1963). Lewis is most famous today as the author of the Christian-allegorical Fantasy series about Narnia (from *The Lion, the Witch and the Wardrobe*, 1950, to *The Last Battle*, 1956); but his SF 'cosmic trilogy' (sometimes called the 'Ransom trilogy') pre-dates his fantasy writing and is in some ways more interesting. In *Out of the Silent Planet* (1938) the protagonist Elwin Ransom is kidnapped by two evil scientists and flown to Mars (or 'Malacandra' as it is called). This journey reveals the solar system to be fully pre-Copernican, imbued with divine grace in a palpable manner – all except Earth, known to the rest of the System as 'the silent planet' because cut off by its dominating demons. Travelling through space reveals it not as 'a black, cold vacuity' but as the very idiom of God: ' "Space" seemed a blasphemous libel for this empyrean ocean of radiance through which they swam' (Lewis, *Cosmic*, p. 26). This elation lessens as they descend to the surface of Mars, but the point of the intricate alien civilisation Ransom discovers there is that it is unfallen. It has needed no Christ to redeem it (with the subsequent confusion of a multiplicity, or even a Bruno-esque infinity, of Christs). In *Perelandra* (1943) Ransom travels to Venus (the title is the aboriginal name for that world) and observes a Venusian Adam and Eve in a prelapsarian paradise being tempted by a particularly irritating and dull Satan, taking the form of Weston. He successfully defeats Weston's plans, again preserving Lewis's solar system against the conceptual unravelling of a plurality of redeemers. Finally, the trilogy circles round, finishing as it began (as the pre-Copernican logic of the whole demands it must) on Earth. *That Hideous Strength* (1945) is a less effective 'fairy tale for grown-ups' about Oxford University, a research organisation called N.I.C.E which is actually the front for a Satanic group, and the resurrection of Merlin from Arthurian myth. In fact the downsizing of the saviour figure from Christ to Arthur in this novel is an attempt to sidestep the same theological problematic. But in one sense it is wrong to read Lewis's deliberately medieval SF in a post-Copernican manner. The main thrust of *That Hideous Strength* makes plain the central project of the whole trilogy: to argue that materialism is not only incompatible with ethics, but must be eliminated root, branch and very concept (Lewis calls it 'objectivism' and presents it straightforwardly as an invention of Satan). For Lewis spiritual realities are true: the material world is a kind of aberration, and dedication to it (as, for instance, by modern scientists) is mere blasphemy: 'the physical sciences, good and innocent in themselves, had already, even in Ransom's own time, begun to be warped ... if this [development] succeeded, Hell would be at last incarnate. Bad man, whilst still in body, still crawling on this little globe, would enter that state which, heretofore, they had entered only after death' (Lewis, *Cosmic*, p. 560). Lewis's science fiction is one with his theological disquisitions: indeed, *That Hideous Strength* is in effect a fictionalisation of the moral of his *The Abolition of Man* (1943), a rather one-sided attack on philosophical relativism that ends with a fictionalised future dystopia in which 'objective value' has been discarded. If we see SF as a dialectic between materialism and spiritualism, between I–It and I–Thou understandings of the relationship of consciousness to cosmos, then Lewis appears at the extreme end of the latter approach.

Yevgeny Zamiatin

The Russian writer Yevgeny Zamiatin (1884–1937) was a Bolshevik revolutionary who later incurred the displeasure of the Stalin orthodoxies of Soviet Russia and ended his days in Paris, an exile hated by both left and right. His masterpiece dystopia *My* ('We') was circulated in manuscript in 1920 and translated into English in 1924, although a Russian-language edition was not published – and then in America – until 1952. It did not appear in Zamiatin's homeland until after the Soviet Union had fallen.

My depicts a totalitarian state premised on the belief that privacy, personality and especially free will are the causes of unhappiness. The citizens, or 'Numbers', of 'OneState' are controlled with mathematical precision. Everybody lives in a glass apartment, so that all are on display to all; everybody's day is regulated by a rigid timetable. Indeed, in some respects even Zamiatin is unwilling to think through to the logical conclusion of this schema: his Numbers are permitted two personal hours a day, and they discretely lower blinds in their glass rooms in order to have sex. A society in which privacy was so anathema that there were no private acts is perhaps too corrosive of the conventions of dramatic fiction for Zamiatin's purposes; although it would surely be the more likely logic of OneState. More familiarly Soviet is the fact that everybody lives under the rule of the Benefactor, a 'Big Brother' figure, and his regime. The narrator-protagonist, called 'D-503', describes his work as part of the team building 'The INTEGRAL', a spaceship, whose purpose is to reach and convert any extraterrestrial civilisations to the happiness of the OneState way. He becomes involved in a resistance group that seeks to do away with the Benefactor. The system responds by insisting that all Numbers have an 'Operation', something like a prefrontal lobotomy, that will render opposition impossible. The narrator eventually succumbs to the Operation, but only after struggling to find an idiom in which to understand, let alone perform, revolt. The novel ends bleakly, although with hints, via D-503's purified consciousness, that things may be changing ('in the western quarters there is still chaos ... unfortunately, quite a lot of Numbers who have betrayed reason ... I'm certain we'll win. Because reason has to win', Zamiatin, *We*, p. 225). Not only does Zamiatin cast Enlightenment reason as cruel and, in a literal sense, dictatorial; he equates that oppressive rationality explicitly with Christian religion. In a meeting with the Benefactor, D-503 is told that their state is that paradise when the burden of free will is removed from humanity by God ('the one who slowly roasts in the fires of Hell all those who rebel against him'). According to the Benefactor, the very cruelty of God is inherent in the concept of atonement:

> A true algebraic love of mankind will inevitably be inhuman, and the inevitable sign of the truth is its cruelty ... remember: in paradise they (human beings) have lost all knowledge of desires, pity, love – they are the blessed, with their imaginations surgically removed (the only reason why they are blessed) – angels, the slaves of God .... (Zamiatin, *We*, pp. 206–7)

The poetry and humour that leaven this otherwise unremitting dystopia save it from being too dismal; and despite the obstacles that were placed in the way of its publication it exercised an enormous influence on dystopian writing across the century. George Orwell, for instance, read a French translation of it in Paris, and worked from its premises to write *Nineteen Eighty-four* (see below p. 208).

Karel Čapek and Mikhail Bulgakov

This chapter must close with two 'High Modernist' writers who do not illustrate the more one-dimensional anti-machinism of the authors discussed above.

The first is the Czech writer Karel Čapek (1890–1938), whose 1920 stage play *R.U.R. (Rosumovi Umělí Roboti)* ('Rossum's Artificial Robots', usually translated into English as 'Rossum's Universal Robots' to preserve the acronym) turns many Modernist pieties upside-down. The drama is set in a factory, located on an island in the South Pacific, which is manufacturing synthetic humanoids. This is the place where the word 'robot' was coined (*robota* is Czech for 'drudgery' or 'servitude'), although Čapek's robots are not metallic, but fleshy. Similarly the company title 'Rossum' is a play on the Czech word *rozum*, which means 'reason' or 'intellect'. The set-up is only too evidently that of a hypostatised 'mind/body', or 'masters/workers' binary.

The robots have been manufactured in order to free humanity from the drudgery of labour, but have therefore become an oppressed underclass themselves. The play begins with idealistic Helena Glory pressing the factory manager Harry Domin to free the robots. He believes they have no souls; but there is never a doubt in the play of the essential humanity of the robots, despite their rather reserved manner; and their rebellion against servitude is inevitable. They storm the stage, killing all humans except Alquist, clerk of works and the only human who still works with his hands.

But without human help they cannot reproduce. The play ends on a queasily religious note, with two modified robots, one male and one female, renamed 'Adam and Eve' by Alquist, sent out into the world to breed without the stigma of original sin. As a socialist allegory the play is too obvious; but as SF it is doing interesting things with the theological anxieties at the core of the genre. Something similar is true of Čapek's other, less well-known SF works. The novel *Továrna na absolutno* ('The Factory of the Absolute', 1922) is premised on a new technology of power generation by the annihilation of matter, the Karburator. In fact, this device works by siphoning the 'absolute', or God, into the real world. The myriad releases of God trigger miracles, and a mess of competing versions of these divine revelations lead inevitably to a devastating religious war. In *Vàlka s Mloky* ('War with the Newts', 1936) mankind discovers, and quickly exploits, a race of giant intelligent newt of the genus *Andrias scheuchzeri*. They are initially used as aquatic workers, co-opted into human armies as marines, and bought and sold; but this exploitation and oppression leads to a Newt uprising. Obtaining explosives from human armies, the newts sink large stretches of land (including much of Europe and Asia) beneath shallow waters. The end of humanity appears inevitable, until, in the last chapter, the author steps into his own narrative to forestall the extinction and consider alternatives. If this summary makes the book sound like an over-obvious satire, then it does not capture the texture of Čapek's novel: in part narrative, in part a scrapbook of all manner of different accounts, texts, fonts, alphabets and illustrations, *Vàlka s Mloky* considers its satirical premise from so many different (human) angles that the Newts acquire a thoroughgoing roundness even as the conventional form of the novel is caramelised. Once again, Čapek's vision is properly science fictional in the sense that it mediates a religious premise through its technological and biological discourse. Facing the apocalyptic flood, one character notes that 'the sea covered everything at one time, and it will do again. That will be the end of the world ... I think the Newts were the cause of it that time too' (Čapek, *Newts*, p. 338). The Fish-Men of the novel become strange re-imaginings of the original fish-man-god, Christ himself.

A similarly expert skill with imaginative satire found in the work of the Russian writer Mikhail Bulgakov (1891–1940). His story Роковые яйца ('The Fatal Eggs', 1924) is a brilliant short pastiche of Wells's *Food of the Gods* (one of the characters goes so far as to reference Wells's novel in order to describe the events he experiences). Set in the near future of 1928, in a neon-lit Bolshevist Moscow, 'The Fatal Eggs' concerns the scientist Professor Persikov – a typical Bulgakov character, hilariously plagued by Soviet bureaucracy and his own ridiculousness and inability to function as a decent human. He makes the chance discovery of a 'ray of life', a light beam capable of accelerating the growth of an organism, produced when the lens of a microscope is twisted in a certain way (the science is spurious, which is to be expected in Bulgakovian satire). Almost immediately, his discovery is taken from him by government agents, who use it to mass-produce poultry in farm collectives in order to counter a deadly chicken plague that threatens to destabilise the Soviet Union. But things turn really sour when the imported eggs used for the hatching produce giant monster chickens and ostriches, and especially when one particular batch of eggs turns out to be not chicken, but anaconda eggs; the Soviet Army engages in desperate combat with these latter. The story relentlessly satirises irrational bureaucracy, short-sighted dogma, the rapid and enforced change of nature, mindless research and the deification of technology, qualities that characterised both capitalist and communist Modernist society. In the end, nature redresses the balance, as the snakes, like the Nazis, are eliminated by the Russian winter, and rioting crowds murder the scientists.

Olaf Stapledon

Olaf Stapledon (1886–1950) was, despite his Scandinavian name, a British writer, who took a PhD in Philosophy from the University of Liverpool in 1925. Inspired by the great nineteenth-century philosophers of the Will, particularly Schopenhauer, he wrote a number of imaginative fictions that dramatised the longest time-scales. The first of these, *Last and First Men* (1930), is couched as a future-history, starting in the manner of Wells (a writer with whom Stapledon corresponded, and to whom he acknowledged a profound debt). But very soon the logarithmically structured storyline sweeps us into a far future; Homo sapiens has evolved into a new species, whose manner of living on the planet is described. This happens repeatedly in the book; evolution (conceived of as a repeatedly punctuated equilibrium, the punctus usually being the near-destruction of all life) produces eighteen distinct races of men, the last a solar system spanning a set of telepaths who are none the less ultimately to be wiped out by a cosmic collision. A sequel, *Last Men in London* (1932), explains how the original novel came to be written, with one of the last men experiencing the (to him) strange life of a twentieth-century Londoner via temporal telepathy. *Odd John: a Story between Jest and Earnest* (1935) takes seriously the possibility of 'supermen', rendering their superiority in spiritual and intellectual terms. John Wainwright and other members of Homo superior found a utopian society on a South Sea island; but they kill themselves rather than run the risk of destroying Homo sapiens. A pendant to this serious and considered work is *Sirius: a Fantasy of Love and Discord* (1944) written from the point of view of a super-evolved dog.

But Stapledon's masterpiece is *Star Maker* (1937), a novel for which even the most extravagant superlatives are insufficient. The events of *Last and First Men* covered several billion years; a tiny micro-fraction of the enormous time-scale of the later novel. A narrator standing on an English hill projects his consciousness through the cosmos, seeking out alien life – at first humanoid in form, later more radically different – with which he psychically combines. There are no spacecraft involved; travel is by the pure will of the narrator, but nevertheless the reader never doubts the premise (unlike the more usually religious 'spiritual voyages' of which we have seen many examples) because it relates at all points minutely and thoroughly to the book's overarching thesis: that behind the veil of reality is nothing but will itself.

One joy of the book is the controlled fountaining of Stapledon's invention, myriad alien forms of life and society described in compelling detail. Moreover, despite lacking a conventional narrative the book acquires a hefty momentum as it moves closer and closer to the ultimate revelation of the nature of the Star Maker at the end. Billennia are traversed; the very stars are revealed as partaking in consciousness; the whole of time and space is encompassed, the whole universe – and then revealed as merely an infinitesimal fraction of the complete sheaf of actual universes.

Viewed from this vantage point, not quite *sub specie aeternitatis* but as close as makes no odds, the Modernist animadversion against 'machines' seems rather pettifogging and irrelevant. Except, of course, anti-machinist Modernists were not opposed to machines as such (machines like beds, pens, pots and pans), but against *technê*, against sophistry and artificiality and in favour of a notional 'authenticity'. In this regard Stapledon is a profoundly and imaginatively episte-mological author. The purpose of his books is not to show off inventiveness for its own sake (although he was a marvellously, extraordinarily inventive writer) but to dramatise the ongoing accumulation of knowledge. And the novel builds towards a tremendous, terrifying climax when knowledge becomes almost overwhelming. The Star Maker has been, we discover, creating an endless series of cosmoi of which ours is only one, and not a very successful one. Most of these He (She? It?) has discarded, as ours is about to be discarded in favour of another one. With 'anguish and horror, and yet with acquiescence, even with praise' the narrator comprehends the inhuman chill of the Star Maker's consciousness:

> Here was no pity, no proffer of salvation, no kindly aid. Or here was all pity and all love, but mastered by a frosty ecstasy. Our broken lives, our loves, our follies, our betrayals, our forlorn and gallant defences, were one and all calmly anato-mized and assessed, and placed. (Stapledon, *Star Maker*, p. 248)

The Star Maker is not wholly bereft of sympathy, or even love; but neither quality is absolute; 'contemplation was ... the cold, clear, crystal ecstasy of contemplation'.

What this ecstasy actually is (the word, from *ec-stasis*, means a standing outside of oneself) is not clear, unless it is the multiple universes of the Star Maker's creation itself – our reality and all other realities. Why has the Star Maker created these things? We do not know, except that (we are told) 'a creative urge possessed him' (Stapledon, *Star Maker*, p. 242). In its radical uncertainty, as much as in its bold attempt to reconfigure the metaphysics of creation, ethics and eschatology as a properly cosmological business, *Star Maker* is an unprecedented and unsurpassed masterpiece.

Fascism

There is a question as to the extent to which these various High Modernist insistences on the alienation of technological society, the validity of a nostalgic and 'authentic' existence, and the mystic and mythic 'communion' of peoples fed into the malign political developments of the 1930s gathered together under the rubric 'fascism'. For some critics (I am one) the consonances between the two render each complicit with the other. Other critics point out that although some leading Modernists had fascist sympathies (Pound and Wyndham Lewis most notably), many repudiated the movement and indeed fought against it in the world war.

It would be nice to argue that the rise of fascism in Europe in the 1930s, and the Second World War to which it led, forced artists to see through the glamour of Futurist brutalism or, indeed, conservatism. Katherine Burdekin's (1896–1963) *Swastika Night* (1937) looked forward into a much darker future. The cover of the first edition laid out its premise for an audience not familiar with such speculative future histories: '**It is the seventh century of the Hitlerian era**. The Nazi empire extends over the whole of Europe and of Africa ... And for centuries civilisation has been dying.' Burdekin's pre-war story reads as horribly prescient, and its feminist emphasis (women are reduced to the status of breeding stock, and Hitler is worshipped as 'Jupiter the Thunderer') provides a very valid critique of fascism. It is also aware of the dangers of using SF speculation merely as wish-fulfilment. Although not strictly an alternative history (in 1937 Burdekin's imaginary future was all too likely actually to come true) *Swastika Night* nevertheless stands at head of the prolific sub-genre of counterfactuals known as 'Hitler Wins'. But one reason why Hitler has drawn so many SF writers into the malign gravity of his posterity is that there *is* something Hitlerian in the manifold power-fantasies of so much twentieth-century SF. The American writer Norman Spinrad (b. 1940) understands this. His alternative history *The Iron Dream* (1972) imagines Hitler becoming an SF illustrator and novelist rather than a dictator, pouring his fantasies into the sorts of novels (the bulk of the book is given over to the text of a novel called *Lord of the Swastika*) with which readers of Pulp SF are only too familiar. Spinrad offended many in the SF community with the book's implicit accusation that SF is in some sense complicit in fascism; but his satire is spot-on nevertheless. Just as High Modernism was often beguiled by the elation of a Nietzschean Will, so Pulp fiction sometimes articulated a similar, if simplified, intoxication:

> Let Adolf Hitler transport you to a far-future Earth, where only FERIC JAGGAR and his mighty weapon, the Steel Commander, stand between the remnants of true humanity and annihilation at the hands of the totally evil Dominators and the mindless mutant hordes they completely control. *Lord of the Swastika* is recognised as the most vivid and popular of Hitler's science fiction novels by fans the world over, who honoured it with a Hugo as Best Science-Fiction Novel of 1954. (Spinrad, *Iron*, p. 7)

Spinrad's *fortissimo* pastiche has less to do with the textual strategies of the *avant-garde* than the other texts mentioned in this chapter. It depends on the mixture of affection and disdain contemporary readers feel for the world of Pulp SF: the subject of the chapter that follows.

References

Baker, Robert S., *Brave New World: History, Science, and Dystopia* (Boston, MA: Twayne 1990)

Čapek, Karel, *The War with the Newts* (1936; transl. M. and R. Weatherall; introd. Ivan Klíma; Evanston, IL: Northwestern University Press 1996)

Carey, John, *The Intellectuals and the Masses: Pride and Prejudice among the Literary Intelligentsia, 1880–1939* (London: Faber 1992)

Cunningham, Valentine, *British Writers of the Thirties* (Oxford: Oxford University Press 1988)

DiBattista, Maria, *High and Low Moderns: Literature and Culture 1889–1939* (Oxford: Oxford University Press 1996)

Foucault, Michel, *Discipline and Punish,* transl. Alan Sheridan (New York: Vintage 1979)

Griffin, Roger (ed.), *Fascism* (Oxford: Oxford University Press 1995)

Huxley, Aldous, *Brave New World* (1932; with an introduction by David Bradshaw, London: Flamingo/HarperCollins 1994)

Huxley, Aldous, *Brave New World Revisited* (1958; London: Flamingo/HarperCollins 1994)

Huxley, Aldous, *The Doors of Perception, and Heaven and Hell* (1954, 1956; Harmondsworth: Penguin 1963)

Huxley, Aldous, *Music at Night and Other Essays* (1931; reprinted to include 'Vulgarity in Literature', London: Grafton 1986)

Huxley, Aldous, *Point Counter Point* (1928; Harmondsworth: Penguin 1972)

Huxley, Aldous, *Those Barren Leaves* (1925; Harmondsworth: Penguin 1967)

Jameson, Fredric, *The Political Unconscious: Narrative as a Socially Symbolic Act* (London: Routledge 1981)

Lewis, C. S., *The Cosmic Trilogy* (*Out of the Silent Planet*, 1938; *Perelandra*, 1943; *That Hideous Strength*, 1945) (London: Pan Books 1989)

Lindsay, David, *A Voyage to Arcturus* (1920; Edinburgh: Canongate 'Canongate Classics 47', 1992)

Spinrad, Norman, *The Iron Dream* (1972; Edgbaston: Toxic Books 1999)

Stapledon, Olaf, *Star Maker* (1937; London: Gollancz 'SF Masterworks 21' 1999)

Zamiatin, Yevgeny, *We* (written 1920–1; transl. Clarence Brown; Harmondsworth: Penguin 1993)

9
Early Twentieth-Century Science Fiction: The Pulps

The previous chapter argued that early twentieth-century Literary Modernism was fascinated by the cultural discourse of science and therefore produced a large amount of SF, but tended to dramatise scientific and technical developments as a negative *technê* rather than an open-ended *epistēmē*. But at the same time as self-consciously elitist Modernists were yearning for a mystical shibboleth they believed lost, a wholly new mode of literature was flourishing, one that found in technologies a liberating epistemological Will-to-Power: mass culture.

The creation of a mass literary culture was tied to a number of social and cultural changes we can date from the end of the nineteenth century. In Britain, for instance, the Education Act of 1870 enormously increased levels of literacy and created a market for a rapidly expanding number of popular newspaper, magazine and book titles. Book publishing in particular saw profound changes. Earlier in the century the dominant form of book publication had been expensive, three-volume hardback editions, known as 'triple-deckers'. Few individuals could afford to buy these books, and most borrowed them from lending libraries. Some 'cheap' editions of more popular writers such as Walter Scott or Charles Dickens were issued, but in general books were very expensive, and the number of new titles appearing in any one year could be counted in the hundreds. As Peter Keating's classic account *The Haunted Study: A Social History of the English Novel 1875–1914* (1991) shows, this world changed very rapidly under the pressures of increased demand from less well-off readers. Triple-deckers went out of fashion and in their place books were published either as one-volume hardbacks or else, increasingly throughout the twentieth century, as cheaper paperback titles.

These latter, as they appeared in the nineteenth century, are often called 'Dime Novels' by historians of publishing, and they were originally more akin to magazines than to mainstream hardback novels. They are, with the benefits of hindsight, appropriated into the broader history of what specialists call the 'Pulps', cheap magazine formats for SF, western, crime or romantic adventures. Edward S. Ellis (1840–1916) wrote several hundred Dime Novels, mostly in the idiom of western adventures, and his *The Steam Man of the Prairies* (1868) took this genre in an odd new direction with the introduction of the title device, a steam-engine in the shape of a giant man which jogged about the Midwest (but had no reverse gear) carrying its passengers into various adventures. This idea was plagiarised by more than one author.

Harold Cohen (1854–1927), a prolific hack-writer, appropriated the device for *Frank Reade and his Steam Man of the Plains* (1876), following the success of this with a number of other adventures centred on the fictional 'Frank Reade', many involving steam-driven technology. In 1882 the original hero was replaced by his son, 'Frank Reade Jr', and in 1892 a Pulp magazine began publication: 'The Frank Reade Library'. This ran to nearly 200 weekly (later biweekly) issues, at first reprinting the various Frank Reade stories published in the 1880s, and later commissioning new work. Many critics see 'The Frank Reade Library' as the archetypal Pulp (Everett F. Bleiler and John Clute describe it as 'the earliest serial publication devoted solely to SF, with more issues than all of Hugo Gernsback's magazines put together', Clute and Nicholls, p. 450). The mix of adventure, energy and can-do heroics has an undeniable kinetic fizz about it, although the racism, sadism, casual slaughter and lazy stereotyping of world and premise make many of the stories unpalatable reading today. But the crucial thing about Frank Reade, in one sense, is the way he mediates wish-fulfilment. His various devices express not a disinterested appreciation of the possibilities of technology, but his own Will-to-Power.

Mike Ashley dates the 'decline' of the Dime Novel to the turn of the century, noting that it would 'survive until the First World War', at which point 'it gave way to the immense popularity of the pulp magazine' (Ashley, p. 21).

Pulps

'Pulp' is a phrase used to denote a particular type of story printed in a series of niche-marketed magazines. The stories (which were not expensive for editors to buy) were written by prolific hack-writers and printed on cheap paper manufactured from treated wood pulp (whence the name) rather than more expensive traditional papers. The point was to keep costs low, sell cheaply and widely, and thereby make money. By the 1920s there were Pulp magazines catering to a variety of genre tastes, including general Adventure Pulps (often aimed at a more juvenile market), Crime Pulps, Western Pulps, Pulps specialising in romantic love stories and non-fiction Pulps.[1] The essence of Pulp SF is not its magazine format, but its cheapness. That said, for much of the early part of the twentieth century it was the periodical form that was most affordable. The Pulp idiom, and its huge popularity, remade SF. Within this widely distributed form, stories were published that appealed to an increasingly socially diverse readership: the emphasis was on eventful narrative, strong characters, a binary ethical code of good and evil, and (especially in SF Pulps) exotic and wonderful locales.

The first Pulp is often identified as *The Argosy*, an American magazine published from 1886 that carried a variety of fiction, including SF. (The title was taken from a rather different British magazine which had been published since 1865.) British magazines such as *The Strand Magazine* (monthly, 1891–1950) and *Pearson's Magazine* (monthly, 1895–1939) carried a great deal of SF between the 1890s and the first decades of the new century. *Strand* serialised Wells's *First Men in the Moon*

(1900–1) and Conan Doyle's *The Lost World* (1912); *Pearson's* ran Wells's *War of the Worlds* (1897) and much else. As important as the stories was the fact that these magazines carried many illustrations, black-and-white engravings to illustrate the story and later colour pictures on the cover.

How can we explain the strange appeal that Pulp SF continues to hold over the genre? Its limitations are too obvious and too undeniable to need much elucidation: it was, by and large, a puerile and aesthetically limited literature, aimed at the lowest denominator, often ideologically reactionary, rarely more than a literature to pass the time, a literature of distraction. Yet there is something more. Flaubert once said that he liked tinsel better than silver because it possessed all the qualities of silver plus one more – pathos. As the number of twentieth-century writers who wrote self-consciously and sometimes self-congratulatory 'highbrow' literature increased, so did the breadth and vigour of those literary traditions whose rationale was purely popular.

Of all Pulps, SF Pulp is the most tinselly: partly in the sense that its content was more dazzling, starry, most likely to lift its readers' eyes, metaphorically, to the brilliances above us; but partly also in the sense that it was aware and even revelled in its own cheapness. Crime Pulps, liquorice-literature, established a narrowly conceived but pungent mood; Western Pulps were a kind of candy cigarette literature, inviting readers into a sweetened miniature version of Malboro Country; Romance or Love-Story Pulps were more obviously fixated on wish-fulfilment fantasy and had the highest sugar content; but SF Pulps, tinsel-literature, had the greatest impact on a major literary tradition (SF proper); its very crudeness correlates often to an aesthetically significant vigour, the luminous fire as well as the noxious exhaust fumes of a genre powering up to escape velocity.

There is another significance of the Pulp era, something linked to its leading editorial light, Hugo Gernsback, after which it is sometimes called the Gernsback Era. Gernsback hoped to found a new literature on strictly scientific-didactic terms; to reshape, in other words, SF by purging all mystical or magical elements from the science-mysticism dialectic that initially formed it. But, despite his influence through the 1920s and 1930s, he failed. By 1930, as Mike Ashley notes, Gernsback had been forced to change the terms of his project, changing the name of the magazine *Science Wonder Stories* to *Wonder Stories* on the grounds that 'the word "Science" has tended to retard the progress of the magazine', a position 'that would have been anathema to Gernsback five years earlier' (Ashley, p. 71). His audiences wanted more than just science. It is probably true, as Gary Westfahl argues, that 'Gernsback made it possible to believe in science fiction; and that belief, more than the literary quality of his initial offerings, enabled his idea of a genre to grow' (Westfahl, p. 27): but the genre that he touched developed rather differently from the one he anticipated in his 1926 manifesto for the genre.

Hugo Gernsback

For many critics and fans Hugo Gernsback (1884–1967) remains what the American critic Sam Moskowitz (b. 1920) called him: 'the Father of Science Fiction'. Born in

Luxembourg, he came to the US in 1903. His interest in electricity, and the invention of electrical devices, led him into a number of business ventures, one of which, a magazine called *Modern Electrics* (the name was changed to *Electrical Experimenter*, and then to *Science and Invention*), was founded in 1908. *Modern Electrics* carried, among other things, SF stories, not least among these Gernsback's own novel *Ralph 124C 41+: A Romance of the Year 2660* (serialised 1911–12). This digressive yarn details the wonderful machines of the twenty-seventh century, most of which, as we might expect from the founder of *Modern Electrics*, are electrical in one way or another. It is a deeply clumsy novel, poorly structured with a limply unengaging narrative clogged with examples of what would later be called 'infodumps' (expository technical and scientific passages inserted into the text regardless of fit). It has its defenders, however, though less on aesthetic grounds and more because it establishes many of the principles that Gernsback would later codify as definitive of science fiction as a genre.[2]

But it was as an editor, not a writer, that Gernsback made his major contribution to the genre. His magazines often contained SF stories; and the August 1923 edition of *Science and Invention* was wholly given over to them. Convinced of the demand for such outlets, Gernsback announced in 1924 that he planned to found a magazine dedicated to SF stories, to be called 'Scientifiction' (Gernsback's ungainly neologism, with its choppy rocking-horse rhythm, thankfully did not catch on). In the event, the first dedicated SF magazine did not appear until April 1926, and then went on sale under the title *Amazing Stories: the Magazine of Scientifiction*. The suggestion that the primary function of the stories in this magazine would be to 'amaze' their readers seems interestingly at odds with Gernsback's often expressed view that SF was as much a didactic as an entertaining mode. Although he always stressed the importance of the latter quality, it is the prescriptive presence of the former that most strikes readers. In his editorial 'A New Sort of Magazine' from the first issue of *Amazing Stories* he insisted that:

> Not only do these amazing tales make tremendously interesting reading – they are also always instructive. They supply knowledge that we might not otherwise obtain – and they supply it in a very palatable form. For the best of these modern writers of scientifiction have the knack of imparting knowledge and even inspiration without once making us aware that we are being taught. (quoted in Ashley, p. 50)

Brian Aldiss thinks this 'instructive' imperative acted as a straitjacket on the SF imagination, 'introducing a deadening literalism into the fiction' (Aldiss, p. 204). It is certainly hard to deny that the fiction published in *Amazing*, and in Gernsback's subsequent SF imprints, tended towards the non-Feyerabendian side of the 'science fiction' dialectic. It was an early form of 'Hard SF', and influenced one major manifestation of the genre.

Gary Westfahl sees Gernsback not only as the inventor but as the first theorist and historian of SF, a man whose work 'launched, anticipated and encapsulated the entire genre' (Westfahl, p. 35). John Clute, on the other hand, has a low opinion of 'the dire weird farcical philistine Hugo Gernsback'. He did 'one good thing' (founding *Amazing*), but 'over and above that ... Gernsback's influence was disastrous':

> The personality which bestrode the tiny not-yet field for almost a decade was humourless, didactic, pedestrian, leaden, fatally prone to the advocacy of rotten science ... SF between 1926 and 1936 was, as a consequence, humourless, didactic, pedestrian, leaden and fatally prone to the advocacy of rotten science. (Clute, *Scores*, pp. 221–2)

Clute is, of course, deliberately overstating matters, but he is by no means alone in his animadversion ('Gernsback was arguably one of the worst disasters ever to hit the SF field', Aldiss, p. 82). There is a sense in which these and likeminded rabbit-punches aimed metaphorically at Gernsback's chin are motivated by the conviction that he was responsible for the deplorable juvenilisation of a genre that ought to have grown into a profound, philosophical and above all adult mode of art (similarly heated denunciations have been occasioned by the 1977 film *Star Wars*). But this is to misunderstand both the genre itself and the force of its development into a mass cultural phenomenon in the twentieth century.

In 1929 Gernsback became bankrupt, very possibly as a result of the machinations of the American publisher Bernard MacFadden (1868–1955), who bought up *Amazing*. Gernsback's response was to found not one but four new SF magazines: *Air Wonder Stories* and *Science Wonder Stories* in 1929, which merged a year later into *Wonder Stories* (issued monthly 1930–36, when it was sold, and the title changed to *Thrilling Wonder Stories*); and *Science Wonder Quarterly* and *Scientific Detective Monthly*, which did not last more than a year. What he discovered was that the public, though interested and willing to be dazzled by 'Super-Science', also wanted a leavening of a less hard-line (we might call it a less materialist) approach. Gernsback's success encouraged the creation of a large number of other magazines, many of which soon folded. Most notable among these was *Astounding Stories of Super-Science* (founded in 1930; from 1933 onwards the title was simply *Astounding Stories*). *Astounding* was to become one of the most important Golden Age SF publications, but began its life by downplaying Gernsback's insistence on didactic science and instead featuring stories in which the emphasis on high adventure, excitement and exoticism was given priority over science.

The 'Magazine Era'

So influential was the magazine outlet as a shaping force in the production of SF during the middle of the century that Brian Attebery goes so far as to label the whole 'period of SF history from 1926 to 1960' the 'Magazine Era' (Attebery, p. 32). A list of magazine titles gives some sense of the fecundity and correlative popularity of this mode of prose SF, as well as giving a flavour of the huckster-chic that informed the choice of magazine names – the more attention-grabbing and exclamatory the better.[3] The later development of SF magazine publishing has mostly seen the abandonment of this Astonishing! – Astounding! – Fantastic! idiom in favour of a modish obliqueness (with titles such as *Analog, Omni, Interzone* and so on); but there is a charming appositeness in the original titular gushiness.

Through the 1920s and 1930s and even, to an extent, into the 1940s and 1950s scores of magazines provided individual stories or serials of high adventure in space to an eager public. The two most significant authors to come to prominence through magazines in this period were Edgar Rice Burroughs (1875–1950) and E. E. 'Doc' Smith (1890–1965), who are treated in a little more detail below. But from among the hundreds of writers and many thousands of novels and stories a number are especially noteworthy. Jack Williamson (b. 1908) published prolifically in various Pulps. *The Legion of Space* (serialised in *Astounding* 1934 and followed by several sequels) is an enjoyably full-throated space-opera in which the hero John Star and beautiful Aladorea Anthor battle the monstrous alien Medusae. In the unconnected *The Legion of Time* (serialised in *Astounding* 1938) Williamson achieved something lasting: an ingenious time-travel concept in which alternative futures fight with one another down the time-lines to try to ensure their own survival. Of the many hundreds of novels Williamson has written (I know of no complete bibliography of his work) most adapt themselves to the demands of his audiences for narrative and spectacle. For this he is sometimes deprecated by critics. Gary Westfahl praises Williamson's story 'The Prince of Space' (*Amazing*, 1931) because it 'astonishingly depicts the first true space habitat in science fiction – an immense cylinder with parks and houses on its inner surface' only to deprecate the story as a whole for using this piece of hardware merely as the background to a story about a space pirate who fights 'a Martian race of vampire plants' (Westfahl, p. 153). The urge to pick selectively among the vast storehouse of Pulp SF for nuggets that can be judged by later criteria of interest or excellence is indeed hard to resist; but the point of Pulps is precisely the raw excitement of the vampire plants rather than the occasional anticipation of a more sober style of 'engineering feat' SF.

Stanley Weinbaum (1902–1935) published relatively little, although his portrait of human spacemen encountering well-realised anarchist Martians ('A Martian Odyssey', *Wonder Stories*, 1934) is regarded as a classic of the genre. Catherine Moore (1911–1987; her writing name 'C. L. Moore' may have been designed to veil her gender in the predominantly male world of early century SF; she also used the mannish pseudonym Lawrence O'Donnell) published her early work in 1930s Pulps, but brought a distinctive fluency with the emotional and sensual aspect of her narratives.

Edgar Rice Burroughs

Will is central to the Pulp vision of Edgar Rice Burroughs (1875–1950), an American ex-Cavalry officer who came late to writing. His first publication, *Under the Moons of Mars*, appeared in the little-known *All-Story Magazine* in 1912: it later appeared in book form as *A Princess of Mars* (1917), the title by which it is now known. This adventure sees the heroic man-of-action (and ex-Cavalry officer) John Carter travelling to Mars by Will alone:

> I closed my eyes, stretched out arms towards the god of my vocation and felt myself drawn with the suddenness of thought through the trackless immensity of space. There was an instant of extreme cold and utter darkness ... I opened my eyes upon a strange and weird landscape. I knew I was on Mars. (Burroughs, *Princess*, p. 20)

Carter, strong and super-competent, fights his way past various alien foes: blue-skinned monsters and green-skinned warriors (the latter immensely tall and with six limbs), as well as red-skinned humanoids (Carter marries a princess of the latter species and fathers children, despite the fact that these Martians are egg-laying). The tenor of the book is fast-paced adventure with the emphasis on stylised but violent combat. Indeed, the title of the book's 26th chapter epitomises the trajectory of the whole: 'Through Carnage to Joy'. It is possible to say that Burroughs simply transferred 'Frank Reade'-style adventures from the Wild West to a more exotically rendered Martian prairie; but there is something more to his books – still in print today, still (despite an inevitably dated savour) enthralling readers. Partly it is the quality of distilled exoticism, the utter genuineness of Burroughs' portrait of 'Barsoom' (as his Mars is known to its natives); something difficult to fake, but easy to fall for. Partly, on the other hand, it is the single-minded vision of John Carter as a cartoon-like but charismatic *Übermensch*: the reinvention of the Will-to-Power as action hero.

A great many sequels followed, tracing the further adventures of Carter and his family, three serialised in *All-Story* (*The Gods of Mars*, 1913; *The Warlord of Mars*, 1913–14; *Thuvia, Maid of Mars*, 1916) and five published in other journals or as stand-alone volumes (*The Chessmen of Mars*, 1922; *The Master Mind of Mars*, 1928; *A Fighting Man of Mars*, 1931; *Swords of Mars*, 1936; *Synthetic Men of Mars*, 1940). Two further, much weaker, tales were serialised in *Amazing*: *Llana of Gathol* (1941) and *John Carter of Mars* (1941–43). The quality of individual titles in this series does vary, although some are very good, and one or two manage not only entertainment and exoticism but something more. In particular, I've always felt that *The Chessmen of Mars* manages, perhaps despite itself, to say some genuinely interesting things about the relations between master–slave and questions of control.

Shortly after the first of the 'Barsoom' stories, Burroughs published the first in what was also to become a long-lived series of adventure tales, one that was to demonstrate far greater cultural penetration: *Tarzan of the Apes* (serialised in *All-Story*, 1912) postulates an English baby raised, Romulus and Remus-like, by African apes, and growing into an idealised noble savage: as strong and ready-to-rumble as John Carter, if limited to the Earth. Burroughs went on to publish 26 novel-length sequels to *Tarzan*, and the series was even more popular than John Carter. Two further multi-book series have remained popular: *At The Earth's Core* (1914) was the first of six 'Pellucidar' books, whose titular location is within the Hollow Earth familiar from the theories of John Symmes (1780–1829); and the 'Venus' books, charting the adventures of Earth-spaceman Carson Napier, beginning with *Pirates of Venus* (1932).

Burroughs' huge fame encouraged imitations of his action-based heroic adventures: as Mike Ashley observes, 'Burroughsian fiction would dominate pulp science fiction for the next 40 years', and, indeed, a great many later writers were directly influenced by him, especially following extensive reprinting of his titles in cheap paperback formats in the 1960s and 1970s (Ashley thinks him 'the most influential writer in the field outside of Verne and Wells' (Ashley, p. 36)). Burroughs' attitude to the rubrics of science is (appropriately for a horseman) cavalier; his emphasis was on a particular sort of narrative in which a male figure proves his courage, strength and what we might as well call his Will-to-Power through hardship, endurance and most especially through combat. It is as a kaleidoscopic representation of the fetishisation of Will (masculinised and configured in warrior terms) that Burroughs' fiction is most effective.

E. E. 'Doc' Smith

Will is also central to the work of another American Pulp writer of note, Edward Elmer Smith (1890–1965). Smith was quite correctly known as 'Doc' (he possessed a PhD in food science), and became one of the most important Pulp writers of the first half of the century. His thundering, unsubtle space-operas garnered a wide audience, some of whom later grew ashamed of their adolescent taste in these primitively written and characterised books.

Smith's first novel, *The Skylark of Space*, was written in the later 1910s, although not published until its serialisation in *Amazing Stories* (August–October 1928). This first book in the 'Skylark' series (the sequels are: *Skylark Three*, serialised in *Amazing*, 1930; *Skylark of Valeron*, serialised in *Astounding*, 1934–35 and *Skylark DuQuesne* serialised in the post-war magazine *Worlds of If*, 1965) initiates the adventures of a clean-cut hero and a black-bearded villain, and is named for the hero's spaceship, a 40-foot sphere (later, in keeping with Smith's increasingly inflating aesthetics of scale, the *Skylark* is rebuilt as a sphere with a diameter of 1,000 miles). Flying through interstellar space Smith's plots orchestrate a variety of alien species, some good and some bad: a high-adventure cosmos of space-pirates, kidnappers, law-breakers and upholders.

More famous is Smith's second sprawling series, the Lensman novels (dates are of first magazine serialisation, all of which was in *Astounding* except for the first title, which was serialised in *Amazing*): *Triplanetary* (1934); *Galactic Patrol* (1937–38); *Gray Lensman* (1939–40); *Second-Stage Lensman* (1941–42); *Children of the Lens* (1947–48). A later novel, *First Lensman* (1950), fits, chronologically, between the events of the first and second book in the series. The premise for the books is a caste of upstanding humans whose telepathic abilities are focused through certain mysterious 'lenses' worn on the wrist, who do battle with various baddies and forces of evil. With each new chapter in the adventure, Smith reveals that the various species and organisations of the previous novels are in fact fronts for larger, hidden forces. The Black Fleet, against which the lensmen battle, turns out to be a facet of 'Boskone', a galaxy-wide conspiracy of evil. Behind Boskone, it is later revealed, is the wicked 'Thrale-Onlonian Empire', although the ultimate source of evil in the Galaxy, the heart of the conspiracy, is finally revealed to be a hidden race of aliens from the planet Eddore. Against the wicked Eddorians, we eventually learn, is balanced the force of the equally powerful and equally mysterious but virtuous 'Arisians'. Everything in cosmic history is revealed to have been manifestations of the primal war between these two godlike alien races: the lenses, for instance, are quasi-magical Arisian technology. (The whole series was later published in book form between 1948 and 1954, at which time Smith revised his earlier titles to make this overarching cosmic duality more apparent.) The effectiveness of this series of staged revelations, each one apparently bringing the reader closer to the secret at the core of the cosmos, is hard to deny; although the books are crude, rather manic and irreducibly adolescent. Indeed, it is now routine for critics of the genre, even (or perhaps especially) male critics who read the books with enthusiasm in their youths, to rubbish Smith's writing. Russell Letson puts it well when he describes the *Skylark* and *Lensman* series as 'the literary acne of SF&F, a painful and embarrassing adolescent memory' (Letson, p. 1).

There have been attempts by some critics to defend Smith's excesses; Joseph Sanders, for instance, suggests that the stepped revelations and continuing heroics and battles of the series dramatise a 'moral struggle for contact/growth' (Sanders, p. 60). But to posit a moral seriousness in the one-dimensional characters and relentless, if inventive, heroics of *Lensman* is surely to miss the point. The extravagance and excess is, in an important sense, the *whole* of Smith's space adventure tales; these are books that embody as well as articulate excess. Smith's writing strives always to be emphatic and is accordingly almost always overwritten, fatally drawn to superlatives (particularly superlatives of size: 'titanic', 'colossal', 'enormous', 'gigantic'), melodramatic and gushing. 'Yet behind the awkward prose and embarrassing dialogue', suggests Edward James, 'lurks an ability to inspire awe and wonder. Smith knew that tremendous size and power were the key to awe' (James, p. 47). Certainly, the first thing that strikes a reader about Smith is his love of the large: billions of years, and of light years, are covered; moon-sized spaceships or planet-sized 'negaspheres' (spheres of 'negative reality' with enormous destructive capability) become involved in huge battles, empires clash, myriad alien worlds and species are encountered – what Brian Aldiss nicely terms 'the glamorous disease of giganticism'. But as Aldiss adds, there is a pasteboard quality to Smith's gigantic stage-sets: his 'banal hearty style ... conveys no visual experience and does not make his immense distances real' (Aldiss and Wingrove, pp. 209–11).

In fact, the persistent appeal of Smith's SF has less to do with this sublime of size, and more with his wholehearted entry into discourses of Will. This, as I have been arguing more generally, is one of the shaping discourses of nineteenth- and twentieth-century SF, ideas that have percolated through from Schopenhauer and Nietzsche. Providing compensatory fantasies of telepathic empowerment for thwarted or low-esteem adolescents is, in other words, only part of what Smith is doing. However banal, his access to the 'sense of wonder' depends on a valorisation of Will over technology. This is particularly evident in the *Skylark* series.

Telekinesis is there called 'the Talent', and, latent in many, can be developed extensively. As a correlative to those things that tend most to preoccupy male adolescents, this 'Talent' mediates sex ('it works best with complementing male and female poles of power', Ellik and Evans, p. 246) and provides otherwise powerless individuals with a world-shattering potency. At the climax of the series the clean-cut hero Richard Seaton teams up with his human arch-enemy Marc C. DuQuesne to repel an invasion by the malevolent alien Chlorans. This resolves into a battle of psionics, a team of humans against 'rabid Chloran attackers ... minds that thundered destruction at them'. The result is a genuinely startling holocaust of alien life, reported in an even more startlingly offhand manner: Seaton and DuQuesne move whole stars (50,000 million of them) across millions of light years, colliding them together to turn them into weapons: the Chlorans 'died in uncounted trillions ... their halogenous flesh was charred back and desiccated in the split second of the passing of the wave front from each exploding double star'. Humanoids are spared: since for each sun destroyed 'an oxygen-bearing, human-populated planet was snatched through four-space into the safety of Galaxy B' (Smith, *Skylark*, pp. 244–7). The book finishes with DuQuesne declaring his love for, and being accepted by, the beautiful, jutting-breasted, narrow-waisted nuclear physicist Stephanie de Marigny: there is no backward glance at the stupendous genocide of Chloran life.

Of course, this takes the fantasy of the individual empowerment by Will to a hyperbolic and extraordinarily distasteful extreme; but the borderline-psychopathology of this adolescent fantasy of mind-power also expresses the depth to which Smith's writing is rooted in a fundamentally theological conception. It is not the world of matter, 'the space–time continuum of the strictly material' that is real: the *true* dimension is 'the demesne of The Talent ... known to some scholars as psionics and to scoffers as magic or witchcraft' (Smith, *Skylark*, p. 240). Similarly in *Lensman* the power of the various titanic and enormous spacecraft is as nothing compared to the power of Thought (defined as 'a mind-generated vibration in the sub-ether', Smith, *Gray Lensman*, p. 41). Ethics in the series is linked specifically to the division between spiritual and material. The evil Eddorians live a thoroughly and disgustingly physical existence (on a dense planet whose liquid atmosphere is corrosive and deadly to human life), enhanced with a great deal of technology. The ideal Arisians, on the other hand, have long since abandoned corporeal form, living as pure thought or Will. This scheme is not Manichean, for the quasi-theistic Arisians are always superior to the evil physical Eddorians in the crucial area, Thought, which is to say Will.

European Pulps

Pulps were especially popular in the US, where they helped create a readership of self-consciously science fiction aficionados. By the end of the 1930s, says Edward James, SF had become 'fully recognised' with 'its own specialist magazines and its own specialist readership', although 'since SF in the United States was largely restricted to the Pulps, it enjoyed none of the literary prestige that was grudgingly bestowed on some SF in Europe' (James, p. 53). Much of the SF we looked at in the previous chapter on 'High Modernism' was indeed European; but this is not to say that there were no European Pulps. On the contrary, a mass audience for SF was catered for across the continent.

In France the nineteenth-century popularity of SF continued up to the Second World War, although it changed character thereafter. Popular Pulp serials included *Sâr Dubnotal* (1909–10; the title character was a superhero who had been trained by Hindu Yogis with magical skills) and *Nyctalope*, a series of Dime Novels by Jean de la Hire (1878–1956) about the interplanetary adventures of a superhuman avenger (titles include *L'Homme qui peut vivre dans l'eau*, 'The Man Who Could Live under Water', 1908; the Mars-set *Le Mystère des XV*, 'The Mystery of the 25', 1911; *Lucifer* (1923); *Le Roi de la nuit*, 'King of the Night', 1923; and the explicitly religious *L'Antéchrist*, 1927). Jacques Spitz (1896–1963) wrote a series of high-concept SF novels with popular appeal: two are *L'Agonie du globe* ('The Agony of the Globe', 1935) in which the Earth breaks into two pieces, and *La Guerre des mouches* ('War of the Flies', 1938) in which mutant flies take over the world and destroy almost all of humanity.

In Germany a series of 165 magazine-like Dime Novels under the general title *Der Luftpirat und Sein Lunkbares Luftschiff* ('The Sky-Pirate and his Navigable Sky-ship', 1908–11) enjoyed considerable popularity. The *luftpirat* is Captain Mors who travels the world in an airship of his own construction fighting evil. Hans Dominik (1872–1945) became a bestselling German Pulp author with *Die Mach der Drei* ('The Power of the Three', 1922), the first of sixteen titles that mixed kinetic adventure with a scientific-magical exoticism. *Die Spur des Dschingis-Khan* ('The Legacy of Ghengis Khan', 1923) and *Atlantis* (1925) confirmed his popularity, although the hyperinflation of interwar Germany wiped out his wealth. An interest in the legends of Atlantis was catered for by the Pulp magazine *Sun Koh, der Erbe von Atlantis* ('Sun Koh: Heir of Atlantis', 1933–36) written by Paul Alfred Müller (1901–70). Otto Willi Gail (1896–1956) was better known to the Anglophone world because two of his titles (*Der Schuss ins All*, 'The Shot into the Universe', 1925 and *Der Stein von Mond*, 'The Moonstone', 1926, both tales of rockets constructed according to correct technological principles as then understood) were translated and appeared in Gernsback's *Science Wonder Stories* 1929–30. Karl Hans Stroble (1877–1946) founded Austria's first SF Pulp, *Der Orchideengarten*, in 1920; although it lasted only 24 issues, it published a wide range of SF, reprinting classic stories and commissioning new ones. In Sweden Otto Witt (1875–1923), the so-called 'Swedish Hugo Gernsback', founded a seminal SF magazine, *Hugin* (1916–20).

In Russia following the Bolshevik Revolution there was an explosion of popular science fiction, seen by the authorities until the later 1930s as a legitimate vehicle for stories expressing the superiority of communist ideals over capitalist ones. Magazines such as *Vsemirnyi sledopyt* ('World Pathfinder') and *Mir priklyuchenii* ('World of Adventure'), which often carried SF, enjoyed enormous readerships, in part because Russia was (and is) a very populous and bibliophile nation. Writers such as Aleksandr Belyaev (1884–1942) mediated the style of Wells and Verne into a Russian idiom. Belyaev's first story *Golova Professora Douellia* ('The Head of

Professor Douellia', 1925) vividly imagines the *stasis* of the head of the titular professor, severed but kept alive by organ-transplant technology. *Zvezda KET* ('The Star KET', 1940) takes its titular initials from the space-travel pioneer Konstantin Eduardovich Tsiolkovski (1857–1935). The novel celebrates and dramatises Tsiolkovski's ideas of reaction rockets. Belyaev is still in print in Russia today, and is regarded as a classic author.

Visual texts

Almost as important as the stories was the visual look of the Pulp magazine; indeed, we can go further and identify the visual component of Pulp SF as in many ways more important than the prose component. Magazines placed a striking, bright, four-colour image on their covers to attract the eye of potential purchasers; and the stories inside were illustrated in black-and-white. Although the quality of artists employed varied, especially when judged by representational criteria (some could do realistic-looking people better than others, for instance) it is a mistake to judge the dozens of SF artists from this period on these criteria. The achievement of Pulp SF art was not representational: it lay in the creation of a wholly original mode of visual representation, highly varied and yet immediately recognisable, that still correlates to SF today.

Despite the enormous variety of images published between 1920 and 1950 most people have a sense of what constitutes a 'typical' Pulp magazine cover. The subject was most likely heroic humans or monstrous aliens posed in mid-action, or monumental futuristic technology; the style, though usually aspiring to 'realistic' representation, was less than photorealist, but was full of energy and vigour. The typical rendering utilised bright primary colours printed in cheap coal-tar dyes. Compositionally, most Pulp covers worked with a pronounced horizontal and vertical arrangement, usually utilising at least one strong and sometimes two crossed diagonals to give the image vigour, as well as (often) a prominent curve. Graphics, especially the title of the magazine but also the other verbal details (of stories and authors inside), contributed to the effect: brightly coloured, thick decorative fonts, sometimes worked into the picture – such that a rocket might shoot *in front of* the magazine's title – as if these words were actually part of the visual field.

Four artists in particular are associated with this distinctive SF magazine cover look. The most famous is Frank R. Paul (1884–1963), born in Austria but a US citizen by the time Gernsback employed him to provide cover and interior illustrations for the first *Amazing* in 1926. That initial cover image, of a chubby silver rocket standing on the launch platform against an orange-and-yellow sky, is characteristic of his work: the emphasis is on giant machinery and architecture rather than human beings, although the overall effect is beautiful. The American painter Howard V. Brown (1878–1945) was particularly associated with *Astounding* but also worked for *Thrilling Wonder Stories* and *Startling Stories*. He specialised in paintings of odd alien life, making them more striking by working with a more restrained colour palette than many of his fellows. Another American, Leo Morey (?1890–1965), worked especially for *Amazing Stories* from 1930 into the 1940s and created more colourful images, with a more imaginative use of composition. The fourth figure was Hans W. Wesso (b. 1894; date of death unknown). Born in Germany as Hans Wessolowski, he emigrated to America in 1914 and worked in a variety of magazines. Although less popular than Paul, some critics prefer his more artistically sophisticated work; Jon Gustafson and Peter Nicholls, for instance, praise his 'more open', less cluttered compositions, which 'seem more concerned with the overall design of each piece ... creat[ing] an almost abstract beauty out of the conventional icons of Space Opera' (Clute and Nicholls, pp. 1316–17).

The connection with abstract art is an interesting one. It is hard to deny that there is an abstract quality about much of this artwork, representational though it purports to be – more so in this period, paradoxically, than magazine covers from the 1960s when actual abstract images were sometimes used. The whole effect is actually in tune with the main currents of American art in the late 1920s and 1930s. The best SF cover art combines the virtues of the two artists Robert Hughes calls the 'outstanding American painter[s] of the 1930s', the careful, unpolished architectural realism of Edward Hopper (1882–1967) and the bright, quasi-journalistic energy of Stuart Davis (1894–1964). Hughes describes the art of the latter as 'extroverted, loquacious, witty, and optimistic, with a strong bent towards pictorial journalese' (Hughes, p. 430); he could be talking about SF art of the period.

In addition to the Pulps, the 1930s saw the emergence of SF 'comic strips', works in which the visual component was even more important. Newspapers often ran brief daily strips, in black and white, of the space-heroes 'Buck Rogers' (syndicated to daily newspapers from 1929) and 'Flash Gordon' (syndicated to Sundays and dailies from 1934), for instance, but these, though popular, were limited: Dick Calkins (1895–1962), the principal artist for the 'Buck Rogers' strip, produced rather mannered black-and-white line drawings that seem crude today; but Alex Raymond's art for the 'Flash Gordon' strip was more vital, punchier and better drawn. But specialist magazines ('comic books') soon emerged. The company 'Detective Comics Inc.' (which, as its name suggests, specialised in crime-story Pulps; it is better known today as a publisher under its initials 'DC Comics') published the more SF *Action Comics* in 1938. This was the venue for the first 'Superman' comics, by writer Jerry Siegel (1914–1996) and artist Joe Shuster (1914–1992). The alien humanoid with superhuman strength and a fondness for rescuing humans in danger became so popular that by 1939 he had his own publication, *Superman Comics*. Even more than John Carter, Superman is the *Übermensch* as action hero, since he turns Will to raw Power. His gravity-defying prowess, initially presented in the form of mighty leaps, evolved into a capability for *bona fide* high-flying, solely on the force of his will. More importantly, he does this in a new, visual medium: readers could 'see' his feats performed before their very eyes, rather than merely fantasise about them based on prose descriptions. Furthermore, Superman was a deliberate answer to the fascism-associated *Übermensch*: as the creation of two American Jewish youths, he was the reverse of the Aryan Overman, and as a child of Roosevelt's New Deal ethos, he used his powers to champion the oppressed, initially against even the corrupt authorities. Imitators followed, including 'Captain Marvel', who first appeared in *Whiz Comics* in 1940, and a number of other 'superheroes', beings whose supernatural powers made them the contemporary pictorial equivalent of ancient gods and mythological heroes, frequently with an SF rationalisation of these powers to make them more believable for twentieth-century audiences (see also the next chapter). In Europe, American comics were often published in translated form, although a number of domestically-produced comics were also published. *L'Avventuroso* in Italy carried the bang-bang space opera *Saturno contra la Terra* from 1937 to 1943, in which Saturnians threaten to destroy the Earth. The French magazine *Junior* serialised *Futuropolis* (1937–38), with elegant artwork by René Pellarin (1900–1998) and a storyline about the last city on Earth. A sequel, *Electropolis*, followed in 1940. The tradition of Robida is here made more complex and involved, and therefore more aesthetically satisfying. In England the bicep-heavy action-hero Garth began appearing in the *Daily Mirror* from 1943. Although these texts installed the 'superhero' as the dominant topic of SF comics, it was not until later in the century that the comic-book and graphic-novel achieved the level of significant art. But their emergence in the 1930s and 1940s does point the way to the increasingly visual bias of the genre.

The significance of the Pulps, in other words, was twofold: first, they increased SF readership, particularly in the US (while simultaneously corralling 'SF fans' into a self-created enclosure). But second, and perhaps more significantly, they created for the first time a *distinctive* SF visual style, and played their part in the broader cultural shift of SF from a verbal to a visual form of art, something that was to become increasingly the case as the century progressed.

Science fiction cinema: the silent era

With hindsight it seems clear that the most important mode of visual SF is cinematic; but this was not evident from the earliest developments of the genre. The first SF films were whimsies; short representations of pantomimic oddity designed as diversions. One of the earliest is *Le Voyage dans la lune* (1902), directed by the French pioneer Georges Méliès (1861–1938). Lunar explorers reach the Moon in a Vernean cannon-fired spaceship to discover Wellsian insectoid aliens before returning to Earth (Méliès claimed to have been influenced by both Verne's *De la Terre à la lune* and by Wells's *The First Men in the Moon*). There is a rather strained wackiness to this film, a side-effect of Méliès' past as an illusionist and stage fantasist: chorus girls in frilly knickerbockers load the spacecraft into its cannon; umbrellas, planted in the lunar surface, blossom and bloom. But there are redeeming features in the visual imagination. In particular the landing on the moon is first shown – in a brilliantly iconic image – as a projectile landing in the eye of the pudgy, cratered face of the man in the moon. This image, still current (which is to say, still liable to be recognised by many), is extremely eloquent: it seems to compact a set of assumptions about man's exploitation of nature, and about alienness, in a wittily amusing manner.

There is another significance in Méliès' image. It emblematises the way SF cinema was largely to make its greatest aesthetic impact: as an idiom of striking and beautiful images, rather than as a medium of, say, narrative, character or even, particularly, spectacle. The key moments in the SF of the last half-century are in essence poetic moments: the resonance and mystery as well as the beauty of a poetic image is what makes luminous (as it might be) the ape throwing its bone into the sky to metamorphose into a spacecraft; or the star-drenched sky of the final paragraph of *Nightfall*; or Wyndham's unsettling Midwich children; or Carrie-Anne Moss suspended in mid-air kung-fu as the camera sweeps all the way around her; or the eerie silences of the first two books of *Years of Rice and Salt*. There are hundreds of examples from the best SF, and they all work precisely as poetic images work. But some of the most resonant and beautiful images come from SF cinema.

It is true to say that narratives in early cinema were rudimentary; the emphasis was rather on spectacle, on giving audiences striking visual novelties. In the US short *The ? Motorist* (1905) a couple in a car slip from the Earth into space, and (among other things) drive round Saturn using its rings as a ring-road. By the second decade of the century, however, audiences wanted more than just striking images.

Searching for compelling stories, SF filmmakers turned to the prose classics of the genre as source material for more ambitious cinematic narratives. Many versions of Stevenson's *Strange Case of Dr Jekyll and Mr Hyde* (1886) were made: one in 1908, two in 1910, one in 1912, three in 1913, four in 1920. A kinetoscope version of *Frankenstein* (J. Searle Dawley, 1910) has been mostly lost, although stills and a brief sequence of the American actor Charles Ogle playing the Monster in wild-haired whiteface survive. More ambitious and much longer films followed. The German director Otto Rippert's *Homunculus* (1916) ran for six episodes and over six hours. The story concerns a scientist who creates a physically perfect artificial man (played by the handsome Danish actor Olaf Fønss), who discovers he cannot love and turns to evil. From the same year came the American Verne adaptation *Twenty Thousand Leagues under the Sea* (Stuart Paton, 1916); the two-hour film's narrative is sketchy, but the underwater photography (filmed from a specially designed steel 'photosphere' suspended beneath a barge) is very impressive.

After the First World War, cinematic narrative tended to fall into more conventionalised structures, but special effects and the visual aesthetic of novelty and spectacle they fostered, developed rapidly. In French director René Clair's *Paris qui dort* ('Sleeping Paris', 1923) a scientist has invented an invisible ray that stops all human action. He turns it on Paris. The only Parisians who escape its effects are those airborne (either on top of the Eiffel Tower or in aeroplanes), who return to Earth to wander around the frozen city. The plot is less important than the striking scenes of moving characters walking through the motionless tableaux of Parisian life. Indeed, one of the things that makes this film so interesting is its tacit awareness of its own medium, the way it inverts the premise of its own novel medium; in Garrett Stewart's words, it dramatises 'the virtual *de*cinematizing of the world's continuous action' (Stewart, p. 166). Science fiction, so often about new technologies, was now being embodied in a new technology, cinema – a circumstance that encouraged a meta-textual cross-fertilisation. New visual media (television, computers) provided similar injections of creative energy later in the century.

Aelita (Yakov A. Protazanov, 1924) emerged from newly Communist Russia. Once again the rather dislocated plot (Russians travel to Mars; there is some flirtation with the beautiful humanoid daughter of the leader, and there is an attempt to provoke a popular uprising; but at the end the whole Martian trip is revealed as a dream) is less significant than the overall look of the film; although unfortunately I have only seen the American print of the film which was drastically cut for its US release in 1929 (perhaps it has more coherence in the original). From its large sets and cast to its stylishly Constructivist costumes and rather mannered acting it makes a striking visual experience.

Metropolis (dir. Fritz Lang, 1926) is probably the single most iconic pre-war SF film. Set in its titular futuristic city, divided vertically between the workers (toiling in subterranean factories) and the ruling class (living on the top of the giant skyscrapers and exercising in sunny Olympic fields). Freder Fredersen, son of the city's chief aristocrat, falls in love with Maria, a saintly woman who ministers to the oppressed workers. For reasons that are not initially clear, Fredersen Sr has the eccentric scientist Rotwang create a robotic duplicate of Maria whose erotic dancing incites lustful rioting among men, and who quickly provokes a citywide uprising. Workers abandon their machines which precipitate a flood that threatens the workers' wives and families. The flesh-and-blood Maria saves the day, and the film ends with industrial relations being put on a more rational, although evidently still unequal, footing.[4]

It is not the inchoate and ideologically naïve story that makes this a great film. *Metropolis* was, it is true, severely cut for American distribution, and the chopped-about version does make less narrative sense than the longer version which premiered in Berlin; but even the longer version – and the novelised version by Lang's wife Thea von Harbou – is disjunctive and fatally schematic.[5] The rulers represent 'the head' and the workers 'the hand' in a clunking allegory; and with a breathtaking banality the film's final shots try to pretend that the moral overall is that 'head and hand must work together, and not fight one another'. If Lang had actually succeeded in dramatising so idiotic a cliché, then the film would have been deservedly forgotten. Metropolis itself is a futuristic urban space; but Rotwang lives in an ancient Gothic mansion and he fights Fredersen on the roof of a medieval cathedral. The film provides no rationale for these conflicting visual styles; indeed, it is possible to argue that the disruptive juxtaposition is part of the point of the film. The fundamental juxtaposition of the movie is the human and the machinic.

This seems as true of the acting as it is of the special effects. It is as if Lang as director had not at this point grasped that cinema requires a different, less projected style of acting than theatre; accordingly, all the performances in *Metropolis* are distractingly over the top, melodramatic to the point of the comical, with distorting exaggerated facial expressions and flinging operatic bodily gestures (silent films as a norm, because of the absence of sound, featured over-projected acting; but the acting in *Metropolis* is hyperbolic even by silent standards). But this is only to say that the human agents in the film perform machinically. It is for this reason that the most effective acting in *Metropolis* is not by a human but by a manikin; the beautiful Art Deco stylings of the robot Maria before she is transformed into the simulacrum of Maria. Standing on its podium while hoops of shining light pass up and down its body, this robot's very stillness makes for a powerful screen presence. Michael Benson thinks 'it has the haunting face of a Mycaenean death mask' and that it 'oozes sensuality' (Benson, p. 23).

Lang's next film, *Die Frau im Mond* ('The Woman in the Moon', 1929), has its supporters, particularly for the earlier sections in which a rocket is assembled and launched (apparently Nazis later suppressed the film and destroyed its special-effects models, for fear that it would give away the secrets of their own V2 programme). The rocket's crew are Moon-bound to find gold; there is a stowaway and various adventures. But there is nothing in the film to rival the iconic impact of *Metropolis*'s cityscapes or the original robot, Maria. It was in such visual moments, although they were often only tiny portions of much longer visual narratives, that cinema made its greatest contribution to the developing art of SF.

Science fiction cinema in the 1930s

Sound changed the way cinema worked in more than just the obvious sense that the aural was added to the visual. Silent films could be made in any country and played in any country, with inter-titles translated. Sound cinema, although it may be dubbed or augmented with subtitles, is much less portable: it bears the impress of the country and culture that makes it to a much greater degree. One consequence for SF cinema was that from the 1930s on, most significant SF films were made in Anglophone countries. Good SF cinema was of course made in other languages as well, but in a twentieth-century world dominated by English, they often failed to achieve the cultural penetration of Anglophone cinema.

The first film to be released with a pre-recorded soundtrack was the historical romance *Don Juan* (Alan Crossland, 1926). The soundtrack for this picture was only an orchestral musical accompaniment; but *The Jazz Singer* (Alan Crossland, 1927) added occasional recorded dialogue. The immense success of this latter ensured that all Hollywood studios had converted production to sound by the early 1930s.

The first significant sound SF film was an expensive misfire: the 1980-set musical comedy *Just Imagine* (David Butler, 1930). The future New York was created with some very impressively realised models and sets, and resembles a cleaner, better Metropolis (it lacks the dystopian lower levels). The film also includes a trip to Mars. Poor reviews and box office sank this film, and it seemed to some at the time as if SF sound cinema was to be stillborn. But a series of excellent, popular adaptations of classic SF books were produced over the following years, films still popular and current today. These included *Frankenstein* (James Whale, 1931), in which the monster is not created as a *tabula rasa*, as is the case in Shelley's novel (a mix-up means that he is given the brain of an executed criminal); but the mix of terror and pathos in his representation is unforgettable. Much of the credit for this must go to the actor Boris Karloff (born William Henry Pratt in Dulwich, England, Karloff adopted his Russian-sounding stage name purely to make himself sound more interesting to Hollywood casting agents), who played him with remarkable subtlety and effectiveness under extensive make-up (the application of which left the actor with scars). But Whale's relatively novitiate grasp of the grammar of film was impressive.

The Island of Lost Souls (Erle C. Kenton, 1932), a version of Wells's *Island of Doctor Moreau*, also trades off the star quality of its lead, Charles Laughton, who plays Moreau with a sinister, slightly effete sadism. Wells disliked the film, because he thought it untrue to his conception of Moreau as an idealist, but it is none the less an extremely powerful horror picture that understands how much more effective it is to *suggest* nastiness than to show it. *Dr Jekyll and Mr Hyde* (Rouben Mamoulian, 1932) remains for many the best version of this much remade title. The actor Fredric March won an Oscar for his jerky, expressive portrayal of the *doppelganger* title characters, but it is Mamoulian's deft handling of the range of cinematic effects that leaves the strongest impression on the viewer: from the use of montage in the first transformation from Jekyll to Hyde to point-of-view shots and moving camera. Whale returned after the enormous success of *Frankenstein* to direct *The Invisible Man* (1933) based on the Wells novella. Claude Rains' portrayal of the title character is limited (except for a brief shot at the end) to his voice alone, but he maintains a striking visual presence throughout, with his face swathed in bandages and eyes covered by sunglasses, and conveys the macabre megalomania of the invisible scientist Griffin with elegant cruelty and a chillingly fascistic rationale. (He tells his unwilling assistant Kemp, 'We'll start off with a few murders: small men; great men. Just to show we make no distinction', and rants, Hitler-like, about the new status quo he will impose, with people trembling in horror beneath his invisible might.)

Enduring though these four films have proved, the most notable SF film of the decade was not an adaptation of a pre-existing book. *King Kong* (Merian C. Cooper, Ernest B. Schoedsack, 1933) remains the most famous 'monster film'. Kong is a titanic, ape-like creature who lives on a remote island. An American film crew who have come to the island to film an adventure documentary discover Kong, and Kong discovers the beautiful if (to him) diminutive Ann Darrow (Fay Wray) who is with them. The beast falls in love with the woman and carries her away. The Americans gas Kong and ship him to New York to display him as 'The Eighth Wonder of the World'. But Kong escapes, grabs Darrow from her hotel room and climbs to the top of the city's tallest skyscraper, before he is gunned down by warplanes and killed. In a sense, it is the Lilliputian episode from *Gulliver's Travels* imagined *from the point of view of the Lilliputians*.

This film has established a wealth of powerful visual icons that have permeated twentieth-century culture: the savage island, on which terrified natives have constructed a huge wall to keep Kong out; Fay Wray in the palm of Kong's gigantic hand; Kong, on display in New York, startled by the flashbulbs of photojournalists and breaking his chains in rage; Kong atop the Empire State Building as biplanes buzz round him. It is a film with a deftly paced narrative, strong drama and a powerful premise; but nevertheless it is hard to shake the sense that it is these images (created by the best special-effects work then seen in Hollywood) rather than anything else that have powered the film into popular consciousness. It is in the nature of the best visual images to resist too precise a definition of their effectiveness and so it is in the case of this film, although critics have been prolific in their various interpretations.

Certainly, what all these films have in common (apart, that is, from their enormous and enduring popularity) is that they configure SF around a monstrous threat. More specifically each of them can be read, reductively, as a fable preaching against mankind's interference in 'Nature'. As such their influence pulled SF towards one pole of the dialectic that defines it: 'Nature' in these works is not predicated on a properly scientific objective 'I–It' relationship, but on a pseudo-religious 'I–Thou' relationship. It is by way of defining a specifically Pulp SF that Scott McCracken offers the suggestion that 'at the root of all science fiction lies the fantasy of alien encounter. The meeting of self with other is perhaps the most fearful, most exciting and most erotic encounter of all' (McCracken, p. 102). It is hard to deny that the various monstrous 'Others' of these films mediate fear, excitement and eroticism; and an allotropic reading of the genre can provide compelling social and cultural contexts to SF texts. What I mean by this is that during the 1930s a number of obnoxious political ideologues enjoyed widespread power partly because of a systematic demonisation of 'the Other' as something monstrous that the 'healthy' body-politic needed to expunge. The most notorious instance of this was in Nazi Germany, where not only Jews but Slavs, homosexuals, Gipsies, Blacks and various others were identified as pathological aliens and scheduled for destruction: but, to varying degrees of extremism, versions of this distressingly effective ideology were evident across the globe. America was perhaps more inclusive than some other nations, but founded as it was by Puritans (for whom the separation of humanity into sheep and goats was a point of religious truth) there was for most of the century a deep-rooted prejudice against people of African ancestry (particularly) as also against Native Americans and various other ethnic groups. American Blacks in particular have borne the oppressive weight of White symbolisation, whereby they are made to 'stand for' a series of negative human characteristics, especially a bestial sexuality, violence and inferiority. It is, for instance, impossible (I would argue) to read the 1932 *Dr Jekyll and Mr Hyde* except in a racial context, with Fredric March as Hyde made up to look distinctly negroid as a tacit libel on black manhood: the film, in other words, tapping into white American anxieties about race, and more specifically about the proximity and danger of Black America. Some critics read *King Kong* in similar terms.

But what is less often noted by critics is the crucial ambiguity of these monsters: in all these films, but especially in the most enduring of them (*Frankenstein, King Kong*), the monster is simultaneously terrifying *and* appealing, evoking a complex of fear and pathos. When Kong is slain the mood is very far from triumphalist.

1930s film serials

Above all, it is as a popular idiom that SF cinema thrived in the 1930s; more particularly in the form of movie serials that were shown week by week. The appropriateness of this format to later television broadcast kept many of the early film serials in currency through the 1960s and 1970s.

There were, of course, attempts to make more deliberately 'serious' or 'highbrow' films, but these proved much less successful: *Things to Come* (dir. William Cameron Menzies, 1936) was a big-budget version of Wells's *Shape of Things to Come* (see above p. 152) that failed at the box office. Academics have been kind to this ponderous, arthritic, unconvincing movie ('one of the most important films in the history of SF', Clute and Nicholls, p. 1219), but it is hard to see why. It takes a complex if preachy prose text and reduces it to a simplistic yet still preachy three-act drama of (first) near-future decline, (second) mid-future anarchy and (finally) neo-fascistic rescue by an elite of aeroplane flying scientists. The film-makers possessed so nugatory a sense of dramatic tension or trajectory that attempts to generate excitement are injected artificially at inappropriate moments (such as an instantaneous riot near the end of the movie, as a rocket is launched to the stars), which adds conniption gracelessness to an otherwise funereal film. J. P. Telotte, in a more intelligently considered account of this film than is to be found in the work of many critics, nevertheless suggests that *Things to Come* is 'one of the more visually impressive films of its time' (Telotte, p. 151) and that is true; although as Telotte himself notes, the special effects, especially towards the end, are 'monumental' in a chilly and rather distant way. Above I listed a handful of iconic visual moments from *King Kong*, moments which would be recognised by most people today; there are no such moments in *Things to Come*.

The first and most successful of the space-opera film serials was *Flash Gordon* (13 episodes; Frederick Stephani, 1936), a film version of the adventures of the popular comic strip hero. The heroic Earthman Flash (Larry 'Buster' Crabbe) and his girlfriend Dale Arden (Jean Rogers) fly off in the spaceship of their friend Dr Zarkov (Frank Shannon) to the planet Mongo to prevent it colliding with Earth. There they battle with the evil emperor Ming (played with pantomime panache by Charles Middleton) and various hawk-men, lion-men, shark-men, horned gorillas and the like. Where *Things to Come*'s special effects were immaculately produced, monumental and lifeless, *Flash Gordon*'s are astonishingly primitive: spaceships hang from visible wires, and rocket exhaust puffs upwards like pipe smoke, aliens are patently actors in suits (moreover, some of them were lifted directly from *Just Imagine*). Yet this very crudity of the cinematography is emblematic of the redemptive eagerness and energy of the whole. It is of course not an ideologically neutral text; the representation of Ming in particular draws on essentialist notions of 'oriental tyranny and decadence' that are racist, and which intensified in 1940 when the US went to war with Japan (although it might be added that, whilst the Mongonians in the comic strips were drawn as racially Far Eastern, the movie version styles them as racially Caucasian; after the success of the serial this was adopted by the comic strip as well).

This serial was so successful that a number of sequels were made, including *Flash Gordon's Trip to Mars* (15 episodes, Ford Beebe and Robert F. Hill, 1939) and *Flash Gordon Conquers the Universe* (12 episodes, Ford Beebe and Ray Taylor, 1940). Other serials were hurried into production, most notably the interplanetary *Buck Rogers* (12 episodes, Ford Beebe and Saul A. Goodkind, 1939). *Undersea Kingdom* (12 episodes, B. Reeves Eason, 1936), starring the Flash-homophonic 'Crash' Corrigan, transferred the plotline of *Flash Gordon* undersea, with Atlantis instead of Mongo threatening the Earth, led by the Ming-a-like Unga Khan (Monte Blue).

Of course, populist SF serials of the 1930s were also often wooden and prone to a lurchingly inorganic sense of narrative form and shape; but they had no pretensions to anything other than narrative momentum and were accordingly much more successful. *Things to Come* purported to be about the future, but its relative motionlessness was incompatible with an age in which the movement of humanity into the future was palpably picking up speed (something many commentators at the time noted). The onward rush of the SF serials, on the other hand, captured precisely the hectic sprint of time towards the unknown.

Orson Welles's *War of the Worlds* (1938)

A radio adaptation of H. G. Wells's *War of the Worlds* by the tyro Orson Welles (1915–1985), which was broadcast in America in 1938, stands as a fitting coda to this chapter – symptomatic of the degree to which SF was achieving a societal penetration, the extent to which many people at least in the US were starting to believe that the topoi of SF were real rather than fantastical. Famously, Welles's radio play, which dramatised the novel in the form of newscasts of an ongoing invasion, sparked mass panic when many people (perhaps over a million) believed it to be an actual news report of an actual alien invasion. Listening to this punchy dramatisation today it is hard to believe that it deceived people: it contains many deictic indicators that it was fiction not reportage (including announcements that it was fictional, a non-realist time-scale that collapsed days into hours and breaks for advertisements). But the panic it occasioned has become a tenet of popular mythology, and has itself been the subject of books and films, such as *The Night that Panicked America* (dir. Joseph Sargent, 1975). It expresses the truth that SF, its assumptions and icons were now part of the mental furniture of most Americans – and most Europeans as well. There are good reasons why the decades that followed, the 1940s and 1950s, are known as 'the Golden Age'.

References

Aldiss, Brian, *The Detached Retina: Aspects of SF and Fantasy* (Liverpool: Liverpool University Press 1995)

Aldiss, Brian, with David Wingrove, *Trillion Year Spree: the History of Science Fiction* (London: Gollancz 1986)

Ashley, Mike, *The Time Machines: the Story of the Science Fiction Pulp Magazines from the Beginning to 1950* (Liverpool: Liverpool University Press 2000)

Attebery, Brian, 'The Magazine Era: 1926–1960', in Edward James and Farah Mendlesohn (eds), *The Cambridge Companion to Science Fiction* (Cambridge: Cambridge University Press 2003), pp. 32–47

Benson, Michael, *Vintage Science Fiction Films, 1896–1949* (Jefferson, NC and London: McFarland 2000)

Burroughs, Edgar Rice, *A Princess of Mars* (1912; New York: Ballatine Books 1979)

Clute, John, *Scores: Reviews 1993–2003* (Harold Wood, Essex: Beccon 2003)

Clute, John and Peter Nicholls (eds), *Encyclopedia of Science Fiction* (2nd edn., London: Orbit 1993)

Ellik, Ron and Bill Evans, *The Universes of E. E. Smith* (Chicago: Advent 1966)

Hughes, Robert, *American Visions: the Epic History of Art in America* (London: Harvill Press 1997)

James, Edward, *Science Fiction in the Twentieth Century* (Oxford: Oxford University Press 1994)

Letson, Russell, 'Something to Think About: Joseph Sanders' *E.E. Smith*', http://www.depauw.edu/sfs/birs/bir44.htm (accessed November 2004)

McCracken, Scott, *Pulp: Reading Popular Fiction* (Manchester: Manchester University Press 1998)

Sanders, Joseph, *E.E. Smith* (Mercer Island, WA: Starmont Press, 1986)

Smith, E. E., *Gray Lensman* (1939–40; New York: Pyramid 1965)

Smith, E. E., *Skylark DuQuesne* (1965; London: Panther Books 1979)

Stewart, Garrett, 'The Videology of Science Fiction', in George E. Slusser and Eric S. Rabkin (eds), *Shadows of the Magic Lamp: Fantasy and Science Fiction in Film* (Carbondale, IL: Southern Illinois University Press 1985), pp. 159–207

Telotte, J. P., *A Distant Technology: Science Fiction Film in the Machine Age* (US: Wesleyan University Press 1998)

Westfahl, Gary, *The Mechanics of Wonder: the Creation of the Idea of Science Fiction* (Liverpool: Liverpool University Press 1998)

10
Golden Age Science Fiction 1940–1960

To describe the science fiction published in the 1940s and 1950s as 'Golden Age' is – obviously – not to use a neutral or value-free description. Coined by a partisan Fandom, the phrase valorises a particular sort of writing: 'Hard SF', linear narratives, heroes solving problems or countering threats in a space-opera or a technological-adventure idiom. Another approach at definition would be to link the Golden Age to the personal taste of John W. Campbell (1910–1971), who played a larger role than anyone else in disseminating prescriptive ideas of what SF ought to be.

Campbell began his SF career as a writer of Gernsbackian Pulp, and some of his stories are quite good, particularly his most famous, 'Who Goes There?' (1938), which was later twice filmed (as *The Thing from Another World*, 1951, Christian Nyby, and as *The Thing*, John Carpenter, 1982). But it is as editor of *Astounding* (which he renamed *Analog* in 1961), a post he took up in 1938 and which he retained until his death, that Campbell made his greatest impact on the genre. He was a proactive editor, with very definitive ideas of what constituted a good story, unafraid to press authors into revisions, to revise their work himself without their say-so, or often simply to reject, in the service of a Platonic ideal SF story. A thumb-nail definition of Golden Age SF might be 'that period when the genre was domi-nated by the sorts of stories that appeared in Campbell's *Astounding* from the late 1930s into the 1950s'.

The sorts of stories that Campbell liked were idea-fictions rooted in recognisable science (and later in his long career, in pseudo-sciences such as telepathy); can-do stories about heroes solving problems or overcoming enemies, expansionist humano-centric (and often phallo-centric) narratives, extrapolations of possible technologies and their social and human impacts. Campbell himself talked about a marked change in generic emphasis from Pulp SF to a new form of the literature. In 1946 he conceded, 'to most people SF seemed lurid, fantastic, and nonsensical trash'; the new age required 'stories of people living in a world where a Great Idea, or a series of them, and a Machine, or machines, form the background. But it is the man, not the idea of the machine, that is the essence' (quoted in Westfahl, p. 182). Fans bicker pleasantly among themselves over the dates most properly connected with this 'Age' (the consensus is that it starts in 1938–39; though whereas some say it ends with the Second World War, others say that it lasts into the 1950s). But calendar precision is not the most pressing business of the critic; on one level the production of SF between 1940 and 1980 was almost inconceivably variegated, without any body that could claim cultural authority (as it were, an *académie scienfictionnaise*) or govern the sprawling textual production. Cultural and ideological discourses, of course, determined some of the broad currents of the SF that was being written; but there are myriad counterexamples that contradict the critics attempt to insist on a 'movement', such as a notional Campbellian 'Golden Age SF'. It is for the convenience of critics and cultural historians, then, who are compelled to organise this mass of material into chapters, that 'Golden Age' can be invoked: a shorthand, best thought of as a Hard SF yang that is always balanced, containing and itself contained in an SF yin quite opposed to Campbellian ideals, although often produced by Campbellian authors.

If we take a slightly coarser perspective then the period of the 1940s and 1950s, although it contains many masterpieces of SF, is less interesting than the 1960s–1970s. Both periods saw an unprecedented kind of cultural prominence for the genre; but only in the latter did a viable dialectic alternative to the restrictive Campbellian vision achieve critical mass (the so-called 'New Wave'), and it is during that time that Russellian and Feyerabendian SF interacted in abrasive, fertile ways to produce the century's greatest prose masterpieces. By the 1980s the success of *Star Wars* and its successors had moved the genre's centre of gravity from prose to visual texts, and the continuing Fan-debate about the proper form of prose SF becomes much less important.

That older generation of critics, who grew up reading material from this period, can of course be forgiven their partiality for it; but 'Golden Age' is not the whole of SF, much of it is not even especially close to the taproot of the genre. In Westfahl's neat summary, Campbell 'makes writing a sort of thought-experiment, in which the author carefully creates a set of hypotheses regarding future events and lets the story grow out of those hypotheses' (Westfahl, p. 185). It is not merely that this represents a constrainingly narrow prescription, although I tend to think it does (others may disagree); it is that it simply does not describe SF in the broadest sense, as it was being developed throughout the century. Indeed, I am tempted to argue that even the best Campbellian SF carried within it parasitic anti-Campbellian elements and that it is these as much as the more obvious elements that makes it great. Fans invoke a 'Golden Age' because gold is highly valuable and ornamental. But it is also inert, unwieldy, and once removed from the symbolic exchange (in which it is coveted solely because people covet it) it has surprisingly few uses. SF, on the other hand, takes its Feyerabendian suppleness precisely from its universal reactivity.

Isaac Asimov

Isaac Asimov (1920–1990) has a good claim to be the century's most famous SF writer. Born in Russia but raised and resident in America, he was a fantastically prolific writer who began to be published at the end of the 1930s, and whose output increased the older he got, as if his career were operating according to a literary principle of anti-friction. Most of his later output was non-fictional, and he wrote virtually no SF between 1958 and the 1980s. The non-fiction is always professional and enlightening, and always worth reading; the novels written in the 1980s are mostly feeble, less interesting in themselves than as symptoms of a dotage-hatched plan to synthesise and unify all his separate earlier universes. But it was the work he created during the 1940s and the early 1950s, at the very heart, chronologically and culturally, of Golden Age SF that has sustained his reputation.

One of his greatest achievements was also one of his earliest, the short story 'Nightfall' (1941). It tells the story of a civilised world that orbits multiple suns, so that there is always at least one source of light in the sky and 'night' is unknown. Archaeologists are puzzled by the fact that the world's previous civilisations seem all to have ended in catastrophic destruction every 2,000 years. Astronomers realise that, once every two millennia, all the suns set at once and night falls. The story builds expertly towards its inevitable conclusion: the darkness finally comes, revealing an unimagined plenitude of stars in the night sky, which sight collapses the minds of the population. This short tale is often cited as the 'most popular' or 'the greatest' SF story ever published; and it has earned its place chiefly on the strength of its 'sense of wonder' ending, the starry splendour that reveals the vastness of the cosmos to a people who had not understood the true scale of things. In other words, this deft little story recapitulates the conceptual crisis that had generated science fiction in the first place: it collapses the Copernican revolution into a single night. This is why it has such resonance for readers steeped in the genre – and why it is so little known outside the genre.

Also written in the short-story idiom, although later assembled into three connected novels, is Asimov's *Foundation* series, a self-conscious translation of Gibbon's *Decline and Fall of the Roman Empire* onto an interstellar stage. The Galactic Empire is on the verge of disintegration, although the only person to have foreseen this is the scientist Hari Seldon, whose own invention, 'psychohistory', can accurately predict the future for large populations (although not for individuals, where there are too many variables). Under the guise of an organisation compiling the *Encyclopedia Galactica* he establishes a 'Foundation' to guide the galaxy through the inevitable millennia of anarchy, materially shortening the time required for civilisation to re-establish itself.

The *Foundation* series is much beloved by many SF fans, although it suffers from a ubiquitous and debilitating dryness of tone. It is almost entirely composed of dialogue, often of an expository or explicatory nature; there is little description, a fact which renders the sequence visually inert; and the characterisation is rudimentary. Why do so many fans love it? In part because it deals in big ideas: the logic governing history, the possibility of a properly scientific prophecy, what role, if any, individuals have in larger historical circumstance. These are good and important matters, and Asimov interrogates them very well: lucidly and thoughtfully. But they have much less purchase today than they did in the 1940s and 1950s; not because history has ceased to matter, but because the development of Chaos Theory in the 1980s has finally put to rest the old Positivist philosophical chimera of a science so comprehensive that it can wholly predict the future. History evidently is a chaotic system. And whilst we cannot expect Asimov to have anticipated Chaos Theory, it is the case that in the *Dune* sequence, one of the great achievements of 1960s–1970s SF (for which see the next chapter) Frank Herbert intuitively felt towards an understanding that history was governed by irrational rather than rational laws. *Dune* stands up well today in ways that the blind faith in science of *Foundation* does not.

But this is not to snipe at Asimov's centrality to twentieth-century SF. Rather, it is to shine the spotlight away from *Foundation* and onto the other genuine contribution Asimov made to SF, and to a lesser extent to culture more generally, in his 'robot' stories. Where robots had previously been, almost exclusively, insensate or dangerous embodiments of the threat of technology, Asimov imagined artificially constructed and intelligent robots as not only humane, but in many ways as *more* humane than humanity. The many robot short stories, and the dozen or so robot novels, all share one deeply felt focus: the exploration of ethical questions. In them, much more compellingly than in the Comtean-Positivist rigidity of the *Foundation* series, Asimov manages to make the subtle, persuasive and dramatic exploration of moral questions vital and compelling.

An example is Asimov's first robot novel, *The Caves of Steel* (1954), a brilliant hybrid SF whodunnit. It is set in a future of two fundamentally differentiated human societies, on the one hand the crowded, dirty mass of life on overpopulated Earth, who are confined to Earth by their own fears and prejudices; and, on the other, the aloof, patrician, wide-open-space-conditioned manners of the Spacers, who live affluent lives on spacious colonised planets. The plot is straightforward; an eminent Spacer scientist has been murdered on Earth. Given the hostility between the two populations, the investigation is a sensitive one. The Earth detective, Lije Bailey, is given the job of investigating the crime, assisted by a Spacer robot called R. Daneel Olivaw. Unlike Earth robots, which are metallic and obviously artificial, Olivaw is constructed to resemble exactly a human being. Bailey shares the Earth prejudice against robots, and is initially hostile to his new partner, even to the extent of publicly accusing him of the crime. Such an accusation is demonstrated to be absurd: no robot could murder a human being, because all robots are constrained by the 'three laws of robotics' (Asimov in fact formulated these celebrated rules in collaboration with John Campbell):

1. A robot may not harm a human being, or, through inaction, allow a human being to come to harm.
2. A robot must obey the orders given to it by human beings, except where such orders conflict with the First Law.
3. A robot must protect its existence as long as such protection does not conflict with the First or Second Law.

It is amazing how much dramatic and conceptual variety, across dozens of books and scores of stories, Asimov was able to generate out of these three little rules. The robot books are a fugue on issues of logic, identity, difference and resemblance.

The binaries the book apparently establishes are in fact in the process of collapsing into a unity, just as Asimov himself spent his last two decades attempting to synthesise a larger 'identity' out of the variegated imaginary universes of the Foundation books, the robot books and the 'Eternals' of *The End of Eternity*.

At root all the robot stories are ethical fictions in which Asimov positions a race of properly Kantian ethical beings (the robots) against the much more nebulous ethics that characterise actual human activity. For Kant, ethical questions were judged as absolutes. 'I am never to act', that philosopher declared in *Fundamental Principles of the Metaphysics of Ethics* (1785), 'otherwise than so that I could also will that my maxim become a universal law'. What this means is that before I commit a murder or tell a lie, I must first consider whether my acting in this way could conceivably apply as a universal law – how things would be if *everybody* murdered, if *everybody* lied. In other words, ethics is a question of consulting not my personal advantage, but a universal moral code. 'There is', Kant insisted in the same work, 'an imperative which commands a certain conduct ... this imperative is Categorical'. This Categorical imperative runs, in many cases, contrary to human impulses: were a murderous madman with a gun to demand of me where my friend is hiding, I might consider it not only prudent but morally justified to lie; but for Kant, lying contravenes the Categorical imperative of morality, and the proper thing to do would be to tell the truth, even in this extreme case. It is this absolutism that renders Kantian ethics unpalatable to many thinkers.

With his robots Asimov created a race of sentient, thoughtful beings in whom the Kantian moral imperative is internalised; robots do not consult their conscience when faced with an ethical dilemma, they obey the three laws that absolutely govern their behaviour. The genius of the invention is that the resulting race of beings is not *absolutely* determined; Asimov's robots do not run on rails, their behaviour is not necessarily predictable, they are not morally clockwork figures. Indeed, the great theme of nearly all the robot novels and stories is the working-out of the implications of what it would be like to live life under this trefoil categorical imperative. Kant talked about, in his famous phrase from the beginning of the *Critique of Practical Reason*, 'the starry heavens above me, and the moral law within me'. Asimov dramatises precisely this philosophical position. Terry Eagleton's account of Kantian ethics is also, unwittingly, a precise delineation of the strengths of the Asimovian robot:

> To act morally for Kant is to set aside all desire, interest and inclination, identifying one's rational will instead with a rule which one can propose to oneself as a universal law. What makes an action moral is something it manifests over and above any particular quality or effect, namely its willed conformity to universal law. What is important is the act of rationally willing the action as an end in itself. (Eagleton, p. 78)

By following this logic Asimov casts light on the ethical dilemmas of ordinary human life; and in particular, the notion that a 'science fiction' ought to embody the values of science in its ethics – according to *rational will*. This is a belief (widespread among

SF writers) that is as unashamedly Enlightenment as is Kant's work itself. Bailey, initially distrustful of R. Daneel, refuses to believe that he/it possesses a 'drive for justice'. 'Justice', he insists, 'is an abstraction. Only a human being can use the term.' His interlocutor, the Spacer Fastolfe, agrees, but invites Bailey to ask R. Daneel what he understands by 'justice'.

> 'What is your definition of justice?'
> 'Justice, Elijah, is that which exists when all the laws are enforced.'
> Fastolfe nodded. 'A good definition, Mr Bailey, for a robot ... [but] humans can recognise that, on the basis of an abstract moral code, some laws may be bad ones, and their enforcement unjust. What do you say, R. Daneel?'
> 'An unjust law,' said R. Daneel evenly, 'is a contradiction in terms.' (*The Caves of Steel*, pp. 83–4)

When Fastolfe insists that 'you mustn't confuse your [human] justice and R. Daneel's' he is making the distinction between 'justice' as an ethical absolute and as a subjective relativism.

It is easy to underestimate Asimov as a writer: the dry, flat, prose style, the underdeveloped characters, a lack of visual and descriptive flair, these things can distract the critical reader from his considerable artistic strengths. Nor have favourable critics been able, effectively, to frame a defence of Asimov's importance except in terms of an entertainer, a puzzle-setter and humorist, a polymathic populariser and so on. SF writer Christopher Priest taught Asimov's *The Caves of Steel* to students in the University of London in 1974. 'Enjoyed reading it', he recalled, in an interview with John Brosnan: 'great stuff, but there was nothing whatsoever to *say* about the book. It was all plot and seemed to have no content at all' (quoted in Ruddick, p. 46). Nicholas Ruddick sees this response as typical; one strand of newer, more 'literary' SF writers in reaction to the old depthless, plot-driven, empty enjoyment of older Golden Age writing. But there is a great deal more to Asimov than this; he used the form to create brilliant fictions of a materialist ethic and great imaginative power.

Early Heinlein

Heinlein towers over many critics' histories of SF.[1] He was aware of his importance to the genre himself, towards the end of his career, and derived, it seems, pleasure from it; for in Heinlein's writing there is an elision between the concepts 'author' and 'authority'. As precious, inert and beguiling as gold, Heinlein's writing is the most representative of Golden Age fiction. If that seems too much of a caricature, let me restate it: more than one generation (mine included) grew up reading Heinlein's fiction as a sort of archetype of what SF should be: forceful, thought-provoking narratives written with a winning fluency and approachability, characters that travelled thrillingly about the solar system or through possible Earths without even losing plausibility or likeability. It may prove that his place in twentieth-century SF becomes seen as less central; he is certainly less generally read today than once he was. But his SF vision was much closer to the Campbellian ideal of 'Golden Age' SF even than Asimov, and his best writing retains its unmistakeable, unique heft.

Heinlein's career passed through three stages. The early phase, before *Starship Troopers* (1959), contains what many see as his best work: what Brian Aldiss identifies as an 'ability to extrapolate', an 'eye for social quirks', a 'simple poetry' and above all an ability to be 'genuinely innovative' (Aldiss, *Trillion*, p. 389). The middle period is often seen as surrendering these virtues to a strident, even desperate 'puppet-master' authorial persona, which harps incessantly and sometimes unpleasantly on a narrow range of ideological concerns: the importance of individual liberty conceived in the American libertarian mode, with a pendant mistrust of 'government' and a fetishisation of authority as such. The final period, from *The Number of the Beast* (1980) through to his death in 1988, is stingingly dismissed by Darko Suvin in a single word: '*senilia*' (Suvin, 1988, p. 262). But this is to underestimate the sheer strangeness of Heinlein's biggest success, and masterpiece, *Stranger in a Strange Land* (1961); a success made all the more difficult to place given that all Heinlein's many subsequent books premised their continued fascination with the status of the messiah on a much more narrowly conceived authoritarianism. By the time of *Time Enough for Love* (1973), a dryly corpulent novel (1,000 pages or so) about the interminably extended life of a Heinlein-mouthpiece character called Lazarus Long, the authority figure has been more precisely defined as the patriarch, and Heinlein's studied non-conventionalism shrinks to a more straightforward sexist ego-fantasy.

But it would be wrong to allow hindsight to diminish the achievement of the earlier work. Golden Age Heinlein almost never put a foot wrong, fictively speaking. He possessed a deep understanding of the ways in which a populist idiom could be used to make serious points. He began writing shorter fiction for the magazine market, usually Campbell's *Astounding*: 'The Roads Must Roll' (1940), set in a future America in which people travel not in individual cars on stationary freeways but on vast networks of rolling roads (on which they can stand, or even enter travelling cafés and eat lunch). Heinlein, never one to be overly beguiled by an ingenious piece of future-furniture, used these roads for a brilliant little tale about the power of collective labour. The workers who service the rolling roads synecdochally represent 'the worker'; and in the trope of 'rolling' Heinlein found a marvellously expressive trope for 'revolution' itself – it is revolution, in its political as well as its literal sense, that is the real theme of the story. The short story 'Universe' (1941), followed quickly by 'Common Sense' (1941) (they were assembled into a novel called *Orphans of the Sky* in 1964), remains one of the classic generation-starship stories; the descendants of voyagers on an interstellar journey lasting many centuries have forgotten that they are even on a spaceship. The story 'Waldo' (1942) about a disabled inventor who uses artificial remotes to do the work his crippled body cannot, gave the language the word 'waldo'.

The Puppet Masters (1951) is a nicely pitched alien-invasion story that shunts its narrative along so effectively, and creates so compelling a mood, that it is only after the book has been closed that the reader starts to doubt the coherence of the whole. The short novel *Double Star* (1956) combines needle-sharp adventure and witty political satire, as an actor hired to impersonate a Solar System politician finds himself stuck in the role. *Have Space Suit – Will Travel* (1958) is a nippy, flawlessly constructed 'young adult' novel in which an ordinary kid gets caught up in wholly believable galactic adventures and returns to school wiser. *The Door into Summer* (1957) rattles through ingenious back-and-forth time travel adventures (a businessman who invents a robot and is double-crossed by his fiancée), articulating a can-do, individualist, free-enterprise ideology on the way.

This last feature (which became increasingly prominent as Heinlein's career progressed) is crucial. In so far as it is meaningful to make the distinction, Asimov was an *ethical* writer where Heinlein was a *political* one. Both positions clearly are ideological; but Heinlein's work and life took place much more deliberately in the arena of politics. In 1938 he campaigned (unsuccessfully) for the Democratic nomination for a California assembly seat, and had connections with a (by US standards) radical left-wing group called 'EPIC'. Later in his life his political allegiance changed completely to a right-wing, militaristic libertarianism. This *volte-face* also involved the suppression of his radical youth, the story of which wasn't unearthed until the 1990s by Thomas Perry, and reported to a wider audience only in Thomas Disch's 1998 book *The Dreams our Stuff is Made Of*. Disch sums up Heinlein's postwar output unsparingly but accurately:

> The main thrust of Heinlein's SF in the Cold War years was to advocate the perpetuation and growth of the military-industrial complex. ... [He] spoke out against restrictions on nuclear testing in 1956. At a World SF Convention in 1961, he advocated bomb shelters and unregulated gun ownership. He was a hawk in the Vietnam years ... These positions, and others more extreme, may easily be inferred from the SF he wrote in the same period. No hawk could boast sharper talons. (Disch, p. 165)

Starship Troopers (1959) is one of the most hawkish SF novels ever written; a novel absolutely in love with all trappings of military existence: boot camp, military training, the transformation of slovenly teenagers into disciplined professional soldiers – in particular the ability to perform the business of ordinary living (washing, eating, moving to and fro) 'on the bounce!' instead of with that sluggish dumb insolence so typical of the modern teenager. The troopers are needed to fight an implacable alien insectoid foe, with whom treaties would be meaningless and who ruthlessly exterminate human life wherever they find it. Within this artificially black-and-white moral framework Heinlein's gung-ho certainty is difficult to resist; but taken out of its artificially pumped-up premises the novel is at least *quasi*-fascistic. Understandably, Heinlein resented accusations of fascism from the first (such accusations were made almost as soon it was published). In the world of the novel only veterans have the vote, and many readers have taken this to mean that the franchise is limited to ex-soldiers; Heinlein countered that 'nineteen out of twenty veterans' in the novel 'are *not* military veterans ... [but] what we call today "former members of the federal services" '. The book, he says, celebrates state service *in toto* rather than just military service. He goes on to insist that, in any democracy, voting should be something earned rather than a right (he is witheringly dismissive of democracy as the term is presently understood, the vote 'handed to anyone who is 18 years old and has a body temperature near 37° C') (Heinlein, *Expanded*, pp. 398–9). There's an element of disingenuousness here; *Starship Troopers* presents the reader with almost no characters other than military ones, and the tone is so gung-ho and celebratory that not one reader in a thousand would see it as anything other than a paean to the military (moreover, as James Gifford shows, 'by the text of the novel, Federal Service *is* entirely military' (Gifford, p. 11)).[2] But to call Heinlein a fascist is to suggest that his books preach conformism to a militaristic national *volk* or subordination to a 'leader', which quite misrepresents his particular brand of ideological reaction. Whilst always a patriotic American, Heinlein was ideologically invested in neither racial nor geographical ideals, and his books consistently advocate a studiedly responsible 'Contrarian' position, especially with respect of the US government. In other words, he was a right-wing libertarian, and his books preach a libertarian gospel. The question is how central this ideologically freighted 'libertarian' problematic lies to the core of Golden Age SF itself. It seems to me the answer is 'very much central', although others might disagree.

North American Golden Age writers

The Golden Age was dominated by white, male North American writers, a fact which can tempt historians into seeing a cohesiveness about the dozen or so major writers of the time that did not actually exist. There were, it is true, unofficial alliances or loose 'clubs' like the Futurians (writers of more liberal political affiliation in the New York area who would meet from time to time during the early 1940s). There were also fan conventions, which many writers attended (the first 'Worldcon', or World SF Convention, took place in New York in 1939), although these did not become today's notorious giant jamborees until decades later. In general, however, the key names did not form a coherent league.

Jack Vance (b. 1916) is one of the underappreciated giants of twentieth-century SF, and indeed of twentieth-century literature more generally. He belonged to no team or club, and his beautifully composed, prolifically produced books seem to argue no particular ideological agenda. His idiom is Romance, in the generic sense of tumbling and exotic adventures that rattle through alien landscapes. Perhaps his many books, when read in bulk, seem to resolve into variations of the same tale (a solitary, rather terse hero makes his way through precisely rendered, ingeniously multifarious cultures and opponents); but it is a mistake to read Vance for the narrative, beguilingly readable though those narratives are. He is a world-builder, an imaginary anthropologist, and above all a stylist; and it is the conjunction of the unflagging fertility of his imagination and the elegant, chill mannerisms of his prose that generate the distinctive Vance quiddity. His first novel, *The Dying Earth* (1950), portrays the ripe complexities of its thoroughly decadent culture in a prose that is icily forensic. It is an absolutely intoxicating book, and had a deep impact on the continuing development of 'Last Man' fictions, a sub-genre of SF that (as we have seen) goes back a long way into the genre's past. But where a book like Grainville's *Le dernier homme* (1805) conceives of the end of the world in religious terms, Vance's amoral, vivid, slightly strange characters acknowledge no superior spiritual authority except their own ingenuity. (There were many sequels to this seminal book, most of them written many decades later.)

Big Planet (1957) is spicy adventure on the oversized and metal-bereft planet of the title. *The Dragon Masters* (1963), like *Dying Earth*, treats Fantasy tropes in an SF manner. *Emphyrio* (1969) may be Vance's most perfect novel; its setting is a claustrophobic world in which artists and artisans work in a complex collective to produce works of art under the oppressive rule of certain lords, and Vance renders every tittle and jot of this society with the clarity of a steel engraving. The adventures of Vance's hero, Ghyl Tarrok, lead him to an understanding that his world, made familiar by Vance's detailed descriptions, is actually radically alien and strange. This, indeed, is the kernel of Vance's genius: his carefully mannered prose slips effortlessly from familiar to alien, treating both with the same disinterested precision. The effect is the witty yet chilling delineation of what alienness is actually like. In this passage Ghyl is travelling through open land with some of the 'lords', who are unused to such hardships. Here is the moment when it dawns on Ghyl, and us, that these aristocrats, whom we had assumed to be decadent humans, are actually alien beings, with radically inhuman perceptions and attitudes:

> Ghyl built a fire on the old stone hearth, which irritated the lords.
> 'Need it be so warm, so bright, with all those little whips and welts of flame?' complained Lady Radance.
> 'I suppose he wants to see to eat,' said Ilseth.
> 'But why must the fool toast himself like a salamander?' demanded Fanton crossly.
> 'If we had maintained a fire last night,' Ghyl returned, 'and if the Lady Jacinth had used my advice to climb high in the tree, she might be alive now.'
> At this the lords and ladies fell silent, and their eyes flickered nervously up and down. Then they retreated into the darkest corners of the shack and pressed themselves into the walls: a form of conduct which Ghyl found startling. (Vance, *Emphyrio*, p. 149)

This goes beyond merely satirising the vacuity of the upper classes; it is a means of apprehending a radical otherness. Vance's other major achievements are all sequences of novels. The 'Planet of Adventure' series (a very uninspiring title for some of Vance's most inspired, marvellous books) is made up of *City of the Chasch* (1968), *Servants of the Wankh* (1969) and *The Dirdir* (1969), and holds familiarity and genuine alienness in perfect balance. The 'Durdane' trilogy (*The Anome*, 1973; *The Brave Free Men*, 1973; *The Astura*, 1974) is set in a world fractured into innumerable cantons, each with their own laws and customs, linked only by fear of the justice meted out by the sinister 'faceless man'. The 'Demon Princes' series (*The Star King*, 1964; *The Killing Machine*, 1964; *The Palace of Love*, 1967; *The Face*, 1979;

and *The Book of Dreams*, 1981) is a revenge tale, in which five wealthy super-villains are dispatched, one per novel, by the aggrieved protagonist. The series cannot avoid a certain repetitiveness, although it is as ingenious and complex as anything Vance wrote. More recently Vance's output has not flagged.

The Canadian A. E. Van Vogt (1912–2000) has few champions among academic critics of the genre; many dismiss as merely incoherent his intricately but discontinuously plotted novels. They are undeniably rococo, often implausible and certainly demonstrate an adolescent infatuation with the compensatory fantasy of superheroes misunderstood by a boorish world. It used to be said that Van Vogt was too slippery and non-commonsensical a writer for the straightforward technical rationalism of American Golden Age SF – that he was, in fact, a European *avant-garde* author born on the wrong continent. And it was certainly the case that some of his books enjoyed much greater popularity across the Atlantic: for instance, French readers were much more receptive to the complex multi-reality shenanigans of *The World of Ā* (1948; the last element in the title is pronounced, and is printed by some publishers as, 'Null-A') than American ones. But there is nevertheless something ineluctably North American about Van Vogt's fantasies, not least his fondness for guns. For instance, the right to bear arms (admittedly, SF guns that know, magically, whether their wielders are firing in self-defence or in criminal assault, and only permit the former) becomes symbolic of all personal freedom against tyranny in *The Weapon Shops of Isher* (1951), and is, in the oblique, associative manner of Van Vogt's writing, somehow also tied in with nothing less than the creation of the entire universe, the Big Bang which ends the book and commences everything else. But the most characteristic Van Vogt book is *Slan* (1946), the story of a 'slan' or mutant Homo superior called Jonny Cross, with tiny horns growing from his head, two hearts and vast intellectual capacity. Cross is persecuted by normal humans, and we first encounter him on the run from a murderous world. But the story bifurcates, or multifurcates, in a bewildering series of directions, and at the end is governed by an almost dream-logic in the assemblage of its various elements. For many this only increases the novel's oblique potency. Indeed, many SF fans have taken the 'slan' as a straightforward metaphor for 'the SF fan' (the slogan goes: 'fans are slans'), persecuted by a cloddish world yet not only inherently cleverer and better, but also (as Van Vogt's peculiar story works through its associative tangles), somehow, obscurely, the secret rulers of the world. A more pertinent connection to make, with regard to this story and much of Van Vogt's output, is the religious one. This is not only because Van Vogt, with his almost automatic-writing aesthetic, intuitively tapped into the buried taproot of SF, although I think he did. It is also because his later career literalised the theological subtext of much of his writing in a personal commitment to the new religion being founded by fellow SF author L. Ron Hubbard (1911–1986). Van Vogt was caught up with Dianetics, although he was not as involved in Hubbard's later religious cult, Scientology. But it seems clear that this yearning towards the mystic complexities of Hubbard's invented religion chimed with a quasi-Gnostic sense of 'significance' that lurked behind the Pulp trappings of his work.

SF fandom holds the memory of the American writer Cleve Cartmill (1908–1964) particularly dear for one Golden Age fluke: his short story 'Deadline' (published in Campbell's *Astounding* in 1944) described the atom bomb in detail, a year before the actual bomb was ready: agents from the US Army 'Counter Intelligence Corps' investigated the *Astounding* offices convinced that military security had been breached; but in fact Cartmill had simply extrapolated from scientific information in the public domain. Campbell later boasted about this as symptomatic of SF's ability to see into the future, although – nice though the story is – it has proved uncharacteristic of the genre.

The prolific and estimable American writer Poul Anderson (1926–2001) once seemed among the most important writers of SF ('he is central to the SF field', opined John Clute in the 1990s, 'although sometimes – this is the downside of reliability – he tends to be taken for granted', Clute, p. 150). Now he is little read, which is a shame, since his best books are not only excellent, but eloquently expressive of the core dialectic of SF itself. His first novel, *Brain Wave* (1954), takes as its premise that the Earth has been moving through a zone of cosmic radiation that has artificially retarded our intellectual development. As we move out of this area intellect suddenly develops exponentially, and everything changes – it is a novel, in other words, about transcendence, touching on the same quasi-religious keynotes as Clarke's *Childhood's End* (see below p. 213) although in a less apocalyptic manner. Anderson went on to publish nearly 100 books, many of them military-based space-opera, always professionally and entertainingly done. Many of them are haunted by a theological-ontological anxiety. In *Tau Zero* (1970) a malfunctioning spaceship picks up more and more velocity, approaching the speed of light and (as Einstein predicted) expanding to almost infinite size, pushing galaxies out of the way. Time dilation means that the crew survive the end of the universe and eventually fly into a reborn cosmos. Most of the problems they face are practical and technical, but everything they do is haunted by what crew member Ingrid Lindgren calls 'that question. What is man, that he should outlive his God?' (Anderson, *Tau Zero*, p. 171).

In a similar way, it is hard to say why the American (of Lithuanian descent) writer Algis Budrys (b. 1931) is no longer read today. In their time his novels had many admirers, and they still read very well: *Who?* (1958) concerns a man rescued from near-death by extensive technical prostheses; turned into a cyborg, he may or may not be a Cold War spy. The novel (though not the rather simplistic film made from it (*Who?* dir. Jack Gold 1974)) manages to coax from its creaky premise an interesting meditation on existence and alienation. *Rogue Moon* (1960), in which a series of *doppelgangers* of the protagonist are killed trying to penetrate a lethal alien labyrinth on the moon, works both as an exciting narrative of tense expectation, and also as a psychodrama to do with the pull of death, mystery and the alienated condition of humanity.

New York-born Alfred Bester (1913–1987) published relatively little science fiction, but his impact on the genre has been enormous. Single-handedly, it sometimes seems, he invented both 'New Wave' and 'cyberpunk'. His first SF novel, *The Demolished Man* (1953), follows the attempts by Ben Reich to avoid arrest for the murder he committed in a world of telepathic 'espers' who can detect most crimes before they are even committed. It is written in a sparkling, brilliantly evocative and laconic manner; 'hard-boiled' doesn't do justice to Bester's command of his idiom, which is never conventional or derivative. His best novel, *Tiger! Tiger!* (1956; in the US published as *The Stars My Destination* 1956) is claimed by some as the single greatest science fiction novel of the century. It is a *Bildungsroman* staged against a brilliantly realised backdrop of a war-torn, hyper-kinetic future solar system. Gully Foyle begins the novel as barely human, a grunting illiterate 'Mechanic's Mate 3rd Class'. He survives the destruction of the spaceship *Nomad* and when a passing ship refuses to rescue him he becomes gripped by a thirst for revenge, which motivates him to escape his prison and pursue those who abandoned him, transforming himself in the process into a highly intelligent, educated, resourceful individual, and ultimately into a sort of saviour of mankind.

It transpires that the fate of the *Nomad* was bound up with its secret cargo, a mysterious war-material called PyrE that functions brilliantly as a plot McGuffin. But the story of this novel, breathless and gripping though it is, is less important than the mood and the sheer exuberance of Bester's world-building. The solar system of Bester's 25th century is one in which people teleport instantaneously, although teleportation in this universe is a function of pure will, unmediated by technology. Once a person realises how, and provided he or she can visualise the destination, they can pop instantaneously from place to place, a practice called 'jaunting' ('a term which', as Scott Bukatman points out, 'describes not only the teleportation technique, but also Bester's hyperactive narrative structure and "jaunty" style', Bukatman, p. 352).

The only barriers to 'jaunting' are unknown destination (which means that security becomes a matter of building elaborate mazes), and more fundamentally the vacuum of space, which cannot be jaunted across. As Foyle grows across the trajectory of the novel's plot, we come to realise that he has, without realising it, cracked the trick of jaunting through space, and so the novel's upbeat American title closes the story in resounding, messianic fashion. Critics have noted that the book is modelled on Jacobean revenge dramas; but Bester is not really interested in the limited theological cause-and-effect speculation of the revengers (God has a severely reduced place in his imagined solar system, and organised religion has been outlawed). Instead, this is a book about 'will', imagined with positively Nietzschean fervour. The book's two central conceits elaborate this key theme. 'Jaunting', for instance, is the externalisation of the desire to travel from place to place simply by thinking about it, the symbol that links 'wish' and 'destination'. PyrE, similarly, is a form of explosive matter detonated by thinking about it. 'Through Will and Idea,' explains Presteign, the patrician villain of the book, 'PyrE can only be exploded by psychokinesis. Its energy can only be released by thought. It must be willed to explode and the thought directed at it' (Bester, *Tiger!*, p. 217). Which is to say, PyrE is the concrete externalisation of mankind's will to destruction, the symbol that links 'wish' to 'destruction'. In this sense Bester's novel, which had an incalculable influence on New Wave and cyberpunk writing, carries within it the germ of a Nietzschean triumph of the Will. That, in a nutshell, is the true dream of the Golden Age: the technological manifestation of the Will of the (masculine) hero/reader.

The postwar British scene

American Golden Age SF was, more often than not, bullish, can-do and outward-looking. British postwar SF had, and to some extent has maintained, a rather different tone; more introverted, downbeat and pessimistic. But this is what we might expect; America in the 1950s and 1960s was an expansive nation; Britain was shrinking. It is certainly hard to overstate just how profound was the cultural shift experienced by Britain, and especially England, after the end of the Second World War. A nation which had once ruled an empire encompassing a fifth of the globe's population began an inevitable process of diminution; a retreat from Empire and a world role into a more proportionate position as an island on the margins of Europe. As the grimness of postwar life (food rationing, for instance, continued for many years after the war) filtered through into a series of more or less downbeat, pessimistic works of art. SF was a mode that allowed the inflation of this sense of loss – an obscure and infuriating sense, since 'we' had after all won the war, but a palpable loss nevertheless – onto global or cosmic arenas. The masterpiece of this particular depressive-pessimistic idiom is George Orwell's *Nineteen Eighty-four* (1949).

Orwell's dystopia is an extrapolation from the totalitarian regimes of the 1920s and 1930s. His future world is divided between Oceania (the old US, Britain, Australasia and South Africa), Eurasia (the old Russia and continental Europe) and Eastasia (China and East Asia). The remaining global territory is continually being fought over in a never-ending world war. Each of the three world-states is governed by absolutist regimes: the 'Party' of Oceania, with which the novel is mostly concerned, ruling according to the principles of 'Ingsoc' or English Socialism, a Stalinist state-system. The main point of the novel is to anatomise this sort of political reality, both its actual ideological rationale and the miserable quality of life it imposes on its subjects.

The novel is concerned with Winston Smith, an inhabitant of Airstrip One (the old British Isles) who lives a shabby, worn-down, miserable existence characterised by chronic shortages of necessities, rationed supplies of poor quality food and low-grade gin drunk at all hours. The population live under constant surveillance by the Party: all citizens have a telescreen in their room, a two-way device that broadcasts propaganda at them but also allows monitors to observe them at all times. Loyalty to the Party is focused on the notional figure of one man, 'Big Brother', whose face is on posters everywhere: 'a man of about forty-five, with a heavy black moustache and ruggedly handsome features ... BIG BROTHER IS WATCHING YOU the caption beneath it ran' (Orwell, *Nineteen Eighty-four*, p. 5). Through the system of telescreens, this slogan is literalised. The narrative of the novel details Smith's dissatisfaction with the world in which he lives. He keeps an illegal diary in which he records his dissident views; he has an illegal affair with a fellow worker called Julia – sex, except for the purposes of reproduction, being a criminal act in Oceania. This affair gives Smith hope that, although Big Brother is too powerful to oppose on a social scale, it remains possible for people to be free in their hearts, as in their purely personal connections. But later in the novel the secret affair is uncovered, Smith and Julia are arrested by the Thought Police, tortured and wholly broken. In an extended session with a senior Party official, O'Brien, Smith is taught the hard lesson that reality itself is only the footstool of Power. O'Brien puts it in these terms: 'there is nothing we [the Party] cannot do. Invisibility, levitation – anything. I could float off the floor like a soap bubble if I wish to.' Smith challenges him:

'But the whole universe is outside us. Look at the stars! Some of them are a million light-years away. They are out of our reach for ever.'

'What are the stars?' said O'Brien indifferently. 'They are bits of fire a few kilometres away. We could reach them if we wanted to. Or we could blot them out. The earth is the centre of the universe. The sun and stars go round it.' (Orwell, *Nineteen Eighty-four*, p. 213)

In a manner of speaking this passage encapsulates *Nineteen Eighty-four*'s uneasy relationship with science fiction. It is a deliberately pre-Copernican SF novel. In some ways it does utilise science fictional tropes: the telescreens, rocket-bombs, the advanced technologies of war-making. But in another sense, the regime of the novel has no use for technology. The slogan about controlling the future is wholly circular: there *is* no future in Oceania, only a continual present of party power. Even the date of the title reinforces this notion; Orwell, writing in 1948, simply reversed the last two numerals to provide his fictive date. Unlike the tradition of future fiction (books such as Bellamy's *Looking Backward 2000–1887*) which it ironically deconstructs, *Nineteen Eighty-four* is not future fiction, because, very precisely, there is no future toward which the inhabitants of Orwell's society can progress.

But nevertheless Orwell's fantasy here *is* science fiction. The model is less the technological SF of 1920s and 1930s Pulps, but rather more the theoretical and subtle SF of Stapledon. We start to have a sense of this in the final section, during Smith's detailed interrogation by O'Brien. These pages are unlike a conventional interrogation in one sense: Smith has little to say, for there is little he can say – the Party knows everything before the interrogation starts. Instead, O'Brien talks at eloquent and chilling length. Some critics see this as a flaw in the book, and admittedly it is not obvious, at first reading, why O'Brien takes such extraordinary pains over Smith. What makes Smith, a wholly insignificant individual, worthy of this special treatment? But in an important sense this novel cannot be read according to the logic of a character-novel in the nineteenth-century sense; it overthrows the logic of character. The SF-ness of the book is not its purported future setting. Rather, it is a world in which the individual has been wholly superseded by the corporate identity of – in this case – The Party. The Party is disillusioned Marxism, Orwell's grim satire on the very notion of a Homo superior; 'it' is what humanity becomes. We can read the book as a grim evolutionary romance. O'Brien himself is straightforward in regarding The Party as a new form of immortal being. 'Can you not understand that the death of the individual', he tells Smith, 'is not death? The Party is immortal' (*Nineteen Eighty-four*, p. 216). It is also all-powerful, all-knowing, a form of secular god that has grown out of humanity. This is what makes *Nineteen Eighty-four* so important for the development of the twentieth-century novel is the way it gestures towards a novel without 'character' at all. It is a much more avant-garde work than most people realise.

'Cosy catastrophe' and alien children

Brian Aldiss's memorable critical slogan 'cosy catastrophe' has taken deep root in the academic loam as a description of the dominant style of postwar British SF. While American SF (the argument goes) explored increasingly expansive possibilities of global, solar and galactic adventure, British SF projected an increasingly insular aesthetic. The situation was not as clear-cut as this suggests; but it cannot be denied that a number of significant novels were written in which middle-class Britons faced disaster or apocalypse; just as real-life middle-class Britons faced the Alice-like shrinkage of the UK from a world power into a caterpillar-sized nation. What makes these catastrophes 'cosy' is their lack of any genuine sense of threat; they fictionalise a sort of adventure playground rather than the terrors of ontological extinction: 'the essence of cosy catastrophe is that that the hero should have a pretty good time (girls, free suites at the Savoy, automobiles for the taking) while everyone is dying off' (Aldiss and Wingrove, *Trillion*, p. 254). There is some truth in this thumbnail of postwar British SF, but only some. In fact, the catastrophes of John Wyndham (1903–1969), Arthur C. Clarke (b. 1917) and John Christopher (b. 1922) are more properly unsettling than this allows, and address eschatological thematics in a fertile and profound way.

There is, for instance, little that is 'cosy' in the harsh post-catastrophe landscapes of Christopher's famine-threatened *The Death of Grass* (1956), icebound *The World in Winter* (1962) or the seismological upheavals of *A Wrinkle in the Skin* (1965). A well-regarded American version of the same thing is George Stewart's (1895–1980) only SF novel, *Earth Abides* (1949), an unusually considered and elegiac treatment of the post-disaster story. Wyndham's *The Day of the Triffids* (1951) constellates – slightly implausibly – two catastrophes: a near-universal blinding of the human race and the rampage of the man-sized walking plants with poisonous flails of the book's title. *The Kraken Wakes* (1953) is an alien invasion tale given added punch by virtue of the fact that the alien threat is never seen: they colonise our deep oceans, attack shipping and eventually start melting the icecaps to flood the world.

But Wyndham's most interesting book is not strictly 'catastrophic'. Indeed, there is a case that the 'cosy catastrophe' genre reached its peak with books concerned not with global mortality, but nativity. Wyndham's *The Midwich Cuckoos* (1957) is one of the greatest achievements of this period. All the inhabitants of the English village of Midwich fall into a trance; when they awake the fertile women are all pregnant. The children born to them grow rapidly, evidencing strange powers of telepathy and a group-mind sensibility. Ostracised and attacked for their oddness, they respond by telepathically compelling villagers to commit suicide, or to turn on one another. Realising the threat they pose, Gordon Zellaby, one of the villagers who has been overseeing the children's education at a house called The Grange, gathers them together and explodes a bomb, killing himself along with all the children. So bald a summary cannot convey the brilliantly chilling opposition of English ordinariness and the compelling oddity of the children with which the book works.

The alienness of children is one of Wyndham's more general themes, as several critics have noted. David Ketterer points out that, from *Planet Plane* (1936) through *The Chrysalids* (1955) and *The Midwich Cuckoos* (1957) to *Chocky* (1968) and a number of short stories (among them 'It's a Wise Child' and 'A Life Postponed'), Wyndham returned again and again to the trope of 'the Different Child', beings both superior and alien in the form of children, a fascination that Ketterer convincingly relates to Wyndham's own psycho-biographical circumstances (Ketterer, pp. 154–5). What makes *The Midwich Cuckoos* different from *The Chrysalids* or *Chocky* is that in this novel Wyndham does not recruit our sympathy for the children. Indeed, we can restate the central premise of the novel in these terms: under what circumstances would it be not only possible but necessary and even heroic to murder a group of 58 children? Thinking of the novel in these terms, I think, makes it a much more obviously postwar work than has been generally acknowledged. By postwar I mean, of course, post-Holocaust; more specifically, I am suggesting that *The Midwich Cuckoos* calls out to be read as an early example of the genre of holocaust fiction. In 1956, when Wyndham was writing, barely a decade had passed since the liberation of the concentration camps. Wyndham himself had been in the army at the end of the war, as had his novel's narrator, who recalls his war, 'the beaches, the Ardennes, the Reichswald, and the Rhine' (*Midwich Cuckoos*, p. 32), placing him with those same armed forces that liberated the first concentration camps. The camps were that place in the war where a people had systemised the murder of children (along with women and men) on the grounds of 'defending civilisation and the human race' against a menace from within. In one sense the novel is an attempt to enter imaginatively (via SF fable) into the mindset of a people who could commit such an act.

This final commission of mass murder is, when one considers it, a startling and even monstrous conclusion to the book. Ketterer points out that not only does Zellaby downplay his 'heroic act of suicide', implying 'that he has not long to live anyway because of a heart condition', but moreover:

> Harris [Wyndham's real name] evades the full horror of his conclusion by having it occur offstage. Furthermore, Harris has simplified or, perhaps more accurately, fudged and sidestepped the moral issues involved by ensuring that no apparent member of Zellaby's own family is among the destroyed Children. (Ketterer, p. 165)

It is striking how *racially* conceptualised the novel is. The alien children, according to army man Bernard Wescott, represent 'a racial danger of a most urgent kind'.

The Russians have destroyed one population of such children in one of their own towns by invoking the extreme step of obliterating the entire town by 'atomic cannon'. The Russian government, Wescott reports, 'calls upon all governments everywhere to "neutralize" any such known groups with the least possible delay'. Another character, Leebody, insists that killing them is not murder, since 'they have the *look* of the *genus homo*, but not the nature'. Zellaby sums up, just before his suicide bombing: 'it is our duty to our race and culture to liquidate the Children' or 'their culture ... will extinguish ours' (*Midwich Cuckoos*, pp. 191, 158, 208). These queasily evasive euphemisms ('liquidate', 'neutralise') are painfully familiar. Moreover Wyndham's description of the children seems almost egregiously Semitic: they all share 'browned complexions', 'dark golden hair' and 'straight, narrow noses', and they carry the sense of 'foreignness' about them (*Midwich Cuckoos*, p. 148). The momentum of the novel exists, we might say, in order to barrel us along the line of reasoning and emotional force, until we reach the same place that the Nazis occupied: these children must be killed to protect our race. This is not in the least to suggest that the humane, considerate Wyndham was in any way a Nazi; indeed, something the reverse. His novel is a brilliantly contrary ethical cipher: it interpellates us as a camp guard: 'here, you must not only kill these children' (or not kill, but *liquidate*, *neutralise*) 'but believe that you are doing good'. Wyndham's satire on the western ideology of 'civilisation' and 'race' still has contemporary bite. There's nothing cosy about this catastrophe.

This trope of 'the child as alien' was a popular one in the 1950s, on both sides of the Atlantic. There was, for instance, a plethora of SF stories that dramatises human children in league with aliens, or as aliens themselves: we could mention Van Vogt's *Slan*, Henry Kuttner's 'Mimsy Were the Borogoves' (1943, in which two children utilise alien educational aids to escape their parents) or Ray Bradbury's 'Zero Hour' (1947, in which human children side with alien invaders). More famously, perhaps, are a number of other books by Wyndham: particularly *The Chrysalids*, with its telepathic superchildren who represent a new evolutionary development over their limited and infanticidal parents is one such; and *Chocky*, in which a human child has an alien 'brother' living in his mind. Something, we might say, was working itself out in the speculative culture of the immediately postwar years: some widespread cultural anxiety about the nature and status of children – some apprehension, in fact, of the uncanny aspect of the child.

But the laureate of the Alien Child is British-born author Arthur C. Clarke (b. 1917). Sometimes saddled with the title 'century's greatest SF writer', Clarke's reputation is founded on his 'Hard SF' fabulation. His *Collected Short Stories* is filled with expert tales in which a technological novum is developed in ingenuous and striking ways. Some of his most enjoyable books are dramatisation of the technological demands of space travel: *The Sands of Mars* (1951) paints what was, according to the knowledge of its day, a thoroughly realistic colonised Mars; *Islands in the Sky* (1952) follows the adventures of a sixteen-year-old in a future of Earth-orbital space stations; *A Fall of Moondust* (1961) sinks a moon-bus into a sea of dust in recognisably 'disaster-movie' style, with the victims' attempts to survive and escape rendered in a perfectly plausible manner. Clarke's own background in mathematics, electronics and engineering always shapes his fiction with a studied scientific rationale. In the short story 'A Slight Case of Sunstroke' (first published in *Galaxy*, 1958), an entire football crowd focus little mirrors on an unpopular referee thereby incinerating him; with another writer the reader might suspend disbelief, grant the author licence to create metaphoric effect. With Clarke, however, the reader never doubts that he has done all the calculations and that incinerating a human with the combined pocket-mirrors of a football crowd is a scientific plausibility. Eric Rabkin notes that one single short story, 'Jupiter Five', required 'twenty or thirty pages of orbital calculations to insure that everything reported in the narrative was true to classical mechanics' (Rabkin, p. 19). This mechanistic Russellian underpinning of his fictional universe is, more often than not, handled unobtrusively by Clarke. But what is perhaps more interesting is the extent to which this rationalist, atheist, Enlightenment writer was drawn to fundamentally religious tropes of transcendence.

His early novel *Childhood's End* (1953) locates the end of humanity in its children. At the book's beginning immensely powerful aliens come to Earth, put an end to war and suffering, and establish a benign dictatorship. Under their secretive rule (the aliens do not initially show themselves to humanity; it later transpires that physically, though not ethically, they resemble devils) a new Golden Age begins. But by the novel's end we discover that these aliens have in fact been acting as shepherds, watching over Homo sapiens as a generation of children transcend physical reality and join the 'overmind'. This transcendence, although a marvellous thing, is also a catastrophic one: it marks the terminus of humanity and indeed of the Earth itself. A generation of parents watch their children become something strange and alien, unable themselves to follow. The transcendent ascension of the children destroys the Earth utterly. Edward James notes that it is 'the apocalyptic and visionary ending of *Childhood's End* is no doubt what has earned it the position in most polls since then as one of the top three of the "Ten Greatest SF Novels of All Time" '. But James also astutely picks up on the barely buried contradictions of this novel: 'it is notable (though perplexing) that Clarke's novel carried this message on the copyright page: "the opinions expressed in this book are not those of the author" ' (James, p. 78). Clarke later explained that the opinion in question was the Overlords' insistence that 'the stars are not for man'; but the fertile self-contradiction runs much more deeply through Clarke's work than that.

An obvious way of reading *Childhood's End* would be to see the initial uncertainty about the Overlords' motives resolved straightforwardly: their intention was always benign towards humanity, the transcendence of humanity into the overmind which the Overlords chaperone is a Good Thing. But as Peter Fitting notes, 'although many critics describe [*Childhood's End*] as a version of the theme of the benevolent alien, it is possible to make the opposite case, viewing it instead as example of the alien invasion narrative. The coming of the Overlords ... brings about the end of the human race and the destruction of the Earth.' As Fitting notes, reading this conclusion as a positive development 'implies that we agree "father knows best", that the Overlords are acting in our best interests even if we cannot appreciate it' (Fitting, pp. 143–4). Fitting draws parallels with British Imperialist paternalism; but it is the psychic and emotive investment in its attraction to/repulsion from 'Father' (or more precisely, with 'the Name of the Father') that, I think, gives the novel its direct impact with readers.

Something similar happens with the Starchild of the last frames of Clarke/Kubrick's *2001* (discussed below, p. 269). In that movie what is most uncanny is not the appearance of the Starchild itself, but rather the scenes that immediately precede that appearance – Bowman stranded in a Louis-Quinze drawing room at the other end of a cosmic wormhole, haunted by himself, and ageing visibly before our very eyes. In other words, the uncanny thing (the newness of the Starchild, and the change, perhaps destructive, to the Earth it forebodes) is transferred in inverted form onto Bowman, the 'parent'. It is his accelerated ageing that is the spooky thing, because it reinforces the trajectory towards death and supersession that is implied by the figure of the Child in the first place. This, in turn, says something interesting about *Childhood End*'s relationship to science fiction as a whole, and illuminates Peter Nicholl's interesting observation of what he calls 'the ACC paradox', that 'the man who of all SF writers is most closely identified with knowledgeable, technological hard sf is strongly attracted to the metaphysical, even to the mystical' (Nicholls, 'Clarke', p. 230). Clarke's particular mediation of the original determining dialectic of SF seems to bias the 'materialist' or Protestant strand; but in fact that mystic/Fantastic 'Catholic' strand is strongly present in subconscious form. But that subconscious is an orphan, and more than an orphan: a parent-slayer, an alien and monstrous embodiment of *jouissance*. One of Clarke's most famous stories is 'The Nine Billion Names of God' (first published 1953): two computer scientists are only too happy to sell an advanced and expensive computer system to a Tibetan lama, to aid in the (as far as the scientists are concerned) futile task of listing all nine billion of God's names. Once this task is completed, the lama believes, the world will end; the scientists keep their mocking condescension to themselves so as not to jeopardise their sale. They hurry away on ponies after the sale to avoid facing the lama's disappointment when his task is complete and the universe continues to exist; and they are almost at the airstrip when the world does indeed end, in the beautifully understated final passage.

'Wonder if the computer's finished its run. It was due about now.'

Chuck didn't reply, so George swung round in his saddle. He could just see Chuck's face, a white oval turned toward the sky.

'Look,' whispered Chuck, and George lifted his eyes to heaven. (There is always a last time for everything.)

Overhead, without any fuss, the stars were going out. (Clarke, *Stories*, p. 422)

The 'sense of wonder' evoked by this story stands in interesting contrast to the stellar superfluity of Asimov's 'Nightfall' (above p. 197). Both tales work in similar, yet diametrically opposite ways: where Asimov dramatised the Copernican revolution in his characters' conceptual breakthrough as to the immensity of the cosmos, Clarke achieves the same effect by, paradoxically, revealing the universe to be *pre*-Copernican, with stars more or less equidistant from a cosmically central Earth, and thereby able to wink out in unison. Edward James recalls reading Clarke's story at 'age 13 or 14': 'an almost religious sense of awe (or wonder) was created in me, as I tried to perceive the immensity of the universe, and contemplate the non-existence of God' (James, p. 107). This is exactly how Clarke works.

Golden Age 'religious' fictions

The religious dialectic that set modern SF in motion did not retreat as an increasingly secular twentieth century proceeded. On the contrary, more and more SF writers explored religious discourse: writing books about religious figures themselves (Christ in Moorcock's *Behold the Man*, 1969; the god 'Sam' in Zelazny's *Lord of Light*, 1967), or books set in religious communities, or in societies dominated by fundamentalist religious stricture. This latter category is by far the larger, and includes a number of SF masterpieces, amongst them Wyndham's *The Chrysalids* (1955), Blish's *A Case of Conscience* (1958), Atwood's *The Handmaid's Tale* (1985), Sheri Tepper's *Grass* (1989), Simmons' *Hyperion* (1989) and Gene Wolfe's work (the last four titles are discussed in chapter 13 below).

Walter Miller's (1922–1996) *A Canticle for Leibowitz* (1959) begins a few hundred years after a devastating nuclear war. Civilisation has collapsed, and in this new age of simplicity, a few books and relics of scholarship are preserved by Catholic monasteries. Over the course of the several centuries the novel traces the monastery is a still point in a brutish world. Eventually civilisation recovers to pre-war levels, but political tensions again threaten nuclear war. The Church organises a spacecraft to travel to the new colony at Alpha Centauri, and so escape the inevitable holocaust.

Miller's prose is careful and graceful, and his narrative line is expertly paced and cadenced, neither rushed nor saggy. But, presumably because of Miller's own deeply held Catholic faith, this novel takes a different approach to the more transcendental SF of some of Miller's contemporaries. They avoid the political or social application of specific religious belief; not so Miller, who in this novel runs the risk of ideological obtuseness by interpreting nuclear annihilation as a function of mankind's original sin, rather than as a political or technological dilemma. Looking back on our time from a post-holocaust perspective, a visiting scholar asks, 'How can a great and wise civilization have destroyed itself so completely?' 'Perhaps', replies a monsignor, 'by being materially great and materially wise and nothing else' (*Canticle for Leibowitz*, p. 139). In practical terms this amounts to a tacit assertion that only by injecting religion into 'materialist' political discourse can disaster be averted; a belief known as 'religious fundamentalism', or 'the Christian Right', or by a variety of other labels. Nevertheless I am tempted to argue that America's Reaganite/Bushite recent history suggests a ruler who believes in absolute religious certitude is more, not less, likely to bring on Armageddon.

What is interesting here is not Miller's own political sympathies (shared, alas, by millions) but the openness of SF to this sort of theological fiction. The tension between material and spiritual is still, in the 1950s, at the heart of SF. It is not only the issue of Original Sin; it is the call – insinuatingly attractive to those living in the monastery of SF Fandom – that a *vocation* validates an otherwise marginalised life. In Miller's short story 'Crucifixus Etiam' (1953), Manue Nanti takes a job working on a Martian terra-forming project. He intends to collect his wages and return to Earth, but the device he uses to oxygenate his blood in the thin Martian atmosphere eventually results in the withering away of his lungs; unable to return to Earth he none the less finds inner peace in the knowledge that he is part of a larger plan to make the world habitable, even though the project will take 800 years. 'Some sow, others reap,' his supervisor tells him; 'if you can't be both, which would you rather be?' Manue finds consolation in his sense of a vocation larger than himself: 'he knew now what Mars was ... an eighth-century passion of human faith in the destiny of the Race of Man.' That this secular vocation has a religious aspect is indicated by the title of the story. 'What man ever made his own salvation?' asks one of the characters (Miller, *Best of*, pp. 67–8).

Some of the very best SF from the 1950s was interrogating, in surprisingly unevolved a fashion, the theological anxieties that had given birth to the genre in the early 1600s. The American writer James Blish (1921–1975) is a case in point. The priest-protagonist of his *A Case of Conscience* (1958) is wracked by doubt because the aliens of the planet Lithia, living sinless lives in a terrestrial paradise, have no concept of 'God' or 'soul'. In a preface to a reprint of the book, Blish notes that he received letters from 'theologians who knew the present [i.e. 1958] Church position on the problem of the "plurality of worlds" ', and quotes the opinion of Gerald Head: 'If there are many planets inhabited by sentient creatures as most astronomers (including Jesuits) now suspect, then each one of such planets ... must fall into one of three categories':

(a) Inhabited by sentient creatures, but without souls; so to be treated with compassion but extra-evangelically.
(b) Inhabited by sentient creatures with fallen souls, through an original but not inevitable ancestral sin: so to be evangelized with urgent missionary charity.
(c) Inhabited by sentient soul-endowed creatures that have not fallen, who therefore,
  (i) inhabit an unfallen, sinless paradisical world;
  (ii) who therefore we must contact not to propagandize, but in order that we may learn from them the conditions ... of creatures living in perpetual grace. (Blish, *Case*, p. 9)

Blish adds the following comment to Head's assessment: 'the reader will observe ... that the Lithians fit none of these categories'. Ruiz-Sanchez, the Jesuit protagonist of the novel, comes to believe that the rational, civilised Lithians have in fact been created by the Devil in order to tempt Earth into disaster. At the novel's end (in a passage it is hard for a non-believer to read in any way other than a monstrous celebration of genocide) Ruiz-Sanchez exorcises the entire world literally *out of existence*, his rite of exorcism happening to coincide with a nuclear chain-reaction set in motion by earthly workmen exploiting Lithian natural resources. The enormous violence of this conclusion picks out a buried strain of hostility to the very notion of a plurality of inhabited worlds. It also mediates E. E. Smith's *übermensch* Pulp vision of alien holocaust, on the one hand, and C. S. Lewis's agonised theological quibbling about number-of-aliens-dancing-on-the-head-of-a-pin, on the other. Blish's novel effectively adds a new category to Gerard Head's theological analysis of alien life (which might read something like '(d) Inhabited by sentient creatures without souls who have been produced by Satan to try and damage God's creation'). But this, by pointing up other 'omissions' from the orthodox Catholic analysis, necessarily suggests another possibility: that other worlds may be inhabited by creatures that have nothing to do with, and were not created by, the God of the earthly Bible. Since this same logic can be applied to Earth as well, it is corrosive of theological certainty.

Blish's 'Cities in Flight' sequence (comprising, according to the books' inner chronology: *They Shall Have the Stars* (1956); *A Life for the Stars* (1962); *Earthman, Come Home* (1955); *The Triumph of Time* (1958)) seems, on first reading, a much more straightforward Hard SF project. Anti-gravity devices, 'spindizzies', powerful enough to lift whole parishes, enable Earth's cities literally to fly into space away from an economically exhausted and politically claustrophobic home world; and to embark on various *voyages extraordinaires* through deep space. Anti-ageing technology allows Blish to keep continuities of character across the very long time-scales involved. The early novels are dominated by a rather grim vision of the way unfettered free-market economics (which govern the future galaxy; the cities must trade to survive) can lead to pinchingly persistent hardship even for the hard-working. But having navigated out of our troublesome galaxy altogether and settled on a likely planet, the city of New York (Blish's main protagonist) and its hero-mayor Amalfi find themselves faced with a transcendental rather than an economic crisis. *The Triumph of Time* matches a Shelleyan pessimism to its Shelleyan title: the universe, it transpires, is dying much sooner than anticipated, and has only a few years left. Rather than see everything vanish utterly, Amalfi and a few others work out a way to position themselves at the epicentre of the cosmic catastrophe, such that – and despite their own deaths – there will be matter from which a new universe can be born. Most of the novel is clogged with Hard SF techno-babble seminar room disquisitions, and equations of the order:

$$\frac{d^2G}{dx^2} + \frac{d^2G}{dy^2} + \frac{d^2G}{dz^2} = a^2\frac{dG}{dt}$$

But at the end, the book reverts into a more recognisably theological creation fable. As Amalfi and his companions, floating in a literal nothingness, press the button on their spacesuits that will turn them into the big bangs of wholly new cosmoi, we are reminded that the only thing they take with them (and which will, presumably, shape the deep structure of the new universes) is love. 'I don't feel deprived,' Amalfi tells the others, in the face of their imminent deaths. 'I loved you all. You have my love to take with you, and I have it too.' 'It is', reflects somebody else, 'the only thing in the universe that one can give and still have' (Blish, *Cities*, p. 595). The whole technological sense of wonder Golden Age text bears down on this final revelation of love as the logos. *Cities in Flight* ends vatically with the sentence: 'Creation began'. It is a testimony to Blish's talents that this last statement carries the weight of all that has gone before, rather than being (as it might have been) an irritating affectation, after the sorts of romantic comedy movies that replace their 'The End' credit with one that reads 'The Beginning …'

Love in this quasi-theological sense (together with its shadow, loss) is at the heart of the considerable literary achievement of Ray Bradbury (b. 1920). His most famous title, *The Martian Chronicles* (1950), is a collection of loosely linked stories that detail a future human colonisation of Mars not in terms of the nuts and bolts (technological or social) of such an undertaking, but as an almost dreamlike inhabitation of empty spaces still haunted by the enigmatic presences of the original Martians. Written in a beautifully judged prose that is poetic without ever being pretentious or showy, the book reads its imagined future via the remembered small-town American past of Bradbury's own childhood. There is this blending of nostalgic solace and childhood night terrors in all Bradbury's best work; certainties evaporate, and alienness is simultaneously externalised and internalised. *The Illustrated Man* (1951) also collected linked stories, this time encapsulated in the magic tattoos of the titular figure. The short novel *Fahrenheit 451* (1953) seems to be a satire on the totalitarian desire to suppress free thought: in Bradbury's future, citizens live in fireproof cities and the job of the 'fireman' protagonist is to burn books; a task he comes, as he grows in knowledge, to regret. But to describe the book in these terms does nothing to capture the weird, under-the-skin oddness of Bradbury's poetry. Rather than being a novel in the *Nineteen Eighty-four* mode of outraged anger, this is a gentle, moving fable about the ways, and the extent to which, it is possible to transcend one's quotidian limitations and become more than one was. Literature, in the fullest sense, represents for Bradbury this potential for growth. It might seem odd to bracket Bradbury here with more obviously religiously inclined writers like Miller and Blish. Yet, running through his fables, there is an almost millenarian yearning for escape. In 1979 Bradbury described space travel as 'exciting and soul-opening and as revelatory'. The latter two terms seem to me the important ones for his writing ('because', he goes on, 'we can escape, we can escape and escape is very important, very tonic, for the human spirit'; quoted in Disch, pp. 72–3). The 'escape', and the terrors, of his writing are linked in this way, I think, to a distinctively American narrative of the soul.

European science fiction of the 1940s and 1950s

Continental Europe suffered acutely during the war years 1940–45, and the SF of this period is very much a product of those upheavals. In the French writer René Barjavel's (1911–1985) *Ravage* ('Devastation', 1943), a futuristic high-tech world suddenly finds its electricity cut off and collapses into anarchy. Out of the disaster a new, old-fashioned and, the book says, much superior agrarian culture develops. Indeed, the book's anti-technological thesis is developed in a bluntly superstitious manner: 'tout cela', says one character, 'est notre faute. Les hommes ont libéré les forces terribles que la nature tenait enfermées avec precaution ... ils ont nommé cela le Progrès. C'est un progrès accéléré vers la mort' ('it's all our fault. Man has freed terrible forces that nature had safely hidden away ... they called this *Progress*, but it's only a progress towards a rapid death!' Barjavel, *Ravage*, p. 85). The pastoral idyll with which the book closes is also rather self-deluding and banal, a phallo-centric fantasy of heterosexual gratification and egotism. Ravage's patriarch enjoys as many women as he pleases ('les générations nouvelles', we are told, 'ont accepté la polygamie comme une chose naturelle'; Barjavel, *Ravage*, p. 298) as he breathes the clean air and eats the good food, sending his many sons out to colonise 'un monde vide', a world picked clean. This vision is a sort of anti-Futurism; dis-daining a Marinettian love of the machinic whilst embracing a similarly anti-life enthusiasm for the great cull. Less totalised and rather more playful is Barjavel's time-travel novel *Le voyageur imprudent* ('The Imprudent Traveller', 1944). Two scientists experiment with travelling through time; when one dies the other attempts to go back and undo the accident, getting tangled up in the 'grandfather paradox'. Barjavel enthusiasts sometimes claim this as the first treatment of this famous theme, although there are several prior examples in American Pulps.

The Bohemian-born writer Franz Werfel (1890–1945) escaped to California during the Second World War. His last novel was *Stern der Ungeborenen* ('Star of the Unborn', 1946), a far-future utopian mediation on alienation and suffering. Resurrected from his own death, the protagonist (who shares the author's name) is shown a far-future society of tremendous technological achievements; but by con-quering suffering humankind has sabotaged its chances for salvation. Werfel's conclusions are not dissimilar to those of Barjavel's *Ravage*, although he reaches them in a much less crude manner.

In the intelligent space-operas of the Soviet Russian writer Ivan Yefremov (1907–1972) the 'official' Soviet view of alien life is clearly articulated: any alien race sufficiently advanced to have mastered interstellar flight must be Communist: for Capitalism (inherently divided among itself by competition over the means of production) could never marshal the enormous collective effort required for so vast an achievement. The aliens encountered in *Cor Serpentis (Serdtse Zmei)* ('Serpent's Heart', 1959) fit this bill. Now that the twentieth century is over and the dreams of that century's SF writers for the rapid colonisation of space have dis-solved into nothing more than a few commercial satellites and a few under-funded robot probes, we may wish to take time to reflect on whether Yefremov was not right. It is certainly difficult to say whether a non-Marxist would find *Tumannost' Andromedy* ('Andromeda', 1958) as beguiling as a reader with Marxist sympathies (such as myself) does: but its compelling portrait of a socialist Earth in the fourth millennium exploring space and making contact with aliens from the Andromeda galaxy is enormously inspiring and powerful.

Films

Historians of SF cinema often dismiss the 1940s as a dud decade, characterised by shabby, cheaply made cash-in sequels, which stumble into horror and comedy without the delicate touch of their progenitors: films such as *The Ghost of Frankenstein* (Erle Kenton, 1942), *Frankenstein Meets the Wolf Man* (Roy William Neill, 1943) and the unfunny *Abbott and Costello Meet Frankenstein* (Charles T. Barton, 1948). However, some populist serials or one-off movies carried themselves off with panache and a febrile inventiveness. Examples include the barmy miniaturised humans adventure *Dr Cyclops* (Ernest B. Schoedsack, 1940), or the *King Kong* imitation *Mighty Joe Young* (Ernest B. Schoedsack, 1949). But it is probably true to say that it was not until the 1950s that we can meaningfully talk of Golden Age SF cinema.

Destination Moon (Irving Pichel, 1950), based on Heinlein's novel *Rocket Ship Galileo* (1947), is well regarded by fans for its 'realistic' portrayal of space travel; Heinlein worked a technical consultant on the project. But it is a dull film, with unmemorable characters and a treacle-slow storyline. In keeping with Heinlein's libertarian political views it is private enterprise rather than governments that gets the rocket aloft. The well-publicised development period created a deal of public anticipation (the posters for the film carried the tag-line 'TWO YEARS IN THE MAKING!'), and an unscrupulous cash-in film was hurried through production to steal its thunder. Indeed, a legal challenge meant that *Rocketship X-M* (Kurt Neumann, 1950) had to include a disclaimer sent to exhibitors stating 'this is not *Destination Moon*'. *Rocketship*'s story abandons scientific credibility (a Moon-bound rocket flies off-course and lands on Mars, discovering a post-atomic war society of troglodytic humanoids); but is nevertheless a better SF film than *Destination Moon* because it understands its kinetic, restless idiom better.

Throughout the 1950s film-makers experimented with the balance between po-faced scientific plausibility and Pulpishly belief-stretching high adventures. On the former side of the fence is the end-of-the-world movie *When Worlds Collide* (Rudolph Maté, 1951); a movie whose earnest elaboration of the attempts to build a space ark to carry a few human survivors to a better world is rendered with a gaudy deadness; although the spaceship (designed by SF artist Chesley Bonestall (1888–1986)), perched atop its ski jump-like launch pad, still has iconic power. On the other, there were films like *The Thing from Another World* (Christian Nyby, Howard Hawks, 1951), a very effective thriller, in which a vegetable alien who must feed on human blood to survive terrorises an isolated Arctic research station. Eventually the humans electrocute the creature, and the film ends with a line that has become a slogan for SF Fans and UFOlogists: 'Watch the skies! Keep watching the skies!' The movie version of Wells's classic *The War of the Worlds* (Byron Haskin, 1953) also plumps for a populist, spectacular idiom and succeeds brilliantly in rendering the Martian machines and the destruction they create: of the film's $2 million dollar budget, $1.4 million was spent on special effects.

The Day the Earth Stood Still (Robert Wise, 1951) strikes the ideal balance between 'serious' and 'Pulp'. The humanoid alien Klaatu (played with dignity by Michael Rennie) and his giant silver robot land a huge flying saucer in Washington DC to warn the Earth that we must change our belligerent and self-destructive ways. He is shot accidentally by a jittery soldier, but recovers, later living among ordinary people under the Christ-like pseudonym 'Carpenter'. After performing 'miracles' to demonstrate his power (for instance, stopping all electricity on the world for several hours) he is killed by humanity. But, again like Christ, he returns to life. Here he reveals that it is the robot rather than the humanoid, who is the real power; a member of a sort of interstellar police force which will 'reduce this Earth of yours to a burnt-out cinder' if we do not give up our violence. Those who consider this threat out of keeping with the otherwise liberal open-heartedness of the film should remember that Christ was also given to threats of Jerusalem's temple being utterly demolished, or the world itself destroyed and the unrepentant sent to Hell. The theological allegory of the film detracts, for some, from the nicely understated direction of this still classic tale; but its presence is symptomatic not so much of contemporary religious discourse as of the theological subtext underpinning the genre as a whole.

The later 1950s saw a distinct boom in 'monster movies', in which either specific humans or humanity as a whole were menaced by various hideous or bestial creatures, most of which are spawned by nuclear testing and rogue radiation. One of the most influential was the Japanese *Gojira* (Inoshiro Honda, 1954) mistakenly rendered into English as 'Godzilla' (there have been 27 sequels to date; the most recent being *Gojira: Fainaru uôza* ('Godzilla: Final Wars'), Ryu Kitamura, 2004). This dinosaur-like giant creature, awoken by atomic testing, rises from the ocean depths to attack Tokyo. Its name, apparently, is a portmanteau of 'Gorilla' and *kujira* (which means 'whale'). Many critics have followed contemporary audiences in reading the film as an allegory of the nuclear destruction of two other Japanese cities less than a decade earlier; but across the sequence of 'Gojira' films the status of the monster is harder to pin down. Often he becomes an ally of man against some other monstrous giant: Mothra, a giant moth; Ebirah, a colossal crab; Hedora the 'Smog Monster', and so on. The American film *Them!* (Gordon Douglas, 1954) also uses nuclear testing as the trigger for gigantic mutation, this time of ants in the Mojave desert. The whole is handled in a low-key, non-hysterical manner by Douglas, which leads to some powerful moments. Less effective but still interesting is *The Creature from the Black Lagoon* (Jack Arnold, 1954) (its sequels are *Revenge of the Creature* (Jack Arnold, 1955) and *The Creature Walks Among Us* (John Sherwood, 1956)), with its iconic human-sized monster swimming menacingly through the waters or lumbering slimily over the land. In *Tarantula* (Jack Arnold, 1955) scientists invent a compound that causes deforming acromegaly in humans but makes animals grow to enormous size; they aim to solve world hunger, but instead create a colossal spider that escapes to the desert preying on people and cattle. It is eventually destroyed by the military. But I shan't go on: a complete listing of 'monster movies' would take up a whole chapter; and most of the output is inconsequential. In the words of Peter Nicholls:

> The boom climaxed with a veritable eruption of monster movies in 1957 … the cascade continued in 1958 with variations on the theme becoming more knowing … but generic rigidity soon degenerated into decline and fall. More monster movies were made 1959–62 than in the whole of 1951–8, but almost without exception they were low-budget exploitationers of no real quality aimed at the teenage drive-in market. (Nicholls, 'Monster Movies', p. 817)

A number of big-budget special effects heavy titles were made in the second half of the decade. *This Island Earth* (Joseph Newman, 1955) is a weirdly unbalanced film. Two-thirds of it is a well-paced, coolly realised mystery, as eminent scientists recruited to work in a secret laboratory in Georgia try to uncover who their mysterious employers are. It transpires they are aliens from the dying world of Metaluna, who are hoping human ingenuity can come up with a way of saving their world. But then the movie shifts gear; the handsome male and beautiful female protagonists are whisked off to Metaluna; they battle with a hideous mutant monster; there are explosions as the planet dies and the humans escape in the nick of time. It is as if this film shows more glaringly than any other the faultline between serious pretension and the crowd-pleasing Pulp extremism that determined so much SF cinema from the period.

Forbidden Planet (Fred McLeod Wilcox, 1956) remains one of the best-loved SF films from this period. Despite artistic pretensions (it deliberately recasts Shakespeare's *Tempest* in an SF idiom) it never sacrifices its Pulpish energy and verve. An Earth spaceship visits the alien world of Altair on which Earth scientist Dr Morbius lives with his daughter Altaira. The Altairans have long gone, but Morbius is (secretly) using their mind-amplification technology, unleashing what the script rather quaintly calls 'Monsters from the Id!' that stomp about invisibly leaving huge taloned footprints and occasionally killing people.

More obviously a product of that socially paranoid period in American recent history known as the 'McCarthyism' is *Invasion of the Body Snatchers* (Don Siegel, 1956). Dr Miles Bennell (Kevin McCarthy) discovers that alien *doppelgangers* are killing and replacing all the inhabitants of his small town with emotionless but otherwise flawless copies of humanity. The creeping paranoia of the film builds to a powerful crescendo; the first cut of the film ended with Bennell screaming at cars on a highway like a lunatic 'You're next, you're next!' (to the detriment of the film, the studio insisted on a more positive ending).

Also noteworthy is *The Incredible Shrinking Man* (Jack Arnold, 1957), a film whose sensationalist title does not do justice to its genuine and acute poignancy. It is the story of Scott Carey, who begins to shrink after being exposed to a radioactive mist and who gets smaller and smaller. In earlier sequences he finds temporary happiness with a female dwarf, but as the relentless process makes him smaller and smaller he ends up alone, battling a spider, and facing ultimate annihilation. The slow-building despair of this remarkably subtle film is unlike almost anything else in 1950s popular cinema, an effect not overturned by a deflating final invocation of God (to whom, we are told, size does not matter). Infinitely sillier, though still memorable, is *The Fly* (Kurt Neumann, 1958), in which a scientist inadvertently swaps heads and an arm with a fly via a prototype teleporter accident. That the scientist is lumbered with a gigantic fly-head (which he covers, discretely, with a cloth) rather than a proper fly-sized one, or that the fly gets a miniaturised human head still capable of speech (in one nicely nightmarish moment we see the fly shouting inaudibly 'Help me! Help me!' as it is trapped on a spider's web) makes, of course, no sense. But there is, for all its silliness, a Poe-like rightness to its nightmare.

Visual artists

Just as SF cinema was, with an increasingly sure touch, creating visual icons of enduring rightness, power and beauty, so the artists of the SF world were developing. Perhaps the most significant contribution in this sense was made by Golden Age SF comics (see below); but the many SF magazine titles and, increasingly, book publications that generated commissions for artists also represented a forum for aesthetic experimentation and achievement. Edmund Alexander (known as 'Emsh'; 1925–1990) created a wide variety of visual styling for his extensive magazine cover work; from quasi-realist brushstroke art to near-abstract pieces (such as the Ben Nicholson/Max Ernst-like representation of spacemen in a capsule which adorned the cover of *Galaxy Science Fiction*, August 1951; a beautiful image that creatively dissociates the visual field into flowing compartments of red, purple, grey and black). Frank Kelly Freas (1922–2005) won eleven Hugos and many other awards for his deftly painted images. Freas' work is always inventive, often witty and only occasionally over-coloured or too gaudy. Several of his images have a wide recognition factor even today; an image of a giant metal robot holding a broken human body in its huge hand, with a look of compassionate grief on its metal face, is unusually spare in its composition and surprisingly powerful (it was originally an *Astounding* cover from 1953; it was recycled as the cover to an album by the pop group Queen in the 1970s). More technically gifted, although rather less lively, were the nearly photographic renderings of Chesley Bonestell (1888–1986); images in *Life* magazine and elsewhere in the late 1940s of famous sites in the solar system created a stir with their verisimilitude (they were collected in the book *The Conquest of Space*, 1950). Richard Powers (1921–1996) adopted a more deliberately surrealist visual idiom (he described himself as 'an American surrealist'), one which works brilliantly for the estranged visions of SF. His book and magazine covers rarely related directly to what they purported to illustrate; the logic was associative, dream-like and articulated a subconscious rightness.

Comics

The 1940s and 1950s was the great age of superhero comics. The enormous and continuing popularity of Siegel and Shuster's Superman character led to a large number of superhero comics, most of which are still being published today. These superheroes have enjoyed significant cultural longevity and penetration, partly through a largely successful latter-day translation into the cinematic idiom.

They came, broadly speaking, in two kinds. First, there were superheroes whose powers or abilities are greater than human – the contemporary demigods and action heroes like Superman himself. The most notable of these action *Übermenschen* is 'Captain Marvel', who first appeared in *Whiz Comics* in 1940, but was soon given his own publication, *Captain Marvel Adventures* (1941–53), and thereafter appeared in various formats. Young Billy Batson is able to transform himself into the superhero 'Captain Marvel' by uttering the magic word 'shazam!' (an acronym derived from 'Solomon-Hercules-Atlas-Zeus-Achilles-Mercury'). Legal pressure by the owners of the copyright on Superman resulted in suspension of the title in 1953. In Britain, where Captain Marvel adventures were reprinted with great success after the Second World War, this suspension compelled the creation of a copycat British character named 'Marvelman' (who used the new magic word 'kimota!' – 'atomic' backwards, more or less). Of the other human-to-superhuman transforming characters, the most SF-oriented were:

- The Poseidon-like 'Sub-Mariner' (created by Bill Everett), a half-human, half-Atlantean mutant, and the 'Human Torch' (created by Carl Burgos), an android that burst into flame when exposed to air (both from 1939).
- The nationalistic icon 'Captain America' (from 1941; created by Joe Simon and Jack Kirby) with his Stars-and-Stripes costume, who began by fighting Nazis, and who was transformed from a sickly youth into a superhero by a combination of 'super soldier serum' and irradiation by 'vita rays'.
- 'The Flash' (from 1940; created by Gardner Fox and Harry Lampert), a college athlete called Jay Garrick who, on inhaling 'hard water vapor', becomes a Hermes-like super-speedster.
- The Aladdin-like 'Green Lantern' (1940; created by Martin Nodell and Bill Finger), an engineer named Alan Scott whose possession of the titular lantern-shaped power source and a 'power ring' charged by it, gave him the ability to shape anything he imagines into concrete forms (and a power over any substance except wood). The Green Lantern was reinvented in a more explicitly SF idiom in 1959 by John Broome and Gil Kane as test pilot Hal Jordan, who receives the ring of power from a dying alien member of the galaxy-protecting 'Green Lantern Corps'.

As mentioned in the last chapter, the mythological and mythopoeic overtones of all of these characters are unmissable (even the Human Torch, who is not reminiscent of a specific god, is an anthropomorphic personification of an element, fire). Yet even more interesting for our purposes is the manner in which they develop. When, for instance, Captain Marvel was transmuted to Marvelman in Britain his new magic word, his new mantra, is no longer derived from ancient gods and heroes, but instead is a pun on atomic energy, the 'new god' of the post-war period. Equally fascinating is the case of The Green Lantern's reinvention as an SF hero and a member of the Green Lantern Corps. Conceived by John Broome (one of the first postmodernist comics creators), post-Sputnik, at the dawn of the Silver Age of comics and SF, the Corps is a direct homage to E. E. 'Doc' Smith's Lensmen. Yet by having the members of this intergalactic police force wield the tremendous 'power rings' (essentially, Aladdin's lamps in ring form), Broome not only transforms an Aladdin rip-off into a Smith rip-off, thus replacing an old 'religion' (mythology) with a new one (SF), but projects the superhero as a creature of Will-to-Power on the grandest scale, the iconographic scale. Aided by Gil Kane, an amazingly versatile artist who was granted the honorary title 'imagineer' late in his career, Broome cemented the true, iconic appeal of comic superheroes and visual SF. Just as each power ring's capabilities was dependent only on the sheer will power of the bearer, SF comics' potential for visual sense of wonder was limited only by the imaginations of comics' creators. In visual SF, anything can happen, because it is a product of artistic Will-to-Power.

On the other hand were those superheroes who were never anything more than human, although humans trained and equipped to an advanced degree. The prototype for these is Doc Savage (whose adventures ran from 1933 to 1949) the prose Pulp hero who had trained his body and mind to near-perfection, but possessed no supernatural powers. But the key icon here is Batman, an aggrieved urban vigilante created by Bob Kane and Bill Finger who first appeared 1939 (the *Batman* comic began in 1940). William Moulton Marston (1893–1947) and Harry G. Peter (?1890–1958) created 'Wonder Woman' (who first appeared in *All Star Comics*, 1941) as a deliberately female corrective to the masculinity of Superman. Hoping to provide women with a positive role model, Marston and Peter presented Wonder Woman as an 'Amazon' armed with various items of technology (special bracelets which she uses to deflect bullets, for instance) but without any magical abilities, at least in her earlier incarnations.

Why have Anglophone SF comics been so overwhelmingly dominated by superheroes from the 1940s to the present day? The argument of the present study is that this vogue expresses in a popular cultural idiom one of the root concerns of SF: the role of the saviour and the status of atonement in a modern, scientific post-Copernican cosmos. In other words, the 'superhero' trope expressed concerns that, for local, cultural reasons (*angst* at the bloodletting of the Second World War, fear at possible nuclear catastrophe, guilt at increasing material affluence, and uncertainty as to what form a saviour would take to relieve these worries), reproduced the same cultural anxieties that lay at the origin of SF as a genre. To put it another way, SF's longstanding mediation of the dialectic between 'material' (scientific) and 'spiritual' (religious) understandings of the cosmos is also behind the various inflections of the superhero in 1950s comic art: some are more material, the human component of the saviour archetype; some are more supernatural, gifted with quasi-magical powers. This is not to say, exactly, that these superheroes are mere 'types' of Christ (rather, I am suggesting that they responded to similar cultural anxieties to those that also generate public receptiveness to religious discourses), although the parallels between the key superheroes and the Christian archetype are sometimes hard to ignore. Roger Sabin, the most respected historian of comics, quotes Siegel's comments on the creation of Superman: 'all of a sudden it hits me – I conceive a character like Samson, Hercules and all the strong men I heard of rolled into one', adding:

> It is clear that the writer [Siegel] must also have been aware of the analogies with Jesus: Superman was similarly a man sent from the heavens by his father to use his special powers for the good of humanity. (Sabin, p. 61)

In place of a largely exhausted religious idiom (a divine saviour combats sin and atones for our inherent sinfulness with a sacrifice) SF comics provided a new popular cultural logic that addressed the underlying anxieties (a superhero battles various incarnations of evil, usually only defeating them at great cost to himself).

This did not happen overnight. After flourishing in the 1940s and early 1950s comics were losing ground by the later decade. But in the early 1960s, which ushered in the 'Silver Age' of comics, one publisher (Marvel) and two authors – Jack Kirby (1917–1994) and Stan Lee (b. 1928), perhaps the two most significant figures in twentieth-century comics – enjoyed new success. In Sabin's words, 'the success of Marvel had the effect of kick-starting the entire industry back into life' (Sabin, p. 74). In short order Kirby and Lee created the most iconic comic figures of all (after Superman). Marvel's first big success was *The Fantastic Four* (1961–present), a team of elemental superheroes including the super-pliable, stretchable hero Mr Fantastic (the element of water), a revamped version of the Human Torch (fire), the Thing (a monster seemingly composed of orange rocks, with enormous strength, thus an analogue for earth) and the Invisible Girl (air). In *Spider-Man* (1962–present), created by Stan Lee and the artist Steve Ditko (b. 1927), a young man bitten by a radioactive spider and given certain arachnoid special powers, Marvel posited the first psychologically 'realistic' superhero, a youth as volatile, sensitive and insecure as any person his age, whose superpowers come at a personal cost and always end up causing him as many problems as they solve. Spider-Man is, at point of writing, the world's most famous superhero, largely due to the success of two excellent film adaptations (*Spider-Man* (dir. Sam Raimi, 2002); *Spider-Man 2* (dir. Sam Raimi, 2004); the same director's *Spider-Man 3* is slated for release in 2007). Two further Kirby/Lee creations have also been successfully filmed: *The Incredible Hulk* (1962–present) was a comic reworking of Stevenson's *Jekyll and Hyde* combined with the Karloff incarnation of the Frankenstein Monster, for the Hulk, the huge, green, monstrous, yet innocent monster of the id lurking inside mild-mannered scientist Bruce Banner, does not express the brutish sinfulness of Stevenson's 'unaccommodated man'. X-Men (since 1963) was another team of superheroes, avatars of a new evolutionary stage in human development in which mutants with special powers are common. The most obviously messianic Kirby/Lee creation is The Silver Surfer (first appearance 1966; series from 1968). The series posits a godlike, planet-devouring entity called Galactus. Norrin Radd, an alien from the idyllic world of Zenn-la, offers himself as a sacrifice to save his home world from Galactus' depredations; Galactus accepts and transforms Radd into the Silver Surfer. After serving him for an unspecified time as a scout of worlds suitable for consumption, the Surfer comes to Earth, where (moved by the nobility of some of the creatures he encounters) he rebels against Galactus, fighting to prevent him from devouring our planet. In this he is successful, but as punishment he is confined to Earth. The tone of the series, as is often the case with Lee's rather wordy and operatic idiom, treats this fanciful scenario with a high seriousness; and at its best does express something important about the role of intermediaries between Earth and cosmic forces that also recapitulates the core dialectic of SF.

European comics

Comics also enjoyed a boom in Europe. In 1950s Britain the *Eagle* ran the very popular 'Dan Dare: Pilot of the Future' comic strip (created by Frank Hampson,

1917–1985). Lantern-jawed, stiff upper-lipped Dare, together with his subaltern Digby (and, depending on the adventure, various other characters) flew about the solar system and beyond, having a variety of adventures. In particular they often clashed with an 'evil genius' alien, 'the Mekon', a mega-cephalic, micro-corporeal green-skinned Venusian who floated about on what looked like a giant dinner plate. Hampson and his team of artists used photographs and scale-models to achieve a high degree of clear-line, bright-coloured verisimilitude; and it is the overall look of this comic, rather than its often rather derivative storylines, that is its most significant achievement. Dan Dare also enjoyed popular success as a British radio serial (broadcast weekly on Radio Luxembourg, 1953–56).

In France comics such as *Espace* ('Space', 1953–54) and *L'An 2000* ('2000 AD', 1953–54) proved relatively short-lived. But postwar France was to become one of the most important forums for the creation of comics, and later graphic novels. A distinctive home-grown aesthetic was developed, in part because of a hangover from the German occupation of the 1940s: 'a Nazi edict banning American comic characters like Mickey Mouse and Flash Gordon was never repealed, giving a home-grown tradition the chance to flourish' (Rambali, p. 145). Indeed, after a period of postwar depression Europe emerged into the 1950s keen to establish a distinctive home-grown tradition of SF.

Fantax ran from 1946 into the late 1950s, and was created by Pierre Mouchotte (1911–1966). Fantax was a superhero, derivative of certain American models; supposedly an English nobleman with the frankly improbable name Lord Horace Neighbour who dons a hood by night and fights crime.

From 1946 to 1947 the magazine *Coq Hardi* ran *Guerre à la Terre* ('War against Earth'), an intricately drawn black-and-white comic in which megacephalic Martians invade our world and are repelled by an heroic pan-European resistance. In a contemporary touch the Martians ally themselves with disaffected Japanese soldiers. *Les Conquérants de l'espace* ('The Conquerors of Space') ran monthly from 1953 to 1964 in *Meteor* magazine, detailing the proto *Star Trek*-esque adventures of the crew of the starship *Space Girl*. It was popular enough for several imitations (such as 1958's *Aventures en l'espace*) to attempt cash-in publications. The magazine *Super Boy* was founded in 1949, although the first adventures of the hero Super Boy (not related to DC's Superman) did not appear in the magazine until 1958. But thereafter the strip, drawn in American style by Félix Molinari (b. 1931), became very popular, running into the 1980s.

French radio hosted the long-running and very popular serial *Signé: Furax* from 1951 to 1960 (with gaps). This inventive comic drama follows the bizarre adventures of super-villain Furax, who later changes sides and joins the forces of Good, where he (for instance) thwarts an attempt at world domination via a mind-control ray and travels by rocket ship to the planet Asterix. One series concerned alien body snatchers that lived inside Swiss cheese, such that anyone eating the cheese became infected with the aliens, which gives a sense of the flavour of the work. The serials were all novelised by the scriptwriters Pierre Dac (1893–1975) and Francis Blanche (1921–1974), and further spin-off novels were written by others in the 1970s.

In Japan, Osamu Tezuka (1928–1989) began drawing the comic serial *Tetsuwan Atomu* ('Mighty Atom', translated in the West as 'Astroboy') for the juvenile magazine *Shonen* in 1952. The titular protagonist is a boy-robot with the power of flight who, Pinocchio-like, yearns to be human, acting as a miniature metal Superman rescuing people from monsters, mad scientists and industrial accidents. Massively popular in Japan, Tezuka supervised the creation of a TV animated series *Tetsuwan Atomu* (1963–present) which helped spread the popularity of the icon.

Gold

For many readers, certainly of a certain generation, 'Golden Age SF' is the real thing, the heart of SF, the paradigm to which definitions of the genre should adhere. It might be thought arguable to what extent these grandiose tech-drunk space operettas and luminous stories-of-ideas can be seen as exemplifying what this present study has argued are the deep roots of SF: the dialectic between the scientific-materialist and the religious-spiritual discourses. Yet, as Alexei and Cory Panshin have argued, the main impulse of Golden Age SF was a 'quest for transcendence'. Campbell's later slide towards mysticism, Dianetics, telepathy and all that bag-and-baggage was not an aberration; it is something included within the fundamentally dialectical logic of SF itself.

References

Aldiss, Brian, with David Wingrove, *Trillion Year Spree: the History of Science Fiction* (London: Gollancz 1986)
Anderson, Poul, *Tau Zero* (1970; London: Gollancz 2000)
Asimov, Isaac, *The Caves of Steel* (1954; London: Grafton 1987)
Asimov, Isaac, 'The Bicentennial Man' (1976), in *The Complete Robot* (London: Grafton 1982)
Barjavel, René, *Ravage* (1943; Paris: Folio 2000)
Bester, Alfred, *Tiger! Tiger!* (Harmondsworth: Penguin 1974)
Blish, James, *Cities in Flight* (1956–62; London: Gollancz 'SF Masterworks' 1999)
Blish, James, *A Case of Conscience* (1958; London: Gollancz 'SF Masterworks' 1999)
Bukatman, Scott, *Terminal Identity: the Virtual Subject in Postmodern Science Fiction* (Durham, NC and London: Duke University Press 1993)
Clarke, Arthur C., *The Collected Stories* (London: Gollancz 2000)
Clute, John, *Science Fiction: the Illustrated Encyclopedia* (London: Dorling Kindersley 1995)
Disch, Thomas, *The Dreams our Stuff is Made of: How Science Fiction Conquered the World* (New York: Simon and Schuster 1998)
Eagleton, Terry, *The Ideology of the Aesthetic* (Oxford: Blackwell 1990)
Fitting, Peter, in David Ketterer, op. cit.
Gifford, James, 'The Nature of "Federal Service" in Robert A. Heinlein's *Starship Troopers*', www.nitrosyncretic.com/rah/ftp/fedrlsvc/pdf (accessed December 2004)
Huntington, John, *Rationalizing Genius: Ideological Structures in the Classic American Science Fiction Short Story* (New Brunswick, NJ: Rutgers University Press 1989)
James, Edward, *Science Fiction in the Twentieth Century* (Oxford: Oxford University Press 1994)
Ketterer, David, ' "A part of the … family(?)": John Wyndham's *The Midwich Cuckoos* as Estranged Autobiography', in Patrick Parrinder (ed.), *Learning from Other Worlds: Estrangement, Cognition and the Politics of Science Fiction and Utopia* (Liverpool: Liverpool University Press 2000), pp. 146–77

Nicholls, Peter, 'Arthur C. Clarke', in John Clute and Peter Nicholls, *Encyclopedia of Science Fiction* (2nd edn., London: Orbit 1993), pp. 229–32

Nicholls, Peter, 'Monster Movies', in John Clute and Peter Nicholls, *Encyclopedia of Science Fiction* (2nd edn., London: Orbit 1993), pp. 816–18

Panshin, Alexi and Cory Panshin, *The World Beyond the Hill* (Los Angeles: Jeremy P. Tarcher 1989)

Rabkin, Eric, *Arthur C. Clarke* (2nd edn., Mercer Island, WA: Starmont 1980)

Rambali, Paul, *French Blues: a Journey in Modern France* (London: Minerva 1989)

Ruddick, Nicholas, 'Out of the Gernsbackian Slime: Christopher Priest's Abandonment of Science Fiction', *Modern Fiction Studies* 32 (1986) 1: 43–52

Sabin, Roger, *Comics, Comix and Graphic Novels: a History of Comic Art* (London: Phaidon 1996)

Suvin, Darko, *Positions and Suppositions in Science Fiction* (London: Macmillan 1988)

Vance, Jack, *Emphyrio* (1969; London: Gollancz 'SF Masterworks' 1999)

Westfahl, Gary, *The Mechanics of Wonder: the Creation of the Idea of Science Fiction* (Liverpool: Liverpool University Press 1998)

Wyndham, John, *The Midwich Cuckoos* (1957; Penguin 1975)

11
The Impact of New Wave Science Fiction 1960s–1970s

1957 is a watershed of sorts for SF: the successful launch of the artificial satellite Спутник (*Sputnik*, the Russian for 'satellite') by the Soviet Union turned space travel from an imagined future to a present reality. John Clute puts it well:

> There may have been a time, in the morning of the world, before Sputnik, when the empires of our SF dreams were governed according to rules written out in the pages of *Astounding*, and we could all play the game of a future we all shared, readers, writers, fans ... But something happened. The future began to come true. (Clute, *Look*, p. 17)

The trajectory of man's space adventure traced what Thomas Pynchon would later call 'gravity's rainbow', the path of a ballistic rocket up, elliptically over and down again. In the late 1950s, and especially with the manned orbital missions and the NASA Apollo mission to the moon in 1969, there was enormous excitement and hope; many people, particularly in the SF community, nurtured on the expansive dreams of Golden Age Fantasy, did believe that the future was coming true. But it did not. By the 1970s it had become clear that space travel was (whisper it) a bit dull. Funding bled away; the Apollo programme was curtailed; space travel shrunk to only commercial and military satellites, augmented by the occasional robot probe. No amount of political barnstorming – as in President George W. Bush's 2004 promise of a manned mission to Mars, a promise few in the world of space travel believe – can recapture that initial transcendent excitement.

Reality let SF down. Golden Age optimism became harder and harder to maintain as the 1970s went on. The science-fictional response to this was complex; a Hard SF denial, or an insistence on looking beyond NASA's limitations by some; a reconfiguring of the logics of the genre by others, a process referred to by the short-hand term 'New Wave'.

'New Wave' science fiction

Critics use the term 'New Wave' to describe a loose affiliation of writers from the 1960s and 1970s who, one way or another, reacted against the conventions of traditional SF to produce avant-garde, radical or fractured science fictions. All these labels for literary movements are problematic; but the label 'New Wave' is more so than most. The phrase itself appropriates the descriptor given to a movement from French cinema, the *nouvelle vague*: but the parallels between 1960s SF and the modish, jump-cut exercises in egregious contemporaneity of directors such as François Truffaut and Jean-Luc Godard are very imprecise.

As Damien Broderick notes, New Wave was 'a reaction against genre exhaustion' which was 'never quite formalised and often repudiated by its major exemplars' (Broderick, p. 49). The term itself was probably coined by Christopher Priest in the days when he was a young fan (rather than the major novelist he would become). It was initially associated with the London magazine *New Worlds*, which had been published, with many interruptions, since 1946, but which was reconfigured as a venue for experimental and unconventional fiction in the 1960s, particularly under the editorship of Michael Moorcock (b. 1939) from 1964 to 1971. Moorcock himself contributed a guest editorial to *New Worlds* before taking over officially as editor in which he called for a more passionate, subtle, ironic and original form of SF, picking out four writers as promising templates of the new style: J. G. Ballard (b. 1930), E. C. Tubb (b. 1919), Brian Aldiss (b. 1925) and John Brunner (1934–1995). In the words of Edward James, 'Moorcock gathered around himself a group of talented British writers', whilst also recruiting 'a new generation of American writers, such as Samuel R. Delany, Thomas M. Disch and John Sladek, all of whom came to London to live, to share in the excitement of those years' (James, *Science Fiction*, p. 169). This perhaps gives too much of a sense that New Wave was a product of the 'swinging sixties' of London counterculture: it is better to think of New Wave SF as part of a broader international interest in experimental and avant-garde literary techniques. Also in *New Worlds*, Ballard called for a comprehensive rejection of SF cliché:

> Science fiction should turn its back on space, on interstellar travel, extra-terrestrial life forms, galactic wars and the overlap of these ideas that spreads across the margins of nine-tenths of magazine s-f. Great writer though he was, I'm convinced H. G. Wells has had a disastrous influence on the subsequent course of science fiction ... similarly, I think, science fiction must jettison its present narrative forms and plots. (quoted in James, pp. 169–70)

New Wave is often taken to be a deliberate attempt to elevate the literary and stylistic quality of SF, which to a certain extent it was; but what Ballard's remarks make plain is the extent to which it was also a reaction to the sedimentary weight of the genre's backlist which new writers were beginning to feel burdensome. By the 1960s so much SF had been published, so many ingenious ideas developed and fleshed out, that thinking of something new, bringing novelty to the SF novel, was becoming harder and harder. What the New Wave did was to take a genre that had been, in its popular mode, more concerned with content and 'ideas' than form, style or aesthetics, and reconsider it under the logic of the latter three terms.

For many fans this was nothing less than a betrayal of what SF was all about. In characteristic curmudgeonly form, the author and SF fan Kingsley Amis

(1922–1995) declared 'the effects of the New Wave' to have been 'uniformly deleterious':

> The new mode abandoned the hallmarks of traditional science fiction; its emphasis on content rather than style and treatment, its avoidance of untethered fantasy and its commitment instead to logic, motive and common sense ... [instead] in came shock tactics, tricks with typography, one-line chapters, strained metaphors, obscurities, obscenities, drugs, Oriental religions and left-wing politics. (Amis, p. 22)

This is, of course, a travesty of the movement; and the self-satisfaction with which Amis announced that 'by 1974 or so the New Wave was being declared officially over' was misplaced. For fans of Amis's persuasion it would be truer to say that the Golden Age never went away. Science fiction continued to be written according to the protocols against which the New Wave was reacting: Murray Leinster (1896–1975), Edmond Hamilton (1904–1977), Clifford D. Simak (1904–1988), Mack Reynolds (1917–1983), Gordon Dickson (1923–2001), Fred Saberhagen (b. 1930), Ben Bova (b. 1932), H. Beam Piper (1904–1964) and various others produced a great quantity of tungsten-hard, mechanically literate and often militaristic SF. Their spiritual home was Campbell's *Analog* (the name for *Astounding* after 1960), and their writing was enthusiastically consumed by many SF fans.

Indeed, on a head-count of novels and stories the bulk of SF written in the 1960s (and since) has been 'Hard SF' of this sort. Many people consider this the best kind, the sort they enjoy most and which therefore they feel most characteristic of the genre. I cannot argue with what many people enjoy, but it is difficult to deny that the major fictional achievements of 1960s SF are much less concerned with the props and protocols of Golden Age Hard SF. The half dozen most important texts from this period are fascinated with one subject: notions of the messiah. Although they approach the topic with a variety of technical and formal innovations, and enjoyed varying degrees of contemporary success, certain key novels from the 1960s and early 1970s have endured the prolonged exposure to the heat of posterity that cremates most of the books published in any given year (even the good ones), and to have become classics: Heinlein's *Stranger in a Strange Land* (1961), Frank Herbert's *Dune* (1965), John Barth's *Giles Goat Boy, or the Revised New Syllabus of George Giles Our Grand Tutor* (1966), Michael Moorcock's *Jerry Cornelius* sequence, the first book of which (*The Final Programme*) was published in 1968, and the novels of Philip K. Dick's great period, particularly *The Three Stigmata of Palmer Eldritch* (1965), *Do Androids Dream of Electric Sheep?* (1968) and *Ubik* (1969). I do not select these texts simply because, by sharing messianic tropes, they happen to support the thesis of this study; they are all massive and influential SF works of the 1960s, certainly among the most important literature from that period. A similar work (published earlier, and outside the scope of the present history, but which came to massive cultural prominence in the 1960s) was J.R.R. Tolkien's *The Lord of the Rings* (1951–53), a deeply sacramental work fascinated to an almost obsessive degree by the theological questions of atonement, free will and incarnation. We see the same messianic impulse in Stanley Kubrick's film *2001: A Space Odyssey* (1968; discussed below p. 269).

There are probably many explanations that might be offered for this persistent fascination with the messiah. The 1960s, after all, was that time when the Beatles declared themselves to be on an equal footing with Jesus (the case could certainly be argued that, at least in terms of fame, they were correct in this); in which alternative religions and cults flourished, that time which many diagnosed as end times, the coming of the Age of Aquarius. There was, it is true, a sense that human technology had finally caught up with the apocalyptic imagination of previous generations of end-of-the-world prophets; and much 1960s SF grappled – usually in rather clumsy ways, marked by a blinkered anxiety – with the fear of nuclear annihilation.[1] But this possibility has not gone away in the twenty-first century (indeed, our weapons are regrettably much more destructive than theirs) without provoking a similarly messianic literature. Besides, it is not just that there was a deal of cultural production on the topic of the messiah, it is that there was a series of masterpieces on that theme, and that those masterpieces took the form of SF.

To make myself plain: I think this messianic turn is connected for deep reasons to the determining logics of the genre itself. While space travel (the ur-narrative of SF) remained something to look forward to in an imagined future, that future event inflected the idiom of SF as transcendence, a metaphor for a more literal escape velocity. But actual space travel quickly revealed itself (inevitably, of course) to be a mundane business even while moving beyond the *mundus*. Nuts-and-bolts accounts of space travel in a fictional idiom seemed less enthralling; space travel as mystical passage chimed more with the spirit of the age. In place of transcendence, SF reverted to one of its core, originary anxieties: all we have learnt, all of our new science and technology, all that we know now about the cosmos – does this not fatally degrade the uniqueness and effectiveness of the very idea of the messiah? Chiliastic (or, strictly, di-chiliastic) concerns may have been part of it, but the fact of 1960s SF was that it was a looking back rather than a looking forward. The determining problematic of the genre was finding this way to work its way through.

Heinlein, Herbert, Barth, Moorcock

It is a mistake to think of New Wave SF as a minority or merely avant-garde interest. As John Huntingdon notes, by the 1960s SF 'had ceased to be the literature of an intensely devoted minority. The broad popularity ... of *Stranger in a Strange Land* and *Dune* is a phenomenon quite unlike the comparatively select popularity' SF had enjoyed before (Huntingdon, p. 2). One reason for the commercial success of *Stranger in a Strange Land*, *Dune* and, incidentally, *Lord of the Rings* is that these titles became campus books, bought and avidly read by hundreds of thousands of students as countercultural accoutrements, or even manifestos. Certainly, the mysticism and the presentation of psychotropic drugs as gateways to transcendental transformation in *Dune* insinuated it into the affections of mystically-inclined drug-taking youngsters; although there is much more to the novel than that. An even more notorious case (in its day) was *Stranger in a Strange Land*.

The story centres on Valentine Michael Smith, a human raised by Martians before being returned to Earth with mystical powers. He founds what is in effect a religion, accumulating many disciples, although his cult is a deliberately rhizomatic and freeform one, reproducing the logic of Martian 'nest' by sharing water and having a good deal of sex.[2] Smith eventually dies, or 'discorporates', at the hands of an angry human mob; although being discorporeal is not a big deal for a Martian, who exists equally contentedly on the material or the spiritual plane. One concept he brings is that of 'grokking', a word fans adopted with such enthusiasm that it has now entered the language (OED: '**grok**, *v. trans* (also with obj. clause), to understand intuitively or by empathy; to establish rapport with'). The knowledge one obtains by 'grokking' is super-rational, total and quasi-mystical; but it is also elusive. It is only three-quarters of the way through the book, after establishing his wildly successful religion and performing many miracles, that Smith can tell his main girlfriend, 'I grok people now, Jill ... I grok "love", now, too' (Heinlein, *Stranger*, p. 127). By the end *what* the novel groks comes close to banality: asked whether he considers himself God, Smith replies, 'with unashamed cheerfulness. "I *am* God. Thou art God, and any jerk I remove is God too ... when a cat stalks a sparrow both of them are God, carrying out God's thoughts"' (Heinlein, *Stranger*, p. 421). A stoned mind might construe this as mystical wisdom; a sober-headed rationalist might find more alarming implications in its muddle. Brian Aldiss diagnoses 'a sinister blurring of fact and fiction' in the 'irrational quasi-mysticism' that haunts the book and quotes with approval the Heinlein critic, Alexei Panshin (b. 1940) to the effect that 'the religious premises of Heinlein's novels are untrue, and super-powers do not exist':

> Without these anyone who attempts to practise the book's religion (which includes mass sexual relations) is headed for trouble. In other words, the religion has no point for anybody. (quoted in Aldiss and Wingrove, p. 290)

On the other hand, it requires a peculiarly narrow-minded literalism (something which is, I concede, often found in SF Fandom) to read the book in these terms: *Stranger* is a novel, not scripture. Far from being pointless, its religion is an eloquent, if sometimes inchoate, re-articulation of the core anxieties that determined the birth of SF in the seventeenth century. On one level Smith functions as a parody-Christ, preaching Martian baptismal 'water sharing' and encouraging cannibalism – Martians, we learn, eat their dead; Jubal Harshaw (Heinlein's mouthpiece in the book) reminds shocked earthlings that 'symbolic cannibalism' plays a 'paramount' part in Christian liturgy (Heinlein, *Stranger*, p. 127). But the book reverses the force of Christian Incarnation and works to contradict the underlying equivalence principle of atonement. A proper 'grokking' enables one to grok 'wrongness' in certain people, a state of affairs which reduces the entire Christian ethical framework to a simple binary, although one with serious consequences. We learn that the Martians grokked 'wrongness' in an entire planet and accordingly destroyed it, leaving only its rubble as the asteroid belt; there is the threat in the book that they might do the same with the Earth. Ultimately, the 'Thou Art God' conclusion of the novel apprehends a resolutely 'I–Thou' understanding of the nature of the universe, although the book is always saved from mere piety by its characteristic, iconoclastic Heinleinian vim.

Frank Herbert (1920–1986) took a rather different perspective on the figure of the messiah in his masterpiece *Dune* (1963–65). *Dune* vied with *Stranger* in its day as the key SF novel of the decade, and is probably the more famous of the two now, thanks to a more thorough cultural saturation and the creation of a *Dune* mega-text, comprising five sequels (*Dune Messiah*, 1969; *Children of Dune*, 1976; *God Emperor of Dune*, 1981; *Heretics of Dune*, 1984; and *Chapter House Dune*, 1985), a film (dir. David Lynch, 1984), two TV mini-series, video games, authorised pre-quels and other para-textual material. The most obvious aspect of *Dune* is that it is an environmental novel: the planet of the title is covered by a world-spanning desert, water is a precious commodity and life is hard; but Dune's sandworms (alien beasts that live and swim beneath the sand) produce a *pharmakon* known as 'spice' or 'melange': an addictive drug that also grants visions of the future to some, and which is vital – exactly in what way Herbert does not say – to the pilots who guide starships through hyperspace. Spice is manufactured only on Dune, which makes that world a valuable piece of real estate. When the family of the protagonist Paul Atreides are given the world as a fiefdom (the politics of Herbert's imagined cosmos is medieval, with an Emperor and a rigid hierarchy of castes beneath him) it prompts Machiavellian manoeuvres by the evil Baron Harkonnen, who assassinates Paul's father to grab the world for himself. Paul escapes with his mother into the desert where the indigenous Fremen (modelled on Arab Bedouins) take him in, and to whom he becomes a messiah figure, the Muad'Dib. Eventually, Paul recaptures Dune from Harkonnen, leading a Fremen rebellion, and becoming Emperor himself.

Paul's mother is a member of a women-only cult known as the Bene Gesserit, modelled (as Herbert conceded) on his memories of the Jesuit Order of the Catholic Church in which he had been raised. They have been conducting a secret breeding programme for many generations, in the hope of producing a messiah of their own, the *kwisatch haderach*. Paul's daughter was supposed to be this individ-ual, although in the event Paul himself short-circuits their plans and assumes the mantle. We might go the whole hog and call the book 'Catholic' (the official reli-gion of the Galactic Empire is a combination of Protestantism and Roman Catholicism, based on 'the Orange-Catholic Bible'). It is at the very least a novel that connects with a particular aspect of the traditions of SF: anti-technological, mystical and transcendent. In Herbert's Galactic Empire computers are interdicted by religious fiat ('Thou shalt not make a machine in the likeness of a man'); specially talented humans called mentats, capable of computer-like rapidity of calculation and thought, have taken their place. There are spaceships and some items of technology, but in general life is lived according to a pre-Industrial Revolution logic, and Herbert uses the desert setting to explore the two great 'desert' religious traditions: the Islamic human saviour ('Muad'Dib', we are told, means 'desert mouse'; but reminds us rather of 'Mahdi') and the Judaic-Christian divine messiah.

The book shares with most mainstream SF a dialectical understanding of the relationship between the technical-rationalist and the mystical. 'Spice' gives Paul transcendental powers of future-sight and inner wisdom; but he can defeat Harkonnen only by deploying atomic weapons (which ordnance has, very sensibly, long been proscribed by a sort of strategic limitation agreement: Paul's use of them is presented as a piece of brilliantly unorthodox generalship, like Hitler ordering his tanks to invade France through the Ardennes). More, one of the book's greatest strengths is its detailed and plausible rendering of the political context; a much more impressive fictional achievement than the sketchy and error-filled representation of Dune's extreme environment.[3] Herbert later said his idea for the novel

> began with a concept: to do a long novel about the messianic convulsions which periodically inflict themselves on human societies. I had this idea that superheroes were disastrous for humans. (quoted in O'Reilly, p. 38)

Less than two decades after the Second World War (the culmination of the mega-death convulsions occasioned by several self-appointed political 'supermen') Herbert could very well have written a 'long novel' about a Hitler or Stalin figure. But Paul Atreides is a political leader and also the founder of a religion. The title of the book's sequels makes the theological bias apparent: *Dune Messiah* (1969) in which Paul, blinded in an assassination attempt, martyrs himself by walking alone into the desert; *God Emperor of Dune* (1981), the fourth in the series, in which Paul's son Leto mutates into a gigantic worm who rules as tyrant for 1,000 years. Herbert wrote no *Führer of Dune* or even *President of Dune*, and there's certainly no *Separation of Church and State on Dune*. Rather, as the series progresses it deepens and and makes more complex its initial insight. In *Dune* the messiah proves 'disastrous for humans' simply in terms of the political upheaval he causes – war, uncertainty and so on – but this is the kind of 'disaster' any conventional political leader can inflict. By the time of *God Emperor* Herbert's understanding of this 'disaster' is much deeper. Leto, monstrously embodied as a giant worm yet maintaining his essentially human consciousness, is both ruler and god; his tyranny goes far beyond the practical oppressions of totalitarian rule. Because he is a god his total knowledge of the cosmos hems in humanity in a far more metaphysically constrictive way. The problem here is the viability of free will, and therefore of the disclosure (rather than enclosure) of human vitality. Only by willing his own death and sending humanity on a mass exodus into the unexplored galaxy can Leto break the deadlock.

Herbert's achievement, in other words, was to render the coming of the messiah in an accurately observed political context, noting as he did so how close the messianic impulse is to the fascistic (*God Emperor of Dune*, with its powerful central image of the dictator as a monstrous worm, may be one of the most effective satires on fascism yet written).

Unlike Herbert's anti-messiah, and very unlike Heinlein's alien *grokmeister* messiah, the messiah of the American novelist and academic John Barth (b. 1930) is deliberately goatish. *Giles Goat-Boy, or The Revised New Syllabus of George Giles Our*

Grand Tutor (1966) tells the story of a human raised as a goat on a farm who comes to believe that he is the messiah. Barth sets his novel in a future world divided between two great universities, parodically translating cold war geographies into the idiom of the campus novel. (Several other SF books from the period imagined the world as a giant campus, most notably *The Last Starship from Earth* (1968) by John Boyd (b. 1931), which is also an archly complex story about a messiah saving the world.) The West Campus is ruled by a supercomputer known as WESCAC (just as the East Campus is ruled by EASCAC), and the titular goat-boy may or may not be the son of WESCAC, the first 'programmed' man. He aspires to the role of the messianic 'Grand Tutor', preaching a gospel of 'Pass All Fail All'. This compendious book (over 350,000 words) packs in a great deal of contemporary satire and word-play, although the division between 'soul' and 'body' is rather over-schematised. Giles translates the slogans of Christ into the campus idiom ('passéd are the flunked!') but is all goat, not only in terms of eating straw but in a relentless and rather tedious randiness that sees him copulating with other goats and with an array of depressingly uncomplaining women; in fact, the novel deals so blithely in so many rape scenes that it goes beyond satire into distress. The sex, and the oblique blasphemy, might have seemed more shocking in the 1960s than they do today, although there is much more to the book than iconoclasm.

The centrepiece of the novel is a lengthy modernised version of *Oedipus Tyrannus* in slangy inventive rhymed couplets. The point is less to parallel Giles and Oedipus, and more to point up the origin of the Greek word 'tragedy' (which is derived from τραγοϛ, *tragos*, 'goat' + οιδε, *oide*, 'song'). Giles' goatish tragedy is in his repeated failure to achieve or even understand what is required of him as saviour; he even fails in his attempt to martyr himself by hanging. The religion that grows up around him cannot even distinguish between Giles and his adversary, the diabolic Stoker (a preface declares the messiah to be 'one Stoker Giles, or Giles Stoker – whereabouts unknown, existence questionable', Barth, *Giles*, 7). But by the end of the book we understand that this confusion is the crux of the novel. Confronting WESCAC Giles short-circuits it (' "I made a short circuit," I admitted ... "but I don't think WESCAC's damaged" ', Barth, *Giles*, p. 780). Barth roots his messianic satire in goatishness not to shock, but rather to invert the premises of mysticism: opposites meet, passing and failing are rendered the same, but in a sub- rather than a super-rational idiom, in the volatile stew of goatish urgings, lusts and confusions that structure the book. Barth's SF messiah is important precisely because he out-materialises the materialist idiom of SF itself. Giles, locking horns with the World Computer, is the bestial saviour in an intellectual cosmos.

Jerry Cornelius, the invention of the prolific British writer Michael Moorcock (b. 1939) inflects the Jesus Christ type (whose initials he shares) in yet another new way. Cornelius is messiah as pataphysicist, an ostentatiously motley character, if indeed the term 'character', with its traditional associations of psychological coherence and verisimilitude, can even be applied to him.

Moorcock began writing at an early age, producing a string of generic Fantasy adventures from the early 1960s to, more or less, the present day. Jerry Cornelius, a rather different sort of figure, first appeared in *The Final Programme* (1968), a book written, Moorcock would later claim, in nine days. This patchwork novel, its text littered with sketchy illustrations and diagrams, the narrative crazy-paved and jolting, fizzes with energy. Cornelius appears as a kind of surreal James Bond, driving and flying a succession of fancy machines across war-ravaged European landscapes, killing without remorse, indulging in drink and sex to excess. The ambiguous villain, Miss Brunner, uses a huge computer called DUEL ('Decimal Unit Electronic Linkage') to create a programmed hybrid of herself and Cornelius, a transgender post-human ('it was hermaphroditic and beautiful ... [its] skull contained the sum of human knowledge', Moorcock, *Cornelius Quartet*, pp. 138–40); it is this hybrid that destroys Europe. In *A Cure for Cancer* (1971) Cornelius is a photographic negative of himself; black-skinned and black-teethed, white-haired. He gets tangled in a bewildering series of non-linear adventures across a war-ravaged Europe, meeting again characters we now realise are archetypes: his doomed sister, Catherine, his evil brother, Frank, his lover, Una Persson, and so on. Dozens of short stories were also published across the decade, some written by Moorcock and some by his friends (Brian Aldiss, M. John Harrison and Norman Spinrad were only a few of the contributors). *The English Assassin* (1972) starts with Cornelius dead and follows his slow, uncertain resurrection. The novel's subtitle, 'A Romance of Entropy', provides one pointer to the focus of the kaleidoscopic series: the only thing treated more casually than sex in the Cornelius books is death, meted out to millions in a paragraph, or struggled back from by Cornelius. In fact, Cornelius is not just a counterculture messiah, but a postmodern Shiva, simultaneously demiurge and destroyer. *The Condition of Muzak* (1977) ranges widely across Moorcock's time-fractured Europe and also recapitulates the earlier books in what Moorcock, with uncharacteristic pretentiousness, called 'something approximating sonata form' (Moorcock, *Cornelius*, p. 144). These last two Cornelius books are somewhat in love with an elegant Edwardian version of their SF world-making, something that anticipates the subgenre that would come to be called steampunk. Various other Cornelius short stories and novels followed, too many to list here (they were collected in the omnibus editions *The Cornelius Chronicles Volume II*, 1986; and *The Cornelius Chronicles Volume III*, 1987). Cornelius also appeared, with a slightly different emphasis, in other Moorcock books. The whimsical far-future sequence *The Dancers at the End of Time* (*An Alien Heat*, 1972; *The Hollow Land*, 1974; and *The Ends of All Songs*, 1976) features 'Jherek Carnelian' as a decadent post-human dandy, whose time-travelling adventures do damage to time itself as he courts the respectable nineteenth-century housewife Mrs Amelia Underwood through his and other universes.

Indeed, Moorcock at about this time, or possibly earlier, began conceiving of his works as all taking place in the 'multi-verse', a neologism that describes the infinite network of possible alternative universes which mesh in complex and unpredictable ways. Cornelius, in this way of thinking, became one with Moorcock's other 'hero' characters: the High Fantasy warriors Elric and Hawkmoon, for instance: all facets of one archetypal 'Eternal Champion', fighting to keep Chaos and Order in balance. But Jerry Cornelius figures as something much more interesting

Jerry Cornelius: a pataphysical Christ (illustration from *A Cure for Cancer*, 1971)

than this; an attempt to figure the messiah as a being who goes beyond order and chaos, a pataphysical incarnation of satire, pastiche and excess. Read with the proper attention, the Cornelius books represent a nicely shocking experiment in messianic thinking. Rather like Christ, Cornelius comes not to uphold the order/ chaos yin-yang, but to overthrow it; and even in his anti-entropic selfishness to overthrow the principle of overthrowing.

Moorcock certainly tried harder to shock than most of his contemporaries, and many moments remain shocking even in our less inhibited age (the 1971 short story 'The Swastika Set-Up' begins with Cornelius's committing incest with his mother: she 'lying on the bed with her well-muscled legs wide apart, her skirt up to her stomach, her cunt smiling', Moorcock, *New Nature*, p. 264). *Epater les bourgeois* textual tactics such as this (and the Cornelius books are full of them) have a clear connection to the disarrangement of conventional narrative linearity and typography of most of the Cornelius books. Moorcock was later to play down this feature of his writing: 'style and technique was merely a means to an end – frequently a very moral means to some very moral ends' (Moorcock, *New Nature*, p. viii); but this doesn't ring very true. The point of Cornelius goes beyond soapbox politicking, even on such wholly creditable positions of anti-Nixon, anti-war and anti-bourgeois hypocrisy. The novels' importance as experimental fiction is not limited to, or even especially encapsulated by, their modishly fractured layout: it lies rather in the thoroughness with which Cornelius himself deconstructs notions of 'saviour', redefining 'atonement' in entropic terms and embedding a brilliantly unsettling arbitrariness into redemption. 'Why', Cornelius wonders in *A Cure for Cancer*, 'was resurrection so easy for some and so difficult for others?' (Moorcock, *Cornelius*, p. 313). The novels suggest there may be no answer to this profound question.

Philip K. Dick

But one key writer from the decade went even further than this in his intervention into the matter of the messiah: the American Philip K. Dick (1928–1982). Dick must be treated at greater length than other authors in a study such as the present one, for he may very well be the most important SF writer of the twentieth century. In his day he was a typical if unusually gifted writer of Pulp New Wave SF: publishing where he could, writing rapidly, prolifically and in a variety of modes to maximise his precarious income (with a consequent roughness of texture in his work: many of his books and stories are fairly crudely written, and read as, let us say, a trifle *under*-revised), winning few awards and attracting little attention. Since his death, however, Dick's reputation has grown enormously. He has proved especially dear to academic critics of SF – indeed, and rather strikingly, much more so than to Fandom. Although he remains in print, and Hollywood continues to excavate his books for movie ideas, Dick has never won a fan base commensurate with his genius, except among that specialist group known as academic critics, who have by and large been egregious in their endorsement.

Dick is most celebrated for the complexity and thoroughness with which he interrogates the notion that reality might not be what it appears. His best novels take thoroughly quotidian characters, often suburban, usually unexceptional, and rake through their (and our) preconceptions about the world around them. Reality and selfhood depend on perception, says Dick; and perception is radically unreliable. Drugs, external catastrophe, inner trauma can all unsettle it. The traditions of the visionary writer (Christopher Smart, Blake, Burroughs) – in which it is proper to place Dick – has often posited a 'real' or base reality hidden behind the veil of appearances: but Dick takes this insight one step further. No reality has ultimate primacy in his hectic imagination; nothing can be trusted. If this leads to a degree of paranoia (an emotion Dick seems to have regarded as fundamentally appropriate and even healthy), and indeed of mental instability, then this is also part of the beguiling flavour of Dick's writing.

But Dick almost always inflects his contingent and proliferating 'realities' in theological ways. This is true as early as *Eye in the Sky* (1957): the novel's sketchy premise (a 'bevatron' field) results in eight unconscious characters finding themselves in the private reality of whichever character happens to be returning to consciousness, in which that character is literally 'god', the fearful Eye in the Sky which the other characters must fear. The book's atmosphere of paranoid *angst* is the distilled essence of Dick. What is striking is not only the freedom but the *precision* with which his imagination addressed doctrinal questions with which seventeenth-century divines would have been fascinated. *The Three Stigmata of Palmer Eldritch* (1965), possibly Dick's best novel (there are many masterpieces that contend for that title), does more than figure its title character as a messiah to a futuristic solar system; and it does more than gesture vaguely at the notion that psychedelic drugs have 'religious' or 'transcendental' qualities. Christ-type Eldritch claims to have found his drug Chew-Z on Proxima Centauri. Intoxication with the drug leads into what seems to be an actual, rather than hallucinatory, world, where Eldritch is the dominant presence. Joe de Bolt and John Pfeiffer get to the heart of the novel when they summarise its premise as 'transubstantiation is actual, not merely symbolic'. They claim the book is a 'weave of the major symbols of Judeo-Christian belief, first in parody, then in a reintegration of them refreshed in meaning' (De Bolt and Pfeiffer, p. 178).

Do Androids Dream of Electric Sheep? (1968) is probably Dick's most famous novel, a celebrity it owes largely to *Blade Runner*, the film made from it by Ridley Scott in 1982. Indeed, the success of the film may cause us to misremember the book, for not only is Scott's rich and multilayered visual text very different from the pared down and rather underwritten source-novel, but the conceptual emphasis is radically altered as well. The novel tells the story of Rick Deckard, one of the dwindling population of a largely abandoned Earth given over to detritus and rubble ('kipple' is Dick's term). His job is retiring runaway androids, which resemble humans in every respect except the empathetic. But even the human characters are almost wholly alienated from their own emotions, relying instead on synthetic emotions generated by 'mood organs'; a premise handled with characteristic Dick wit; before he leaves for work at the start of the novel Deckard dials 'a creative and fresh attitude to his job' for himself, and setting 594 for his wife ('pleased acknowledgement of husband's superior wisdom in all matters') (Dick, *Androids*, p. 10). Religion is also machine-mediated and commodified; the messiah an elderly man called Wilbur Mercer. Humans connect with his experiences via an 'empathy box', experiencing what he experiences as he climbs a hill, is stoned to death by mysterious persecutors, descends into the 'tomb world' and is resurrected to climb the hill again. This curious and deliberately *ersatz* religion balances the 'kipple' of the novel's imagined world (one character puts it in those terms explicitly: the entropic kipple 'reproduces itself ... drives out nonkipple'. 'No one can win against kipple', we are told, '... except of course for the upward climb of Wilbur Mercer', Dick, *Androids*, p. 53). Mercer is assumed to be an alien messiah ('an archetypal superior entity, perhaps from another star', Dick, *Androids*, p. 158); but he turns out to be a sham, a human actor playing a role. In a complex dénouement it transpires that the journalist who exposes Mercer as a fraud is actually an android, hoping to destabilise human society; but recognition of Mercer as a sham will not affect the religion. Deckard meets Mercer, and even fuses with him in some sense, to hear this ultimate lesson:

> 'I am a fraud,' Mercer said ... 'I am an elderly retired bit player named Al Jarry. All of it, their disclosure, is true ... they [the androids] will have trouble understanding why nothing has changed. Because you're still here and I'm still here ... I lifted you from the tomb world just now and I will continue to lift you.' (Dick, *Androids*, p. 162)

It is as if Dick has emptied the notion of 'messiah' of all transcendental, religious or even practical content only to find that it still works (Mercer's real name, Jarry, alludes, of course, to the ur-postmodern pataphysical writer Alfred Jarry, 1873–1907, whose aesthetic informs Dick's theological conception). Dick's bleakly witty vision of a collapsed postmodern world constructed out of nothing but surface, simulation and detritus nevertheless coalesces a genuine atonement.

Ubik (1966) is based on two premises. One is a future in which those individuals with 'psi' talent, telepaths and precognitives, are routinely used by corporations for industrial espionage. Glen Runciter makes a living in this world managing a corporation that hires out *anti*-psi personnel to neutralise the psi danger. The second (adapted from Dick's 1964 story 'What the Dead Men Say') is that dead people are maintained in a cold storage 'half-life', where their brainwaves can be read and the living can communicate with them. This half-life slowly decays towards 'full' death, but so long as it lasts the dead are not wholly dead. But the story moves in a startling direction from these fairly standard SF premises. Runciter is assassinated and his team put him into cold storage; but the survivors soon discover the real world is decaying in strange ways; recently bought coffee is weeks old and mouldy in the cup; technology reverts to 1930s levels (one character's TV 'devolves' into a valve radio); Runciter's face appears on coinage and banknotes. It appears that Runciter is trying to communicate with his employees from beyond the grave by leaving bizarre messages in various places. It slowly becomes clear that although Chip and his fellows assumed that they had survived the bomb blast and Runciter had died, in fact it was the other way around. As Runciter puts it in Situationist style, his words appearing as crayon graffiti in a bathroom mirror:

JUMP IN THE URINAL, AND STAND ON YOUR HEAD,
I'M THE ONE THAT'S ALIVE. YOU'RE ALL DEAD.
(*Ubik*, p. 109)

The whole group is, it seems, dead and in cold storage in Switzerland, with the living Runciter attempting to communicate with them. The entropy that preys on individual members can be countered only with a mysterious product 'ubik', which appears in the form of an aerosol spray. Spraying the contents on oneself produces a new vigour and energy, whereas lack of ubik leads to exhaustion and death.

In other words, *Ubik* is a book about death. Almost all the protagonists are dead for most of the novel, and in their post-mortem existence they decay further towards a deader form of death. It is also, in the manner of Dick's strangely but potently constellated thematic quantities, a novel about commodity culture; as if 'dying' and 'consuming' were somehow the same thing. Joe Chip, the novel's protagonist, says something like this late in the book: that life is consumption.

Metabolism, he reflected, is a burning process, an active furnace. When it ceases to function, life is over. They must be wrong about hell, he said to himself. Hell is cold; everything there is cold. The body means weight and heat; now weight is a force which I am succumbing to, and heat, my heat, is slipping away.

And, unless I become reborn, it will never return. This is the destiny of the universe. (*Ubik*, p. 158)

Entropy is here figured as a form of *consumption*, rather than the tendency of patterns to disintegrate or of energy to run down to a low universal constant. If in *Androids* the saviour was an artificial man, in *Ubik* it is quite literally commodified: Christ as aerosol spray.

These three novels, *Stigmata*, *Androids* and *Ubik*, come closest to encapsulating Dick's wayward genius. But he was an immensely prolific writer, and while almost all his works contain tremendous strengths, some become tangled up in their own cat's cradle of con-trick realities, Chinese boxes and veiled conspiracies, and can be rather bewildering to read. The situation was complicated in 1974 by Dick's own religious epiphany: under various stresses, he experienced a series of 'visions' or 'communications' from a being he called the Vast Active Living Intelligence System, or VALIS. He saw pink light that communicated to him in dense bursts of information. He heard God in the Beatles' 'Strawberry Fields' ('God talked to me through a Beatles tune,' he wrote in 1980. 'A random assortment of trash blown by the wind & there is God'; quoted in Sutin, p. 225). He believed himself to be the reincarnation of Simon Peter, Christ's apostle; and also to be Buddha. The experience, which he shorthanded by date as '2–3–74', formed the basis for a process of private writing that bordered on hypergraphia: the aggregation of over a million words in a journal he called *Exegesis*. It also formed the raw material for several novels: *VALIS* (written 1978, pub. 1981); *The Divine Invasion* (1981); *The Transmigration of Timothy Archer* (1982). We might want to call the experience of '2–3–74' a nervous breakdown, and Dick himself was aware of the possibilities that his visions were hallucinatory; but he preferred a more complicated explanation. 'In my opinion Holy Wisdom herself took over my life and directed me ... I lived a wild, unstable, desperate, Quixotic life, & would soon have died. Hence it is not accident that Holy Wisdom came to me; I needed her very badly' (Sutin, p. 217). At other places in the *Exegesis* this saviour is called Christ, the Holy Spirit, the Greek god Apollo, his twin sister Jane (who had died in childhood), technologically advanced beings from the future and by various other names. What is clear is not only the prime importance of these models of salvation to Dick's subconscious, but also – and more crucially – the extent to which Dick's Christ is a Christ of indeterminacy. This is the main insight of his work: that embracing a vision of the world as radically indeterminate was not only radically corrosive of rationality, but also of certainty and even coherence. To go right back to the seventeenth-century theological controversy about the plurality of inhabited worlds: as William Empson put it, the Church denied this plurality precisely because, if accepted, then 'Christ was crucified on Mars too; indeed, on all inhabited planets' and 'his identity in any one appearance became precarious' (see above p. x). Dick makes the startling step of accepting this at face value: the appearance of his Christ is indeed precarious; but not only does this not negate his messianic potential, in fact it is revealed as the *very ground* of his power to save.

Ursula Le Guin

Ursula Le Guin (b. 1929) is a writer of genius whose entire career has been a dialogue between the mystic and the materialist idioms. Her first novels display the influence, in about equal measure, of Tolkien and Hard SF (*Rocannon's World*, 1966; *Planet of Exile*, 1966; *City of Illusions*, 1967). By her own account, '[I] got my pure fantasy vein separated off from my science fiction vein' in 1967–68 'by writing *A Wizard of Earthsea* and then *Left Hand of Darkness*, and the separation marked a very large advance in both skill and content. Since then I have gone on writing, as it were, with both the left and right hands' (Le Guin, *Language*, p. 23). Her left-handed 'Fantasy' writing has been without equal: The *Earthsea* series (*A Wizard of Earthsea*, 1968; *The Tombs of Atuan*, 1971; *The Farthest Shore*, 1972; *Tehanu*, 1990; *The Other Wind*, 2001) may be the single best High Fantasy series written; spooling out a traditionalist imagined archipelago in which celibate male wizards wield a carefully balanced magic, only in the latter two books elegantly to revise the whole from a powerfully realised female point of view. But her SF is also amongst the most highly regarded in the field.

The *Left Hand of Darkness* (1969) is perhaps her most notable book. On the icy world of 'Winter' the humanoid population has no fixed gender, moving from asexual into either male or female depending on circumstance. An off-world ambassador, Genly Ai, lives among these people for several years, noting the different ways society is inflected without the pressures of fixed gender. There is something solid and appealing in the society Le Guin portrays, although at the same time it is intensely conservative, the frozen landscape it inhabits an externalisation for its inner *stasis*. The novel also incorporates a mystical aspect, one the rationalist Genly finds hard to assimilate. Travelling from the quasi-western Karhide to the quasi-Communist totalitarian country of Orgoreyn, Genly eventually makes his way over the glaciers back to Karhide with his companion Estraven.

A similar ideological divide (based on the cold war era Western and Eastern blocs), and a similarly circular pattern of journey from one to the other that leads to a journey back home, is found in *The Dispossessed: An Ambiguous Utopia* (1974); a novel that remains one of the most mature and intelligent analyses of the utopian impulse written. The planet Urras (wealthy, capitalist and unequal) is circled by the arid moon Anarres, which is settled by anarchists; one of these, the brilliant scientist Shevek, finds the surreptitious but inescapable social pressures of status and competition unbearable (in a society in which everybody is supposedly equal) and flees to Urras. There he realises the authorities wish to exploit him, and he returns, circuitously, home.

The *Left Hand of Darkness* is often discussed, and indeed taught, as a vehicle for thinking about gender; and it performs that function admirably. But there is much more to it than that; and there is a rather dangerous gender essentialism inherent in the assumption that Le Guin, because female, must have subordinated her aesthetic project to feminist proselytising. The truth is that Le Guin's writing is always much more balanced than that; and indeed, that balance as such forms one of her major concerns. Both *Left Hand* and *The Dispossessed* balance form to theme, of symbol to narration, flawlessly.

One way of looking at Le Guin's two 'hands', her 'Fantasy' and her 'SF' impulses, is to configure them as reflective observations about the creative artist. A parable such as *The Lathe of Heaven* (1971), in which the main character finds that his dreams overwrite reality for everybody else, turning his fantasies into actual reality, can be read as a commentary on the power and danger of the creative imagination. On the other hand, more recently Le Guin has become much less *esemplastic* (to use Coleridge's term for the creative, shaping imagination), and much more observational. There is an anthropological coolness to *Always Coming Home* (1985), which reveals the balanced society of the Kesh in a post-apocalypse California Napa Valley, through a variety of documents, stories, poems, and so on. But it is not that Le Guin is interested in meta-textual game-playing. Rather, as Warren Rochelle argues, she is interested in the socially embodied principle of communication, the currency of communities: she 'argued for a true human community, one of the heart, in which human life can be lived with worth, honour and value' (Rochelle, p. 173). A character from *The Dispossessed*, Odo, puts it tremendously well:

A child free from the guilt of ownership and the burden of economic competition will grow up with the will to do what needs doing and the capacity for joy in doing it. It is useless work that darkens the heart. The delight of the nursing mother, of the scholar, of the successful hunter, of the good cook, of the skilful maker, of anyone doing work and doing it well – this durable joy is perhaps the deepest source of human affection and sociality as a whole. (Le Guin, *Dispossessed*, p. 207)

Brian Aldiss

The British writer Brian Aldiss (b. 1925) began publishing in the 1950s and has remained a constant and dependable producer of excellent fiction (in various modes) to the present day. The timing of his birth meant that he served in the Second World War, in Burma and Sumatra, an experience he has fictionalised powerfully in the Horatio Stubbs saga (*The Hand-Reared Boy*, 1970; *A Soldier Erect*, 1971; and *A Rude Awakening*, 1978) and then observed the dismantling of the British Empire and the changes of the 1960s. Older than some of the other New Wave writers, Aldiss was in a position to make one of the most significant contributions to the creative evolution of Prose SF in the 1960s.

Hothouse (1962) is a novel set far in the future on an Earth that has ceased revolving, and whose daylight side is dominated by a titanic banyan tree, among whose branches the diminutive devolved descendants of Homo sapiens scamper and scurry. Giant spider-creatures float through the sky, travelling as far as the Moon, which is also teeming with life. A summary cannot do justice to the brilliant, crowded, sparkling *oddness* of this book. Although conventionally written and plotted, it is a book that deconstructs notions of character: its main humanoid,

Gren, is almost unthinking – indeed consciousness as we understand the term is in this world a deadly kind of fungoid parasite. Aldiss deliberately overloads the more coherent conventions of 'adventure narrative', hurling a seething mass of imagery and novelty at the reader, and using Gren's humble consciousness as a means of emphasising the mental crush of sensation.

> Gren sank to his hands and knees among the painful stones at the mouth of the cave. Complete chaos had overtaken his impressions of the external world. Pictures rose like steam, twisting in his inner mind. He saw a wall of tiny cells, sticky like a honeycomb, growing all about him. Though he had a thousand hands, they did not push down the wall; they came away thick with syrup that bogged his movements. … The mirage fogged over and vanished. Miserably he fell back against the wall, and the cells of the wall began popping open like wombs, oozing poisonous things. The poisonous things became mouths, lustrous brown mouths that excreted syllables. (Aldiss, *Hothouse*, p. 162)

This sort of hallucinatory writing works because much of the novel is so precise in its delineation of its future environment that it reads as real. This vividness depends, in large part, on the expertly realised childishness of the post-human consciousnesses Aldiss creates: indeed, though violent, sexual and bizarre, *Hothouse* seems to me one of the great evocations of what it is to be a child. Seized by 'the tribe of the True World' a group of humans is brought before a group of deformed elders held captive in large urns: 'one had no legs. One had no flesh on his lower jaw. One had four gnarled dwarf arms.' The human response is a straight-forward childish disgust; which is met with a childish logic:

> 'You are too foul to live!' Haris growled. 'Why are you not killed for your horrible shapes?'
>   'Because we know all things,' the Chief Captive said … 'To be a standard shape is not all in life. To know is also important. Because we cannot move well, we can think.' (Aldiss, *Hothouse*, p. 34)

Later in the novel, an intelligent fungus called 'the morel' attaches itself to Gren's head, enslaving him while also enhancing his power of thought. Gren's mate, Yattmur, catches it in a gourd, thereby freeing him: 'he stared down at the still-living morel. Helpless and motionless now, lying like excrement in the gourd' (Aldiss, *Hothouse*, p. 179). In this novel *intelligence itself* – that fetish of SF (as 'literature of ideas') – becomes excremental, parasitical, deformed. Aldiss deconstructs thought itself; a startling and brilliant strategy that goes beyond the avant-garde literary tinkering with conventional plot or prose style to an undermining of the underpinnings of literature itself. As this novel dramatises, most of life is unthinking, and there is nothing intrinsic to 'thought' that means it ought to be privileged: it is only one evolutionary strategy among many. (Stephen Baxter makes the same point in his 2002 novel *Evolution*.)

Perhaps the most significant feature of Aldiss as a writer is his very restlessness. Some of his early stories were assembled into a patchwork 'future history' narrative, but taken as a whole his work refuses to settle into or reify any given form. Throughout the 1960s Aldiss continued to push the boundaries of fiction with a number of experimental or challenging novels. Good though these are, their very showiness means that they are less successfully unsettling as the fecund thought-lessness of *Hothouse*. *Report on Probability A* (1968) is a paranoid story of multiple voyeurs spying on one another, written in a deliberately dislocated and estranging style. *Barefoot in the Head* (1969) is an exercise in exuberant drug-influenced surrealism. It begins fairly conventionally with its protagonist Colin Charteris travelling across a future Europe; but the fallout from a recent war in which psychedelic drugs were used as weapons gradually infects not only the character but the narrative itself. Peter Stockwell notes the book's increasing 'typological and graphological fragmentation, with the narrative prose interrupted with poems and song lyric' as well as 'the technique of lexical blending and deviant syntax, which increases in deviance and complexity as the narrative progresses and Charteris leads a Messianic convoy of acid-heads through a series of massive motorway pile-ups and riots' (Stockwell, p. 63). David Pringle and John Clute argue that the presiding theme that dominates much of Aldiss's writing is 'the conflict between fecundity and entropy, between the rich variety of life and the silence of death' (Clute and Nicholls, p. 11). The fact that fecundity often manifests itself in Aldiss pathologically (cancer and fever are common themes) gives his writing a certain darkness. A novel like the superbly controlled *Greybeard* (1964), set in a future in which humanity is sterile, the population ageing and facing death, treats its *memento mori* topic gracefully and movingly. The darker tone, howsoever wittily and inventively treated, is a distinctively Aldissian feature. A vicar in one of his most recent novels, *Affairs at Hampden Ferrers* (2004), confronts, in the size and hostility of the cosmos, the same spiritual crisis as a thinking religious person from 1600. The universe, he argues, is an entity: 'this entity is non-living. It is a sort of process, a cancer on a tremendous scale.' Asked 'where does God come into all this morbid cosmology?', the vicar replies that God 'prevails only over this planet Earth, and perhaps over the other planets of the solar system, but probably not over planets of distant stars' which, he concedes, makes God 'so insignificant' (Aldiss, *Affairs*, pp. 230–1). This dark vision is more than mere pessimism; it is a restatement of the root anxieties at the heart of the birth of SF. Several of Aldiss's best novels rework classics of the genre in ways that would later become characteristic of postmodernism: *Frankenstein Unbound* (1973) mixes inadvertently time-travelling Americans with Mary Shelley and her monster in a story that fractures and tessellates the original *Frankenstein* in enormously creative ways, even as the world itself is broken by time-quakes (the fallout from a future war). By the end Frankenstein's monster becomes almost a type of Christ ('My death', he tells his killer, 'will weigh more heavily upon you than my life ... though you seek to bury me, yet will you continuously resurrect me'; as he dies he announces that he is off to harrow hell (Aldiss, *Frankenstein*, p. 156)). Two books in a similar idiom followed: the Wellsian *Moreau's Other Island* (1980) and the Stokerish *Dracula Unbound* (1991). After the energetic reinventions of the New Wave these later titles are a tacit acknowledgement that the weight of pre-existing SF was becoming increasingly clogging, something with which writers of new fiction *had to deal* in some way.

Anglophone science fiction

The English writer and composer Anthony Burgess (1917–1994) was a prolific fabulist of enormous skill and range who, for reasons of cultural and religious snobbery (his lower-middle-class background and his Catholicism), was largely ignored by his native country. But *A Clockwork Orange* (1962) is a masterpiece of focus, a deeply Catholic science-fictional novel. Burgess establishes his setting (a state-controlled future UK greatly influenced by Soviet Russia) less through description, and more – brilliantly – through an invented future argot, an idiom of slang, Americanisms and Russianisms in which the first-person narrator Alex tells us his tale. He is a thug, bully and rapist; he is apprehended and brainwashed by the state, transformed into 'a little machine capable only of good' (Burgess, *Clockwork*, p. 122). Alex's crimes are very bad, but his conditioning is much worse; this is the meaning of the oblique title, the inherently monstrosity of the cyborg collision of organic and mechanical ('the attempt to impose on man, a creature of growth and capable of sweetness, to ooze juicily at the last round the bearded lips of God, to attempt to impose, I say, laws and conditions appropriate to a mechanical creation', Burgess, *Clockwork*, p. 21). The main flaw in the novel is Burgess's grumpy refusal to believe that pop music could be anything other than pap: Alex, so believably teenage in his thuggishness, nurses an improbable love for Beethoven, whose choral symphony whips him into a state of violent excitement. Thrash metal (or, in 1962, rock and roll) would surely be more like it. But the novel's excellence is linked to the unswerving way it elaborates its point: that it is wholly better for humanity to possess free will, even if some people use it to behave badly, than for that freedom to be taken away and humanity conditioned, even if the conditioning results in everybody behaving well. It is because of, and not despite, this explicitly Catholic thesis that *Clockwork Orange* is a great SF novel.

Other titles from the decade now seem less significant, despite being praised extravagantly in their own day. The British author John Brunner's (1934–1995) *Stand on Zanzibar* (1968) is a lengthy disquisition layered over a sort of spy plot, set in a monstrously overpopulated world. But its choppy, 'experimental' style, lifted directly from the work of the American Modernist John Dos Passos (1896–1970), seems second-hand and over-boiled, and the premise of the novel has a phlogistonic lack of contemporary bite (overpopulation had not brought the world to a standstill by the start of the twenty-first century, and will not do so by the start of the twenty-second either). Of course, Brunner was not alone in thinking his premise sharply relevant: many writers in the 1960s and 1970s adopted positions of Malthusian glumness on the subject of overpopulation; a better treatment of the theme than Brunner's (better because rooted in a Pulp terseness rather than a High Modernist prolixity) is Harry Harrison's (b. 1925) *Make Room! Make Room!* (1966).

A much more enduring perspective on impending catastrophe is found in the perfectly dislocated and obsessive SF novels of the British writer J. G. Ballard

(b. 1930). Ballard placed his near-future apocalypses somewhere between literal precision and symbolic generality. Ballard's four early disaster novels construe air, water, fire and crystalline earth; but isolating thematic patterns like this does not capture the unique strangeness of the texts. In *The Wind from Nowhere* (1962) a gigantic gale blows Britain to destruction. In *The Drowned World* (1962) melting icecaps have submerged much of the Northern Hemisphere; but the novel's focus is psychological rather than meteorological. In *The Drought* (1965) a polluting film on the ocean's surface prevents evaporation and provokes a global drought. There is a slightly didactic whiff about this book (man tampers with the world to his peril) and the lack of such implicit moralising makes *The Crystal World* (1966) a much better novel. The African rainforest, and everything and everybody in it, is being transformed into crystalline form by some strange cosmic effect. The novel might be called 'surreal', if that term didn't suggest an inappropriate correlation of imagery to subconscious affect. Ballard's best writing cannot be reduced to any simple schema; they work in the poetic rather than rationalist idiom. His various stylistic and formal experiments from later in the decade (collected in *The Atrocity Exhibition*, 1970) strain perhaps too hard to exterminate the rational reader-response, although some of them – for instance, the bizarre meditation on the violence of 1960s America entitled 'The Assassination of J. F. K. Considered as a Downhill Motor Race' – haunt the mind. In the 1970s Ballard moved away from the more recognisable conventions of SF and became fascinated with the actuality of urban desolation. If there is nothing 'realist' in his version of London it is because Ballard does not consider the lives people actually live in major cities to partake of realism: something eloquently if sternly worked out in his car crash-erotica fusion *Crash* (1973), the Beckettian isolation of his protagonist on a traffic island of *Concrete Island* (1974), and the atavistic tower-block ghetto novel *High Rise* (1975). It can be argued that Ballard has since lost something, although he certainly gained a much larger readership and made a deal more money by turning to a more recognisable world in autobiographical fictions in the 1980s.

One consequence of the successes, and the cultural currency, of certain New Wave writers was to point ambitious writers in the direction of genre. The British poet and novelist D. M. Thomas (b. 1935) plundered what he called 'myths suggested by science fiction stories by Ray Bradbury, Arthur C. Clarke, Damon Knight' and others for a range of poems that conspicuously cross-bred mainstream Modernist poetry with Pulp tropes. 'Missionary' (1968) rewrites T. S. Eliot's 'Journey of the Magi' ('A harsh entry I had of it, Grasud; / the tiny shuttle strained to its limits ...') with its alien messiah as both Magus and messiah ('I loved them, and they / killed me', Thomas, pp. 91–3). In 'A Dead Planet' a tentacled alien captain lands his ship on the 'broken-pillared plain' of a dead Earth, and instructs his crew to resuscitate one of the human corpses. This "Man"/ – Such was the thing called' revives with joy that 'his faith was not in vain':

> '*Dear Christ!* ... *how blissfully Thou dost abate*
> *The grave's* – ' His gaze took in the plain;
> The ring of orbs devoid of love or hate,

> The ray-guns poised to mow it down again
> When they had sorted out its true estate.
>   (Thomas, p. 110)

Many of Thomas's poems ('Two Sonnets from Drifting Worlds', 'Elegy for an Android') strike this rather mordantly elegiac pose, recording sadness at the superseding of religious mystery by loveless, hate-free science.

The Russian-born Vladimir Nabokov (1899–1977) wrote fables in which reality becomes malleable under the influence of imagination and art, many of which explore precisely the dialectic of science fiction. Yet, perhaps typically of the 'non-SF writer' of this time, he refused the idiom of generic SF sometimes, explicitly and wittily, in his fiction itself. In his 1958 story 'Lance', in which the parents of an exploring astronaut wait anxiously for his safe return, the narrator airily announces with respect to the timeframe of the tale:

> I gladly leave the replacement by a pretentious '2' or '3' of the honest '1' in our '1900' to the capable paws of *Starzan* and other comics to atomics. Let it be 2145 AD or 200 AA, it does not matter. I have no desire to barge into vested interests of any kind. This is a strictly amateur performance. (Nabokov, *Stories*, p. 634)

Yet, whether he was amateur or Starzanic, Nabokov wrote very powerful science fiction. The best is the long novel *Ada or Ardor: A Family Chronicle* (1969). Set in a richly evoked and sometimes oppressively vivid alternative world, 'Antiterra', this combines the formal and stylistic experimentation with the Pulp tradition of 'alternative history'. Nabokov's alternative America has been as largely colonised by Anglophone Russians, electricity is illegal and the knowledge of the existence of our world, Terra, is energetically debated. The story traces the love affair of two cousins, Van Veen and Ada, who later discover they are actually brother and sister; but its densely celebrated eroticism in fact chimes with a series of densely intellectual meditations on the nature of time, memory and what we might call the many worlds hypothesis. I can think of few SF books as rich or challenging as this one.

Yet neither Thomas nor Nabokov's names are ever linked by critics with New Wave SF; and many other writers who might be thought New Wave-y (or who have been classified as such by some critics) either repudiate the identification or are omitted from standard histories. A good example of the former is the stylish and challenging American writer Samuel Delany (b. 1942), who began his career in the 1960s. *The Jewels of Aptor* (1962) and the *Fall of the Towers* trilogy (*Captives of the Flame*, 1963; *The Towers of Toron*, 1964; and *City of a Thousand Suns*, 1965) show some of the gaucherie of youth, together with a poetic complex of imagery and theme; but they are also share many of the trappings of Campbellian Hard SF. Stylistic and formal innovation, as in his quasi-imagistic exercise in SF-*noir*, the short story 'Time Considered as a Helix of Semi-Precious Stones' (1968), and his presence in London at the appropriate time mean that many identify him as New Wave. He disagrees: 'To say that *I* was, somehow, a representative of the New Wave is tantamount to saying that science fiction *has* no history.' He has attacked critics for peddling an 'unresearched and ahistorical myth of the New Wave' (quoted in Luckhurst, p. 162). His belief is, presumably, that if SF '*has* a history', then that history must be fine-grained enough to distinguish the New Wave proper and Delany's own subtle distinction from that orthodoxy. But a Critical History that paid such scrupulous attention to the myriad burgeoning and passing away of microclimates of the genre would stretch to too many thousands of pages to be accommodated by most publishing houses. It is certainly true that Delany, though sometimes accused of literariness and pretension, never abandoned the more visceral delights of the Golden Age – *Babel-17* (1966), for instance, harnesses its rollicking space-opera plot to some serious philosophical inquiry into the extent that language shapes and indeed makes reality; and *Nova* (1968) manages a similar blend of colourful, kinetic space-opera and profounder intensities. But his work shared many strategic and formal attitudes with other New Wavers. In the best of his many jewel-like short stories, he opens trapdoors in generic conventions that drop the reader into strange new territories. 'Aye, and Gomorrah ...' (1967) invents a new sexual perversion, and then treats the 'frelks' who practise it (they are attracted to spacemen who have been neutered by their experiences in space) sympathetically and open-endedly. 'Time Considered as a Helix of Semi-Precious Stones' (1969) is a complex meditation on guilt, time and poetry, set in a stylised but wholly convincing idiom with much in common with the later 'cyberpunk'. Most critics see *The Einstein Intersection* (1967) as his most significant early work: 'short, but still a work of extreme complexity and resonance ... tightly patterned', in George Slusser's words (Slusser, p. 42). Various seemingly archetypal characters interact on an Earth from which humanity has been removed: the messiah archetype finds vigorous and rhizomatic life as Orpheus, Christ and even Ringo Starr. The big novel *Dhalgren* (1975), set in the multifaceted and time-fractured future city of Bellona, has received less favourable critical attention. It sold extremely well (perhaps because it was, for its day, extremely sexually explicit), despite its *Finnegans Wake*-inspired experimental form and length; but some critics have felt that it is too turgid to achieve its considerable ambition. *Triton* (1976) relates the build-up and destruction occasioned by a war between a future solar system's more socially restrictive planets (Earth and Mars) on the one hand, and various smaller satellites with more libertarian politics on the other. This is a book that explores and details the tensions between 'centre' and 'margins' in several senses. The main character, Bron Hellstrom, mediates a series of familiar and (to this critic, at any rate) unfamiliar aspects of human sexuality. Hellstrom has himself surgically altered from male to female, and the novel as a whole works tirelessly to unpick the traditional notions of 'sex' as somehow inherent in biological bodies. The commitment to a thoroughgoing polymorphous perversity, and the unflinching acceptance of the violence this sometimes entails, makes the book a compelling read. Some critics, though, may share Robert Elliot Fox's reservations about this novel: 'in *Triton* Delany castigates "overdetermined systems" – e.g. government bureaucracy – but it strikes me that the compulsive polymorphousness of the sexual relationships he depicts ... is itself overdetermined, like the more extreme varieties of Gay Liberation and Feminism, or the fundamentalism of the so-called Moral Majority' (Fox, p. 49). A better novel is *Stars in my Pocket Like Grains of Sand* (1984), as densely compact and challenging as anything Delany has written: a sort of love story between Marq Dyeth and Rat Korga (a freed slave and criminal). Delany's fondness for rough-trade characters like Korga is evident in all his work. There is sometimes a self-consciously critical-theoretical flavour to Delany's later writing, the product of his immersion in the newer critical theory of the 1970s, particularly the work of the French philosopher-historian Michel Foucault (1926–1984) and the post-Derridean strategies of Deconstruction. In part these informed a series of fiercely intelligent and demanding critical interventions into SF and other cultural discourses, among them *The Jewel-Hinged Jaw: Notes on the Language of Science Fiction* (1977), *Starboard Wine; Some More Notes on the Language of Science Fiction* (1984), *Silent Interviews* (1994) and *Longer Views* (1996). The ferocity and omnivorousness of Delany's intellect makes him ideally positioned to encounter the changing semiotics of SF, as both a creative writer and a critic. Just as his criticism, and much of his later fiction, has become post-structuralist, we might want to identify his classic period not so much as New Wave as post-Golden Age; which might, indeed, be a workable descriptor for the New Wave *tout court*.

There were many other writers of the 1960s and 1970s who, like Delany, declined the New Wave invitation to abandon the puerilities of trad-SF and yet did not sacrifice aesthetic ambition or worth. Harry Harrison (b. 1925) produced a consistent run of highly entertaining space-opera, most in a comic idiom. *The Stainless Steel Rat* (1961) was the first of many adventures for its space-faring anti-hero. Even his tale of desperate survival on an inimically hostile planet, *Deathworld* (1960) (followed by *Deathworld 2*, 1964 and *Deathworld 3*, 1968) is leavened by a witty humanity; and his *A Transatlantic Tunnel Hurrah!* (1972; a much better title than the dreary US version *Tunnel through the Deeps*) is one of the most charming and amusing alternative Victorian history stories in the genre. At the other extreme was the posy humourlessness of Harlan Ellison (b. 1934), one of the period's *enfants* (or *hommes*) *terribles*. The perpetual adolescent *angst* of the story 'I Have No Mouth and I Must Scream' (1967) is precisely captured by its *outré* title (the story's protagonist, tormented by a malevolent computer inside a virtual reality, literalises the title's screaming mouthlessness). The likewise egregiously titled ' "Repent Harlequin!" Said the Ticktockman' (1965) is a slightly more sophisticated account of an over-regulated society. Much of Ellison's work mediates this - ill-tempered Beat-ish aesthetic (think of Ginsberg's *Howl!*, 1956) through the teenage structures of the genre. His most important contribution to SF may well prove to be editing the collection *Dangerous Visions* (1967), which brought the cutting edge of the best genre writers of the period to a wide audience.[4]

The balance between Pulp and Literariness was most perfectly achieved by the American Robert Silverberg (b. 1935), perhaps the most prolific writer in a field not underprovided with prolific writers. Silverberg began work in the worthy tradition of hack SF writer, producing a great many SF novels and other forms of fiction for money in the 1950s, and indeed announced his retirement from SF in 1959. Frederick Pohl, then editor of *Galaxy*, is supposed to have wooed him back to the genre by convincing him that the market was ready for a more literary form of SF. Silverberg (who had stopped genre writing rather than writing altogether) re-emerged as a major figure, perhaps the most technically gifted and prolific author of his generation. Despite a second hiatus (he declared another retirement from SF in 1976, seemingly disillusioned with the insularity of the SF culture, although he was publishing SF again by 1980) he has maintained an impressive rate of production. His 'semi-official' website (www.majipoor.com) lists the titles of 1,200 novels in its bibliography, a number almost certainly incomplete (much of Silverberg's early work, particularly his erotica, was published pseudonymously, and not all the pseudonyms have been recovered). The striking thing about Silverberg's best period (in contrast to most writers nurtured on hack overproduction, where 'good enough' is the main aesthetic criterion) is how well written, expertly handled and technically flawless his books are. Perhaps less surprising, given how completely he has immersed himself in writing, is the frequency with which his best novels become meditations on the protocols of storytelling itself. One of his most fully rendered novels, *Dying Inside* (1972), concerns a receive-only telepath called David Selig who is scamming a living in 1970s New York. The arc of the story traces Selig's gradually waning abilities, but the marvel of the book is the vividness with which Silverberg renders his promiscuous hopping from mind to mind, some rendered in stream-of-consciousness, some more imagistically; even, in passing, slipping into the mind of a bee ('there are no verbal outputs from the bee, nor any conceptual ones ... [but] how *dry* the universe of a bee is: bloodless, desiccated, arid. He soars. He swoops', Silverberg, *Edge*, p. 655). In other words, Selig is an author practising telepathically that empathetic entry into the minds of all manner of consciousness that a good writer must. Similarly, *A Time of Changes* (1971), a poised anatomy of a society on an alien world in which the use of 'I', 'me' and 'mine' is an obscenity (the favoured locution is 'one', as in 'one would wish for warmer encouragement from one's bondbrother'; although the narrator travels to an even more puritanical society where the only decent locutions are in the passive voice), is an excavation of the premise of the 'first-person narrative', as well as being a compelling story of the repression of subjectivity (perhaps more resonant for an English middle-class sensibility, where such repression is also *de rigueur*).

But a broader perspective on Silverberg's varied output must acknowledge the extent to which he modified and reproduced the New Wave fascination with 'the messiah'. *Thorns* (1967) constellates varieties of physical and emotional pain by way of interrogating not only whether such suffering has a purpose, but whether pain beyond a certain level is compatible with love. *Downward to the Earth* (1970) transposes a Conradian fiction about the exploitation inherent in imperialism (Silverberg is frequently Conradian) onto a jungle planet, following a former colonial governor Gundersen's pilgrimage in search of atonement for his formerly oppressive ways. The book ends with a mystical union between Gundersen and the alien aboriginals which, we realise belatedly, transforms him quite literally into the messiah (the book ends with his understanding that 'I am the resurrection and the life. I am the light of the world ... A new commandment I give unto you, that ye love one another'; Silverberg, *Edge*, p. 392). *Lord Valentine's Castle*

(1980) is a readable and fairly swaggering adventure across a Vance-style Big Planet of Majipoor, which involves, amongst many other things, some battling with monsters. By the third in the 'Majipoor' cycle *Valentine Pontifex* (1983) the human inhabitants must make atonement for their crimes against the indigenous life of the world (the 'monsters' of the first book). This involves a rather deadening literalisation of the notion of 'atonement': in the latter book Majipoor's crops won't grow until Valentine atones for the human slaughter of the whale-like Sea Dragons, as if contemporary guilt at environmental abuse could wholly encompass the concept of sin. But in his most notable work Silverberg is better at striking the buried chord of SF than almost anybody. *Son of Man* (1971), perhaps his best novel, resuscitates a twentieth-century man in a future so far distant that the Moon has disappeared, and the natives have never heard of Homo sapiens. Through a series of incarnations as different forms of (in some sense) human life, the protagonist embodies the messianic identity indicated by the novel's title (and reinforced by quotation from the New Testament): making plain a fascination with the problematics of incarnation present throughout Silverberg's work.

The British writer Christopher Priest (b. 1943) is, like Silverberg, a major novelist who has been under-celebrated by the literary establishment. In part this may be because the particular excellence of Priest's writing is hard to isolate: there is (although I don't wish to be essentialist) an Englishness about Priest's work: a control of tone, a Burne-Jones colour scheme, a sense of loss that hovers on the edge of inexpressibility, a particular *timbre*. The piquant, downbeat paranoia of his first novel *Indoctrinaire* (1970) was called by contemporary reviewers 'Kafkaesque', although his unsettling book creates a tone quite unlike Kafka's. The mysteriously oppressive bureaucracy of *Der Proceß*, through sprawling, is at least Germanically (or Czech-ly) efficient. The oppressive atmosphere of *Indoctrinaire* is more ramshackle, enacted habitually rather than consciously, something on the borders of comprehensibility. The closest thing to it in mood is Ballard, but there is an elegiac and attenuated aspect to Priest's prose that is missing from the more aggressive writing of his compatriot. Priest's second novel, *Fugue for a Darkening Island* (1972), is redolent of a peculiarly British 1970s sense of social decay. Set in a near-future England falling into anarchy, it places its rather unpleasant protagonist at the centre of an efficiently dark little narrative, related in a deliberately fractured way. Priest specifically juxtaposes before and after narrative viewpoints, braiding each into each, so that the tale proceeds stereoscopically. Reading through a dialogue of past and future which creates a deeper, more powerfully bleak artistic whole. *Inverted World* (1974) takes inversion as its key conceit: the planet on which we live is a finite world existing within an infinite universe. What would it be like to live on an *infinite* world that existed within a *finite* universe? Priest unpacks his premise ingeniously, building towards one of the great surprise endings in SF. But the greatest achievement of *Inverted World* is its holistic aesthetic, the way premise is intimately woven into every aspect of the novel. The coming of age of the Helward Mann, his forward journey through life, functions as a ratio inferior of the way his whole world (the city in which he lives) must continually move through its strange world, aiming always for an optimum location which itself is constantly moving. The physical reality of the life that this challenge entails – moving an entire city forward on rails through a constantly changing environment, picking up the massive rails from the rear and transporting them to the front, scouting the land ahead, and so on – is sparely but vividly rendered. The first sentence of *Inverted World* strikes the keynote: 'I had reached the age of six hundred and fifty miles.' In other words, this novel is much more than simply a metaphor for life (although it works well as a symbolic fable along those lines). It is about the *spatialisation* of time, the way we so often think of the non-spatial quality 'time' as if it had extension, breadth, depth, as if it were a river or a road. Heidegger has some interesting but indigestibly written things to say about this. Priest's writing is never indigestible. His novel simply, eloquently, takes us into a literalisation of this belief, such that when he provides his *coup de théâtre* ending, we are shaken up in a deeper way than a simple narrative twist in the tale has any right to manage.

The Space Machine (1976) is a recursive confection of H. G. Wells's *Time Machine* and *War of the Worlds* which reconfigures the 1890s into a tale in which Wells himself makes an appearance, taking us from southern England to Mars and back; it is a book that interrogates the nature of fiction: the dialectical integration of created worlds of the imagination and real worlds of experience with great agility. This theme found powerful expression in his next novel, *A Dream of Wessex* (1977), an intensely English, intensely science-fictional, beautifully parsed and formed novel. It is, indeed, something wholly new in fiction: SF mediated through Thomas Hardy. The Hardy connection is flagged up by Priest's evocation of a future 'Wessex', but it runs much deeper than that. It has to do more with an awareness of the ways in which character and destiny shape one another. The plot concerns 39 human minds connected in a consensual virtual reality in which Wessex has been separated from mainland UK. This imaginary world is a possible future (a century and a half in the future in fact), and its inhabitants are unaware that their consciousnesses are determining this world. Some critics have talked of the novel's conceit as representing an early manifestation of 'cyberspace' (seven years before Gibson's *Neuromancer*): but this isn't quite right. The exploration of 'reality' and 'simulation' is much more complex in *A Dream of Wessex* than in the typical matrices of cyberpunk fiction and film. Priest explores the way reality is shaped as much by our unconsciousness as our conscious minds; he described *A Dream of Wessex* as

> a kind of valediction to traditional sf, because it explicitly describes the process of futuristic imagining, then subverts the whole business. It has been described as the novel that predicted virtual reality, but that's because whoever said it hadn't spotted the subversion.

This is not to say that Priest is nothing more than a subverter, a narrative game-player who enjoys leading his readers up the garden path, a dealer in trick endings and nothing more. His novels access a deeper truth. It is not fiction that toys with our consciousness; it is the nature of mind itself. This is Freud's insight, of course; the invention of an unconscious mind to which the conscious mind does not have access, and which indeed thinks *against* the conscious mind in many ways. Something like this is the root ambiguity of Priest's work.

Feminist science fiction

This Critical History has so far avoided ring-fencing any portion of any chapter and corralling within it examples of 'female SF'. This has not been because of any hostility towards feminist criticism or the (massively valuable and important) project of feminism itself; but rather as a reflection that the many women that have written SF since the eighteenth century did not, until the later 1960s, do so out of a sense of gender solidarity, or even as members of gendered groupings seeking to advance women's rights. But one of the most significant cultural facts of postwar European and American social life was the rise of a complexly interconnected series of women's movements across several generations. This inevitably found correlatives in SF, as in all modes of art; and although the fictional articulations of the first generation of feminist activists tended to be couched in realist modes (by way of capturing), many younger thinkers and writers were excited by the imaginative possibilities afforded by a non-realist idiom. In the words of the critic Sarah Lefanu:

> One of the major theoretical projects of the second wave of feminism is the investigation of gender and sexuality as social constructs ... The stock conventions of science fiction – time travel, alternate worlds, entropy, relativism, the search for a unified field theory – can be used metaphorically and metonymically as powerful ways of exploring the construction of 'woman'. (Lefanu, pp. 4–5)

Joanna Russ (b. 1937) may be the most celebrated avowedly feminist writer the genre produced at this time. Her most famous novel, *The Female Man* (1975), presents a four-fold perspective of women's experience of the world, including the women-only utopian planet Whileaway. Each of the four protagonists, Janet, Jeanine, Joanna and Jael, is an aspect of one character; and yet each is very different from the others, determined as they are by very different societies: two set in different 1960s Americas, one on Whileaway, and one during a future war between the sexes. The book is put together in a rather blocky manner, the four elements written in each woman's voice not so much interwoven as jammed up against one another, chunks rather than threads. One positive effect of this is that of a commendable *nouveau roman*-esque dislocation of our assumptions about character, and (the book's project) gender; one negative effect is that the book lacks flow, and some readers find it hard going. Perhaps the most famous 'feminist' SF novel in English, *The Female Man* has received a great deal of respectful criticism. But Gwyneth Jones is surely right when she judges it a relative failure compared with some of Russ's other fiction. The utopian possibilities of Whileaway are juxtaposed with the variously oppressive three other possible worlds (Russ insists on all four as contemporaneous facts). Russ has written about all-female societies elsewhere, most notably in *When it Changed* (1972), but, as Jones points out, the female society there is 'not unreasonably idealised', the constituent women combining the faults and strengths of 'the whole of humanity'. But three years later, in *The Female Man*, the all-woman world 'has been got at. Its inhabitants have become female characters in a feminist science fiction, their vices and virtues bowdlerised and engineered precisely to fit the current demands of sexual politics.' Russ's novel becomes in effect hijacked by a feminist agenda: '*When it Changed* is feminist fiction, *The Female Man* is feminist satire' (Jones, pp. 125–6).

Russ was no bestseller and is perhaps more highly rated by academic critics of the genre than by fans. But there were women SF writers in this period who enjoyed very considerable commercial success. For instance, the American author Marion Zimmer Bradley (1930–1999) created an SF mega-text with her 'Darkover' novels (26 titles from *The Planet Savers*, 1962 to *A Flame in Hali*, 2004, completed after her death by Deborah J. Ross). These entertaining if sometimes one-dimensional adventure stories, all set on the planet Darkover at various stages of its technological development, gained a sizeable fan community, who not only discussed the novels but wrote their own fan fiction set in this universe. The earlier Darkover novels were conventionally masculinist in focus, but during the 1960s Bradley's interests moved in the direction of feminism, and a New Age variety of spirituality, taking her sizeable fan base with her. In 1985 she spoke for many when she declared that 'my current enthusiasms ... are Gay Rights and Women's Rights – I think Women's Liberation is the great event of the twentieth century, not Space Exploration. One is a great change in human consciousness; the latter is only predictable technology, and I am bored by technology' (Bradley, p. 13).

Marge Piercy (b. 1936) has also enjoyed popular as well as popular success with a number of novels that use SF as a means of exploring the cultural logic of oppression (in terms of gender, but also more generally) and the possible strategies for combating this. The ambitious *Women on the Edge of Time* (1976) contrasts, perhaps too schematically, the miserable existence of its protagonist Connie Ramos – a poor, overweight Chicana woman in 1970s New York who has been committed to a mental institution – with the future society of which Connie has glimpses, a gender-equal utopia of the year 2137, where the very pronouns for 'he' and 'she' have been replaced by the all-purpose 'per'. Perhaps Piercy's best novel is the Arthur C. Clarke award-winning *He, She and It* (1991). Set in the America of 2056 (largely dystopian except for a few 'free cities' where life is better), this text juxtaposes a narrative about the creation and programming of a robot by a brilliant, emotionally wounded woman called Shira, with a story-within-story about the creation of a golem by the famous Jewish mystic Rabbi Judah Loew ben Bezalel in the seventeenth century. The appeal, and dangers, of the new cyborg beings is mediated via this expressively science fictional dialectic, the twenty-first-century technological and the seventeenth-century magical.

Arguably the most gifted female writer of this period was Alice Sheldon (1915–1987), who wrote under the pen-name James Tiptree Jr. Indeed, it may be that the mellow and subtle complexities of her (sometimes simple-seeming) stories have less currency among twenty-first-century fans than does the story attached to her writing identity. Adopting her gendered pseudonym (the surname came from a brand of marmalade) and responding to enquiries with many biographical details – such as her stint working in the Pentagon and her role in setting up the CIA – except her gender, she was assumed by many to be male. The revelation of her actual gender in the late 1970s embarrassed two notable figures in SF in particular: Robert Silverberg (b. 1935), who had written an introduction to Tiptree's collection *Warm Worlds and Otherwise* (1975) rubbishing the suggestion that she could possibly be a woman;[5] and Ursula Le Guin (b. 1929), who had prevented Tiptree from adding a signature to a feminist petition on the grounds that she was a man. As an icon of female genius in the face of the inherent masculinist bias in culture in general, Tiptree has been enthusiastically adopted by feminist critics of the genre. And some of her stories do make penetrating points about gender, most famously 'The Women Men Don't See' (1973), in which a group of women are as happy to live in 'the chinks of the world machine' of an alien spacecraft as in the interstices of a male-dominated society. In 'The Screwfly Solution' (1977) the widespread murder of women turns into a global holocaust, as an alien agent, wanting to depopulate the globe prior to moving in, causes men to confuse sexual lust and bloodlust. It is a genuinely chilling and upsetting tale that works as both a caricature of conventional male attitudes and as a properly SF intervention into a plausibly extrapolated world.

But her stories range far beyond gender issues. 'I'll Be Waiting for You when the Swimming Pool is Empty' (1971) is an energetic satire on the 1960s Peace Corps, in which infectious 'American' values overwhelm the native cultures with which it has contact – although the story's comedy saves it from being po-faced. 'The Girl Who Was Plugged In' (1974) nicely satirises corporate commodification and reification; both 'Faithful to Thee, Terra, in Our Fashion' (1968) and 'And I Awoke and Found Me Here on the Cold Hill's Side' (1971) are clever explorations of colonial and post-colonial logics from both sides of the equation. Overall, her unillusioned and powerful vignettes of extrapolated reality are among the most compelling in recent SF.

European science fiction of the 1960s and 1970s

Europe continued producing a torrent of superb SF; although the lamentable observation must be made that, as SF became increasingly commercialised towards the end of century, and especially as it became implicated in the mostly American idioms of mass cultural TV and film, the relative impact made by non-Anglophone writers diminished. As American culture assumed an unprecedented global infiltration, a great deal of good SF became swamped in the Anglophone chauvinism in which the English speaker felt less and less pressure to learn another language, and deigned to notice literature only if translated into English.

After dominating the genre in the nineteenth century, French literary SF was entering a silver age (although French visual SF, and in particular 'bandes dessinées' – comics both serialised and in album form – was coming into its glorious flowering, with the creation of adult, often controversial texts, such as Jean-Claude-Forest's erotic 'Alice in Wonderland' variant *Barbarella* (1962), the publication of the vastly influential anthology title *Métal Hurlant* (from 1975) and the rise of creators such as Jean Giraud (b. 1938, pen-name 'Moebius') and Phillippe Druillet, who used comic book art as a launch pad for archetypal explorations of the collective unconscious, introspective mysticism and existentialist, often surreally and psychedelically inclined philosophising). The reasons for this are not clear; Jacques Goimard, a French journalist, wrote in 1970 that 'la science-fiction française ne manque donc pas d'inspiration. Sa véritable maladie est d'origine économique' ('French SF doesn't lack *inspiration*; its real problem is *economic* in origin'; quoted in Gattégno, p. 32). For whatever reasons, clearly related to the radical social and cultural changes that took place in postwar France, written SF became less noteworthy.

Although René Barjavel (see above p. 218) continued writing until his death in 1985, his later fiction is much less focused and powerful than his earlier works: for instance, the novel *La Nuit des temps* ('The Night of the Age', 1968), in which an over-familiar premise of post-nuclear war doom is treated at too great a length. But even a book like the Belgian Jacques Sternberg's (b. 1923) *Toi, ma nuit* ('You, My Night', 1956), which brilliantly and wittily treats the coming of a new age of sexual indulgence, could find no audience outside France. Some French books did make a larger impact: perhaps the most famous work of postwar French SF is *La Planète des singes* ('The Planet of Apes', 1963) by Pierre Boulle (1912–1994); although the fame is due more to the successful 1968 Hollywood film than to the mordantly witty 'conte philosophique' form of the original novella. The Algerian novelist Robert Merle (1908–2004) is better known for his mainstream fiction, but he produced two powerful and influential books, *Un Animal doué de raison* ('An Animal Gifted with Intelligence', 1967) and *Malevil* (1972); the first dealing with a group of dolphins taught to speak with humans, the second with the aftermath of nuclear war and the precarious process of rebuilding. Once again, extra-continental fame for these books owed more to film versions than the novels themselves (the films were, respectively, *The Day of the Dolphin* (Mike Nichols, 1973) and *Malevil* (Christian de Chalonge, 1981)).

In Germany, by contrast, the 1960s saw a renaissance in popular SF. In large part this was due to the worldwide success of the 'Perry Rhodan' series. The brainchild of Walter Ernsting (b. 1920) and Karl-Herbert Scheer (1928–1991), the first Perry Rhodan novel appeared in 1961; subsequent novels in the sequence have appeared regularly (on occasion as often as once a week, produced by a fast-working team of authors). Rhodan is an American astronaut who discovers a crashed alien spacecraft on the Moon, and becomes involved in a fantastically burgeoning series of adventures that lead him to assume the role of 'Peace Lord of the Galaxy'. The American Forrest J. Ackerman (b. 1916) edited a series of English-language translations of the first 118 novels from 1969 to 1977, and a sizeable Anglophone fan community grew up, with the usual trappings of fan culture: conventions, fan fiction, blueprints of the spacecraft, art, comic books, several spin-off series of books (*Atlan*) and magazines (*Dragon* and *The Planet Series*), and a low-budget Euro movie, *Perry Rhodan: SOS aus dem Weltall* ('SOS from Outer Space', dir. Primo Zeglio, 1974). The American and British markets dried up in the late 1970s.

Nevertheless, the series has run continuously in its native Germany since the first title appeared in 1961. More than 2,200 novels have now appeared, some 80 million words, giving *Perry Rhodan* an unprecedented longevity as an SF presence.

Taking the whole series as a sort of mega-text, an extraordinary extension of the concept of the *roman fleuve* to the idiom of Pulp SF, presents us with an especially fascinating problematic of critical interpretation. Clearly, Perry Rhodan takes to unique extremes one of the most culturally persistent forms of SF: the Pulp serial adventure, 'mega-texts' such as *Flash Gordon*, the *Star Trek* franchise, or the six-title *Star Wars*. But none of these composite works has the sheer scale and complexity of Perry Rhodan. A similar though much smaller success has been enjoyed by the 150 novels in the *Orion* sequence (many by Hans Kneifel, b. 1936), which spun off from a successful *Star Trek*-like German TV series *Raumpatrouille: Die phantastischen Abenteuer des Raumschiffes Orion* ('Space Patrol, the Fantastic Adventures of the Spaceship Orion', 1965–66).

Other European SF authors worked in a more self-consciously literary idiom. The Austrian Herbert Franke (b. 1927) has published a series of very highly regarded novels, beginning with *Das Gedankennetz* ('The Thought-Net', 1961) and *Der Orchideenkäfig* ('The Orchid Cage', 1961), in which astronauts explore a mysterious planet. His earlier fiction tends to portray humans overwhelmed by hostile or indifferent superior forces, but his later writing has been more interested in the possibilities of human supermen, often individuals whose super-potential is not immediately obvious; notable examples are to be found in the short story collection *Zarathustra kehrt Zurück* ('Zarathrustra Returns', 1977) and the novel *Schule für Übermenschen* ('School for Supermen', 1980). The Italian writer Italo Calvino (1923–1985) is famous for his witty and carefully braided meta-texts, stories that reflect the process of story-making and embody themselves. Many of these are fantastical, and some are SF; for instance, the interlinked short stories *Ti con zero* ('T at Zero', 1967) and *Le cosmicomiche* ('The Cosmicomics', 1968), in which an unimaginably old narrator called Qfwfq retells the story of the cosmos from a series of ingenious, comedic perspectives.

The greatest European writer of postwar SF is probably the Polish writer Stanisław Lem (b. 1921), an uncompromisingly European writer who has nevertheless established an enormous international reputation. It has helped that he has been widely translated, and that two highly regarded SF films (*Solaris*, dir. Andrei Tarkosvky, 1971, *Solaris*, dir. Steven Soderbergh, 2003) have been made out of his novel *Solaris* (1961). That novel, Lem's most famous, takes as its premise an oceanic planet that is conscious and intelligent, and which creates simulacra humans that interact and unsettle the human occupants of an observing space station. As a meditation on the tendency of human consciousness to reduce alienness to variations of human sameness the book has rarely been bettered; and the films create some powerful visual correlatives to the book's meditations. But the true flavour of Lem's witty, intensely thoughtful, ingenious writing is poorly captured by cinema. Peter Swirski comments that 'there can be no doubt about the centrality of philosophical and scientific thought in Lem's writings', and adds that 'it would take a polymath equal to Lem himself to engage critically his hypotheses and scenarios' which range over a staggering range of the human sciences and arts. Swirski thinks that this *'embarras de richesse* may partly explain why so few literary scholars to date have penetrated (his) conceptual frameworks in a systematic way' (Swirski, pp. xi–xvi).

The linked short stories *Opowieści o pilocie Pirxie* ('Tales of Pirx the Pilot', 1968) are fantastically imaginative and wide-ranging, while always connecting to a deeply considered meditation on the human condition. Similarly, the tales collected in *Cyberiada* ('The Cyberiad', 1965) constitute a robot-centred mock-epic, which is darkly funny and thought-provoking. Almost as if flaunting his heroically fertile imagination, Lem published a large number of reviews of imaginary books, a notion he took from Jorge Luis Borges: *Doskonała próżnia* ('Perfect Vacuum', 1971) and *Wielkość urojona* ('Imaginary Magnitude', 1973) contain more ideas and perception than is found in the entire careers of most middle-ranking western SF authors. *Katar* ('Catarrh', also known in English as 'The Chain of Chance', 1977) reads like the synthesis and terminus of all detective novels, tracing a baffling mystery through its myriad of essential and accidental circumstances back to a satisfying explanation that reinforces the contingency of human existence.

The Russian brothers Arkady Strugatski (1925–1991) and Boris Strugatski (b. 1931) collaborated on a number of very interesting SF books. Their first major success was *Trudno byt' bogom* ('Hard to be God', 1964), in which human agents from a Communist Earth are working undercover on an alien world helping it develop past its medieval-level technology without falling into fascism. But being regarded as gods by the indigenous people corrupts them, although not in the obvious ways a lesser writer might have been drawn to. The novella *Piknik na obochine* ('Picnic by the Roadside', 1977) is a brilliantly handled mood-piece, dealing with a mysterious zone in Canada where aliens, it seems, have discarded various artefacts. It was filmed by the Russian director Andrei Tarkovsky as *Stalker* (1979), which is either the most magisterially beautiful and profound, or the most constipated and boring, film ever made: it is really hard to decide which.

Japanese science fiction

Hoshi Shin'ichi, perhaps Japan's most famous modern SF writer, has described the 1960s as 'the golden age of Japanese SF' (quoted in Matthew, p. 41). The profound changes that reshaped Japanese society after the war, turning an old-fashioned feudal society into an industrial and consumer powerhouse, unleashed a potent complex of forces; and SF – as that genre best able to mediate and analyse the impact of rapid technological change – grew rapidly. By the 1990s SF represented a major portion of Japanese popular culture; Shibano Takumi estimates that by the 1990s '400 Japanese original and 150 translated SF books are published each year' (Shibano, p. 640). The influence of American culture on postwar Japan has been extensive and much Japanese SF reflects this; but there has been a good deal of aboriginal writing in the genre. 1960 saw the first appearance of *S.F. Magajin* ('SF Magazine'), publishing original Japanese SF; it is still published today.

Hoshi Shin'ichi (b. 1926) recently announced his retirement after publishing 1,000 stories, most of them SF. His story 'Bokko-chan' (1958) is a sharp and ingenious story in which a robot girl, programmed with only a basic level of response, becomes the unwitting agent of destruction for the various lonely men who come to the bar in which she works. It can be read as a satire on the limited social roles permitted women in traditional Japan. Abé Kobo's (1924–1993) *Daiyon Kanpyo-ki* ('Inter Ice-Age 4', 1959) is a complex novel in which a super-computer programmed accurately to predict the future exercises a baleful effect on its programmer's life, dominating and then wrecking it. The computer is called MOSCOW II. Abé, a Communist, was expelled from the Japanese Communist Party in 1962. Komatsu Sakyo (b. 1931) enjoyed considerable success with *Nippon Chinbotsu* ('Japan Sinks', 1971), a disaster novel in which the Japanese archipelago begins to slide into the ocean, and the entire population must be evacuated; over four million copies of this novel were sold. Komatsu's *Nippon Apatchi-zoku* ('The Japanese Apaches', 1964) concerns a tribe of disenfranchised Japanese living on a fenced-in waste land who, by eating steel, become indomitable cyborgs, repelling assault and eventually being the only Japanese to survive a nuclear war.

References

Aldiss, Brian, *Hothouse* (1962; London: Sphere 1976)

Aldiss, Brian, *Frankenstein Unbound* (1973; London: Pan 1975)

Aldiss, Brian, *Affairs at Hampden Ferrers: an English Romance* (London: Little, Brown 2004)

Aldiss, Brian, with David Wingrove, *Trillion Year Spree: the History of Science Fiction* (London: Gollancz 1986)

Amis, Kingsley (ed.), *The Golden Age of Science Fiction* (Harmondsworth: Penguin 1981)

Barth, John, *Giles Goat-Boy, or, The Revised New Syllabus* (1966; London: Penguin 1967)

Bradley, Marion Zimmer, *The Best of Marion Zimmer Bradley*, ed. Martin H. Greenberg (London: Orbit 1993)

Broderick, Damien, 'New Wave and Backlash: 1960–1980', in Edward James and Farah Mendlesohn (eds), *The Cambridge Companion to Science Fiction* (Cambridge: Cambridge University Press 2003), pp. 48–63

Burgess, Anthony, *A Clockwork Orange* (1962; London: Penguin 1972)

De Bolt, Joe and John Pfeiffer, 'The Modern Period, 1938–1975', in Neil Barron (ed.), *Anatomy of Wonder: Science Fiction* (New York: R. R. Bowker 1976)

Disch, Thomas, *The Dreams our Stuff is Made Of: How Science Fiction Conquered the World* (New York: Simon and Schuster 1998)

Fox, Robert Elliot, 'The Politics of Desire in Delany's *Triton* and *Tides of Lust*', in James Sallis (ed.), *Ash of Stars: On the Writing of Samuel R. Delany* (Jackson, MI: University of Mississippi Press 1996), pp. 43–61

Gattégno, Jean, *La Science-fiction* (Paris: Presses Universitaires de France 1971)

Heinlein, Robert, *Stranger in a Strange Land* (1961: New York: Ace Books 1987)

Huntingdon, John, *Rationalizing Genius: Ideological Structures in the Classic American Science Fiction Short Story* (New Brunswick, NJ: Rutgers University Press 1989)

James, Edward, *Science Fiction in the Twentieth Century* (Oxford: Oxford University Press 1994)

Jones, Gwyneth, *Deconstructing the Starships: Science, Fiction and Reality* (Liverpool: Liverpool University Press 1999)

Lefanu, Sarah, *In the Chinks of the World Machine: Feminism and Science Fiction* (London: Women's Press 1988)

Le Guin, Ursula K., *The Language of the Night: Essays on Fantasy and Science Fiction*, ed. with an intro. by Susan Wood (London: Women's Press 1989)

Luckhurst, Roger, *Science Fiction* (London: Polity 2005)

Matthew, Robert, *Japanese Science Fiction: A View of a Changing Society* (London: Routledge/Oxford: Nissan Institute of Japanese Studies 1989)

Moorcock, Michael, *The Cornelius Quartet* (*The Final Programme, A Cure for Cancer, The English Assassin, The Condition of Muzak*) (1968–77; London: Phoenix 1993)

Moorcock, Michael et al., *The New Nature of the Catastrophe*, ed. Langdon Jones and Michael Moorcock (London: Orion 1997)

Nabokov, Vladimir, *The Stories of Vladimir Nabokov* (Harmondsworth: Penguin 1995)

O'Reilly, Timothy, *Frank Herbert* (New York: Ungar 1981)

Rochelle, Warren G., *Communities of the Heart: the Rhetoric of Myth in the Fiction of Ursula Le Guin* (Liverpool: Liverpool University Press 2001)

Russ, Joanna, 'The Image of Women in Science Fiction', in S. K. Cornillon (ed.), *Images of Women in Fiction* (Bowling Green, KY: Bowling Green University Popular Press 1972)

Shibano Takumi, 'Japan', in John Clute and Peter Nicholls, *Encyclopedia of Science Fiction* (2nd edn., London: Orbit 1993), pp. 639–41

Silverberg, Robert, *Edge of Light: Five Classic Science Fiction Novels* (*A Time of Changes* (1971); *Downward to the Earth* (1971); *The Second Trip* (1972); *Dying Inside* (1972); *Nightwings* (1968/9)) (London: HarperCollins 1998)

Slusser, George, *The Delany Intersection: Samuel R. Delany Considered as a Writer of Semi-Precious Words* (San Bernando, CA: Borgo Press 1977)

Stevenson, Randall, *The Last of England?* ('The Oxford English Literary History vol. 12, 1960–2000'; Oxford: Oxford University Press 2004)

Stockwell, Peter, *The Poetics of Science Fiction* (Harlow: Longman 2000)

Sutin, Lawrence, *Divine Invasions: A Life of Philip K. Dick* (Carroll and Graf 2005)

Suvin, Darko, 'Afterword: With Sober, Estranged Eyes', in Patrick Parrinder (ed.), *Learning from Other Worlds: Estrangement, Cognition and the Politics of Science Fiction and Utopia* (Liverpool: Liverpool University Press 2000)

Swirski, Peter, *Between Literature and Science: Poe, Lem, and Explorations in Aesthetics, Cognitive Science, and Literary Knowledge* (Liverpool: Liverpool University Press 2000)

Thomas, D. M., 'Two Voices', in *Penguin Modern Poets 11: D. M. Black, Peter Redgrove, D. M. Thomas* (Harmondsworth: Penguin 1968)

12
Science Fiction Screen Media 1960–2000: Hollywood Cinema and Television

Two major things happen to SF in the last four decades of the twentieth century. Most importantly, it undergoes a transformation, becoming increasingly a genre dominated by 'visual media' and especially by 'visual spectacularism', a special sub-genre of cinema that is predicated primarily on special effects, the creation of visually impressive alternate worlds, the realisation of events and beings liable to amaze. The other shift linked to this is that SF becomes less markedly a 'literature of ideas' and becomes increasingly dominated by an imagistic aesthetic: this involves both more conventional poetic or literary images and more strikingly potent *visual* imagery which penetrates culture more generally (a bone thrown into the sky by a prehistoric ape cuts to a spaceship in orbit ...). It is in the nature of images that they cannot be parsed, explicated and rationalised in the way that 'ideas' – excerpted from a notional 'literature of ideas' – can. Accordingly, there is something oblique about the workings of the best SF of the latter part of the twentieth century; something allusive and affective that can be difficult exactly to pin down.

Linking, as I do here, the rise to prominence of 'visual SF' and the subtle effectiveness of the poetic image may strike some as an odd thing to do in this context. It is more usual to note the rise of SF cinema and TV with regret, as a dilution of the effectiveness and sophistication of the genre. But this, I think, is a mistake. It is true that a new chief mode of SF comes into being, especially after 1977: a form of text known colloquially as the 'Hollywood blockbuster'. Hollywood has now taken on the negative connotations of 'popular cinematic art', and the shorthand is used here. For many, 'Hollywood' denotes lowest common denominator commercialism; but whilst there have been many low-quality, exploitative or reactionary films produced under this cultural logic, there have also been many masterpieces. Moreover, the inherent populism of this idiom has meant that such works achieve a much deeper cultural penetration than was the case with novels or poems.

I am working here from an assumption that human culture is deeply invested in two forms of art in particular: 'story' – which is to say, narrative + characters; and 'lyric' – which is to say, moments of aesthetic intensity that stir and move us, art that captures 'epiphanies' that make the hairs on the backs of our necks stand up.

What I mean by this is that almost all human beings crave stories, and most crave intensities: we find the former in gossip, newspapers, novels, biographies, histories and many other forms; and the latter in what Wordsworth calls 'spots of time' evoked by art or literature, but also and more pervasively in religious experience, in sex and sexual art, and in various other ways.

For much of the last 300 years the dominant mode of 'story' in western culture has been the novel, and the dominant mode of 'lyric' has been poetry. This is, I think, no longer the case. Although there are millions of people around the world today who read novels with great pleasure and indeed more than pleasure, the fact is that most of the global population (even most novel readers) access the *stories* they need primarily through visual media, particularly cinema and TV. I think something similar has happened in *lyric*; the audience for poetry has dwindled startlingly in the last 100 years, but billions of people now find their 'epiphanic' moments of intensity in pop music.[1] This is a rather crude and deliberately over-stated generalisation, but I think it stands as a general cultural backdrop to the rise to prominence of visual media in the history of SF.

The visual has a long pedigree, of course. In previous chapters we have looked at works such as Robida's 'picture-plus-text' confections; the Pulp magazine owed much of its effectiveness to the strong visual component. Professional SF artists, many of them highly talented, became increasingly prevalent as the century progressed. The technology of colour reproduction enabled mass advertising, full-colour magazines and newspapers, which with the eventual ubiquity of TV and cinema has resulted in a super-saturation of the visual in culture as a whole: this is most true in the developed world, but is true to a degree across most of the planet. Indeed, it is the case today that there are many texts in which quite extraordinary visual sophistication and indeed beauty are married to startlingly primitive char-acterisation, narrative and dialogue: let George Lucas's *Star Wars: The Phantom Menace* (1999) stand as an emblematic representation of what I am talking about. In such a culture, the critic who spends his or her energies merely denigrating the text for its insufficiencies – something that's often very easy to do – is embarking, I think, on the less interesting response. The critic's job is to elucidate the extent to which the greatness of the key texts can be explained with reference to its visual potencies. To put this another way: the requirement is for an account of the devel-opment of SF cinema that, rather than seeing special effects as mere ignorable dec-oration, sees them as core to the text: a sense of SF cinema as self-reflexively aware of its own (visual, technological) idiom, and all the more powerful for that.

Early 1960s science fiction cinema

Here is an example of what I am talking about. One of the main successes of George Pal's *The Time Machine* (1960) lies in the use of stop-motion photography to convey the rapid passage of time from the point of view of the time traveller. Indeed, the effectiveness of these scenes renders this colourful film somewhat lopsided, formally speaking. The initial scenes, set as the century turns from 1899 to 1900, see the 'time traveller' (Rod Taylor) telling his friends about his invention of a time machine among nicely realised late Victorian clutter. The film then moves into a brilliant series of vignettes in which time travel is rendered variously with the sun shooting across the sky, with a witty sequence in which the hems of dresses in the shop across the road rise and fall with the annual fashions, and with stop-offs at significant points of future history (including the nuclear bombardment of London in the 1960s). Indeed, so powerfully rendered are these scenes of special effect that they become much more than a mere transition to the far-future of Wells's novella; instead, they dominate the whole. When he arrives among the Morlocks and Eloi there is no time for the complex development and class satire of Wells's original. Taylor's character quickly surmises the nature of the future world, leads the Eloi to freedom against the monstrous Morlock oppressors, and returns to London only to collect books for the continuing job of re-civilisation. Wells's hyper-evolutionary pessimism is wholly missing; but in a way this doesn't matter, for the sequence among the Eloi is the least important in the film. The images that stay with the viewer are of the eons-long accelerated perspective – both the sense of power this grants the viewer (swallowed by lava, Taylor's time traveller merely fast-forwards his device until the rock is picked away by natural erosion and he emerges into a sunny pastoral world) and also the horror (Taylor accelerates into the future after killing a Morlock, and the camera revoltingly but strikingly compresses the decomposition of the corpse into a few seconds). In other words, *The Time Machine* stands at the threshold of a new kind of cinema, a cinema predicated upon the special effects of the camera that reflect upon the medium itself.

It is not that special effects ('SFX') were unknown in earlier cinema, of course. Quite the reverse: *Metropolis* (1927), for instance, incorporates some impressive special effects, but most of these (large, purpose-built sets, a big cast, scaling of models to interact with live-action performance) are, in essence, the special effects of the Grand Theatre of the nineteenth century. They create a sense of spectacle (as did D. W. Griffith's transhistorical epic *Intolerance* (1916), also via the creation of huge theatrical sets), but they do not create a sense of specifically *cinematic* spectacle. SF, always enamoured of technology, finds in those new technologies of the cinema developed in the last half of the twentieth century not only a means of realising its vision, but of embodying its very aesthetic.

Special effects (less sophisticated but still used to striking effect) are also the saving grace of *X – the Man with the X-Ray Eyes* (1963), a low-budget feature by Roger Corman (b. 1926). After a conventional start, in which a scientist experimenting on himself develops X-ray vision, *X* goes on to develop a weird and rather compelling quality of existential *angst*, as its protagonist sees through to the essential hollowness of the universe and goes mad. A similar pessimism, powerful though a little adolescent, also inflects Joseph Losey's B-movie *The Damned* (1961), about a secret British government programme to irradiate children in order to render them invulnerable to atomic fallout.

But, speaking generally, the cultural mood of the early 1960s in the US and UK did not affect SF cinema in the same liberating way that it affected written 'New Wave' SF. Presumably this was because the much higher costs of producing a film-text mean that cultural production is more implicated in corporate and investment interests, and there is less space for the explicitly counter-cultural. Both Corman and Losey were working outside the system of the big studios. The more mainstream SF films of the period tended to reflect the political concerns of the Cold War, the Cuban Missile Crisis (1963), and the on-going fear of nuclear war. Films such as the sombre, low-key *On the Beach* (Stanley Kramer, 1959, based on Nevil Shute's 1957 novel) is set in Australia, the only country not destroyed by a nuclear war, although the approaching cloud of radioactive dust is bringing inevitable annihilation with it; and *Fail Safe* (Sidney Lumet, 1964) which presents the destruction of the world not in terms of spectacular explosions but rather a miserabilist, talky post-mortem of humanity's failings. Non-US films set in post-nuclear environments were often less preachy. In the ambitious Japanese *Dai-sanji Sekai Taisen: Yonju-ichi Jikan No Kyofu* ('Third World War: 41 Hours of Horror', Shigeaki Hidaka, 1960) the US accidentally drops an atom bomb on Korea, leading to a war in which most of the world is destroyed. That the atomic apocalypse many in the 1960s expected did not materialise (or has not yet materialised) does not give us the right to look condescendingly on the very real fear it generated.

Nevertheless, it is Stanley Kubrick's black comedy *Dr Strangelove or: How I Learned to Stop Worrying and Love the Bomb* (1963) that is the only one of this sub-genre not to have been diminished into a period piece by the passage of time. This still hilarious film tells the story of a lunatic American general (obsessed with his own cleanliness and 'precious bodily fluids') who launches an unauthorised nuclear strike against Russia. The frantic attempts of the western authorities to counter-mand his order are related partly through a near-documentary visual style of urgent cutting and handheld camera work, and partly through a more hypertropic and caricatured representation of the US War Room. At one point rival army officers fall to blows, prompting the president (played by Peter Sellers) to reprimand them: 'Gentlemen! You can't fight in here – this is the War Room!' The title character, also played by Sellers, is an ex-Nazi rocket scientist, confined to a wheelchair and with a false arm, presumably intended as a broad-brush satire on the high public profile of the former Nazi Wernher von Braun (1912–1977), then director of NASA's Marshal Space Flight Center and overseeing the Apollo programme. But Sellers' Strangelove is a long way from any actual human nuance; his insane delight at the prospect of the coming disaster, his ridiculous cod-German accent (Sellers manages more comic business with the single mispronunciation 'compuder' for 'computer' than can easily be believed) and the fact that his prosthetic arm seems to have a life of its own, spasmodically jerking into a 'Heil Hitler' salute beyond the control of its master, all contribute to a hilariously burlesque mood which is all the more powerful for being offset by those sections in the film shot in quasi-documentary style. The film was very successful.

Another marginally SF film from the same period, which also relies on caricature, exaggerated sets and props, and a certain black humour, was the first James Bond movie, *Dr No* (Terence Young, 1962). This film loses out to *Strangelove* in positing its super-villain (who also has mechanical hands and a maniacal desire to see the world descend into apocalypse) *outside* the western establishment rather than, as with Kubrick's masterwork, within it. But it proved the first of a desperately popular series charting the repetitive adventures of super-spy Bond, an action hero in the finest SF tradition who uses advanced and occasionally space-travelling technology as a means of augmenting his own super-competent physical strength and Will.

But the films from the 1960s that went on to have the biggest impact on SF-cinema and cinema more generally used the SF mode as a means of working for-mally through the possibilities of the new grammar and idiom of cinema itself. One example is the half-hour black-and-white French film *La Jetée* ('The Jetty', Chris Marker, 1962), regarded by many as one of the best SF films made. Set after a nuclear holocaust in a Paris in which the law of cause and effect is breaking down, this subtle and suggestive time-travel drama in fact externalises the very logic of cinematic montage. Almost all the visual text is made up of a succession of photographic images, held on the screen for varying lengths of time. In an attempt to undo the damage of the nuclear war, scientists are attempting to send subjects backwards and forwards through time. The main subject is chosen because of his peculiar attachment to a particular photograph (the face of a woman). The effect is to concentrate the viewer on the power of 'the image', as well as on the eloquence that can be achieved by the straightforward juxtaposing of images. As it ends, tragically, motion blossoms into the static succession of still photos to very moving effect. The pared-down quality of *La Jetée* is part of its extra-ordinary effect; but the way it compels the viewer to reflect not only on the story being told but also on the medium used to convey it makes it distinctly and bril-liantly SF. Jean-Luc Godard's oblique SF *policier Alphaville* (1965) is preferred by some, although it seems obfuscatory compared to the luminously still complexi-ties of *La Jetée*. (When the American director Terry Gilliam – a very different kind of genius – remade Marker's film in 1996 as *Twelve Monkeys* the replacement of the original's cool stillness with Gilliam's trademark burly energy and hyperactivity eroded the emotional affect, although it did produce an interestingly dislocated SF text.) In retrospect the significance of *La Jetée* lies in the ways it anticipates the aes-thetic and, to some extent, mood of the first undeniable masterpiece of SF cinema, Stanley Kubrick's *2001: A Space Odyssey* (1968).

Mid-1960s SF produced fewer great films. On the one hand, there were a num-ber of Pulp SF films. For instance, in the Japanese *Matango* (Inishiro Honda, 1963; released in the US as *Fungus of Terror*) shipwrecked survivors are obliged to subsist on a strange mushroom. One by one they turn into mushrooms; the only survivor appears to get back to civilisation intact only to discover mushroom growths sprouting on his face: rescuing a silly concept by its brio and energy. On the other hand were more 'art-house' films. *Fahrenheit 451* (1966) is one of French director François Truffaut's least interesting films, although it is the one with which Anglophone cinema-goers are likely to be most familiar. Truffaut, a *nouvelle vague* director with a spectacular gift for capturing the rhythms and textures of ordinary life, stumbled a little over Ray Bradbury's classic novel of a bibliophobic future world. Bradbury's almost evanescent subtlety of effect is transferred into a too obvious anti-authoritarian satire. Then again there were films which straddled this divide. *Barbarella* (Roger Vadim, 1967) was an Italian production, although it starred the American actress Jane Fonda in the title role. A camp, erotic, far-future romp, this film – although it retained a certain cult appeal into the 1980s (the high-profile pop group Duran Duran took their name from one of its characters) – has dated badly: the sex seems fussily peek-a-boo and the SF plotline risible.

2001: A Space Odyssey

Kubrick's monumental film was several years in the making, and finally released in 1968. Kubrick had worked closely with Arthur C. Clarke (see above p. 212), expanding his short story 'The Sentinel' into something more transcendent and mystical. The film is divided into four, unequal sections. The first, set in Africa at 'the dawn of time', wordlessly relates the adventures of a group of pre-human hominids. A strange black monolith, like a colossal black tombstone, appears among them as they sleep, apparently upgrading their intellects; after its appearance they become tool-users. The transition from this prelude to the main body of the film is effected via the most famous cut in the whole of cinema: a hominid, delighted at having killed a rival with an animal's jaw-bone, hurls it into the blue African sky. Kubrick's camera follows it up to its apogee, and as it starts to fall cuts to a twenty-first century spaceship tumbling through outer space in orbit around the world. The implication (that the spaceship, though considerably more complex, is simply another tool, like the bone) is gracefully alluded to rather than preachily insisted on; and, indeed, the effect of the transition is less ratiocinative and more poetic – strangely affective, uplifting and obscurely beautiful. Indeed, by far the most important thing to note about *2001* is simply how beautiful a visual artefact it is.

The second sequence sees a certain Dr Floyd travelling to a lunar base and out onto the lunar surface, where another example of the monolith has been discovered. But the conceptual mystery of the monolith is less compelling than the slow, fluid beauty of Kurbick's *mise-en-scène*, the drawn-out sequence of space travel harmonised by a Strauss waltz soundtrack. The monolith sends a signal in the direction of Jupiter, and the third (longest) section of the film concerns the events of the spaceship *Discovery* travelling to Jupiter to investigate. *Discovery*'s crew are all in hibernation with the exception of Bowman and Poole, and the sinisterly softly spoken computer, HAL. The main event of the movie is HAL's insanity; the computer murders Poole and nearly succeeds in killing Bowman, although he is able to enter the machine's central core and shut it down. Once again, exciting though these actions are, it is the slow, graceful stillness of Kubrick's direction that is the most effective element. *Discovery* is a fully realised environment; and Kubrick the first director to convey that actual space-travel is (relative to the enormous distances that must be covered) achingly slow, that it simply *takes a very long time*. As Hollywood in the 1960s and 1970s became more and more fascinated with speed for its own sake – showcased in the exhilarating acceleration of the automobile of *Grand-Prix* (John Frankenheimer, 1966); *Bullitt* (Peter Yates, 1968) and *The French Connection* (William Friedkin, 1971) – this strikes a gloriously contemplative, melancholic note. It has not beguiled all critics. According to Vivian Sobchack, films such as *Destination Moon* (1950), *When Worlds Collide* (1951) and *Forbidden Planet* (1956) portray their spaceships as 'breathtakingly beautiful' and 'palatial', but *2001* 'gives us in "Discovery" a mechanism which barely tolerates and finally rejects human existence', and a sense of 'entrapment and confinement' also present in *Marooned* (John Sturges, 1969) and *Silent Running* (1972) (Sobchack, pp. 5–6). This may strike many as a strange judgement; partly because Kubrick's emphasis throughout is precisely on the 'breathtakingly beautiful'. But more importantly, it is surely the case that the limitations of the *Discovery* are the limitations of the hermitage. A meditative quality dominates *2001* as a film, a potent realisation of qualities of solitude almost monk-like.

The film's final section, though brief and functionally a coda to the whole, tends to dominate analysis, as if it provides some sort of key for comprehending the 'mystery' of *2001*. Bowman, alone, finds a massive version of the alien monolith in orbit around Jupiter. This opens as a sort of gateway or portal, and Bowman is taken down an interstellar corridor composed purely of trippy, multicoloured images. At the end of this psychedelic transcendence he finds himself in an oddly neon-lit Louis Quinze suite of rooms, where – in, once again, a beautifully elongated and poised set of shots – he grows old, before being reborn as a 'star child', a luminous and apparently enormous foetus in orbit around the Earth. But although many at the time, and since, have attempted to excavate a significantly meaningful if oblique message from the film by way of this conclusion, in fact it is a perfectly straightforward version of a longstanding SF theme – the transcendence of humanity, our uplift with the help of aliens into a higher form of being. Indeed, this is one of the core themes of Golden Age SF, and as such the reductive moral of the film is rather backward-looking. The genuinely new thing that *2001* brought to SF was not in its content, but in its form; and in particular in its forging of a specifically space-age visual lexicon out of which Kubrick fashions a powerfully allusive visual poetry. Some people find *2001* a rather chilly, unlikeable film; and some students (to whom I have had the pleasure, over the years, of teaching it) consider it too long, slow and – devastating word – boring. These sorts of reactions are understandable. The characters in the film are all rather distant, passionless, almost machinic; the 'moral' of the film, should you wish to read the film in that reductive manner, is one of ultimate human passivity: the whole of human history is revealed as nothing but a sort of mental computer virus inserted by off-stage aliens via the monoliths. We can take credit for none of it. But to read the film not as a disquisition on the nature of humanity, and certainly not as manifesto, but rather as poem unlocks its massive elegance. It approaches what Wallace Stevens (1879–1955) famously called 'the Supreme Fiction' – Stevens, as poet, is also sometimes accused of being chilly, a little inhuman, in his work. The three commandments embodied in the poem 'Notes toward a Supreme Fiction' (1947) ('it must be abstract', 'it must change', 'it must give pleasure' (Stevens, pp. 329–30)) are behind the 'vivid transparency' and 'celestial ennui' of Kubrick's masterwork.

TV science fiction: *Star Trek* and *Doctor Who*

The origins of televisual SF can be traced back to the late 1940s and early 1950s with shows such as *Captain Video and His Video Rangers* (1949–55) and *Tom Corbett,*

Space Cadet (1950–55). If, as critic Mark Bould perceptively notes, 'the production values of these shows now look extraordinarily amateurish, it is indicative of the extent to which the very fact of TV no longer seems miraculous' to us today (Bould, p. 88). Often broadcast live, these shows frequently transferred the dramatic conventions of western or crime genre shows into a notional outer space: a heroic male protagonist fighting evil. Throughout the 1950s and into the 1960s umbrella formats such as *Science Fiction Theatre* (1955–57) or episodic unlinked serials such as the low-key, contemplative *The Twilight Zone* (1959–64) and the more *outré* and surreal *The Outer Limits* (1963–65) raised production values and narrative expectations.

Many of these early TV serials have vanished from the collective radar of SF Fandom (for instance, *The Man and the Challenge*, 1959–60; *Men into Space*, 1959–60), sometimes because tapes of episodes have literally been wiped. But many of the shows retain dedicated, if specialist, followings even today. On American screens the animated series *The Jetsons* (1962–63) played the strangeness of a space-faring family against the conventions of American suburban family life with comic intent, sometimes fairly amusingly; *Voyage to the Bottom of the Sea* (1964–68), an all-American underwater adventure series set aboard a futuristic submarine which, though energetic, did not live up to the Vernean pretensions of its title; the concept behind *Lost in Space* (1965–68) was a futuristic space-marooned *Swiss Family Robinson*, and accordingly the minutiae of family interaction overwhelmed the rather campy SF elements. *The Time Tunnel* (1966–67) tossed its hapless protagonists from one memorable historical event to another via the titular device's ongoing malfunction; wherever they ended up, from the Ancient World to the Second World War, the travellers were sure to encounter English-speaking historical celebrities.

British TV SF has been a lesser phenomenon, British TV being a proportionately smaller business than in the US. Nigel Kneale, the writer-creator of the quintessentially English scientist Quatermass, put him in peril from alien invasion and buried monstrosities in three successful dramas, *The Quatermass Experiment* (1953), *Quatermass II* (1955) and *Quatermass and the Pit* (1958–59); the character was resurrected, although with less success, in *Quatermass* (1979). In all cases the show connected with the taproot of SF itself, its various storylines all riffing on (to quote Peter Nicholls) 'Kneale's obsessive 30-year repetition of the science-meets-superstition theme' (Clute and Nicholls, p. 983). Kneale was also the writer of the one-off drama *The Year of the Sex Olympics* (1969) a slightly over-boiled but still intriguing dramatisation of a future world in which the bulk of humanity lives only through TV, even down to their sex lives. Gerry Anderson and Sylvia Anderson were behind a number of very successful puppet-based shows aimed predominantly at children, among them the spaceship-centred *Fireball XL-5* (1962–63), the underwater *Stingray* (1964–65) and the galumphing *Thunderbirds* (1965–66), which concerned a global rescue and emergency service run, apparently not-for-profit, by an eccentric millionaire. *Thunderbirds*, with its boy-friendly rockets, space-station, giant cargo-planes and submarines, was the most successful of Anderson's shows; but arguably two later Anderson shows surpassed it in many ways. *Joe 90* (1968–69) was based on the intriguing premise of a youngster into whom could be downloaded the knowledge and skills of a variety of adults. *Captain Scarlet* (1967–68) pitted the colour-coordinated cadre of Earth-protection agents against the threat of a malevolent, disembodied Martian threat known only as 'the Mysterons'. The slightly arbitrary premise of the show concerned the title character, 'a man whom fate has made indestructible' (the theme song still reverberates inside my head, thirty years on: '*In*-de-*struc*-tible! Captain Scarlet!'), which perhaps gives Earth a rather unfair advantage in their interplanetary conflict. No matter how many times the Mysterons killed off Scarlet he always resurrected. But the vivid, beyond-Bond adventures of these stringless puppets captured something of the kaleidoscopic energy of the 1960s, and still plays well today.

But the most important feature of 1960s TV SF was the development of the two most influential TV serials in science fiction, shows that would demonstrate powers of endurance far greater than their original framers could have conceived: the American show *Star Trek* (1966–69) and the British serial *Doctor Who* (1963–89, 2005–). The extraordinary and, to an extent, continuing success of these two franchises says important things not only about the increasing dominance of TV as a cultural medium (and as the century moved towards its conclusion, TV has increasingly become the world's major narrative mode, enjoyed by billions of human beings nightly, something that cannot be said of any other mode of art) – but also about the broader development of SF itself. By this last statement I mean two things: first, to reiterate what I have already argued, this is symptomatic of an increasingly visual bias of the genre. But I am also making an argument about the changes in the textual focus of SF. To some extent it is the TV serial – a collocation of individual texts subordinated to a premise or particular imaginative identity – that becomes the template for all SF textual production. Instead of producing singular, stand-alone texts, SF writers and creators increasingly produce 'mega-texts', interlinked sequences of texts, often spanning several media. As was noted earlier, a single concept, and single novel, like Frank Herbert's *Dune* (1965) can become, by the century's end, a dozen novels, a cinema film, two TV serials, video games, comics, artwork by a dozen artists, and so on. The same is true of most of the significant SF works of the last four decades of the century, and continues to be the case today; the only difference is that, instead of accreting these multiple textual additions, SF works are planned in advance as mega-texts. So, for instance, *Star Wars* (1977) is now six interlinked movies, but also dozens of tie-in novels set in the 'Star Wars' universe, many comic-books, half a dozen video games, magazines, websites, fan fiction, action figurines and many other forms of cultural production. The *Matrix* (1999) franchise was configured as a trilogy of cinema live-action films, into which (to make the story-arc comprehensible) the true fan needed to add a compilation of six shorter animated films *The Animatrix* and a video game *Enter the Matrix*, all set in the imaginary universe of the text. I discuss this development in the genre in greater detail below (pp. 286–7); the issue is relevant at this point in so far as it elucidates the ways these two TV serials grew into popular consciousness throughout the 1960s and into the 1970s.

Doctor Who is chronologically antecedent, although it has not enjoyed quite the success of the *Star Trek* franchise. 'The Doctor' (he has, it seems, no other name) first appeared as a crotchety elderly man, played by William Hartnell, who travelled through space and time with certain companions in a spacecraft shaped like a Police telephone box. This machine, called the TARDIS, is much larger on the inside than the outside; whilst most TARDISes can metamorphose its exterior to blend in with whatever environment it finds itself, the Doctor's craft has lost this facility. We learn that the Doctor is a humanoid alien, one of a race called the Time Lords who, in a nebulously unspecified way, are supposed to 'police' the timeways. Time Lords enjoy extended life spans by virtue of 'regenerating' their bodies, a device which allowed Hartnell to be replaced in the role by the physically dissimilar Patrick Troughton in 1966. Troughton's more impish doctor was followed by the silver-haired and foppish Jon Pertwee in 1970 (for much of Pertwee's time in the role, the Doctor was confined to Earth as a punishment). Tom Baker took over the role from 1974 to 1981, and remains for many the definitive Doctor, playing the role with an endearing eccentricity always edged with an indefinable sense of danger. The fact that Baker was the youngest and (to use a rather outmoded term) most virile of the actors to take the role also added an effective if buried sexual subtext to his relationships with his, often female and nubile, assistants. While less relevant to the extensive audience of children watching, for the adult audience this added spice to a show that, in terms of plot, was becoming rather repetitive. After Baker's day the show ran into diminishing returns, with Peter Davison, Colin Baker and Sylvester McCoy and (after hiatuses) Paul McGann; but it was successfully relaunched in 2005 with the sexy and eccentric Christopher Ecclestone.

Practically speaking, this premise allowed for an indefinitely extended series of adventures: typically the Doctor and his companion(s) would arrive in some location (perhaps an historically interesting period of Earth's history or an alien world), have a number of adventures, face down an adversary and wrap everything up before departing. The Doctor's anti-heroics (he never wields a gun, preferring to outwit opponents; and the character is played in most of his incarnations with an almost caricature Englishness – polite, good-humoured, a little odd) are well balanced by a series of vivid villains. The first of these, the six-foot bollard-shaped Daleks, first appeared early in the show's career: these conscienceless quasi-Nazi cyborgs, with their grating cry 'Exterminate! Exterminate' are genuinely memorable, and have reappeared many times over the years, even going as far as to monopolise the role of the villain in the Doctor's single two cinematic outings, *Doctor Who and the Daleks* (1965) and *Daleks: Invasion Earth 2150 AD* (1966).

Dr Who ran more or less continuously on British TV until the late 1980s and was sold widely around the world. The quality of individual storylines was variable and often dipped alarmingly low: budgets rarely matched the writer's vision; even when that vision was not especially exciting. But the show gathered a very large and dedicated fan base. It is this that has kept *Dr Who* alive, even beyond its cancellation by the BBC in 1989. What is interesting is the way the constraints of the show, its central character and its distinctive English mood, mix with the radical freedoms of the central premise (which enables adventure to be located anywhere in time and space) to enable a seemingly inexhaustible series of new texts to be generated: nearly 200 spin-off or tie-in novels of *Dr Who* adventures have been published, together with a great quantity of unpublished fan-fiction, circulated among friends and at *Dr Who* conventions; there are magazines, numerous websites and other forms of cultural production. Apprehending this very significant SF mega-text would, for the uninitiated, be the work of many years.

Star Trek (1966–69) was conceived with a similarly accommodating premise, and has resulted in an even more impressively large body of textual production, much of it by fans. The original show followed the travels of the USS *Enterprise*, a capacious starship from the Federation (a human dominated interstellar commonwealth), captained by the charismatic if strenuous and sometimes over-earnest James T. Kirk (William Shatner). The *Enterprise* was on what the title-sequence voiceover called 'a five-year mission, to seek out new life and new civilisation, to boldly go where no man has been before'. That notorious split infinitive seems somehow to capture precisely the naïve energy and slightly clumsy charm of the original series. The ensemble cast was built around a triad of characters: Shatner's alpha-male Kirk and his two friends: the emotionless alien science officer Spock (Leonard Nimoy), and the over-emotional medical officer 'Bones' McCoy (DeForest Kelley). It was the developing interactions of these three characters as much as the various SF premises and alien worlds that won the series so devoted a following – particularly among female viewers. Indeed, some critics suggest that *Star Trek* is more responsible than any other SF text for the increase of female interest in the genre.

Star Trek ran for three series before being cancelled; but in an unprecedented move concerted action by disgruntled fans put enough pressure on the TV studios to have the franchise resurrected. Initially, this took the form of a rather cursory animated cartoon version, also called *Star Trek* (1973–74); but plans for a movie eventually bore fruit in the spectacularly dull and plodding *Star Trek: the Motion Picture* (Robert Wise, 1979) – a work whose subtitle was perhaps a necessary reminder to audiences that they had not stumbled by mistake into an interminably snail-paced piece of *avant-garde* art. The film, though bad, nevertheless made a great deal of money: a testament to the fierce loyalty of 'Trekkies' or 'Trekkers' (as the fans were known). Much truer to the spirit of the original series was the lively space-melodrama *Star Trek II: the Wrath of Khan* (Nicholas Meyer, 1982). Though rather daft in premise and plot, this film did give its viewers a series of exciting space battles, a rip-snorting villain in the genetic superman 'Khan' (played without restraint by Ricardo Montalban), and an effective surprise ending in which Spock dies saving the *Enterprise*. The follow-up *Star Trek III: the Search for Spock* (Leonard Nimoy, 1984) resurrected Spock via an experimental 'Genesis device', and in doing so made manifest a latent 'resurrection' theme that runs through much of the original series: in many original episodes many characters are killed only to be reborn. That there is a danger of critical pomposity in reading this as a secularisation of the Christian mythos should not deter us from recognising it as another manifestation of the buried roots of the genre as a whole.

Subsequent movies were less interesting works, although they were sometimes entertaining – as in the played-for-laughs *Star Trek IV: the Voyage Home* (Leonard Nimoy, 1986), and the played-straight-but-inadvertently-laughable *Star Trek V: the Final Frontier* (William Shatner, 1989).

By this time, however, the franchise had developed exponentially. A sequel TV series, *Star Trek: the Next Generation* (1987–94) began with a stilted first series, although its likeable cast usually made it watchable: the Shakespearian actor Patrick Stewart brought gravitas to the role of Captain Picard, and the emotional triad of the original series was transferred onto his Shatnerian first office Riker (Jonathan Frakes), the emotionless android Data (Brent Spiner) and the passionate Klingon tactical officer Worf (Michael Dorn), complicated by various female characters, including an emotionally telepathic 'ship's counsellor' Troi (Marina Sirtis) who in the later series works through a soap-operaish love triangle with Riker and Worf. Picard's super-ego super-competency floated serenely above all these interactions. Not until the third series, when the nicely sinister alien threat posed by the 'Borg' turned Picard literally into the enemy, did the quality of episodes pick up. By the late 1980s *Star Trek: TNG* was the most-watched TV SF show. Paramount invested in several spin-off series, one set on a Federation space station *Star Trek: Deep Space Nine* (1993–99), one on a female-captained spaceship lost far from home *Star Trek: Voyager* (1995–2001) and another a prequel to the original series called simply *Enterprise* (2001–5). Indeed, the iconic spaceship, designed not as a rocket-ship but as a white sprawling mantis-like structure with a wide saucer for its head, has been one of the most successful features of the show. Not only was one of NASA's prototype space-shuttles named *Enterprise* in honour of the series' popularity, but plans for commercial space flights have recently been announced, also utilising a craft called *Enterprise*.

Perhaps more important than the burgeoning body of TV and cinema texts is the role *Star Trek* has played as the focal point for a vigorous, worldwide fan base. Hundreds of novels have been published, some adapted from scripts but most original fictions set in the Star Trek universe. Despite the fact that critics often dismiss (or simply ignore) the phenomenon of TV/film tie-in novelisations, some of the *Star Trek* books are very considerable works of fiction. Writers as highly regarded as James Blish, Greg Bear and Joe Haldeman have written original fiction set in the Star Trek universe; and with the excellent *Spock's World* (1988), Diane Duane (b. 1952) wrote perhaps the best SF novel of that year. Magazines, comics and graphic novels, several video games, a wealth of privately published or unpublished fan fiction, and critical studies have created an enormous cultural resource. If it would take a newcomer years to apprehend fully the paratextual material associated with *Dr Who*, it would take the same individual a decade or more to apprehend the material associated with *Star Trek*. Even in the widespread 1980s and 1990s culture of the SF mega-text, the detail into which fans of Trek have gone is unusual. One of the most popular alien peoples, the warrior Klingons, have featured prominently, especially in the later series. Fans have picked up on this not only by dressing as Klingons at fan conventions;[2] but by developing an actual and, I am told, working Klingon language: dictionaries and grammars of this invented tongue have been published, and sections of Shakespeare and the Bible translated into it. The amount of ingenuity and effort expended on this (some might say) pointless activity might be seen as an index of the creative possibilities *Star Trek* has opened up to a wide body of non-professional creative writers and artists.

Film from *2001* to *Star Wars*

The financial success of certain SF films (particularly *2001*) and the continued audience appetite for TV SF persuaded many producers that money was to be made out of the genre. Although it resulted in many weak films and many flops, sometimes this faith proved justifiable. One hit, for instance, was *Planet of the Apes* (Franklin J. Schaffner, 1968). Loosely adapted from French writer Pierre Boulle's *La Planète des singes* ('The Planet of the Monkeys', 1963) this movie strands astronaut Charlton Heston on what appears to be a distant planet where eloquent monkeys are the dominant life-form, and human beings are dumb beasts. Unluckily shot through the throat, Heston's character cannot communicate with his ape captors until late in the film when, to simian astonishment, he exhorts his captors to 'get your stinking paws off me you damn dirty ape!' Monkeys are men; men are monkeys; the simplicity of this satiric inversion is well played, although there is perhaps a certain incoherence in the conception: Is this inversion, in which intelligent monkeys maltreat mindless humans, an animal rights satire on mankind's cruelty to the world's fauna? Is it more obliquely a satire on racial prejudice? Is it a satire on the callousness of the industrial-military complex from the point of view of inarticulate hippies? The famous twist ending, in which Heston's character discovers the half-buried Statue of Liberty on a desolate coastline and realises that he is not on a distant planet but rather a far-future Earth, has had its impact diluted by its over-familiarity, although it retains its power as a visual icon. But what is especially interesting about the film is the way it generated not only a whole series of successful sequels, but also a TV series (1974), and nearly a dozen novels, some original fiction set in the 'Planet of the Apes' universe. In *Beneath the Planet of the Apes* (Ted Post, 1969) a new astronaut (James Franciscus) chances on an atom-bomb worshipping cult underneath the ruins of New York, and finally detonates the device destroying the entire world, a twist ending that managed to match the original film's unexpectedness. The uncompromising pessimism and sheer oddness of this film are, if anything, even more powerful than the original movie. Undeterred by the fact that this film unambiguously destroyed the imaginary universe in which the franchise was based, further films followed: *Escape from the Planet of the Apes* (Don Taylor, 1971) has three talking apes return to the twentieth century via time travel, and manages some neat satirical points. *Conquest of the Planet of the Apes* (J. Lee Thompson, 1972) and *Battle for the Planet of the Apes* (J. Lee Thompson, 1973) fill in the story between the third and first movies; by now the appeal of the films was being understood as an allegory of race relations – a hot topic in the US in the late 1960s and early 1970s. Eric Greene argues that the films contain a 'double allegory', one which 'codes white humans as white humans' and 'codes black humans as apes'; and another which 'code white gentile humans as orang-utans, white Jews as chimpanzees, and African-Americans as gorillas' (Greene, p. 55). The series' various meditations on racism and the undesirability of conflict as a means of resolving difference are therefore rather undercut. The premise in fact displaces one SF fantasy (a world in which black humans are the dominant power, along the lines of Heinlein's *Farnham's Freehold*) into another (a world in which *simians* are the dominant power). It is hard to deny that, to quote Greene again, 'the issues of racial conflict and racial oppression' are 'the central issues' of the five *Ape* films; and that their success was in part a result of the way they 'connect[ed] with the individual or collective experience of large numbers of the consuming public' (Greene, pp. 1, 8). The relative failure of the remake, *The Planet of the Apes* (Tim Burton, 2001), though partly due to its narrative and aesthetic incoherence, may also be explicable as a function of the lower register of cultural concern about racism 30 years on.

It is hard to generalise about SF cinema of the early 1970s, except to say that the broader cultural concerns of the decade fed through into the films, which (given the wide range of cultural concerns) doesn't tell us very much. But one thing is noticeable: most of these films were infused with a very grim sensibility. Robert Wise's efficient thriller *The Andromeda Strain* (1971) imagines a rather hit-and-miss but eventually successful governmental-scientific response to a deadly plague from outer space. Douglas Trumbull's good-looking but pompous eco-drama *Silent Running* (1971) is set aboard a spaceship containing the Earth's last greenery, plants not wanted by a flora-free post-nuclear war planet (it is not explained how the home world oxygenates its atmosphere). The slightly hippy protagonist (Bruce Dern) is a junior crewman-gardener on the spaceship; he disobeys his orders to destroy this cargo, and instead sends it into deep space as a sort of message in a bottle. Even though Dern can achieve this only by murdering his crewmates, this is an ending that does not avoid tweeness. Stanley Kubrick's follow-up to *2001* was the ultra-violent *A Clockwork Orange* (1971). By putting visuals to Anthony Burgess's vivid prose, Kubrick apparently went too far; watching Alex murder and rape on screen allegedly provoked copycat crimes, and Kubrick withdrew the film from British (although not continental) cinemas during his own lifetime. It is true that, by jettisoning the book's moralising final chapter and presenting Alex's violence as more or less without consequence the movie achieves a dislocated and disturbing intensity; but it is hard to believe that so brilliantly weird a future vision in fact acted as a blueprint for Britain's own juvenile delinquents.

Andrei Tarkovsky's version of Lem's *Solaris* (1972) was called in its day 'the Russian *2001*', a reflection less of substantive similarities between the two movies and more the fact that it struck western audiences as so very slow and contemplative. In fact, by comparison with some other films by the glacial Tarkovsky it is full of incident; but several extended sequences suggest a deep sense of receptivity to alienness itself. Like the same director's *Stalker* (1978), this is a film nearly impenetrable for the casual viewer, but persistence rewards with sensibilities, insights, and visual poetic compositions of quite extraordinary beauty. Though much peppier, and poppier, Michael Crichton's *Westworld* (1973) can do little with its interesting premise of a quasi-Disneyworld staffed by robots who interact with the human holidaymakers to provide a hyper-real 'western' (or 'Roman Empire') environment, except (when the robots inevitably malfunction) provide a couple of effective shock-moments. The ultra-white bleached-out dystopia of *THX-1138* (1974) was George Lucas's first feature. Lucas was to become more famous for another SF film made a few years later, but in *THX-1138*, constrained by a small budget, he produced a very interesting visual artefact. The sight of white-suited bald actors moving through white corridors and white rooms, all slightly over-exposed, makes a much more sophisticated visual statement than the film does in terms of its derivative and ultimately sloppily humanist story.

But two films from this period did manage what very few SF films have managed; a workable blend of genre and comedy. Woody Allen's *Sleeper* (1973) demonstrated, among other things, that the conventions of SF cinema were now so familiar that it was possible to score comic points off them with ease. John Carpenter's *Dark Star* (1974) is a more rounded piece of film-making, a satirical swipe at human, and more specifically American, insularity in the face of cosmic splendours. It is often very funny indeed; but its satire is more often eerie (as in the scenes where the crew hold conversations with the frozen remnants of their dead captain's consciousness) or even exhilarating (the sentient bomb that decides it is God and explodes becomes almost the nihilist hero of the piece) than pointed.

Other films seemed to be saying slightly incoherent things about western fascination with sport and/or violence (as in the extended rendering of the bone-crunching game of *Rollerball* (1974) in Norman Jewison's film) or the cult of youth. *Logan's Run* (Michael Anderson, 1976), in which beautiful young people live idyllic sensual lives, only to be exterminated before they get old and ugly, is a simplistic and even banal piece of filmmaking. Nevertheless, its blend of beautiful actors and actresses in revealing costumes, and the continual threat of violence, touched some sort of chord: a spin-off TV series and several tie-in novels followed.

Nicolas Roeg's *The Man Who Fell to Earth* (1976) flirts with pretentiousness but manages, just, to stay the right side of the line. It is helped in this regard by the casting of pop-star David Bowie as the alien protagonist: in 1976 Bowie was going through the weirdest phase of his sometimes very weird life; and he captures exactly the combination of oddity and charisma that keeps the film on course. Bowie's mannered, slightly distant, Anglicised alienness colours the whole film (Roeg, another mannered, slightly distant Englishman, finds visual correlatives throughout for this mood).

Although there are aspects of *Tron* (Steven Lisberger, 1982) that seem quaintly dated nowadays, it has been and in part remains an enormously influential movie. The plot concerns a computer hacker who is literally transported inside a computer, where he is forced to battle against a tyrannical Master Control Program. It is the 'clean', minimalist look of the movie (designed by the comic-book artist Moebius), the gleaming red or blue neon lines against a black background, with human faces in pared-down torsos riding various simplified motorcycles, tanks and ships over an apparently endless grid, that is its greatest achievement. It established a visual grammar for the representation of cyberspace that informed much cyberpunk fiction. But all these films, interesting and varied though they are, seem in retrospect mere preludes to the big event of SF cinema, which occurred in 1977 and which changed the cultural logic not only of Screen SF, but of the whole genre.

Star Wars

George Lucas's feisty, juvenile space-opera *Star Wars* (1977) marked a radical change not only in SF cinema, but in cinema *tout court*. Despite his Oscar-nominated pedigree as a director, few believed that there was any profit or even merit in Lucas's pet idea for a high adventure yarn in the SF idiom. In the event, *Star Wars* became the highest grossing film in cinema history. It cost $11 million to make (with an additional $4 million spent on prints, distribution and advertising); to date it has grossed $926 million globally, a figure which is not adjusted for inflation (if that adjustment is made the film has earned nearly $2 billion worldwide; and three subsequent *Star Wars* films have each grossed nearly $1 billion: see the appendix to this chapter for details). This was wealth-generation on a scale unprecedented in cinema.

Star Wars changed everything. In the wake of its success filmmakers flocked to replicate the winning formula, and a very large number of often high-budget (and sometimes highly successful) SF films and TV series were made throughout the 1980s and 1990s. Some of this success spilled back into the more traditional fields of written SF, with readerships increasing and spreading across a number of previously resistant demographics; but the fact of the matter today is that, speaking globally, very few people read SF, whereas very many people watch it. This is the major change in the genre over the century.

For many the juggernaut of *Star Wars'* success is wholly deplorable. Lucas's film is seen as responsible, sometimes single-handedly, for the dumbing down of SF, or even of world, culture. It is indeed a puerile film in the sense that its primary audience was children, and accordingly it does not deal with a number of adult concerns (sexuality, for instance, is almost absent from the first film, and is treated very clumsily in the sequels). Those who deprecate *Star Wars* see it as reactionary escapism infected throughout by a juvenile sentimentality – it is often stabbed and sliced by critics as the murderer of 'proper' SF, which is implicitly taken to be 'a literature of ideas'.[3] It may be true that cinema is poor at communicating ideas; characters verbalising complex intellectual ideas make for a turgid movie-going experience. What cinema is good at is image, action, narration and (to a certain extent) character.

It is probably not true to say that the success of *Star Wars* rewired SF literally single-handedly; but it certainly cemented what was already a tidal change in the genre. From the late 1970s to the present day SF has metamorphosed from a primarily written literature of ideas into a primarily visual idiom of poetic imagery and spectacle. For many critics of the genre (whose personal taste runs to the former) this is a consummation devoutly to be unwished, but it is surely pointless to deprecate such cultural sea changes. SF is now the most popular form of art on the planet *because* it has colonised visual media: of the fifteen top-grossing films of all time only one is not SF or Fantasy (the odd one out is James Cameron's 1997 movie *Titanic*; and even Cameron is primarily thought of today as a director of SF films). Graphic novels and comics, although published in most genres, are overwhelmingly SF, Fantasy and Horror texts. Video games are released in a wider variety of genres (sports simulations, spy adventures and historically-based titles are all popular) but SF is nevertheless the biggest single idiom.

Can all this really be traced back to *Star Wars*? What was it about this particular text that meant it had such a huge impact on the development of the genre? To talk about the film in terms of its manifest content does little to answer this question: the storyline, though engaging, is derivative and rather banal. Luke Skywalker (Mark Hamill) is a young man on a parochial desert planet on the edges of an oppressive quasi-fascist Galactic Empire. He meets the wise old magician Obi Wan Kenobi (Alec Guinness), one of the last of a Galactic order of knights with supernatural powers called the Jedi, who harness a ubiquitous and mystical 'Force'. Through him, Luke becomes embroiled in a plot to free a beautiful woman, Princess Leia (Carrie Fisher), from imprisonment on one of the Empire's Death Star – a moon-sized spaceship with enough firepower to destroy whole planets. The Princess is indeed freed, although in the process Kenobi dies fighting the sinister black-clad and black-masked Darth Vader, sidekick of the wicked Emperor and an embodiment of evil himself (this part was played by David Prowse, but voiced by James Earl Jones). Skywalker and his new friends – Han Solo (Harrison Ford), a wise-guy space pilot, his hairy alien co-pilot Chewbacca, and two comic-relief robots – join the rebellion against the evil empire. An attack on the Death Star, aided by Skywalker's increasing awareness that he too is gifted with the Force, results in its spectacular destruction, although Darth Vader escapes to fight another day.

Lucas was open that his screenplay dealt not in 'characters' in any meaningful sense, but rather in *types*, specifically story archetypes which he drew from the structuralist sub-Jungian musings of Joseph Campbell's *The Hero with a Thousand Faces* (1949). Lucas himself partly credited the film's universal appeal to this fact, although in practical terms it schematises the characters, something Lucas's stiffly unidiomatic written dialogue only makes more pronounced. But as a *visual* artefact the film was unprecedented, and remains stunning. It was not the content of the film that is so praiseworthy (certainly not the ideological content, which is conservative and borders on the racist militarism of US militia groups), but rather the genuinely sense-of-wonder visuals of the piece. The one element of the film's content that enjoyed success apart from the film was the religious-mystical doctrine of the Force.

In the sequel, *The Empire Strikes Back* (Irvin Kershner 1980), the rebels come under sustained Imperial attack, Luke trains as a Jedi under the diminutive green-skinned muppet Jedi Master Yoda (voiced by Frank Oz), and finally fights Vader, thereby discovering – in what was a genuine surprise to original audiences – that Vader is actually his father. The third film, *Return of the Jedi* (Richard Marquand, 1983), is mostly concerned with the ultimate rebel assault on a second Death Star, in the process of being built by the Empire. Luke, now a fully-fledged Jedi, confronts Vader a second time. After much light-sabre swashbuckling, Vader rediscovers his love for his son, assassinates his Emperor and dies. Princess Leia (revealed to be Luke's sister) finds love with Han Solo.

Many fans consider the second film the best, because the darkest, of the three films, and because it takes the narrative risk of ending with the plot still *en l'air*, a fact which, combined with the horrific revelation of Luke's parentage, added a temporary gravitas to the ongoing trilogy. But the potential was largely wasted: there is indeed something too cosy about the revelation in *Jedi* that the whole cosmos-spanning battle between Good and Evil has actually been a sort of hyper-tropic family drama, a cosiness too lamentably epitomised by the teddy-bear-like 'Ewoks' who inhabit the forest moon below the under-construction Death Star – certainly the most throat-clenchingly annoying aliens ever rendered on film (the fact that the explosion of the Death Star into the atmosphere of their Forest Moon will almost certainly wipe out all life on that world is some consolation for the outraged sensibilities of the film's viewers).

It as an obvious point to make about the trilogy that it is highly derivative; and some critics suggest that this alone is enough to explain its enormous success (although there have been very many SF films made that are equally derivative, or more so, that have not replicated *Star Wars* popularity). Certainly the whole Star Wars series is confected out of Lucas's own reading: the planet Tatooine is a version of Herbert's *Dune*, the Death Star derives from E. E. 'Doc' Smith's interplanetary mega-weapons, Coruscant is stolen from Asimov's *Foundation*, the teddy-bear-like Ewoks are close to the creatures in H. Beam Piper's *Little Fuzzy* (1962), Darth Vader is an amalgam of two of Jack Kirby's comic-book characters, Dr Doom and Darkseid (and the Force has many things in common with the Source, from Kirby's *New Gods* series), and so on. But this is not particularly important. More germane are those moments in the films of sheer visual beauty, many of which have attained an iconic status in western culture: the enormous, wedge-shaped Imperial Star Destroyer rumbling into view over the top of the cinema screen at the very beginning of *A New Hope* (an iconic externalisation of the idea of political tyranny); Luke watching twin suns set on his uncle's farm; Darth Vader's black death's-head mask; several of the intricately choreographed space battles; father and son fighting light-sabre battles and many other moments. The level on which Star Wars works most effectively is precisely as *visual* myth.

After *Star Wars*: *Alien, Blade Runner, The Matrix*

The very large number of SF films released after *Star Wars* would, if I attempted to list them all, turn the remainder of this chapter into a tediously extended list of titles and dates. But on the other hand, it is quite hard to select those titles which have proved 'most significant' to the continuing development of the genre from the morass – or, if you prefer a different metaphor, from the bountiful harvest – of 1980s and 1990s SF cinema. This is because of the close association between Visual SF and Fandom, which means that most SF films (even relatively minor ones, ones that lose money or lack critical plaudits) have created around them a cultural microclimate in which they are discussed by dedicated fans, and from which they become the seed-ground of small-scale SF mega-texts. The ubiquity of the internet facilitates this cultural production.

But four titles stand out as unusually successful – generating not only other film sequels and prequels, but a wide variety of paratextual material and remaining culturally alive and significant today. Three of them were noted by Will Brooker, writing in the late 1990s:

> *Star Wars: a New Hope, Alien* and *Blade Runner* were all released between 1977 and 1982; yet almost two full decades later, the narratives of Luke Skywalker, Ellen Ripley and Rick Deckard maintain their grip on popular culture. While texts of the same period like *Jaws, Close Encounters of the Third Kind* and *Saturday Night Fever* are now regarded as nostalgic exhibits in film 'history', these three science-fiction sagas endure vividly in the 'present' of the late 1990s (Brooker, p. 51)

To Brooker's three we may wish to add *The Matrix*, a film which although released much later (1999) has already generated the combination of mass-cultural reach and extensive fan interest needed to establish it as a mega-text of contemporary significance.

Alien (Ridley Scott, 1979) has been treated with a greater respect by critics and academics, by and large, than *Star Wars*: its violence excluded an audience of children, and its monstrous alien is seen as polysemously symbolic of a number of concerns that have chimed well with the interests of research-publishing academics in the last two decades – particularly in the fields of feminist and racial studies. Moreover, it marked a change of emphasis in the representation of the alien in SF film. After the Cold War invasion films of the 1950s and prior to *Alien*, most Hollywood SF films followed *2001*'s lead in treating aliens as benign, if sometimes mysterious, creatures. Two movies by Steven Spielberg marked the apogee of this sort of film. *Close Encounters of the Third Kind* (1977) begins by figuring potential alien encounter as an ominous business, but moves gear easily two-thirds of the way through into its saccharine, fluting, musical climax in which childlike aliens mix and mingle with humanity. Even more successful was *E.T. the Extra-Terrestrial* (1982), a film I discuss below, whose ugly-cute central alien, like a benignly super-intelligent de-shelled tortoise, does create a very powerful sentimental effect. The sentimental *Starman* (John Carpenter, 1984) and the even more sentimental *Cocoon* (Ron Howard, 1985) stand as pendants to this tradition: in all respects lesser works than Spielberg's two movies they were also out of time: by the mid-1980s almost all the aliens in successful movies were very nasty indeed – a trope that can be traced back to the impact made by *Alien*.

Alien was set aboard a cavernous interstellar cargo-ship given the Conradian name *Nostromo*. Woken from their hibernation by a broadcast distress signal, the crew find a crashed alien spacecraft in which numerous leathery eggs have been laid. A swift-moving hand-shaped alien creature leaps from one of these and affixes itself to a crewmember's face, although this being later drops away and dies. The crewman (Kane, played by John Hurt) seems unharmed, but in fact the alien has laid a maggot in his torso. This bursts out of his chest, bloodily, whilst the crew are having lunch – a moment of gut-wrenching visual poetry, and one of the most celebrated and genuinely shocking moments in cinema. The creature rapidly grows into a seven-foot towering alien carnivore, jet-black, with acid instead of blood and a second set of teeth on a retractable tongue inside his already sharply betoothed mouth. One by one the crew are killed, until finally Ripley (Sigourney Weaver) destroys the monster.

Part of the film's appeal lies in its design and look, a two-fold triumph. On the one hand, there was Scott's expert directorial eye which ensured that his spaceship looks lived in, worn-down, battered, distressed, dark (a radical move in a cinematic tradition where spaceships tended to be gleaming white examples of polished technology). On the other hand, there was the contribution made by the Swiss artist who designed the alien itself and many of the sets, Hansruedi Giger (b. 1940). Giger had exhibited and published several books of his distinctive artwork (most notably *H. R. Giger's Necronomicon*, 1977), work which drew him to the attention of the producers of *Alien*. His designs for that film further developed his style: sinister, twisted, black images of weird organic shapes and machines that looked grown rather than constructed, generally rendered in dark inks, acrylic paints, with a high level of surface gloss that highlights the themes of convoluted quasi-biological form and shapes reminiscent of sexual organs, and conveys a palpable odour of death and violence. In the words of Peter Nicholls, Giger 'revolutionized the look of sf cinema to a degree it would be difficult to overstate' (Clute and Nicholls, p. 495). Arguably it is his work that raises *Alien* from being a well-executed but otherwise conventional 'bug-eyed-monster-attacks-humans' B-movie into (a frequently abused but unavoidable phrase) a work of art. The point here, once again, is that it is primarily as a beautiful if unsettling visual arte-fact that *Alien* works. The film's expertly orchestrated string of shock-moments work very well on a first encounter with the text, but inevitably reduce in effec-tiveness in subsequent viewings. On the other hand, as with all good visual art, Giger's designs repay close attention, and make the film worth returning to.

The significance of this art has been widely debated. Feminist critics have read the film as being in some ambiguous sense 'about' woman-ness. Ripley was unusual in late 1970s film in so far as she was an active action-hero rather than a passive wilting adjunct to a man. Barbara Creed points out that in the film 'virtually all aspects of the *mise-en-scène* are designed to signify the female: womb-like interiors, fallopian tube corridors, small claustrophobic spaces'. Creed's argu-ment is that in late twentieth-century culture 'the body, particularly the woman's body, has come to signify the unknown, the terrifying, the monstrous' (Kuhn, pp. 215–16). The success of *Alien* may indeed be explicable in these terms; as a text that tapped into a broad cultural mixed fascination-repulsion with 'the body' as such, and with 'the female body' in particular. The sequel *Aliens* (James Cameron, 1986) swaps Scott's fascinated lingering on the visual magnificence of Giger's designs for a much snappier, fast-paced action-adventure; but even here the sheer rightness of Giger's art gives the film resonance beyond its whiz-bang storyline. Weaver reprised her role as Ripley, this time travelling to a colony world with a band of space-marines to confront a large infestation of the alien creatures. Only one of the colonists has survived, a young girl called 'Newt' (Carrie Henn), and the aliens make short work of the nervous and inexperienced marines. Once again Ripley saves the day, battling the enormous egg-laying alien queen in a narrative that builds relentlessly in pace and excitement, and manages a genuinely climactic set of final scenes. The third Alien film, *Alien³* (David Fincher 1992), crash-lands Ripley on a forbidding prison world populated entirely by sociopaths, murderers and rapists. Whilst a stowaway alien wreaks havoc among this unlikeable crowd, it transpires that Ripley has herself been infected by the beast. The film ends with Ripley plunging into molten metal, immolating herself and her foetus-like alien, which bursts from her chest during the plunge, together. The bleakness and deliberate visual *ugliness* of this film, its unremitting pessimism and its consistently realised aesthetic of dour anti-beauty (the usually handsome Weaver appears in this film skeletally skinny, with shaven head and boils) renders it surely one of the least 'Hollywood' films ever to come out of Hollywood. Presumably only the size and receptiveness of the *Alien* fan-base could have persuaded Fox studios to invest the money in making it. *Alien Resurrection* (Jean-Pierre Jeunet, 1997) contrives a way to bring Ripley back to life, and is in its way an even weirder film than *Alien³*, fascinated with monstrous hybridity and revolting mutation, although it lacks the coherence and focus of the earlier films, since it meshes what could have been a fairly standard Hollywood screen story with a more European directorial approach, replete with joyfully over-stylised visual moments and narrative *non-sequiturs*. The fifth in the series is the feeble and largely incoherent prequel *Alien versus Predator* (Paul Anderson, 2004).

There have been spin-off novels, but fewer of these than for *Star Wars* or *Blade Runner*. This is presumably because the *Alien* franchise has fewer narrative and conceptual possibilities; and indeed, examined too closely, it reveals certain difficulties (the Alien in the first film was primarily interested in eating the humans it encountered; but such humans are integral to alien fertility and can incubate Alien maggots only if alive. It is surely counter-evolutionary for a creature to develop that devours its means of reproduction). But the look of the film has inspired a range of spin-off visual art, including graphic novels (indeed, the *Alien vs. Predator* concept first originated in comic books), computer games, artworks and statuettes of the alien creature. It is as a triumph of design, and an embodiment of a certain style of dark neo-Gothic visual art (art that correlates to threat and violence as well as elegance), that *Alien* lives on.

Blade Runner (Ridley Scott, 1982) is in a rather different category. Where the *Star Wars* films have grossed $3.5 billion (unadjusted for inflation); the *Alien* films nearly $1 billion and the *Matrix* films nearly $2 billion worldwide, *Blade Runner* was not a runaway box-office success (it cost $28 million to make and grossed $35 million on its first release) and no sequel or prequel has been made.[4] Yet it is considered a classic of SF cinema not only by critics but by a large fan base. In part this is because it is seen by many fans as the best adaptation of Philip K. Dick's work to the big screen (and Dick remains a talismanic figure for many SF fans). In part it is because the *look* of the film has proved so enduring.

This is more than simply a question of good design, although the design, special effects and cinematography on the film are all world-class. It is a reflection of the consonance of style and substance. As in Dick's *Do Androids Dream* (see above p. 241) on which the film is based, Rick Deckard (Harrison Ford) is a hunter of androids known as 'replicants'. A group of these beings have escaped from their servitude and come to Earth, hoping to find a way to extend their four-year life-span. One by one Deckard kills these artificial humans, chasing them through the intricately realised multi-ethnic and postmodern city space. On the way he loses his certainty about what distinguishes 'real' and 'artificial'; he discovers that the beautiful Rachel (Sean Young), with whom he has a relationship, is a replicant with implanted memories (she does not realise herself that she is not human). There is the possibility that Deckard is also artificial, although the film never resolves the ambiguity on this issue.

The film is sometimes criticised for failing to capture the subtleties of Dick's original novel, and it is true that there are many differences between the two works. *Blade Runner* lacks the fascination with the notion of an artificial messiah, and its interrogation of the difference between 'simulated' and 'real', between 'surface' and 'depth', takes place under a newer cultural logic, one called 'post-modernism'. (I have taught courses on postmodernism for many years, and *Blade Runner* is perhaps the text most often instanced by critics as a paradigm for that slippery term.) More importantly, *Blade Runner* reconfigures Dick's imagination in visual terms. The transfer from one prose masterpiece to another visual masterpiece enacts the broader shifts of SF.

Blade Runner is just exquisite-*looking*. The film revels in a density of visual affect (Scott himself talked about his approach to directing as the 'kaleidoscopic accu-mulation of detail … in every corner of the frame … [making] a 700-layer cake'); but more than that it is in many ways about 'looking', about the visual. In Scott Bukatman's words, '*Blade Runner* is all about vision. Vision somehow both makes and unmakes the self in the film, creating a dynamic between a centered and autonomous subjectivity (eye/I) and the self as a manufactured, commodified object (Eye Works)' (Bukatman, p. 7). 'Eye Works' is the name of a commercial emporium that sells artificial eyes. The chief replicant Roy Batty (played with a brilliant mix of campness and Aryan menace by Rutger Hauer) visits, and the tech-nician recognises him as a replicant. 'You Nexus, heh?' he says. 'I designed your eyes!' Batty replies: 'if only you could see what I've seen with *your eyes* …'

The temptation is to read the film as a gloss on cinema itself, the artificial eye through which we see wonders (at the end of the film, as he dies, Batty tells Deckard: 'I've *seen* things you people wouldn't believe: attack ships on fire off the shoulder of Orion …', but SF cinema offers the visual artist possibilities that are not available to other film-makers. As the screenwriter Hampton Fancher worked through multiple drafts of the script Scott advised him, to help imagine the look of this future world, to read copies of the French comic *Métal Hurlant*. Elements of *Metropolis* and *The Shape of Things to Come* were appropriated, New York and Los Angeles were jumbled, punk fashions and 1930s gumshoe-chic, everything was mixed into a rough-edged visual melange. Where both *Star Wars* and the *Alien* films elaborated one recognisable visual style, *Blade Runner* succeeded in melding an encyclopaedic visual aesthetic.

The *Matrix* franchise also achieves its greatest effects in the visual idiom. The premise of the original film (*The Matrix*, Wachowski Brothers, 1999) is not original: engineered from Dick, Gibson's *Neuromancer* and several other sources, it concerns an ordinary hero, Thomas Anderson (Keanu Reeves), working in a dead-end office job in what appears to be the present day and cracks his world open to show him dizzying new depths. Anderson meets the hacker and digital terrorist Morpheus and is told that what he takes to be reality is in fact an elaborate computer simulation called 'the Matrix' programmed by malign machine intelligences which have enslaved humanity. Unplugged from this consensual simulation, Anderson becomes 'Neo' and joins Morpheus's rebellion of humanity against the machines in the grimy 'real' world. Morpheus believes Neo to be 'the One', a predicted saviour who will deliver mankind from the machinic bondage. Neo doubts this, but the course of the film sees him return to the Matrix (with *Joe-90*-style upgrades that give him almost superhuman strength and agility) to be killed and then return to life. He saves the day, and accepts his destiny as messiah.

But what struck many viewers when they first saw it was less this narrative and more the extremely stylish, visually gorgeous and arresting manner in which it was realised on screen. The special effects, the cinematography, the design, the whole *cool* of the thing captured fannish hearts. One particularly celebrated cinematographic trick was a shot-sequence invented for this project and called by the filmmakers 'bullet-time'. At several key moments, the frame appears to freeze in time, and the camera angle swoops through 180 (or more) degrees around the static actors. The first 'bullet-time' image is the most famous of the film: inside the matrix, the police attempt to arrest the beautiful Trinity; she kung-fu's her way free. At one point in the fight she leaps in the air, her arms curved away from her body, her legs in mid-kick. The image freezes when she is in mid-air and the point-of-view circular-pans all the way round her, before the image unfreezes, Trinity kicks a cop in the chest and drops back to the floor.

Other key bullet-time moments pick out, or visually highlight, similar moments in action sequences. In one, Neo falls backwards to avoid a barrage of bullets fired by an agent; as he tips backwards the camera spins round his flailing body and the bullets make themselves apparent with lines of aerial ripples. In another, Morpheus escapes the clutches of the agent, being wounded in the leg by a bullet; the camera sweeps along the floor and around as the bullets rake the room through which he is running. A fourth bullet-time moment captures the moment when Neo and Mr Smith, mid-fight, launch themselves at one another, the camera circling their bodies in mid-air. Bullet-time, in other words, seems to be a form of visual italics, a method of underlining moments of particular excitement. 'Excitement' here connotes the conventional climactic moments of the action-thriller genre: guns, fist-fights, and so on. But there is something very beautiful about them too; the term 'action choreography' has never been more appropriate. In fact these moments concretize the Fan's urge to 'get inside' the world of the Matrix; to have a proper look around, rather than simply and passively accept the 2D image.

The huge success of the first film led to two sequels – *The Matrix: Reloaded* (Wachowski Brothers, 2003) and *The Matrix: Revolutions* (Wachowski Brothers, 2003) – and a number of spin-offs, most notably the nine animated shorts set in the Matrix universe collected in *The Animatrix* (Peter Chung, Andy Jones, Yoshiaki Kawajiri, Takeshi Koike, Mahiro Maeda, Kôji Morimoto, Shinichirô Watanabe, 2003). In plot terms the sequels saw the story become more diffuse and incoherent, and were less liked by fans (although they made a great deal of money for the filmmakers). But visually they added a number of potently iconic images to the whole: Neo fighting an army of replicated versions of his arch-enemy Smith; an extraordinary chase sequence driving the wrong way up a crowded freeway; the white-bearded architect of the matrix itself sitting in a strange room every wall of which is covered with TV screens; millions of giant spermatozoon-shaped killer robots pouring into the forward hall of humanity's last city in the real world, and so on.

In all these cases, these films (or more accurately these *franchises*) have endured not only because they provide the usual satisfactions to cinema audiences, but because they offer something more: a fictive SF environment realised well enough to permit fans imaginative entry. In each case it is the imagined world of the film that has fully endeared the films to its fan base. The bulk of paratextual material allows viewers, in effect, to go behind the stage sets and special effects; to wander from planet to planet in the *Star Wars* universe, to fantasise about worlds in which women as beautiful as Sean Young can be purchased from an android store, or in which the grind of a menial office job can be thrown off in a whirl of kung fu – in which, in fact, a detailed knowledge of computers becomes an index not of nerdiness but of super-cool, kick-butt, handsome-as-Keanu-Reeves saviour of mankind potential.

But there's more to it merely than a hospitably escapist concept. The reason why these four franchises struck their chords, as opposed to the hundreds of others (many of them better conceived and better thought through), has to do with their visual power. All four are breathtaking, beautiful works of visual art. It is on that criterion, rather than on the level of character, plot, or even premise, that their greatness must be judged. Each has a distinctive look, a recognisable and potent visual styling; and like the greatest paintings they are artworks in which the viewer can lose herself for long hours.

Science fiction blockbusters of the 1980s

A number of other SF films have enjoyed considerable success and have also generated a great deal of paratextual material and fan activity, without reaching the levels of cultural saturation of these four titles. Chief among these is Steven Spielberg's *E.T. the Extra-Terrestrial* (1982), a film of significant emotional power that manages to be very moving without being too sentimental (a trick which earned it three-quarters of a billion dollars). In part this is because Spielberg is so expert a director of children, the main players in this drama. The plot concerns a runty but appealing little space-alien who gets separated from his spaceship and holes up among the toys of a sympathetic human friend, the ten-year-old Elliot (Henry Thomas). Elliot lives with his brother, sister and single mother in an LA suburb, and forms a supernatural bond with E.T., who possesses psionic powers. The adult world, searching for the alien, rudely intrudes, sealing the house and performing tests on the sickening E.T., who dies. Everything seems lost, but E.T. resurrects himself, and with the aid of the children escapes the adult scientists, rejoining its spaceship and leaving the planet. The film achieves a genuinely numinous mood by mediating that uplift via the open-heartedness of childhood; indeed, it seems to be elaborating the New Testament script about becoming again as a little child as a prerequisite for entering heaven. In several dozen specific points, too numerous to list in detail here, the story echoes the life of Christ, of which his resurrection is only the most obvious. There are good reasons for explaining the film's success in terms of its emotionally potent numinous quali- ties, an example of the way an apparently simple content (the plot summary above will give no sense of the power of the film to those who have not seen it) can achieve a luminous almost spiritual effect as a secularised quasi-religious myth. In this respect *E.T.* is a film in touch with the deep roots of SF as a genre.

Another film that managed a significant impact was *Terminator* (James Cameron, 1984), which has generated two notable sequels: *Terminator 2: Judgment Day* (James Cameron, 1991) and *Terminator 3: Rise of the Machines* (Jonathan Mostow, 2003); and which franchise also includes graphic novels, crossover works, video games, tie-in novelisations (at least one of these – S. M. Stirling's *T2: the Future War* (2004) – is really rather good). The titular terminator is a man-shaped cyborg designed to kill humans in a future world in which mankind is engaged in a world-wrecking total war with machines. On the verge of losing this war, the ruling misanthropic computer system Skynet sends a terminator (played with an appropriately lumpish muscle-bound brutishness by Arnold Schwarzenegger) back in time to the 1980s to kill the mother of the man, John Connor, who has led humanity to the brink of victory. With Sarah Connor dead, John will never be born and the machines will win. Connor sends a lone human soldier back to protect his mother. The film then plays out as a lengthy, and very effective, chase- and-fight action adventure set in 1980s LA; with the relentless Terminator being eventually stripped of his fleshy human covering revealing a gleaming metal skele- ton, a modern refashioning of the old *memento mori*. This theme becomes plainer in the sequel, in which once again a potential assassin and a protector are sent back to Sarah Connor (now with her young son John at her side), the twist being that Schwarzenegger, still a Terminator, is now the protector; the other Terminator (the slender but terrifying Robert Patrick) is a newer model made of 'liquid metal' that can morph into almost any shape. *Terminator 2* is a purer film than the first; jettisoning almost everything except the chase, a strategy which highlights the absolute unstoppable relentlessness of the killer Terminator all the more. This machine is, we realise, a trope for death itself, from which we can run, but which we can never *out*run. Sarah Connor's attempts to change the future and avert the nuclear war that will lead to the rise of the machines appear successful at the end of the second film, but *Terminator 3* (in which Schwarzenegger plays yet another 'good' Terminator, and the role of the liquid-metal opponent is taken by a woman) restores destiny to its world-destructive path. Despite the gravitational pull towards Hollywood 'happy endings' the *Terminator* franchise has enough inertia in its vision of the inherent violence and destructiveness of mankind to avoid being upbeat. Indeed fans are less interested in the resolution of the ethical tensions of the premise in favour of 'good' (non-violent, caring, sharing) solutions, and are drawn more to the figure of the Terminator itself, an effective externalisation of what Freud, in a rather different context, called the Death Drive. In *Terminator 2*, the boy John Connor, whom the good Terminator has been programmed to obey, orders him 'not to kill anybody, OK?' But this apparently peace-endorsing motto is resolutely and exhilaratingly deconstructed by the action of the movie: the Terminator does not kill, but does cripple and terrorise, the humans who stand in his way. Future nuclear war can, it seems, be averted only by massive destruction in the here-and-now; buildings explode, cars and helicopters crash, billions of dollars of damage is done, and, as we learn in *Terminator 3*, all for nothing. This film is in love with destruction, partly as an end in itself (as thrilling cinematic spectacle) but also as a means to another end: the machinic blankness and regularity of death itself.

As with *E.T.* many contemporary commentators noted religious overtones in the movie, from the fact that John Connor (the future saviour of mankind in the war against the machines) shares his initials with Jesus Christ, to the apocalyptic subtitle of the second film. But although there is some marginal mystical discourse – apparently only 'living tissue' can travel through time, on account of some ill-defined 'field' it generates; hence the Terminators' need to clothe their machine skeletons in cloned flesh – in fact these films are wholeheartedly in love with machinery: they valorise technology as precise, impermeable, efficient, cool-looking and utterly desirable.

These visions of monstrous and violent otherness made much more of a cultural impact than Cameron's other big-budget SF film from the 1980s, *The Abyss* (1989). Here mysterious aliens at the bottom of the sea contact the human crew of an underwater research establishment. They prove, after some tension, to be benign, but friendly aliens were out of step with what the movie-watching public wanted in the 1980s and 1990s. More successful was Australian director George Miller's (b. 1945) trilogy of near-future anarchy and social collapse, *Mad Max* (1979), *Mad Max 2* (1981), *Mad Max: Beyond Thunderdome* (1985) (a fourth, *Mad Max: Fury Road*, is forthcoming). In the first film a traffic policeman Max Rockatansky (Mel Gibson) becomes brutalised by the murder of his family and pursues an ingeniously violent revenge on the gang responsible. The modest success of this low-budget movie enabled the much weirder sequel, in which rival gangs fight over petrol in the post-civilisation Australian outback. The third film is odder still, with an imaginative and creative oddity, spending less time with the crunching road-battles that had made the reputation of the first two films, and more with an intriguing society of children in an oasis into which the now middle-aged Max stumbles. He is treated by these kids as a literal messiah, who will lead them to paradise.

Another successfully iconic nasty alien was the titular *Predator* (John McTiernan, 1987), who has comes to Earth (in this film, to the jungles of Central America) to hunt humans (in this case a team of US commandoes led by Arnold Schwarzenegger), using various high-tech weapons and disguises. The Predator kills all the commandos but Schwarzenegger, who kills it. In the sequel *Predator 2* (Stephen Hopkins, 1990) the alien goes hunting in crime-torn LA, killing many bad guys and some good ones before being dispatched by policeman Danny Glover. Since this franchise is patently based on the *Alien* films, it is fitting that a successful series of comics and then a less successful 2004 film pitted the two creatures together. One of the most interesting things about *Alien versus Predator* is the fact that the human characters, caught in the middle of this battle, choose to ally themselves with the latter. Technology links us; which in turn suggests that the Predator is in some sense a projection of human self-image outwards.

An even less pleasant alien appeared in John Carpenter's (b. 1948) superb remake of *The Thing* (1982). A *tour de force* of revolting and startling special effects, the titular alien (when not impersonating humans) changes shape into a nightmare torrent of monstrous forms.

The Dutch director Paul Verhoeven (b. 1938) has made a number of notable SF blockbusters. *RoboCop* (1987) is a darkly witty and ultra-violent satire on zero-tolerance policing: a murdered policeman is brought back to life as a clunking cyborg, with chrome prostheses and a robotic voice; he takes inevitable revenge on his murderers, but the sparkle of the film is found in its interstices: pastiche adverts and TV-spots, moments of deft extrapolation. The film's sequels (none of which were directed by Verhoeven) have been, without exception, lamentable. Verhoeven's broad-brush adaptation of Philip K. Dick's 'We Can Remember it for you Wholesale', *Total Recall* (1990) worked less well (Dick's first name was, ominously, misspelt 'Phillip' in the opening credits): Arnold Schwarzenegger was miscast as a regular working Joe who is swept up in spy-mystery adventures on Mars, or else only thinks that he has been. There are varieties of disbelief it is impossible to suspend, and the notion of monstrously muscled Schwarzenegger as an ordinary guy is one such. Verhoeven's invisible man film *The Hollow Man* (2000) similarly flagged, despite impressive special effects. But his version of Heinlein's *Starship Troopers* (1997) comes within a bicep's width of being a masterpiece. Heinlein's original novel is a lively but horribly earnest endorsement of militarism and the army as the fount of all virtue. Verhoeven reads his source text brilliantly against the grain, casting a series of beautiful-looking, plastic soap-opera starlets as the army grunts, and then delighting in the battle scenes where many of them are sliced into little pieces. This is a film that cries out to be read as funhouse satire on American TV mores. The bones of a conventional military movie are still visible, just (tough sergeant whips unpromising group into tight fighting unit), but they are overlaid with such visual panache and so many witty touches that the viewer ceases to pay attention to the reactionary froth and is caught up in the visceral, deconstructive fun.

The late 1980s and 1990s saw a flurry of superhero movies. Superman was transferred from comic book to big screen (*Superman*, Richard Donner, 1978; *Superman II*, Richard Lester, 1980; *Superman III*, Richard Lester, 1983; *Supergirl*, Jeannot Szwarc, 1984; *Superman IV: the Quest for Peace*, Sidney J. Furie, 1987; *Superman Returns*, Bryan Singer, forthcoming). Batman followed (*Batman*, Tim Burton, 1989; *Batman*

Returns, Tim Burton, 1992; *Batman Forever*, Joel Schumacher, 1995; *Batman and Robin*, Joel Schumacher, 1997; *Catwoman*, Pitof, 2004; *Batman Begins*, Christopher Nolan, 2005) and then, even more successfully, Spider-Man (*Spider-Man*, Sam Raimi, 2002; *Spider-Man 2*, Sam Raimi, 2004). The Incredible Hulk lost his superlative (in more than one sense) for *Hulk* (Ang Lee, 2003); while the ensemble X-Men (*X-Men*, Bryan Singer, 2000; *X2*, Bryan Singer, 2003) enjoyed both popular and critical success. As with the continuing interest in superhero comics, the box-office dependability of these films connects with the broader western cultural anxiety about the status of the 'saviour' in the modern West. More, the shift away from Superman (an alien saviour gifted with supernatural powers) to Batman and Spider-Man (more ordinary humans, with human foibles) tracks a similar shift of cultural fascination, from messiah as god to messiah as man.

A number of Japanese *animé* or 'animated' SF films also achieved global cultural penetration in the 1980s and 1990s. The most famous is still *Akira* (Katsuhiro Ôtomo, 1988); based on an enormous *manga* (comic-book) series (1982–90), this tells a rapid and, to the casual viewer, incoherent story about New Tokyo, a gang of kids on enormously souped-up motorcycles, and the coming of a meta-human mutant with massively destructive powers. But as with all the key SF films of this period, it is the visual look of the whole that is where its greatness lies.

Similarly, if one attempts to understand *Kōkaku kidōtai* ('Mobile Armoured Riot Police', Mamoru Oshii, 1995; the film is known in Anglophone countries as 'Ghost in the Shell') as a meditation on the mind–body split based on its textual story and dialogue alone, it would appear clogged and jejune. But the film manages to be such a meditation, thanks to its beautiful, multilayered visual text, each image of which speaks volumes about the film's theme. Like many *animé* films, it becomes understandable, even poetic, on an intuitive, nonlinear, associative level. The sequel *Innocence: Kōkaku kidōtai* (Mamoru Oshii, 2004) is more pretentious (one character declares, 'Life and death come and go like marionettes dancing on a table, once their strings are cut they crumble easily') although it similarly achieves a sometimes numinous beauty.

1990s

Spielberg's 1990s also saw a number of significant SF films. *Jurassic Park* (1993) adapted a Michael Crichton (b. 1942) novel about genetically recombined dinosaurs overrunning a near-future safari park and eating people. It was extremely, perhaps extraordinarily, successful. Its sequel, *The Lost World: Jurassic Park* (1997, also directed by Spielberg) reworks *King Kong* with dinosaurs instead of a giant ape terrifying westerners on a distant island before being transported to San Diego and running amok. As a deliberately homage to Kong the film has a degree of postmodern charm, although audiences were no longer liable to be impressed simply by the sight of CGI dinos. *Jurassic Park III* (Joe Johnston 2001) was a thin retread of the first movie, and it looks as though *Jurassic Park IV* (due out in 2006) will be similarly uninspired.

A.I.: Artificial Intelligence (Spielberg, 2001) was based on a Brian Aldiss story, 'Supertoys Last All Summer Long', about a robotic boy-child designed to have feelings and intelligence who becomes the surrogate child of a couple whose real child is in cryogenic suspension with an incurable disease. Stanley Kubrick had been developing the film for a dozen years before his death, passing the project on to Spielberg. The earlier part of this film achieves a genuine power, as the real son recovers and the family has no further use for the robot son, whose programmed love for his 'mother' can nevertheless not be unwritten. There is a chilly, beautiful mournfulness to some sections of this picture which build towards a significant emotional power, although this build is ruined by a tacked on and wholly unbelievable happy ending that spoils the whole. Kubrick delayed making the picture because he hoped that technology would develop a real robot which he could then cast as the boy; and to work properly the project needed to be bleached of almost all human emotion. But Spielberg, often expert at modulating sentiment precisely, demonstrated a less than characteristic deft touch in this case. Mind you, given that the title was changed from *A.I.* to *A.I.: Artificial Intelligence* because US test audiences assumed the two initials referred to 'A1', a popular brand of American steak sauce, the world into which the film was being released might not have been ready for a fully Tarkovskian movie-going experience. Better was *Minority Report* (Spielberg, 2002), based on Philip Dick's story of the same title: a 'pre-crime' division, which uses precognitive data to stop crimes being committed in the first place, has made mid 21st-century Washington DC virtually crime-free. But the leading 'pre-crime' policeman John Anderton (Tom Cruise) is accused of the future-murder of somebody he has never even heard of. The future world is brilliantly realised (Spielberg gathered a team of sixteen futurological experts in Santa Monica to brainstorm the year 2054); the gadgetry is very convincing and the action-adventure element well handled. But the film as a whole is actually about 'free will', and as such it connects directly with the theological taproot of SF as a whole. 'Anderton' (in the original screenplay the character's name was the more obviously son-of-mannish 'Anderson') struggles with the theological dilemma of whether future-omniscience is compatible with freedom of will. Spielberg, rather unconvincingly, wants in this picture to suggest that it is.

Many SF blockbusters from the 1980s and 1990s worked, as did *A.I.*, with these sorts of theological constructs. Where the religious element was taken literally it often backfired badly: an example is *Event Horizon* (Paul Anderson, 1997) in which an experimental faster-than-light drive system accidentally opens a gateway to an unreconstructed medieval Hell, condemning the entire crew to Bosch-like torments. The spaceship *Event Horizon* was deliberately modelled on the Parisian cathedral of Notre Dame and is full of crucifixes, but the overall effect is simply silly. The idea of Hell as a literal location no longer has mainstream cultural purchase. Instead, the important questions of atonement, of the status and problematic of the messiah, has been translated into metaphoric and visual-materialist terms: as the Christological Neo in the *Matrix* films; as the superhero (as Superman and Spider-Man); and even in comic terms as the mechanical alien *The Iron Giant* (Brad Bird, 1999), or the wittily profound and thrilling pastiche of the Fantastic Four, *The Incredibles* (Brad Bird, 2004). The shaping forces at the heart of SF are still evident, although they have grown in SF cinema into a rich and strange visual iconography.

These are the forces that lie behind the first significantly successful SF TV texts of the twenty-first century. *Battlestar Galactica* (1978–79) was originally a cheesy *Star Wars* imitation (the pilot episode was released into cinemas as a film in 1978): after the destruction of their home world, humankind is searching for a mythical paradise planet ('Earth') on which to settle, travelling in a convoy of spacecraft protected by the titular starship against repeated depredation by the 'Cylons', the genocidal machines that are persecuting them. It was remade, brilliantly, in 2004 as a series of thirteen episodes, produced by David Eick and Ronald D. Moore; visually spectacular, well-written and acted, and with a modish jittery camera style redolent, despite the lavish SFX, of hand-held or documentary TV. The original series contained a number of barely coded allusions to the belief structure of the Church of Jesus Christ of Latter Day Saints ('the Mormons') of which the show's producer, Glen Larson, is a member. These echoes have, in the remake, assumed a more general Christian flavour, with the Cylons themselves seemingly possessed of a religious mania in which the destruction of humanity, and the creation of a hybrid human-machine race, are not only divinely sanctioned but necessary for the atonement of the various failings of both human and machine.

Appendix: All-Time Top Fifteen Movies by Global Box Office (gross in dollars)

1. *Titanic* (1997)	1,835,400,000
2. *Lord of the Rings: Return of the King* (2003)	1,117,02,779
3. *Harry Potter and the Philosopher's Stone* (2001)	975,800,000
4. *Star Wars: The Phantom Menace* (1999)	925,800,000
5. *Lord of the Rings: The Two Towers* (2002)	922,986,073
6. *Jurassic Park* (1993)	920,100,000
7. *Harry Potter and the Chamber of Secrets* (2002)	869,400,000
8. *Lord of the Rings: Fellowship of the Ring* (2001)	867,683,093
9. *Finding Nemo* (2003)	844,400,000
10. *Shrek 2* (2004)	840,581,107
11. *Spider-Man* (2002)	821,700,000
12. *Independence Day* (1996)	813,200,000
13. *Star Wars* (1977)	797,900,000
14. *Harry Potter and Prisoner of Azkaban* (2004)	781,767,207
15. *E.T. the Extra-Terrestrial* (1982)	775,913,554

These figures are not adjusted for inflation. When this adjustment is made (across the last century), the top five titles are:

1. *Star Wars* (1977)	981,848,794
2. *E.T. the Extra-Terrestrial* (1982)	810,536,150
3. *Titanic* (1997)	774,918,846
4. *Star Wars: The Empire Strikes Back* (1980)	573,323,886
5. *Star Wars: Return of the Jedi* (1983)	560,754,164

Note: These inflation-adjusted figures are for US box-office takings only; these numbers should be roughly doubled to give an indication of global box-office takings adjusted for inflation. In other words (assuming one considers *Finding Nemo* to be a Fantasy film) every film on these lists is SF or Fantasy with the single exception of *Titanic*.

Source: www.the-numbers.com/movies/records/#world [accessed November 2004].

References

Aldiss, Brian, 'Speaking Science Fiction: Introduction', in Andy Sawyer and David Seed (eds), *Speaking Science Fiction: Dialogues and Interpretations* (Liverpool: Liverpool University Press 2000)

Bould, Mark, 'Film and Television', in Edward James and Farah Mendlesohn (eds), *The Cambridge Companion to Science Fiction* (Cambridge: Cambridge University Press 2003), pp. 79–95

Brooker, Will, 'Internet Fandom and the Continuing Narratives of *Star Wars*, *Blade Runner* and *Alien*', in Annette Kuhn (ed.), *Alien Zone II: the Spaces of Science Fiction Cinema* (London: Verso 1999)

Bukatman, Scott, *Blade Runner* ('BFI Modern Classics'; London: BFI 1997)

Greene, Eric, *Planet of the Apes as American Myth: Race, Politics and Popular Culture* (Hanover, NH: Wesleyan University Press 1996)

Kawa, Abraham, *Eikonika Vlemmata* (*Virtual Gazes: Postmodern Narrative in comics, film and fiction*) (Athens: Futura 2002)

Kuhn, Annette (ed.), *Alien Zone: Cultural Theory and Contemporary Science Fiction Cinema* (London: Verso 1990)

Sobchack, Vivian, 'Images of Wonder: the Look of Science Fiction' (1997); in Sean Redmond (ed.), *Liquid Metal: the Science Fiction Film Reader* (London: Wallflower Press 2004), pp. 4–10

Stevens, Wallace, *Collected Poetry and Prose*, ed. Frank Kermode and Joan Richardson (New York: Library of America 1997)

Tulloch, John and Henry Jenkins, *Science Fiction Audiences: Watching Doctor Who and Star Trek* (London: Routledge 1997)

13
Prose Science Fiction 1970s–1990s

Two apparently contradictory observations suggest that something peculiar happened to written SF in the last decades of the twentieth century. The first is that, over this period, more and more SF novels and stories appeared annually, among them many significant achievements and some undeniable masterpieces, such that SF grew into one of the most successful branches of publishing. But a second, more value-based observation is that during this period the novel stopped being the prime mode of SF. As visual SF (particularly cinema and TV) increasingly came to dominate the mainstream, prose SF became increasingly sidelined – an energetic sideline with many passionate adherents, but a sideline none the less. Scores of SF novels in the 1980s and 1990s became bestsellers and were acclaimed as classics in their day; but a very small number of those titles are still alive today in a meaningful sense. By 'alive' I mean a book still in print, still the subject of discussion and recommendation among readers (outside small-scale, dedicated fan bases), still influencing new writers, still making a cultural impact.

In other words, SF publishing from 1980 to 2000 embodies an odd paradox: a vivid flourishing that is also a waning away. Some may disagree and insist that they personally read dozens of SF novels each year that are nothing short of genius; that they continue to derive enormous pleasure and meaning from prose SF, and that, far from dying, the SF novel is stronger and more alive now than it has ever been. Fair enough. If people derive pleasure and significance from reading the – admittedly – often brilliant prose SF being written today, good luck to them. But a more than fannish perspective is needed. I am going to argue that, looked at objectively, the SF novel is now a lively but minor cultural phenomenon.

The first thing to say is that this creeping obsolescence of the SF novel (for which I am going to argue) may well be simply a function of the larger obsolescence of the novel *tout court*. By the 1970s some critics were confident that the novel was a moribund art form. Indeed, prognostications about the death of the novel had been popping up for decades, and by the 1970s had become something of a critical commonplace. As early as the 1920s and 1930s the distinguished critic Ortega y Gasset declared:

> It is erroneous to think of the novel – and I refer to the modern novel in particular – as of an endless field capable of rendering ever new forms. Rather it

may be compared to a vast but finite quarry. There exist a number of possible themes for the novel. The workman in the primal hour had no trouble finding new blocks – new characters, new themes. But present-day writers face the fact that only narrow and concealed veins are left them. (Ortega y Gasset, pp. 57–8)

The UK critic Bernard Bergonzi responded in 1970:

At what point was the last piece of territory occupied, the last vein of the quarry exhausted? Inescapably the answer must be that it was in the decades between 1910 and 1930, in the work of Proust and Joyce, whom Moravia has referred to as 'the gravediggers of the nineteenth-century novel.' *À la recherche du temps perdu* and *Ulysses* mark the apotheosis of the realistic novel, where the minute investigation of human behaviour on all its aspects – physical, psychological and moral – is taken as far as it can go, whilst remaining within the bounds of coherence. (Bergonzi, p. 23)

According to Bergonzi, by the 1970s the novel had to accept the fact that it was no longer (as its name tells us it once was) defined by its newness. Novelistic responses to the fact that 'the novel was no longer novel' varied in approach and in effectiveness. One attempt to inject newness into the traditional form was the so-called *nouveau roman*, a self-avowedly 'new' sort of novel that originated in France with writers such as Alain Robbe-Grillet (b. 1922) and Michel Butor (b. 1926), but which was taken up by Anglophone and other writers: my favourite American practitioner of this sort of writing is Robert Coover (b. 1932), although he does not write SF. According to Robbe-Grillet's manifesto *Pour un nouveau roman* (1963) the *nouveau roman* challenges the taken-for-granted assumptions of the 'traditional novel', conventions such as an omniscient and self-effacing narrator, coherence of narrative, believable characters with whom the reader can relate in the same ways s/he relates to the real people in his/her life, and in general the understanding that the novel functions as a window through which 'real life' (in a manner of speaking) can be viewed – all these things are, in effect, lies. They do not articulate the reality of modern experience, which is more alienated, more fractured and much more self-aware than this suggests. Classic *nouveaux romans* present deratiocinated agents moving through dislocated worlds; the emphasis is on the artificiality of the fictional construction, with a highlighting of linguistic effect and a commitment to a form of aesthetic estrangement. This relatively short-lived movement nevertheless fed into a broader cultural phenomenon, which we adduce with the contested label 'postmodernism', to which I shall have recourse to return in a moment.

All this strikes me as very interesting from the point of view of SF. Indeed, I might wonder whether the parallels between 'the mainstream literary establishment' and 'the SF world' during this period aren't alarmingly close.[1] The first thing to say is that many of the textual strategies we find in New Wave science fiction are directly or indirectly derived from the *nouveau roman*. It would not, I think, be correct to say that New Wave SF is simply the SF form of the *nouveau roman*.

Nevertheless, the hostility that the mainstream literary establishment evidenced towards the *nouveau roman* finds its correlative in the hostility of the mainstream Hard-SF community towards New Wave.

Bergonzi is very condescending about Robbe-Grillet's *avant-garde* injunctions to dispense with the contaminated attributes of the traditional novel of 'characters, story, atmosphere … Robbe-Grillet shows a forceful tendency to throw out the baby with the bathwater' (Bergonzi, p. 3). He likes 'characters, story, atmosphere'. Most readers of prose SF over the last three decades have liked them too, and have demanded them (with a sprinkling of techno-novelty on top) from their prose SF.

To make myself plain: it seems that there was another way of making the novel novel again in the last decades of the century, an alternative to the avant-garde extremism of stylistic and formal experimentation associated with the *nouveau roman*. It lay in the cross-fertilisation of 'fiction' with modes of discourse other than the traditional humanist idioms (what Bergonzi terms the 'physical, psychological and moral' attributes of human life), or perhaps merely the extension of those. By the end of the century discourses of science had changed utterly from the ones with which Victorians or Edwardians would have been familiar. A properly science fiction form might have injected vivid new life into the novel. As the present chapter will, I hope, show, some novels were written that attempted to do this, often very well. But, speaking broadly, such hybridisation has not reinvigorated the form of the novel, either in SF or in general.

One index of this transformation is the reception of *Gravity's Rainbow* (1973) the enormous novel by American writer Thomas Pynchon (b. 1937). This work has a plausible claim to be the greatest SF novel of the 1970s. Although lauded by the academic community, it is too long, complex, rebarbative and obscene ever to have enjoyed popular success (that it is still in print today is almost certainly because universities require their students to buy it). But more important than the general reception the novel has enjoyed is the response, or non-response, of the SF world. The two most prestigious awards in SF are the Hugo Award (the premier award of SF Fandom, voted for by people attending the annual 'SF World Convention') and the Nebula Award (voted for by professional writers who are members of the Science Fiction and Fantasy Writers of America). In 1973 the Hugo went to Asimov's limp late novel *The Gods Themselves* (1972). Asimov's book did, in a sense, reflect the concerns of its day (a free source of apparently limitless energy is discovered to be draining power from an alternate universe, with terrible consequences for the alien life in that place) as the energy crisis began to bite in the West. The Nebula went to Arthur C. Clarke's *Rendezvous with Rama* (1973), in which a mysterious alien spacecraft of considerable size flies into the solar system and then flies away again, allowing some astronauts just enough time to encounter indigenous life and wonder about the architects of the craft.[2] Because of the peculiarities of the voting rubrics, and the fact that mass-market paperbacks (issued sometimes a year later) often generate the fan vote that wins Hugos, it is possible for one novel to win the Nebula one year and the Hugo the following: *The Gods Themselves* and *Rendezvous with Rama* did precisely this, the former text winning the 1972 Nebula and the latter the 1974 Hugo.

It cannot be maintained that either of these novels represented the best SF novel of the year. Rather they were novels by writers with major reputations, sustained by very large pre-existing fan bases; as fiction they were deemed *good enough*. It is, of course, in the nature of fan communities that they prize a sometimes doggish loyalty to their tribal shibboleths over a broader view of artistic merit: but it remains, for instance, inconceivable that, had *The Gods Themselves* been published by a first-time author, it would have won either award.

The true significance of the 1973 awards, I think (and I am not the first person to argue this), is that they are symptomatic of the SF community turning decidedly inwards, eschewing not only the wider literary world but also the sorts of more experimental hybrid SF that had come out of the New Wave. SF Fandom from here on tended (not exclusively, but largely) to judge its preferred SF by a restrictive and self-fulfilling rubric. In some senses this rubric was more accommodating than it had been hitherto: for instance, an increasing number of fans were female, and women writers were more often praised, awarded prizes and their cause adopted than had earlier been the case. But in other senses it provided a procrustean bed. One of the things that happened in the 1980s is that some of the most talented writers of prose SF abandoned writing 'genre fiction' and declared themselves to be 'mainstream' or 'non-SF' authors: Michael Moorcock, J. G. Ballard, Christopher Priest, Margaret Atwood and many others. That these (to be frank) irrelevant statements of personal affiliation were taken as monstrous betrayals by outraged Fandom is one index to the partisanship that increasingly afflicted this world.

I earlier argued that, although sometimes sneered at, SF Fandom is, as Henry Jenkins has shown, a tremendously energetic, creative, passionate and significant phenomenon; a feature of popular culture itself worthy of significant study. This continues to be the case. But at the same time, SF Fandom has – in the bulk – manifested a lamentably conservative taste in fiction. SF fans may prize originality of *premise*, and fans often like books that engage with the biggest questions, that dramatise the sheer vastness of the universe, that evoke a 'sense of wonder', that make the head spin. But at the same time, the majority of SF fans prefer these big ideas to be actualised in novels established on (to use Bergonzi's phrase again) 'characters, story, atmosphere'. What I mean by this is that in the books SF Fandom acclaimed during the 1980s and 1990s there is an unmistakable bias towards (a) characters with whom readers can identify and empathise, whom they like and about whom they care; (b) a story which provides the satisfactions (set-up, development and dénouement, with all loose ends tied up) of a nineteenth-century novel, and which doesn't play with chronology or narration; and (c) a transparent and serviceable prose style, rather than one that experiments with language, or affects too 'literary' an idiom. Most of the novels discussed in this chapter test positive for one or two of (a), (b) and (c), and many exhibit all three. It is not that there is anything wrong with such unadventurous tastes *as such*. But the effect of the shaping power of this subculture has been a large number of novels and stories that treat exciting and original premises in dowdy, old-fashioned, limiting and two-dimensional fictional form. The SF novel, by and large, is no longer novel; or more strictly, SF novels that attempt novelty are often ignored or disdained by the majority.

Pynchon's *Gravity's Rainbow*, on the other hand, is a novel of brilliant and triumphant newness. Its characters, though fascinating, are often grotesques, amoral or bizarre; its plot is so peripatetic as to appear to the casual glance formless, although in fact the myriad wonders of the book are organised according to a tightly structuring thematic principle (the parabola of ballistic flight) rather than to the great-aunt conventions of beginning–middle–end. And the style is one of the wonders of contemporary Anglophone writing; endlessly inventive, baffling, obscene and brilliant, shifting registers and points-of-view continually, and managing somehow to adapt itself to the encyclopaedic gamut of the novel.

The novel is set during the last stages of the Second World War and its immediate aftermath. Tyrone Slothrop, an American officer stationed in London, goes on a peripatetic quest across Europe. It seems that mapping Slothrop's sexual encounters produces a map of where the V2 rockets are going to fall. The rocket stands at the symbolic core of the novel: the fact that V2s travelled supersonically (and that, therefore, they would explode into their targets first, and that only *after* that would people hear them coming) stands, for instance, as a symptom of a larger derangement of cause and effect in the world of the novel. The arc of the rocket's flight (from which the book takes its title) also governs the form of the novel. Indeed, the ambition of the book is so enormous, and so fully achieved, that the rocket becomes a prodigiously expressive symbol. In Richard Poirier's words, 'the central character is the rocket itself'; and the 'secret' with which the book is concerned 'is that sex, love, life, death have all been fused into the Rocket's assembly and into its final trajectory' (Poirier, p. 15).

One thing the novel does, very powerfully, is explore precisely the very real dialectic of good/evil expressed by that 'rocket', the same rocket, essentially, that flies through the dreams of SF's interplanetary dreams. The balance is between what one character calls 'a good rocket to take us to the stars, an evil rocket for the World's suicide, the two perpetually in struggle' (Pynchon, p. 727). These are ethical questions of the largest scale: is the American achievement in sending astronauts (by rocket) to walk on the Moon compromised by the fact that their rocket programme drew heavily on Nazi research instituted for the V2 programme under the Nazi Wernher von Braun (1912–1977), who was later willingly appropriated by the Americans to head their own space programme? That there may be an ineluctable darkness at the heart of the bright vision of spaceflight, something tied to the very mechanism of the flight itself, is nowhere else so dazzlingly or brilliantly explored.

Science fiction Fandom

The SF fan community has become very effective at marshalling written literature within the genre. There are, every year, many conferences at which fans gather to hear authors speak, to discuss the merits of various books among themselves, and to buy new and rare SF items from dealers. In addition to the premier SF awards (the Hugo, Nebula, John Campbell, Philip K. Dick and, in Britain, the Arthur C. Clarke Award) there are over 90 more specialist SF awards, some voted for by fans, some awarded by panels of experts, most given annually – there are so many, indeed (more than any other literary genre by a factor of 20) that it becomes quite hard for an author of serious SF not to win one at some point during his or her career.[3] Over the period under consideration a great many amateur fan magazines ('fanzines') were published, similarly discussing in great detail the books being published in the genre; although by nature transitory, some of these publications are still issued today; some of these embodied such high productive values and have so wide a circulation that they are known as 'semiprozines'. Since the advent of the internet, the exchange of opinion and judgement among the worldwide community of SF fans has expanded considerably. I have no idea how many SF websites are now in existence, and do not have time to attempt to count them: it is certainly in the thousands, and the best of such sites contain some of the best criticism (usually in the form of reviews and essays) available on the genre. This situation means that a large, eager, intelligent and opinionated audience awaits the publication of new SF fiction; and that such novels will almost certainly be reviewed widely (this is not true of mainstream fiction, where novels are quite often not reviewed at all). SF fans are very often articulate and knowledgeable – more knowledgeable, sometimes, than salaried academics studying the genre, such as myself (although I am also, of course, a fan). All in all the size and vigour of the fan community is such that it now represents a very significant proportion of the SF world itself; a fact reflected in the fact that some SF awards (such as the Hugo) give an annual award for fan writing. Histories of SF Fandom have been published by various interested parties.[4]

Partly because publishers know that they can rely on this body of fans to buy new SF fiction, and partly too because the fans are so dedicated at disseminating their appreciation of the genre, more SF novels are published annually now than at any time in the past. Of course, a proportion of this output is schlock, and many of the books are clichéd, merely conventional or simply bad. But a great many good, and some masterful, SF novels are published every single year. We might deduce from this that from the late 1970s to the present, SF prose fiction has been in a state of rude health, and has been getting consistently healthier.

But this would be an erroneous deduction. Within the fan community there is an often overenthusiastic endorsement of pet authors and fans too easily trade in the language of the 'classic' – in the sense of 'such-and-such is a classic'. There is a broad current of over-compensation in such claims; since it is a tenet of faith in the world of fans that 'they' – whoever they are ('the literary establishment', for instance) – are deliberately ignoring or suppressing science fiction; that SF has been hemmed into a ghetto by a hostile world, that no SF author would 'be allowed' to win the Pulitzer, Booker or Nobel literary prize, and so on. It seems paradoxical that a mode of literature that is enjoying such rude health (according to one way of looking at things) should also be prey to such delusions of persecution and believe itself harried almost to death. This siege mentality is neither gracious nor helpful. It's very understandable, of course, that readers want to champion books they like, and some SF fans have liked some books very much indeed; but the aggressive hyper-assertiveness of many fans is counterproductive: you don't persuade people to like the things you like by hectoring them. Moreover, it is not clear why it matters so much to SF fans that others sometimes don't like the things they like. The SF fan community is, I think, the largest and most active literary community in the world (outside the artificially maintained factory-lines of school/college literature teaching). Each year it has hundreds of excellent books to chew over; readings to attend; conferences to congregate at; online chat rooms to engage with. It is the very strength of SF Fandom that renders most of the complaints of SF Fandom irrelevant.

There is another consideration, which applies particularly to written SF in the 1980s. It is hard to be objective about books that have had an impact on one's own life, even though those same books may have had very little broader cultural impact. Moreover, as Peter Nicholls wittily puts it, it might be said that 'the Golden Age of SF is 14' (Clute and Nicholls, p. 506): critics who grew up reading the SF of the 1970s and 1980s (I am one) will tend to look more rosily at the fiction of that period than might otherwise be the case.

But the fact of the matter is that very few SF novels published in the 1980s are still living works of literature today. Even titles that enjoyed considerable popularity, perhaps bestseller status, are now, most of them, irretrievably dated and gone. For example, Julian May (b. 1931) enjoyed a widespread popularity and enviable sales with her *Saga of the Pliocene Exiles*, four chunky novels detailing the adventures of time-travellers from our own near future who settle in prehistoric Europe (*The Many-Colored Land*, 1981; *The Golden Torc*, 1982; *The Nonborn King*, 1983; and *The Adversary*, 1984). These books are all examples of highly enjoyable tale-telling, a proficient mix of future technology and magical human ability, with a well-structured plot, addictively readable. Millions read it in its day. But the *Saga* is no longer a current book: which is to say, a book still selling, still in print, still avidly discussed (outside the occasional niche fan forum), a book still influencing new writers, still making a cultural impact. In a sense it reflects no discredit on this book or this author that this is the case; May's work did its job. But like an overwhelming majority of prose SF from this decade, it has become an element in the general mulch of the past.

My argument is that during the 1980s the dominant mode of SF altered from written to visual paradigms, towards cinema, TV and graphic novels; this is the root explanation, I think, for the relative scarcity of SF prose from this period that is still alive. But I am not suggesting that 1980s and 1990s SF is a totally dead zone in the critical history of the genre, like an over-exuberant jungle whose fecundity has utterly stifled itself. Clearly some books from the 1980s are still alive; books including (to cite only the unarguable masterpieces): Gene Wolfe's *The Book of the New Sun* (1980–83); Russell Hoban's *Riddley Walker* (1980); William Gibson's *Neuromancer* (1984) – a book which, quite apart from its own merits, founded a new subgenre called 'cyberpunk'; Margaret Atwood's *The Handmaid's Tale* (1985); Orson Scott Card's *Ender's Game* (1985); Iain Banks' 'Culture' novels, beginning with *Consider Phlebas* (1987); Sheri Tepper's *The Gate to Women's Country* (1988) and *Grass* (1989); and possibly Dan Simmons' *Hyperion* (1989). These works are discussed in as much detail as space allows below. But many other titles, even some highly successful in their day, have been squeezed out. From inside the world of SF Fandom it seemed as if novel *x* or novel *y* was going to change the world. Almost always it did not.

Prose science fiction of the 1980s

The success of *Star Wars* (1977) percolated through into fiction as well as influencing film. A great many new readers were drawn to the genre, looking for *Star Wars*esque adventures to read, and publishers supplied their needs. By 1989 *Locus* magazine was reporting a 50 per cent rise in the number of SF titles being published annually since 1980 (Clute, *Science Fiction*, p. 87). Indeed the majority of SF published during the decade was 'commercial' in this sense, and most of that tended to look back to the formats and conventions of Golden Age SF (to the delight of many, for whom such SF represented the acme of the genre) rather than building on the aesthetic advances of the New Wave. That many of these books were politically right-wing merely reflected a widespread political shift, the dominance of Ronald Reagan's scorched earth Republicanism and Margaret Thatcher's foully monetarist Conservatism.

Larry Niven and Jerry Pournelle collaborated on a number of blockbusters, including the disaster novel *Lucifer's Hammer* (1977) about Earth's attempts to avoid a massive asteroid collision, *Oath of Fealty* (1981) in which wise autocrats crush attack by idiotic ecologists and other liberals, and, most notably, *Footfall* (1985). This last book, in Damien Broderick's words, 'captured the Reagan 1980s even more vividly than the movie *Rambo II*' (Broderick, p. 83): aliens who physically resemble small elephants invade Earth. They are communal beings, ciphers for the Communist threat that right-wing ideologues still bruited as the greatest danger facing the Free World at that time (bruited wrongly; Communism was in fact on the point of collapse). Earth retaliates, and human grit, determination and fighting spirit crush the invaders. This type of bang-bang-bang adventure writing is so gung-ho as to approach self-parody, although its large and generally sympathetic audience read its ideological simplifications and obfuscations as actual insights into the world, which is a worrying thing to contemplate.

Much less crude, although still informed by a right-wing libertarian ideology was the Heinleinian writing of John Varley (b. 1947). Varley began the 1980s with a very successful trilogy: *Titan* (1979), *Wizard* (1980) and *Demon* (1984). Near future exploration discovers 'Gaea', a spoked wheel-shaped space habitat of prodigious size that is also a sentient being, orbiting Saturn. The crew crash into this artefact, and are absorbed through the wall into the inside, some of them undergoing profound changes in the process. The inside, they discover, is a varied and colourful series of interlocking ecologies, all generated by Gaea herself and many of them adapted from Earth's popular culture (which Gaea has been monitoring). Some of these work better than others; the intelligent Centaurs, with their complex sexual protocols (they possess both human and equine genitals; Varley spares us none of the detail) are tiresome; but the point in *Demon* where Gaea herself, now mad, takes the form of a Brobdingnagianly proportioned Marilyn Monroe achieves a powerful weirdness that almost substitutes for satirical punch. The moon colony of *Steel Beach* (1992) is Heinleinian (strictly speaking – some of the colonised have modelled themselves on Heinlein's writings) contains many interesting elements, but the book as a whole does not escape the claustrophobic solipsism of a talented writer looking back to previous novels rather than outward to the cosmos.

This solipsism is not exactly a failing in Varley's work. Rather, it reflects the increasingly clogged and constipated nature of the genre, another function of its success. Keeping the SF novel *novel* (to use Bergonzi's terms) means finding new premises, new technology, as well as new textual forms and strategies. But increasingly SF writers are content to recycle older premises and technologies. This is less plagiarism, more a function of a new cultural logic, usually indexed with the shorthand term 'postmodernism'. Varley's most successful short story symbolically articulates precisely this turning-in: 'The Persistence of Vision' (1978), a clever and moving updating of Wells's story 'The Kingdom of the Blind', ends with a transcendental involution of the narrator away from his external senses into a hinted-at bliss.

The American writer Greg Bear (b. 1951) published his first story in 1967, but it was not until the 1980s that he began to make a name for himself. *Hegira* (1979) is a striking although ultimately unsatisfactory novel set on a vast hollow world that is home to many species. Although barely 200 pages long, it reads like the condensed version of a much longer novel. More effective is *Beyond Heaven's River* (1980), which treats a standard SF premise (of mysterious, godlike aliens playing games with human pawns) with considerable power and originality. Kawashita, a pilot in the Japanese air force during the Second World War, is plucked from certain death by an alien spacecraft; he spends centuries living in an artificial environment on a distant world acting out scenes from Japan's history. Eventually he is rescued by star-travelling humanity, and faces the dual challenge of reintegrating into a human world and trying to understand who abducted him and for what alien reason. Again, though a small book, this is replete with big ideas (probability spheres, living oceans, hallucinogenic pollen, million-year histories dropped into a paragraph and so on). But there is something even older than Golden Age SF about the aesthetic ambition. Bear dedicates the novel to the spirit of Joseph Conrad, and it's not an empty gesture. Worthy of Conrad is the expertly realised representation of an individual, a sailor from the East, encountering degrees of strangeness that spiral up and beyond all expectations.

Bear enjoyed greater success with *Blood Music* (1985), a wonderfully spooky tale in which nanotechnology assimilates all of humanity (almost) into a global-sized, living, sentient mass of sludge. The trillions upon trillions of molecule-sized computers of which this is composed, each of them intelligent, eventually creates a noosphere of such thinking density that the Earth itself passes into a new dimension. Like *Childhood's End*, with which the novel has things in common, this is a book wholly implicated in the SF dialectic, expressing itself through a rigorously applied scientific framework yet shading into the mystical ('magic') transcendence at the end. Bear enjoyed even more success with the more conventional space opera of *Eon* (1985) and its sequel *Eternity* (1988), with hollowed-out asteroids travelling through space and then time; but his best book is certainly *Queen of Angels* (1990), a novel of genuine richness and depth. The depth isn't immediately evident; the showy, well-handled murder plotline keeps one reading, the many futuristic and mind-tickling ideas keep popping up to make one think. But at the core of this book is an exploration of the nature of consciousness as profound as any in literature. Martin Burke is a psychologist who investigates the motivations for a murder in a society, using nanotechnology to explore the Country of the Mind. AXIS is a sophisticated computer that has travelled to a distant planet. The two journeys of exploration are paralleled, but in their different ways they prove relative dead-ends. Bear's masterstroke is surreptitiously to delineate the shift of another sophisticated computer, JILL, from a linear intelligence based on processing data to a self-aware sentient intelligence that is a genuine consciousness. Overall this novel embodies exactly what Prose SF can do well – philosophical investigation into the mystery of consciousness expressed in popular and accessible form.

The American writer C. J. Cherryh (b. 1942) has a large fan base, and writes both Fantasy and SF, sometimes combining the two. Her Arthurian romance *Port Eternity* (1982), for instance, is set aboard a stranded spaceship in which androids (programmed to act out Arthurian personalities) start to assume the roles for real. It is a tightly written, expert novel and may be her best, but it is not typical of her output. Her SF fame derives from the many lengthy space-operas she writes prolifically and which often have a texture rather like the novelistic version of expanded polystyrene. *Downbelow Station* (1981), which won the Hugo for best novel of that year, throws together a great many characters, species, plots and ideas (mostly venerable SF ideas) into a future universe without faster-than-light travel in which spaceships are home to generation after generation of increasingly feudal, bickering space-families. A dozen or so of these titles (brought together under the collective moniker 'the Merchanter novels') connect with another dozen-or-so 'Union-Alliance' novels, set in the same imagined future galaxy. The 'Chanur Saga' brings together an alien–human encounter-centred series of novels, which also fits into the larger future history Cherryh is developing. Everything she has written, it sometimes seems, fits together into a single ambitious future history, reminiscent of Niven or Heinlein. The zenith (or, if you prefer, the nadir) of this love for largeness is *Cyteen* (1988), a huge ziggurat of a novel: well over 1,000 pages of flapdoodle about clones, androids and humans. Books like this bear witness to the general trend, which became increasingly prominent in the 1980s and 1990s, for more and more words, thicker and thicker spines, as if quality can be measured primarily in terms of quantity. But the only formal innovation in Cherryh's books is scalar: bigger and bigger, more and more plots and characters confected. Undeniably the fact that she is able to construct on such a scale, and to link each monumental block into the larger edifice of her future history, is evidence of some technical skill. But her prose style is greyly enervating, her characterisation old-fashioned, and there is little to the novels except the various stories. It is hard to shake the sense that (to appropriate the words of a critic from a different context) a lifetime is too short, and eternity scarcely long enough, to read the total output of C. J. Cherryh.

Much more interesting was Brian Aldiss's mid-career leviathan, *Helliconia*; published as *Helliconia Spring* (1981), *Helliconia Summer* (1983) and *Helliconia Winter* (1985). When these appeared they were treated as one of the great events of SF publishing; a masterful, absorbing, complex and open-ended process of world-building. That they seem to have dropped off the radar lately may owe more to the general under-valuing of Aldiss (one of the handful of genuinely significant twentieth-century writers of SF: see above pp. 245–8) rather than a reflection on their own merits. Aldiss's Helliconia is a planet whose eccentric orbit round several stars gives it mini-seasons (as on Earth) as well as a very long Great Year in which the whole climate shifts as the individual volume titles indicate. Humans and aliens cohabit the world, and civilisation as a whole grows from hunter-gatherer to something analogous to the late nineteenth century. As the story grows we discover that the population of an enervated future Earth is watching the events on Helliconia, beamed back to them via myriad remotes. When I say that this huge novel is a masterpiece I hope I mean more than that it had a juggernaut effect on my own sensibilities when I read it in my late teens. I think there is something genuinely 'classic' about this book, a text which finds interesting new ways of expressing the dialectic between materialism and mysticism. On the one hand, the myriad details are rendered with an extraordinary and vivid sense of verisimilitude, as actual possibilities (James Kneale and Rob Kitchin note that for this novel 'Aldiss took advice from academics, including Jack Cohen, a reproductive biologist who has acted as "consultant" to several SF authors who wanted plausible extraterrestrial life-forms', Kneale and Kitchin, pp. 10–11). On the other hand Helliconia possesses a dimension of soul, or spirit (which Aldiss does not – in this dramatisation – give the Earth).

But the complexity and innovation (subtle but palpable) of Aldiss's big 1980s novel are not characteristic of the subgenre 'Big 80s SF Novels' in general. Most marry innovative premise to traditional backward-looking or reactionary models of plotting and characterisation, producing novels that rarely rise above the unexceptionable. There are, for instance, many admirers of the writing of David Brin (b. 1950) but nobody could claim that he has made any formal innovation of advance in the technical aesthetics of fiction. The many fat volumes of his 'Uplift' sequence (*Sundiver*, 1980; *Startide Rising*, 1983; *The Uplift War*, 1987; *Brightness Reef*, 1995; *Infinity's Shore*, 1996; *Heaven's Reach*, 1998) chug along as if powered by two-stroke engines, throwing humans, genetically engineered intelligent dolphins and chimpanzees, and many alien species together in a plot built around the premise that alien races can be 'uplifted' to higher states of consciousness, a process begun by the now-vanished Progenitors. Reading these novels is a perfectly acceptable way of passing the time, and the scale and invention is large. But they are not especially noteworthy.

John Barnes (b. 1957) achieved something interesting with his second novel, *Sin of Origin* (1987) – a book that, in a sense, reverts to the seventeenth-century theological origins of SF itself (missionaries encounter an alien life-form that lives as three-part symbiosis, and map their own understanding of the Holy Trinity onto what they see – converting the aliens but with malign results).

Russell Hoban (b. 1925) seemed to some to be, in *Riddley Walker* (1980), aping Burgess's *Clockwork Orange*. Like the earlier novel, Hoban's book is set in a future world (in this case a post-nuclear Kent) and is written in an invented language. But where Burgess's idiom reads as forced and artificial, Hoban's is rich and believable, although since much of it consists of odd and variant spellings of familiar words it sometimes reads rather inappropriately as Molesworthian. The protagonist, Riddley, is caught up in attempts to resurrect the mystery of the atom bomb itself, the '1 Big 1' the memory of which is preserved in quasi-religious myth concerning 'the Littl Shynin Man the addom':

> Eusa put the 1 Big 1 in barms then him & Mr Clevver droppit so much barms thay kilt as menne uv thear oan as thay kilt enemes. Thay wun the Warr but the lan wuz poyzen from it the ayr & water as wel. (Hoban, p. 82)

The book's portrayal of a brutish yet genuine future world is convincing, absorbing, and Hoban never lets the mythic and theological aspect of his fiction overshadow the telling. Because, of course (again like *Clockwork Orange*), this resolves itself into a fiction of Fall and Original Sin (figured here as splitting the 'the Littl Shynin Man the addom' and the barminess of putting him/it into a 'Barm' to do so much damage) as well as, in a sense, being about redemption. Once again the buried symbolic drama of SF reasserts itself – the dialectic between a purely religious and a purely materialist understanding of the cosmos. As John Clute notes, the premise of the novel – life struggling back towards civilisation after a nuclear war – 'is a situation much explored in the SF of the latter half of the 20th century', but not only is Hoban's 'penetration of the moral and cultural complexities involved ... acute' but the linguistic experimentation of the novel succeeds brilliantly (Clute and Nicholls, p. 574). The style succeeds in simultaneously making strange (as non-standard English) and making familiar (as a childlike idiom of misspelling and phonetic transliteration).

Gene Wolfe's *The Book of the New Sun* (1980–83) and its sequels

For many critics, the American writer Gene Wolfe (b. 1931) is the most important writer of fiction, in or out of the genre, of the last few decades. Such critics may well be right. By the 1990s it had become clear that, a few short-stories and singleton novels aside, the prolific Wolfe was actually engaged in writing one single enormous novel, rather like Proust – with whom he has occasionally, although erroneously, been compared.[5] *The Book of the New Sun* (which first appeared as four volumes: *The Shadow of the Torturer*, 1980; *The Claw of the Conciliator*, 1981; *The Sword of the Lictor*, 1982; and *The Citadel of the Autarch*, 1983) is set on a very far future Earth (called 'Urth'), where the sun is dying, and human society has bedded down into intricate quasi-medieval structured leavened with occasional touches of high technology (space flight, time travel) but otherwise a world recognisable from Fantasy. Severian, the protagonist, is an apprentice torturer, who leaves the city in which he has been trained to trek across the surface of Urth, encountering many people and having a variety of strange adventures. He renounces his calling and travels further, eventually becoming the 'Autarch' or ruler of Urth. A single-volume sequel to *The Book of the New Sun*, *The Urth of the New Sun* (1987) takes Severian into space in search of the streaming 'white hole' which will be used to revivify the dying Sun. This miniature summary does not even come close to capturing the baroque complexity of Wolfe's novel, its many interlocking narratives and scores of characters, its shifts of tone and emphasis, sometimes written in prose sometimes in dramatic form.

The Book of the New Sun links obliquely but certainly with what became Wolfe's next major project, *The Book of the Long Sun* (comprising *Nightside of the Long Sun*, 1993; *Lake of the Long Sun*, 1994; *Caldé of the Long Sun*, 1994; *Exodus from the Long Sun*, 1996). The hero of this large work is a priest (very Catholic in style although actually in the service of various pagan deities) called Patera Silk, who lives in his modest way aboard a colossal Generation starship, the *Whorl*, a word which sounds more like 'World' if articulated in certain varieties of American accent. This ship with its many inhabitants has been so long on its journey that most of its passengers have forgotten that it is a ship at all – a venerable SF idea, but here treated with a new seriousness and thoroughness as an essentially spiritual fact. A trilogy followed that continued the *Long Sun* story directly, *The Book of the Short Sun* (*On Blue's Waters*, 1999; *In Green's Jungles*, 2000; *Return to the Whorl*, 2001), mostly set on the two planets ('Blue' and 'Green') which were the Long Sun starship's destination, and where it has long been orbiting. Silk and his family are caught in new adventures, travelling between the two worlds and returning, periodically, to the *Whorl*.

Stylistically, Wolfe is a talented writer, although some find his deliberately mannered and archaic idiom a deterrent, and his more recent books rely too much on over-lengthy and expository dialogue. But his great achievement is not stylistic but formal, the creation of a text that construes narrative, character and atmosphere into the ambiguities and complexities of which they are made. Little is straightforward in a Wolfe novel; books can be read and reread again to reveal new perspectives. Like a *nouveau romancier*, or a postmodernist (although Wolfe – a practising and conservative Catholic – may be surprised to hear that he is called 'postmodern') he deconstructs our assumptions about narrative closure, about the description and working of character and about 'meaning' in a series of challenging ways. Although his books are all 'religious', none of them resolves into straightforward allegory, or even symbolism, although all of them are replete with Christian symbols: roses, fishes, the sun, trinities, and so on. But the action of beginning to decipher the symbols creates more rather than less textual insecurity. The effect of these three complex works is perhaps to baffle the reader used to simpler fare.

Because of this, exegesis can become a fetish with fans of Wolfe (several have published lengthy keys to the mystery: Michael Andre-Driussi's *Lexicon Urthus* (1994), Peter Wright's *Attending Daedalus: Gene Wolfe, Artifice and the Reader* (2003) and Robert Borski's *Solar Labyrinth: Exploring Gene Wolfe's BOOK OF THE NEW SUN* (2004) are three worth mentioning). The watchword for these fans is that his lengthy novels must not only be read but *re*read, often several times, before their beauties and depths become apparent: Wolphiles' insistence on this point is sometimes folded into a more generalised grumpiness about the hectic pace of modern living and the virtues of close, careful attentiveness to the text. This is a reasonable, if crankily middle-aged, view to hold. But it is also worth noting that, of all the major writers of SF alive today, Wolfe is the one most thoroughly to divide the SF fan base. Many fans have never read him, or have tried to read *New Sun* but have given up. Internet bookshops such as amazon.com, which permit readers to post their reactions to the books they have bought by grading them from 1 star to 5 stars, provide interesting if partial snapshots of general responses to works. Wolfe's books draw extravagant praise and 5-star assessment and extravagant dismissal or hostility and the lowest rankings in about equal measure.

But before Wolphiles simply dismiss the latter reactions as bred of ignorance and a pig-headed resistance to the beautiful complexities of a deliberately difficult writer, they need to confront the fact that some SF critics of great distinction have shared this dislike. 'I cannot stand postmodernists', announces Darko Suvin, adding:

> I cannot follow the semantic and diegetic contortions of Gene Wolfe, fleeing the Master Narrative ... I shamelessly confess I prefer a good story by Heinlein, Cherryh or Gwyneth Jones to most philosophies, since they show me worlds with actions, resistances and psychozoa for whom both mean something. (Suvin, p. 241)

'Contortions' is a little unfair; although his books *are* complex, serpentine and cat's cradle-like it gives the wrong impression of Wolfe (always an elegant, controlled writer) to call him 'contorted'. But Suvin gives voice to a widespread suspicion, not that Wolfe is a bad writer exactly (Suvin concedes that there are 'some impressive facets to his major series'), but that the kind of writing he practises is a wrong turn in the development of prose SF. We are back, in other words, to mourning the loss of Bergonzi's nineteenth-century criteria: 'characters, story, atmosphere'.

My sense is that Wolfe has more to fear from some of his enthusiasts than his opponents. It is fair enough, if one enjoys puzzles and games, to treat Wolfe's fiction as a gigantic textual box of puzzles and games. But to read only ludically is to miss the main point of Wolfe's writing, which is always serious ('playfully serious' sounds merely oxymoronic, but comes close to the truth). His work is engaged in a genuinely profound excavation of the core dialectic of SF, the relationship between the material and the spiritual. *The Book of the New Sun* is very finely constructed to balance precisely on this knife-edge: it can be read throughout as heroic Fantasy (in which matters are explicable in terms of magic) or as science fiction (in which the various wonders have a technical, material explanation), and is as often categorised with one genre as the other. The point of this is not the categorising itself; it is rather that Wolfe's aesthetic is concerned with these matters because they relate to the world in a genuinely problematic manner. Many of the things humanity used, formerly, to attribute to God can now be explained in ways that leave no space for the divine. If the search is pushed as far as it can go, outwards to the stars and throughout time, do we come to a *pou sto*, an ultimate standing-place that we must concede to be God? Wolfe's dramas usually dramatise this quest, with many creatures that initially appear to be gods revealed later to be artefacts. There are many blind alleys, much misdirection, and the crucial points often appear minor, easily ignored, on first view: the involutions, and evolutions, of this search acts as template for Wolfe's plotting.

The conclusion that Wolfe comes to is that there is an 'Increate' or 'Outsider' whom we might call 'God'. 'Once I believed you three were gods,' says Severian to a number of appearing entities, 'and then that the Hierarchs were still greater gods ... But only the Increate is God, kindling reality and blowing it out. All the rest of us, even Tzadkiel, can only wield the forces he's created' (Wolfe, *Urth of the New Sun*, p. 353). This rather conventional monotheism is even more explicit in the *Long Sun* books, where the gods worshipped aboard the *Whorl* are early on revealed to be the simulated personalities of long-dead aristocrats preserved in a mainframe database, occasionally appearing to the populace through view screens called 'Sacred Windows'. The protagonist Patera Silk is a priest of these gods, devoted to the truth, celibate, absolutely dedicated to making the correct moral choices in a complex and changing world – indeed, he doesn't altogether avoid the charge of priggishness (although he is prepared to steal, or even kill, if he believes it serves a higher purpose). During the course of the text Silk comes to understand that the gods he has served and thought good are not gods, and are not particularly good, although that loss of faith is compensated by the apprehension of the greater God outside the *Whorl*. Indeed, rather boldly, Wolfe begins this lengthy novel with a moment of this divine revelation that becomes comprehensible only after the reader has read a long way into the book. The first sentence of *Nightside of the Long Sun* is: 'Enlightenment came to Patera Silk on the ball court; nothing could ever be the same after that.' We are also told that 'few of these hidden things made sense, nor did they wait upon one another' (Wolfe, *Nightside*, p. 9).

The *Book of the Long Sun* is about revelation. On the one hand, its starship-travelling characters who have unthinkingly assumed that their destination is a long way ahead of them realise that not only have they arrived, but they have been at that destination for a long time. But more important, in a way, is the 'enlightenment' that Patera Silk experiences right at the beginning, which he describes not as the sudden realisation but as a recognition of something he had long known: 'it was as though someone who had always been behind him and standing (as it were) at both his shoulders had, after so many years of pregnant silence, begun to whisper in both his ears' (Wolfe, *Nightside*, p. 9). That 'someone' is the 'Outsider', also called 'Ah Lah'. Personally, I fear this more direct introduction of Wolfe's own ecumenical God into the narrative unbalances the whole. *Long Sun*, although marvellous, is a lesser achievement than *New Sun* in part because of this unambiguous theological bias. It is also less thrillingly written, with too much expository dialogue and not enough descriptive or meditative prose; some of the many conversations in the book come close to interminability. Some critics have not seen this as a weakness: according to John Clute, 'the dialogue, once it is understood as the central artery and pulse of the entire enterprise, seems anything but a diversion. This sound of incessant interrogation is, one might say, the song of the epistemology of revelation' (Clute, *Scores*, p. 144).

In *New Sun* Severian is revealed as a type of Christ – he even rises from the dead – although throughout we see things from his point of view and are therefore privy to his own character flaws, his disbelief in himself. It is one thing to have certain people not believe in Christ; it is quite another to have Christ himself not believe in Christ. Severian is an infinitely more complex and compelling character than Silk, and because the world of Urth is mediated through his consciousness the novel in which he appears is more effectively rendered.

By comparison Patera Silk, though humble and self-effacing to an irritating degree, is actually much more self-assured: possessed (as Severian is not) of a strong moral compass and sense of right and wrong. But the mismatch between this rather simple, grounded character and the relativist, unreliable richness of the plot and mise-en-scène tend to drain the whole work of power, I think; where the creative tension between Severian and the world he travels through in *New Sun* adds immeasurably to it. The fact that Horn, the protagonist of *The Book of the Short Sun* actually changes is a third attempt to work through a similar thematic business, and is the least successful of the three, precisely because it is too literally externalised. But in this late work Wolfe is still chafing the rough borders of 'Fantasy' and 'SF' up against one another. Horn, in *In Green's Jungles*, is regarded as a sorcerer by some, although he insists that he cast no spells, but only marshalled a number of technical tricks. One major species in this trilogy are the 'inhumi', a type of shape-shifting vampire: a type that can be taken as supernatural creatures (as much horror writing does) or that can be explained in pseudo-scientific terms. Wolfe picks out a carefully judged line between these two approaches.

Having said that, it still seems to me that it is *The Book of the New Sun* that is Wolfe's major achievement. The more books that are added to his creation the less ambiguous it becomes, even when those additional books are full, in David Langford's words, 'of tantalizing enigmas … shades of Borges and Escher and strange loops' (Langford also, rightly, talks of 'those passages of blinding simplicity in which this extremely unsimple author loves to mask himself' (Langford, pp. 288–90)). There is in *New Sun* something approaching a radical relativism, a vision of ambiguity going right down to the core of existence, despite the occasional invocation of the Increate. By the end of *Short Sun* we are being told that this ambiguity is not radical, but merely a kind of metaphysical ketchup applied to the ontological dish for the benefit of the cosmos's inhabitants. In the halting words of another character, Patera Remora, 'what all men, and most – ah – females, require is not theophany, not the divine palpability. Tangibility. It is the – ha! – possibility' (Wolfe, *Return*, p. 402). Remora suggests that Horn (or Silk, as perhaps he now is) institute the worship of the Outsider who will not trouble human beings' taste for possibility over actuality by doing something so disruptive as incarnating himself, or revealing himself to his congregation ('He, at least – … will not come to your Window, Your Cognizance. I believe I can assure Your Cognizance of that'). I find this a cop-out. We are too certain of the existence of

'Ah Lah' by the end of the whole sequence, something not true of *New Sun's* Increate. But, to quote John Clute once again, just as *'The Book of the Short Sun* is all about salvation' (Clute, *Scores*, p. 260), so is the entire sequence. Wolfe not only revisits many of the conventions of twentieth-century SF, he goes further back than that, tapping into the deep roots of the genre: interrogating the many ways in which notions of 'salvation' are inflected by our much broader materialist understanding of the cosmos.

William Gibson's *Neuromancer* (1984) and cyberpunk

As the number of commercial prose SF titles increased it became proportionately harder for one writer to make an impact. But one who managed to do so was the American William Gibson (b. 1948); and the book which, arguably, made the biggest splash of any SF novel in the 1980s was his *Neuromancer* (1984).

'Cyberpunk' was a key genre invention of the 1980s. The term was coined by the little-known American writer Bruce Bethke (b. 1955) in a short story 'Cyberpunk' (1983). The first significant novel written in this idiom, according to many, predates the terminology: the American writer Bruce Sterling's (b. 1954) *The Artificial Kid* (1980): set on a distant planet in the far future, it largely concerns the 'kid' of the title, who uses his enhanced martial arts skills to fight for a living, an entertainer in a post-scarcity culture. But this novel, good though it is, is hardly characteristic of the great wash of cyberpunk fiction that was to follow. Sterling's work, while trendily violent and streetwise, is nevertheless expansive, picaresque, colourful and mostly set in the larger landscapes of its imaginary world. More often than not typical cyberpunk books are claustrophobic, dark and wedded to a narrow urban vision: symptomatic of a more pessimistic assessment of the dangers of computing (which was the boom industry of the western world in the 1980s) and the increasingly city-centred focus of contemporary life.

It was Gibson's *Neuromancer* (1984) that established many of the premises of cyberpunk: combining the premise of the movie *Tron* (1982) – a consensual, computer-generated virtual reality or cyberspace, which Gibson calls the Net, or Matrix – with the grubby-chic stylings of *Blade Runner* (1982): a *noir* plot, a degree of violence, and a conceptual breakthrough at the end. Gibson's hero 'Case' is a streetwise hacker, who becomes embroiled in a complex plot to steal protected data from cyberspace. The adventure leads him across the face of the world and then up into a space habitat in orbit, before the heart of the narrative's mystery is revealed as a computer AI that is trying to become self-aware – and which ultimately succeeds. Without this powerful literalisation of the *deus ex machina* trope, Gibson's two sequels (*Count Zero*, 1986 and *Mona Lisa Overdrive*, 1988) are much less successful, derivatively Le Carréan exercises in street near-future mystery spy-adventure. His later novels almost always contain interesting elements but none of them exactly coheres in the same compelling way *Neuromancer* does: *Virtual Light* (1993) makes too little of its intriguing titular metaphor, devolving instead into more chase-and-run adventure. By the time of *Idoru* (1996) Gibson's novels were being defined almost wholly by their sense of déjà-vu, the constraints of the

Gibson brand-name. But he remains a talismanic writer for many in the world of SF, someone who determined a particular style, a post-industrial lightweight postmodern 'cool', and his influence scattered in a wide number of directions throughout the 1980s and into the 1990s.

At the time, many saw 'cyberpunk' not just as another style; but rather as (in Larry McCaffery's words) part of some 'enormously exciting' developments in culture generally, and as a key component of 'postmodernism', which McCaffery sees as a 'complex set of radical ruptures – both within the a dominant culture and aesthetic and also within the new social and economic media system (or "post-industrial society") in which we live' (McCaffery, pp. 1–2). There are as many different definitions of 'postmodernism' as there are of SF; but in so far as it privi-leges surface over depth, a shifting present over the past, collage, quotation and intertextuality over 'originality', and presides over a euphoric emptying out of emotional content ('the waning of the affect') postmodernism does indeed set itself against spiritualist or mystical idioms. Embedded in the glorious rhizomatic multifariousness of contemporary material culture, the typical postmodern text is enthusiastically non-mystical. But we may want to query the extent to which cyberpunk *is* a materialist idiom. On the one hand, cyberpunk authors spent a good deal of attention on the *things* of their world, the quiddity, the textures and flavours of actual experience. But the more cyberpunk authors delved in the *material* of their urban dystopias, the more the other half of the SF dialectic asserted itself in their work. As Dani Cavallaro says in her useful critical study of the cyberpunk movement:

> On one level contemporary technoscience seems to perpetuate the rationalist approach preached by the Enlightenment. On another level, the Gibsonian configuration of cyberspace as a hallucinatory experience alludes to science's involvement with the irrational ... Cyberculture thrives on these ambiguities: rationality and irrationality coexist within its territory ... one of cyberpunk's main contributions to contemporary reassessments of knowledge and agency lies in its fusion of mythological and technological motifs. (Cavallaro, p. 52)

Perhaps this is an inevitable feature of the particular pocket universes of cyberpunk's virtual reality premise: it turns the 'I–It' environment of science into an (at root, religious) 'I–Thou' environment with an almost facetious literalness. Cavallaro quotes both Darko Suvin and Samuel Delany by way of emphasising the religious component of Gibsonian cyberspace: fiction written, she suggests, from a desire to escape the squalor of everyday megalopolitan existence, to forge alternative forms of cohesion among increasingly alienated individuals. According to Darko Suvin, 'a solution logically latching onto cyberspsace, and allowing the reconnecting (*re-ligio*) between disparate people and their destinies ... is then religion'. This reading is corroborated by Samuel Delany who maintains that 'the hard edges of Gibson's dehumanized technologies hide a residing mysticism' (Cavallaro, pp. 57–8; ellipsis in original). This mysticism may redeem, as it were, the otherwise stony grimness of Gibson's books.

Something akin to this mysticism is present, in varying forms, in the most important cyberpunk writers (a few praiseworthy names among a great many authors who jumped on this bandwagon) of the 1980s and 1990s. The American Pat Cadigan (b. 1953) is, after Gibson, the most important practitioner of this sort of writing. Her version of cyberpunk attends more closely to the interface between human and computer precisely in order to be able to articulate this dialectic. *Synners* (1991), probably her best book, articulates the problematic mostly in terms of disease (computer viruses that may be conscious and are implicated in certain human deaths). In *Tea From an Empty Cup* (1998) the dialectic is more clearly dramatised along cultural lines: the premise (a virtual reality that gives the user access to certain ecstatic sexual sensations) is viewed as a proper medium for spiritual revelation and cosmic mythological truth by a Japanese character in the novel, but as a non-magical purely materialist experience by a white character. Another American writer, John Shirley (b. 1953), superheated the conventions of cyberpunk to an even greater intensity, exaggerating both the hardcore materialism and the Gnostic possibilities. His Eclipse trilogy (*Eclipse*, 1985; *Eclipse Penumbra*, 1988; and *Eclipse Corona*, 1990) is especially notable.

Where a writer goes to the lengths of purging his/her cyberpunk fantasy of its buried mysticism, the result can be too overwhelmingly dark for some tastes: the British writer Richard Morgan's (b. 1965) sequence of novels, beginning with *Altered Carbon* (2002), has many of its characters' minds reinserted into new bodies ('resleeving') on death; but this resurrection is a wholly quotidian, non-mystical experience, and the world of the novels is among the most claustrophobic, the darkest and grimmest in the genre. It may be that the dark chocolate pleasures of ultra-violent, dystopian cyberpunk only reflected the anxieties and excitements of the Reagan/Thatcher 1980s. The many such books published during that decade almost all read as painfully dated nowadays; not because our own age lacks anxiety or excitement, but because the nature of those emotions has shifted its cultural logic. The problem is not with sub-genres and conventions as such, but with the calcification of those conventions. Many 'cyberpunk' novels are well written and well realised, but too many are incapable of reaching a conclusion that does not suit their initial prejudices. That is not the case with *Neuromancer* itself, although it is sadly true of several of Gibson's later novels, and it generally characterises the sub-genre to which he gave rise.

Orson Scott Card's *Ender's Game* (1985)

Orson Scott Card (b. 1951) seemed in the 1980s to have become rapidly one of the dominant figures of world SF (his first story was published 1977). That dominance looks a little less assured at the beginning of the twenty-first century; not because his output has flagged but because his later books, whilst always interesting reading, have added little to his earlier work. Everything Card writes is formally conservative, recycling basically nineteenth-century iterations of 'characters, story, atmosphere'. But in *Ender's Game* (1985) he found an ingenious central premise for a novel that was able to dramatise a genuine ethical dilemma. The novel's future Earth fears overwhelming attack by the unfortunately named insect alien species the Buggers. The authorities take promising individuals at a very young age and raise them in a military academy to maximise their belligerent potential. The hero, young Ender Wiggin, is the most gifted strategist, playing increasingly complex games of battle and war in a virtual reality environment. Only at the end of the tale is it revealed that, while Ender thought he was playing war games, in fact he has been guiding the actual human space fleet in a final battle against the Buggers, wiping them out entirely. Two things make this more than an ingenious twist-in-the-tale SF fable: one is the unillusioned but ringingly true portrait of the bullying, jockeying-for-power and danger of the children's world in the academy. The other is Card's follow-up volume *Speaker for the Dead* (1986), where the thrilling military climax of the war against the Buggers – thrilling for the triumphant Ender, but also for the reader, who turns the pages egging Ender on – becomes ashes in his and our mouths. This sequel dramatises the ethical consequences of committing genocide, or more strictly, to quote the title of the third in the Ender series, *Xenocide* (1991): Ender travelling the galaxy trying to make amends for his war crime. The moral point is well dramatised, even if Card does sometimes veer towards the schematic; and what makes it especially interesting is the sense that Ender, though responsible for the crime, was not in a position consciously to will it. Card's fascination with this matter of Free Will has evident Christian resonances.

A Mormon, Card has frequently returned to his faith as inspiration for his writing. His novel sequence *The Tales of Alvin Maker* (*Seventh Son*, 1987; *Red Prophet*, 1988; *Prentice Alvin*, 1989; *Alvin Journeyman*, 1989; *Heartfire*, 1988; *The Crystal City*, 2003; and *Master Alvin*, 2005) reworks the life of the founder of the Mormon religion, Joseph Smith (1805–1844). More impressive is his five-volume *Homecoming* series (*The Memory of Earth*, 1992; *The Call of Earth*, 1992; *The Ships of Earth*, 1994; *Earthfall*, 1995; *Earthborn*, 1995), a sequence that rewrites the Bible as SF, telling an Old Testament story of interstellar Exodus. The planet Harmony was colonised by humans 40 million years ago. Now the orbiting computer, Oversoul, which has governed the planet for all that time, is starting to fail. It selects some humans to make the long trip back to Earth to contact the even more powerful Keeper and find out how to proceed. The story reveals itself to be most interested, once again, in questions of free will, the reasons why God (Card's 'Keeper') grants it and the fact that some people will use it to do evil things. This summary sounds dry, and Card cannot quite escape a deadening piety in his treatment. But in building his novels around directly engaging characters and clear-cut moral dilemmas he always writes a cut above the mass of the market.

'Non-genre writers'

'Non-genre writers have always written SF novels,' observes John Clute (*Science Fiction*, p. 230). It is interesting that even a critic of Clute's excellence can invoke the category of 'non-genre writers' as if SF were a feature of the physical constitution of individual authors, like skin colour, rather than a description of certain sorts of text. Clute's is an attitude often found in SF circles, where a conceptual apartheid separates 'genre writers' (good) from 'non-genre writers' (bad until proven otherwise). More, non-genre writers who trespass into SF more often than not receive a roasting at the hands of SF fans. There are, perhaps, reasons why this is so. Some writers who have built reputations on novels that nobody would call SF attempt SF tropes, and sometimes in doing so display a discreditable ignorance of the lengthy corpus of pre-existing SF treatments. Alternative history, for instance, is sometimes lighted on by mainstream writers with cries of joy as if they have discovered it for the first time. Beryl Bainbridge (b. 1934), in *Young Adolf* (1978), imagines Hitler living in Liverpool in 1913; Philip Roth (b. 1933) in *The Plot Against America* (2004) posits the aviator Charles Lindbergh becoming a quasi-fascist President of the US in 1940 instead of Franklin Roosevelt, with malign consequences for the Jewish family at the heart of the fiction. But there is an aesthetically debilitating caution in both novels, with actual history soon reasserting itself, and in neither case is the trope used in an especially illuminating or original way. The most notorious case of this circumstance is *Time's Arrow* (1992) by Martin Amis (b. 1949), which was celebrated by the literary establishment of its day as a work of stunning originality and depth.[6] In fact the originality (the premise of time running backwards) is better handled in the book where it was first used, Philip K. Dick's *Counter-Clock World* (1967); and the depth (the book tells the story of an old Nazi moving back in time through his experience of the Holocaust) seems glib compared to the complex and moving meditation on precisely the same theme, also rendered via a fractured chronology, of Kurt Vonnegut's *Slaughterhouse 5* (1969). Amis's is a deeply derivative and deeply conservative book (there's no mistaking its implicit thesis that 'the past is better than the present'). That it could have made the splash it did is rather as if an eighteenth-century painter had announced to the art establishment that he had invented the principle of perspective and – what is the really remarkable thing – that *they had believed him.*

But many so-called 'non-genre writers' have written very interesting SF. Maureen Duffy (b. 1933), in *Gor Saga* (1981), treated of the creation of a human–gorilla hybrid. Set in a well-realised run-down future Britain, the hero Gordon ('Gor') has many adventures, enabling Duffy eloquently to dramatise questions of the worth of life, although her thumb is in the scale to a certain (pro-animal) extent. Paul Theroux's (b. 1941) occasional excursions into SF have been marked by a dystopian gloominess. The shallow, consumerist neo-hell America of *O-Zone* (1986) leaves a sour taste in the reader's mouth, but is very powerfully rendered. Indeed Theroux's pessimism has influenced even his travel writing, the mode in which he is more celebrated. In *Sailing through China* (1984) his perspective on overpopulated, treeless, polluted modern China is powerfully expressed via an SF dystopic imagination. 'In a hundred years or so, under a cold, uncolonized moon, what we call the civilized world will all look like China, muddy and senile and old-fangled ... Our future is this mildly poisoned earth and its smoky air' (Theroux, p. 100). The baleful implications of that eloquent phrase 'uncolonized moon' are carried on the cultural buoyancy and ubiquity of science fiction's once-positivist assumptions.

Recently, the Canadian novelist Margaret Atwood (b. 1939) has annoyed many members of the SF community by denying that she writes any such thing as 'science fiction', a Pulp genre she has denigrated as concerning merely 'intelligent squids in space'. But in spite of her denials, Atwood's three best novels are all SF. *The Handmaid's Tale* (1985), her most famous title, won the first Arthur C. Clarke award for its vivid portrayal of a dystopian future US in which Christian fundamentalism has established a repressive and misogynist society based ostensibly on biblical principles. The novel's protagonist, 'Offred' (nameless except in so far as she belongs, as a handmaid, to her master – 'of Fred'), becomes the focus for the book's more generalised outrage at the many ways men objectify and enslave women. And yet, despite many expert touches, the novel ultimately does not succeed. The earlier sections are brilliantly claustrophobic, not only in their representation of the attenuated existence Offred endures, but also the limited life led by Fred himself. Atwood is making a penetrating point about the way powerful individuals can become trapped by the very ideology that has helped them to power, restricted in different but nevertheless choking ways. But two-thirds of the way through the book Fred takes Offred to a secret orgy, in which powerful members of the regime are seen enjoying illicit sex, alcohol and the like. This weakens the whole; Fred becomes a more two-dimensional villain – not the less believable, of course, for many autocrats have paid only lip-service to their supporting ideologies whilst enjoying illicit pleasure in secret; but a less interesting character according to the fictive logic of Atwood's created world. The novel's conclusion, with Offred escaping the limiting worlds of New Gilead, is also a little anticlimactic.

Her tenth novel, *The Blind Assassin* (2000), was very well received by critics and the mainstream literary establishment, winning the prestigious Booker Prize for fiction, although it is only partially successful as a novel. It is a family saga made more interesting by a Russian-doll inset narrative, a story being told by one of the novel's characters from which the novel as a whole takes its title. This latter story is a brilliantly realised SF take, a superior Edgar Burroughs-like planetary romance set on the exotic planet of Zycron. Indeed, so effective is this inset tale that it rather overshadows the frame narrative, putting those predictable permutations of family betrayal and secrets revealed in the shade. The effect of the whole is an unbalanced novel, one that only in its SF sections achieves the imaginative and bewitching fictive power of Atwood at her best. It is this that makes Atwood's *Oryx and Crake* (2003) the most successful of her books: an unembarrassed entry into a dazzlingly realised dystopian imaginary world. Although the subject matter sounds unappealing in summary, its execution is so wonderfully handled as to bring the whole to a glorious life.

Sheri Tepper

The American writer Sheri Tepper (b. 1929) came late to writing; her first book did not appear until 1984. As a woman publishing SF in the 1980s, and on the strength of the gender fable *The Gate to Women's Country* (1988), she is sometimes classified under the multicoloured umbrella of 'feminist SF'. But this is not quite right – not because Tepper's work is inimical to feminist thought (it takes much of feminism as a *sine qua non*), but because her own anger is aroused by what is, at root, fundamentally a *theological* category: pride. Embodiments of the destructive power of pride haunt the protagonists of her river world novel *The Awakeners* (1987). It is pride that treats the environment as merely a resource to be exploited, and pride (mostly in male form) that threatens humanity (mostly in the form of women and children) in the post-holocaust world of *The Gate to Women's Country* (1988), where war-drunk men live in different cities from the rest of the more civilised population. It is for this that the agnostic humanism of Tepper's novels relates medially with a religious awareness that taps into the root of SF itself, something evident in Tepper's masterpiece, *Grass* (1989), a spacious, well-plotted, thought-provoking and wise book that rewrites the 'beauty and the beast' mythos in interesting ways.

The human population of the planet Grass is dominated by the aristocratic 'bons', an elite caste of close-knit families whose life is dominated by hunting – not hunting foxes with horse and hound as on Earth, but hunting a more fearsome alien 'foxen' (a shapeless blur of teeth and claws) with alien 'hippae', velociraptor-like creatures of great cunning and intelligence. The world is close-knit and doesn't welcome outsiders, but it seems to hold the cure for a galactic plague, and so the horse-loving Earthwoman Marjorie Westriding Yrarier and her family travel to the planet to try to insinuate themselves into the world of the 'bons' and learn the secret of Grass's immunity. Westriding's journey towards this knowledge also unlocks the mystery of the circle of life on the planet, and mirrors her own ongoing estrangement from her unsatisfactory husband, via the possibility of a love affair with one of the bons, to a startling but much more satisfying conclusion. The apparently savage foxen are revealed as victims, the persecuting hippae as diabolic. In the end, rather than transforming the Beast into a human with her love, Marjorie leaves her husband and runs off with the Beast, enjoying Beastly congress. Tepper's satire on the conventional Catholicism of the Westridings, and on the stifling quasi-Mormon official religion of her Earth, is a little heavy-handed. In a loose sequel to this novel, *Raising the Stones* (1990), fundamentalist Islam is similarly lambasted. But Tepper is never po-faced in her attacks on organised thought. In *Raising the Stones*, Marjorie and her foxen appear briefly through a wormhole gate before a gathering of people called the Baidee, to whom she imparts some wisdom (for instance, 'there is no sin inherent in any mind save the sin of pride in believing one has seen or been taught the absolute truth'), saving her 'greatest commandment' for last, the profoundly wise and true statement (possibly the single wisest and truest thing ever written in a SF novel): 'even when people are well meaning, do not let them fool with your heads' (Tepper, *Raising*, p. 155). Tepper fast-forwards the narrative several hundred years, and we discover that her words, taken as divine revelation, have become codified, resulting in a society just as restrictive and oppressive as before except that brain surgeons, psychiatrists and hairdressers have been outlawed. The satire on the reification of revelation is pinpoint and hilarious. This, for instance, of Marjorie's commandment 'be not sexist pigs':

> Where the Low Baidee found a prohibition against sexual discrimination in the words of the prophetess ('be not sexist'), the male Scrutators of the High Baidee found a warning against bestial behaviour ('be not pigs'). It was not long until bestial behavior was defined as consorting with the other kind, that is, the Low Baidee. (Tepper, p. 156)

Octavia Butler

The American writer Octavia Butler takes the idea of alien abduction as the starting point for her powerful *Xenogenesis* trilogy (*Dawn*, 1987; *Adulthood Rights*, 1997; *Imago*, 1998). Lilith Iyapo, a well-to-do African-American woman, wakes to find herself in a grey, enclosed room aboard a spaceship. She has no memory of how she came to be there, although she does remember the nuclear war that had destroyed her world, and remembers the death of her family. The aliens whose spaceship it is, the Oankali, have removed the survivors from the aftermath of nuclear conflict and preserved them in suspended animation aboard their orbiting spaceship for 250 years, until the Earth could be made habitable again and humanity can return to it. In return for acting as saviours in this manner they require human genes. This is their mode of existence; travelling the galaxy continually augmenting their own bodies with genetic diversity from other species, a process they can control at a molecular level. The Oankali are, in other words, a symbolic embodiment of the Saviour as the principle of diversity. In the second book of the trilogy, Lilith says as much, talking to her son, a hybrid human-Oankali called Akin:

> 'Human beings fear difference,' Lilith had told him once. 'Oankali crave difference. Humans persecute their different ones, yet they need them to give themselves definition and status. Oankali seek difference and collect it. They need it to keep themselves from stagnation and overspecialisation ... When you feel a conflict (within yourself), try to go the Oankali way. Embrace difference'. (Butler, *Adulthood Rites*, p. 80)

Perhaps this sounds banal, excerpted from its context; but the trilogy as a whole articulates the beauty as well as the necessity of difference with comprehensive power and truth. Butler's Oankali are not blandly utopian ciphers either; they are themselves implicated in seeking their own goals regardless of, or by actively harming, the otherness they claim to revere.

The placement of a black woman at the core of this story inevitably brings us back to issues of racial difference, one of the topics to which Butler often returns in her fiction. We discover that Lilith has been awakened by the Oankali for a particular reason: she is to 'parent' a group of newly awakened humans, to guide them into a position of acceptance of their new position. She doesn't want this job, but that is the very reason why she has been given it: 'somebody who desperately doesn't want the responsibility, who doesn't want to lead, who is a woman' (Butler, *Dawn*, p. 157). In other words, as Jenny Wolmark puts it, this is a novel about the ways a character's 'marginality, articulated in terms of both gender and race, (can) become her strength' (Wolmark, p. 32). As a black woman, Lilith might traditionally be represented as marginal; but Butler's SF context redefines the concept of the marginal with the hybrid space aliens in whose domain the story takes place.

The 1990s

Dan Simmons (b. 1948)'s *Hyperion* (1989) created a tremendous buzz among SF fans on publication (it won the Hugo in 1990). A complex, dazzlingly handled space-opera it and its sequel *The Fall of Hyperion* (1990) are novels expressing the original anxieties of SF in unusually pure form. This compendious two-novel book begins with a story about a Catholic priest, Lenar Hoyt, whose faith has been eroded by the plurality of inhabited worlds of Simmons' future Galactic civilisation. He discovers a race of mute aliens with a glowing cross on their chests which, with delight, he takes as proof of the truth of Christ's Incarnation. In fact, the crosses are a separate form of malicious alien parasite, and the priest is disillusioned; but the trajectory of the two novels actually works to reinscribe certain core Christian values. A being called the Shrike is impaling a great many people on the thorns of a vast metal tree. The Shrike, it is revealed, hopes to tempt forth a universal principle of 'Compassion' that though not explicitly identified with Christ expresses a similar form of divinity. Simmons added to the series with *Endymion* (1996) and *The Rise of Endymion* (1997), set in the same universe, although one in which a rather caricature version of the Catholic Church has become the dominant, tyrannical political power. William Gibson and Bruce Sterling's *The Difference Engine* (1990) also caused a splash on publication with its nicely done alternative history nineteenth century (in which Charles Babbage's 1820 computer, from which the novel takes its title, works). The relocation of cyberpunk shenanigans to the steam age even earned a short-lived sub-generic tagging ('steampunk'), although the limitations of this mode of writing have kept notable examples to a minimum.

Of the many writers and novels from this decade, only a few have created fiction that seems assured of continued life, although (of course) it is very hard to judge these matters via the extremely foreshortened perspective of contemporaneity. The British writer Gwyneth Jones (b. 1952) has published a series of extraordinarily intelligent, subtle and effective novels; too subtly rounded, perhaps, for SF Fandom, which has yet to embrace her wholeheartedly. A book such as *White Queen* (1991) eloquently realises a subtly alien race of humanoid visitors, tracing their multifarious interactions with a near-future humanity. Jones is better than almost any other living writer at meshing a convincingly realised scientific or technical context with a profoundly imagined understanding of the co-dependencies and hostilities of interpersonal interaction. The idiom in which she is most at home is one that balances the two categories of what might, generically, be called 'SF' and 'Fantasy', something clearest in her on-going Arthurian series (*Bold As Love*, 2001; *Castles Made of Sand*, 2002; *Midnight Lamp*, 2003), a powerfully rendered near-future series in which a rock star becomes a quasi-Arthurian ruler of England. By exploring the dialectic between 'science' and 'magic' (in the broadest sense of these terms) this series could not be more central to the core project of SF.

The first of the 'Culture' novels, savvy space-opera from the Scottish writer Iain M. Banks (b. 1954), was *Consider Phlebas* (1987); but Banks's rise to becoming one of the best known writers of SF was a phenomenon of the 1990s. There seemed, at first glance, little new about Banks's inventive star-hopping narratives except, possibly, a nicely judged contemporaneity of tone. Banks's technologically enabled 'Culture' lives a left-wing (as opposed to right-wing) libertarian dream; a society of seemingly endless possibilities for self-realisation that none the less has a social conscience. For many this was a welcome change from the largely right-wing bias of much Golden Age American SF; and Banks brought a tremendous warmth and likeability, as well as a knack for ingenious and satisfying plots, to the mix. There have been many subsequent Culture books (*The Player of Games*, 1987; *Use of Weapons*, 1988; *Excession*, 1996; *Inversions*, 1998; and *Look to Windward*, 2000), some better than others, but all successful. More politically pointed was fellow-Scot Ken McLeod (b. 1954) whose sharply intelligent socialist-libertarian novels have won him positive critical attention.

The closer this history comes to the present day, the harder it becomes to make judgements about which writers and which books will 'last'; but one figure who not only deserves but seems assured of a major reputation is the American Kim Stanley Robinson's (b. 1952). His 'Orange County' books imagine three different sorts of future for Robinson's home state California: a post-nuclear dystopia in *The Wild Shore* (1983), which works brilliantly by refusing to become too grim (the addition of a window to the protagonist's rudimentary hut acts as a surprisingly effective upbeat ending); *The Gold Coast* (1988) extrapolates current Californian pollution, overcrowding and corruption into a well-realised future; and finally and perhaps least successfully as Utopia in *Pacific Edge* (1990) set in a clean, environmentally sustainable future. But Robinson's major achievement in SF so far is his 'Mars' trilogy (*Red Mars*, 1992; *Green Mars*, 1993; and *Blue Mars*, 1995) three spacious, well-orchestrated novels tracing the terra-forming of Mars from a near-future into a longer-term vision of the new world. This is realised with such precision, such verisimilitude, as well as with such aesthetic panache, that it creates a powerfully believable whole. The best of the three books is probably *Red Mars*, a book pregnant throughout with the possibilities of future terra-forming, and minutely fascinating on the details of the job. Robinson handled a large and well-drawn cast of characters with aplomb, and builds to a climactic conflict between Earth and Mars that is wholly rooted in the plausibility of political interaction. In order to keep some characters common to a series of novels that must (by virtue of the time-scales required for terra-forming) take place over several centuries, Robinson introduces a slightly dissonant 'longevity treatment', giving his characters a potential immortality. But although this seems opportunistic in the first two books, Robinson brilliantly puts it centre stage – or more precisely, puts the strain such longevity places on the memories of those who live so long – in *Blue Mars*. This means that the final book in the trilogy, apart from detailing life on the aerated and oceanated Mars, meditates on the nature of memory, managing an almost Proustian penetration at times. None of Robinson's subsequent novels has quite lived up to this marvellous work: the near-future *Antarctica* (1997), though full of interesting facts about the continental South Pole, never quite brings its disparate stories under effective control. The alternative history *The Years of Rice and Salt* (2002), in which all of Europe dies in the Black Death and the world from the Middle Ages to today develops without it, is actually ten linked shorter novels, some of which are more effective than others. But what Robinson brings to the SF card table is a more thoroughly considered approach to what he has himself, in interviews, identified as the 'fact-value' problem; a way of restating the core dialectic of SF.

The Australian Greg Egan (b. 1961) has steadily, and without fanfare, been fashioning something new out of the SF novel by inhabiting its assumptions more thoroughly than other writers. Some find an inhuman chill clings to Egan's tales of artificial intelligence, mathematics and physics, but for others his rigour enables his powerful imagination to extend the possibilities of the genre in ways that more traditionally liberal-humanist aesthetics cannot. In *Diaspora* (1997) most humans have up-loaded their consciousnesses into virtual realities where they live as 'infomorphs'. Even the possible destruction of the material Galaxy cannot threaten the marvellous possibilities of life as information rather than atoms. *Schild's Ladder* (2002) is set in a far-future galactic civilisation in which (as with much Egan) humanity is divided by those who choose to live in their bodies, and those who prefer the possibilities of 'acorporeality', living as information in complex data banks. A scientist unwittingly creating a new variety of vacuum, a novo-vacuum, that spheres outwards from its originary point at half the speed of light. Six hundred years later it has swallowed dozens of stars and inhabited systems, and humanity is fleeing all along its length. But Egan's treatment of this fairly standard 'disaster novel' scenario is bracingly hard-science and brilliantly estranging. The discovery of life inside this expanding sphere presents a genuine alterity.

Not all authors have abandoned the dream of rewiring SF according to a Pynchonesque rather than a Heinlein aesthetic. The American writer David Foster Wallace's massive novel *Infinite Jest* (1996), set in a future America even more thoroughly interpenetrated by commercial concerns than is the case today, configures its subject according to the logic of a sprawling and exploded encyclopaedia of contemporary malaise and emptiness. It is the very excess of this vision that makes it interesting to SF. (Some critics see that excess as deplorable – the American critic Dale Peck thinks that '*Infinite Jest* may well be the first novel to out-*Gravity's Rainbow Gravity's Rainbow*', and he doesn't mean to praise Wallace (Peck, p. 43).) But excess, as an aesthetic correlative of the Feyerabendian possibilities of the 'science' that now governs our life, may be the most fruitful approach to the difficult business of 'novel writing' in our present age. Some of the best SF of the 1990s follows this route, although not necessarily by sprawling, like Wallace, through well over 1,000 pages. The American writer Douglas Coupland (b. 1961) manages to convey precisely this excess in under 300 pages in *Girlfriend in a Coma* (1998), the best end-of-the-world novel published in the 1990s.

Less excessive (although certainly given to writing big, inclusive books) is the British author Stephen Baxter (b. 1957), a figure who epitomises the dialectic of SF for which this study has been arguing. A prolific author of 'Hard SF', grounded always in discourses of scientific plausibility, Baxter is better than almost any current writer at creating the 'sense of wonder'; billions of years of cosmic history, an immeasurable spread of possible alternative realities blooming like a fantastically complex flower from the nubbin of the Big Bang. But Baxter is also a writer deeply informed by a religious buried logic. Born into a Catholic family, Baxter made his name with the 'Xeelee' books (*Raft*, 1991; *Timelike Infinity*, 1992; *Flux*, 1993; *Ring*, 1993; *Vacuum Diagrams*, 1997), which trace out a future history in which the solar system is invaded and humanity enslaved by an alien race; throwing off this yoke, mankind's far future becomes embroiled in universe-scale war with aliens from a different dimension, made of 'baryonic' or 'dark matter' who are trying to chill the stars' fires the better to suit their own lives. This is, as Baxter himself has made clear in interviews, a 'war in heaven' with strongly Miltonic overtones – one of his short stories, 'The Tyranny of Heaven' (1990), is even built around the notion that a quotation from *Paradise Lost* might be the cunningly hidden trigger to a profound human genetic change. More recently his powerful novel *Coalescent* (2003) science fictionalises the Roman Catholic Church itself as a sort of mariological hive-mind, living in a hidden buried city right at the heart of modern-day Rome. It is not – and to say this is to reiterate the whole thesis of this lengthy critical study – that Baxter's novels are religious allegories: nor are they hidden avowals of Catholic faith. But they are aware that what SF does better than other forms of literature is mediate the scientific and the mystical perspectives of the cosmos: rationality and the unnameable; and that it does this because the shaping conditions that gave birth to the genre have also determined its growth. The *underlying* logic of SF still addresses those same anxieties, finding a newer less theological but nevertheless at root metaphysical vehicle in which to do so.

A certain prejudice against SF adheres still to the Anglophone literary mainstream, but it is less evident in the literature of other languages. Indeed, it may be the case that the most enduring SF written in the last few years may prove not to have been created 'inside the genre' at all. Three examples of this will have to stand in for a great many.

The Portuguese author and winner of the 1998 Nobel Prize for Literature José Saramago (b. 1922) is fascinated by the intrusion of scientific or Rationalist disturbances into ordinary life. In *Ensaio sobre a Ceguira* ('Essay on Blindness', 1993) society is afflicted by a strange, contagious form of blindness. The first sufferers are quarantined in an asylum, but eventually the plague spreads across the whole world. Saramago uses this premise as a way of meditating on social health and insight, and it is interesting to contrast his treatment with, for instance, Wyndham's *Day of the Triffids* (1951), which also begins with the widespread onset of blindness. Wyndham rationalises his affliction (the blindness has been caused by military satellite weaponry), and concentrates on the practical struggles of, mostly non-blind, survivors. Saramago, on the other hand, does not explain his blindness; he does not even give us his characters names (they are referred to throughout as 'the Doctor', 'the girl in the dark glasses', and so on). There is a bleakly exquisite blankness of affect in the novel, a match between form and premise. *Caverna* ('The Cave', 2001) plays an Orwellian dystopia against the simple life of a potter. But close to the heart of Saramago's art is a fascination with the humanity of the saviour and the extent to which atonement is implicated in suffering and pain. His alternative sacred history *O Evangelho segundo Jesus Cristo* ('The Gospel According to Jesus Christ', 1991) insists on the materiality of the messiah and the blank face that atonement presents to human mortality (told by God that he must pay for the world's sins with his life, Jesus rebels: 'You are the Lord, and forever taking from us the life You gave us.' 'There is no other solution,' God replies, with chilling practicality. 'I cannot allow the world to become overcrowded' (Saramago, p. 197). There is no rebirth for *this* Christ after the crucifixion.

The gifted Japanese novelist Haruki Murakami (b. 1949) has written science fiction, but his SF novels owe more to the traditions of the ghost story and supernatural fable than to the particularly cultural dynamic that shaped the western genre. In *Supūtonku no koibito* ('Sputnik Sweetheart', 1999) circling sputniks emblematise the alienation and distance of human emotional connection, and a gently handled supernatural element may or may not be explicable in scientific terms. *Umibe no Kafuka* ('Kafka on the Shore', 2000) might similarly be described as 'magic realism', except that Murakami's characters are too passive, in a sense too machinic, and his world too thoroughly immersed in the idiom of contemporary technology. French writer Michel Houllebeq (b. 1958) achieved a *succès de scandal* with his novel *Les Particules élémentaires* ('The Elementary Particles', 1998; translated into English as 'Atomised'. The story follows two brothers, one a molecular biologist almost without a sex-life and the other a libertine; what begins as savagely comic satire moves inexorably into SF dystopia to brilliant effect. In all three of these cases the SF is Feyerbendian, a supple thought-experimental medium via which to penetrate.

The 21st-century science fiction novel

It really would be premature to pass judgement on novels published in only the last few years; although it can be said that published SF shows no signs of withering away from its 1990s hyper-fecundity, and may indeed even be expanding. Attempts to impose manifesto-prescriptive order upon this jungle-like fecundity have not been particularly successful. For instance, in the UK a group of writers pronounced themselves part of the 'New Weird', advancing a darkly knobbly aesthetic based on a love for the fantasies of H. P. Lovecraft (1890–1937), Mervyn Peake (1911–1968) and (especially) M. John Harrison (b. 1945). Harrison's ironically titled *Light* (2002) was ecstatically greeted as a masterpiece by many in the SF world, and the many may well, in this case, be right: its powerhouse assumption of pretty much all the old conventions of space-opera is impressive, although the more lily-livered among its readers may baulk at the way Harrison then puts them through a prism refracting them into relentless cruelty and calculated ugliness. But the book is an immense achievement by any standards. Some New Weird writers have established significant reputations, most notably the marvellously inventive tyro-work of the British writer China Miéville (b. 1972), whose 'New Crobuzon' novels (*Perdido Street Station*, 2000; *The Scar*, 2002; and *Iron Council*, 2004) have rapidly established themselves as contemporary genre classics. But the movement's impact has not reverberated beyond the narrow world of SF itself. More promising in many ways are writers who are less inward-looking. The Northern Irish writer Ian McDonald (b. 1965) brings perhaps the best sensorium and stylistic talent to the business of writing science fictionalised Africa, Mars or India. British novelist David Mitchell (b. 1969) favours complexly interlocking narratives that move smoothly between SF and other genres. An attempt to turn this cross-generic fertilisation into a movement, 'Interstitiality', has also floundered rather than boomed. The US-based Insterstitial Arts Foundation (www.artistswithoutborders.org) promised to champion a commendable catholicity of aesthetic approach, but has not made a noticeable cultural impact. Jeff Vandermeer (b. 1968), himself a talented writer of bizarre fantasy, was in 2003 initially persuaded by the movement's promises ('a real renaissance has come about ... ["interstitial"] has captured not just the moment but the nature of this sea change in literature'); but only a year later was qualifying his support (he says he has 'soured' on the term, arguing it is becoming 'insular and one-sided') (Vandermeer, pp. 44, 50). 'The SF Community', that nebulous, un-unified agglomeration of fans, writers, reviewers and others, has a poor record at communicating its enthusiasm outside its own bailiwick. Insular one-sidedness, too often the currency of our world, serves nobody.

References

Bergonzi, Bernard, *The Situation of the Novel* (1970; Harmondsworth: Pelican 1972)

Butler, Octavia, *Dawn* (1987; New York: Warner Books 1997)

Butler, Octavia, *Adulthood Rites* (1988; New York: Warner Books 1997)

Cavallaro, Dani, *Cyberpunk and Cyberculture: Science Fiction and the Work of William Gibson* (London: Athlone 2000)

Clute, John, *Science Fiction: the Illustrated Encyclopedia* (London: Dorling Kindersley 1995)

Clute, John, *Scores: Reviews 1993–2003* (Harold Wood, Essex: Beccon 2003)

Clute, John and Peter Nicholls (eds), *Encyclopedia of Science Fiction* (2nd edn., London: Orbit 1993)

Kneale, James and Rob Kitchin, 'Lost in Space', in J. Kneale and R. Kitchin (eds), *Lost in Space: Geographies of Science Fiction* (London: Continuum 2002), pp. 1–16

McCaffery, Larry (ed.), *Storming the Reality Studio: A Casebook of Cyberpunk and Postmodern Science Fiction* (Durham, NC and London: Duke University Press 1991)

Ortega y Gasset, José, *The Dehumanisation of Art and Other Writings on Art and Culture* (New York: Doubleday 1948)

Peck, Dale, *Hatchet Jobs. Writings on Contemporary Fiction* (New York: The New Press 2004)

Poirier, Richard, 'Rocket Power', in Harold Bloom (ed.), *Thomas Pynchon's Gravity's Rainbow: Modern Critical Interpretations* (New York: Chelsea House 1986), pp. 11–20

Pynchon, Thomas, *Gravity's Rainbow* (1973; London: Picador 1975)

Saramago, José, *O Evangelho segundo Jesus Cristo* (1991; transl. Giovanni Pontiero, *The Gospel According to Jesus Christ*, London: Harvill Press 1993)

Suvin, 'Afterword', in Patrick Parrinder (ed.), *Learning from Other Worlds: Estrangement, Cognition and the Politics of Science Fiction and Utopia* (Liverpool: Liverpool University Press 2000), pp. 233–71
Tepper, Sheri S., *Raising the Stones* (1990; New York: Bantam 1991)
Theroux, Paul, *Sailing Through China* (New York: Houghton Mifflin 1984)
Vandermeer, Jeff, *Why Should I Cut Your Throat? Excursions into the Worlds of Science Fiction, Fantasy and Horror* (Austin, TX: Monkeybrain Books 2004)
Wolfe, Gene, *The Urth of the New Sun* (1987; London: Futura 1988)
Wolmark, Jenny, *Aliens and Others: Science Fiction, Feminism and Postmodernism* (New York and London: Harvester Press 1994)

14
Late Twentieth-Century Science Fiction: Multimedia, Visual Science Fiction and Others

Comics, graphic novels

Comics, and especially superhero comics, have – as we have seen – been a tremendously important aspect of later twentieth-century SF. Throughout the 1980s and 1990s a new sophistication entered into the world of Anglophone comics. In addition to the traditional form of magazine format comics, there came on the market a new mode known as 'graphic novels', usually issued in a series of part-formats but later bound as a single volume.

In earlier chapters I argued that the vogue for SF superhero icons expressed in a popular-cultural idiom one of the root concerns of SF: the role of the saviour and the status of atonement in a modern, scientific post-Copernican cosmos. Few people nowadays think of these questions in theological terms – and, indeed, it is one of the achievements of SF to have mediated these deeply embedded questions through new, materialist discourses. It is worth remembering this, because any discussion of what is almost certainly the most significant SF novel of the 1980s – Alan Moore and Dave Gibbons' graphic novel *Watchmen* (1986–87) – must concede the extent to which it derives a great deal of its power precisely from this context: the role and nature of the saviour in a nuclear, technologically advanced world: what must be done to save mankind from death and corruption, and the price that must be paid for such redemption. It would miss the point of this complex visual novel to suggest that it can be reduced to a 'Christian allegory' after the manner of C. S. Lewis (it really cannot). The point is not that SF is 'actually' a Christian discourse, hiding conventional religious iconography beneath high-tech exteriors. But these are the questions that determined the modern genre at its origins, and which therefore continue to shape it; because these are the points at which a properly Copernican world-view chafed at the Western *Weltanschauung*. Moore certainly has been fascinated with the deeper problematics of the superhero 'saviour' figure: the contradictions of his/her simultaneous divinity and humanity, and the price s/he pays to atone for human destructiveness. This can be seen in *Miracleman* (1982–88), a satisfyingly layered updating of a 1950s–1960s superhero known successively as 'Marvelman' and 'Miracleman'. Able to transform into a near-invulnerable superman by uttering the phrase 'kimota' Michael Moran finds that

326

'Miracleman' destroys his marriage. 'His emotions are so pure,' he tells his wife. 'When *he* loves you it's gigantic. His love is so strong and direct and clean. When *I* love you, it's all tangled up with who's not doing their share of the washing up, and twisted, neurotic little things like that' (*Miracleman*, 7: 4). A more eloquent summary of the painful disjunction between divine and human love is hard to find. 'Miracleman' is more explicitly a trope for the messiah than most comic supermen: he saves the world from a devilish fate and rebuilds it as a rather bizarre and restrictive utopia.

Watchmen is set in the 1980s in an alternative US (Nixon is still president), and taps into the widespread 1980s nuclear anxiety – one aspect of the 'watch' from which the book gets its title refers to the so-called 'doomsday clock', in which nuclear war was metaphorically registered on a clock face as *midnight*, and the number of minutes to midnight registered the immanence of this apocalypse. During the course of the novel the minute hand approaches ever close to midnight, and the atomic destruction of the world seems unavoidable (as it did to many people in the 1980s).

But the world of *Watchmen* is also one in which 'superheroes' are real members of society. Indeed, the pragmatic treatment of the 'superman' icon is one of the most consistent features of writer Alan Moore's work. He imagines a world in which characters like Batman or Superman actually exist, and then teases out the minutiae of how they might actually function. In our world 'vigilantes' are likely to be people drawn to violence, socially dysfunctional people, and some of Moore's superheroes are like this. Others might be idealists, or performers, drawn to the (costumed) publicity of the role. Moore portrays them with an equal degree of psychological detail and verisimilitude. In *Watchmen* there are two sorts of 'superhero'. Most, like Batman, are ordinary humans who have trained themselves or built themselves machines to enhance their abilities. But at the same time there is one superhero in the novel with supernatural powers: 'Dr Manhattan', a blue-skinned humanoid mutated during a nuclear accident to be able to change his size, teleport himself or others, and manipulate matter at the atomic level by will alone. Dr Manhattan, still residually human although cold and distant, is perhaps the true hero of *Watchmen*: the closest the novel has to a supernatural saviour.

The novel asks after the price of salvation. 'Ozymandias', the superhero identity of Adrian Veidt (the smartest man in the world), prevents nuclear apocalypse by forging the materialisation of an alien life-form in New York that kills tens of thousands (this event, staged to look like the opening salvo of a full-fledged alien invasion, unites the Cold War superpowers in frightened alliance). Many more would have died in nuclear war, but is this salvation too dearly bought? More centrally, Dr Manhattan – the closest the novel has to a god – gives up on humanity. 'I understand', he tells Veidt, 'without condoning or condemning. Human affairs cannot be my concern. I'm leaving this galaxy for one less complicated' (Moore and Gibbons, *Watchmen*, XII: 27). What if Christ not only denied the existence of God ('the world is not made' is his judgement: 'Nothing is made ... a clock without a craftsman', Moore and Gibbons, *Watchmen*, IV: 28) but abdicated the responsibilities of his atonement and left humanity for another galaxy altogether?

The artistic and commercial success of *Watchmen* helped create a micro-climate of serious, adult graphic novels in the late 1980s and early 1990s. Most of these were fascinated with the status of the saviour/superhero in the modern world. The Scottish writer Grant Morrison and illustrator Dave McKean published *Arkham Asylum* (1989), a sophisticated revisioning of Batman as borderline lunatic. This graphic novel recycles *Alice in Wonderland, Psycho* and *Lovecraft* in the tale-within-the-tale of Amadeus Arkham, the founder of the titular asylum. More importantly, though, the novel posits both Batman and the inmates of the asylum – the villains from his rogues' gallery – as embodiments of Tarot and Jungian archetypes, and explores the narrative possibilities of each, through the pretext of having the Batman wander through the underworld of the Asylum corridors Alice-like, facing each one of them successively. Perhaps a man with messianic aspirations does nothing but project scenarios of his mind onto the world, as the *Alice*-influenced villain Mad Hatter points out: 'Sometimes I think the asylum is a head. We're inside a huge head that dreams us all into being. Perhaps it's your head, Batman. Arkham is a looking glass. And we are you' (Morrison, McKean: unpaginated). Japanese artist and writer Masamune Shirow achieved a certain Descartesian sophistication in his cyberpunk graphic novel *Kōkaku Kidōtai* ('Ghost in the Shell', 1991).

Visual art: painting, illustration

SF art, which was largely limited to illustrations for genre magazines from the 1920s through the 1960s, began to cross over into the artistic mainstream in the 1970s and 1980s. The British artist Chris Foss (b. 1946) has developed a very distinctive style of SF painting: using the pneumatic airbrush rather than conventional bristles enables him to create a photorealistic patina for non-realist images. Foss specialises in enormous spaceships, usually represented in flight, bulky, blocky, architectural craft rendered with a brightly coloured precision that is simultaneously monumental and precise. He is not known for his portrait or figure painting (although his sketches of copulating nudes for the very popular *Joy of Sex* (1972) by Alex Comfort (1920–2000) are very convincing) except to provide visual scale for titanic compositions. Foss's work was once seemingly ubiquitous on the covers of SF paperbacks in the 1970s, particularly his cover-art for reissues of Asimov's *Foundation* and E. E. 'Doc' Smith's *Lensman* books. Today he is seen in some quarters as a little cheesy. But this is not a fair judgement: at its best his art achieves a striking fusion of, as it were, the nineteenth-century apocalyptic painter John Martin and the monumental abstracts of Mark Rothko. One sign of his resuscitating reputation is the way mainstream British artist Glenn Brown (b. 1964) plagiarised several Foss images, repainting them and exhibiting them as avant-garde art. Brown was nominated for the 2000 Turner Prize and caused a stir when one of his most powerful paintings *Ornamental Despair (painting for Ian Curtis)* (1994) – a mountain girdled with a vast collar of ice in orbit around the Earth – was recognised by the SF community as a repainting of one of Foss's pieces from his own book *Diary of a Spaceperson* (1990; Foss's original includes a bright red and black spaceship that Brown omits).

Jim Burns (b. 1948), although as popular as Foss, is a more limited artist, whose considerable technical ability is subordinated to an almost Jeff Koons-like fascination with the erotic possibilities of skin, materials and metal. Having said that, at least he never drops to the levels of unctuous soft-porn made popular by the Peruvian-born artist Boris Vallejo (b. 1940). More aesthetically interesting is the English artist Roger Dean (b. 1948), who came to widespread fame through his album covers for groups such as Yes and Asia. Dean's art is instantly recognisable: intricate alien landscapes, very colourful and exotic, with trees and features following helix, sine wave and $x = y^2$ curves and a wealth of filigree detail. The spaces portrayed are more surreal than real, but rendered with a liberating if stylised flow.

Sculpture and performance art

Other artists have explored the somatic possibilities of SF. The Cypriot-born Australian Stelarc (b. 1950s) has worked since the late 1960s with a cyborg-inspired series of performance artworks. His 'obsolete' (his own term) human body is augmented with a wide variety of technical prostheses. Stelarc is interested in information storage and retrieval as much as in purely physical additions; and his website archives his extensive body of work (www.stelarc.va.com.au). The New York artist Rammellzee (b. 1960) creates paintings, sculptures and rap-influenced performance art that feature elaborate collages of organic and metallic materials. As fascinated by graffiti and urban detritus as the machinic, Rammellzee adopts an almost medieval modernity: he describes his graffiti art as 'illumination', and considers that he is following the traditions of 'the monks'. Nevertheless the energetic creation of what we might call (to appropriate Philip K. Dick's term) 'kipple art' does capture the centrifugal social forces of a world hurtling into an urban future. Another SF performance artist, or perhaps it would be more accurate to describe her as a sculptor, is the French artist Orlan (b. 1947; she took her *nom-de-travail* from a variety of Russian spacesuit). Her most famous project has been a prolonged self-fashioning; undergoing plastic surgery to alter her appearance, inserting subcutaneous lumps and bumps to her face and otherwise altering her features. Quite apart from deconstructing notions of 'conventional beauty' (which she does in a very salutary way) she has now reached a stage where she most resembles an alien from a high-budget SF film. Certainly her aim to literalise a variety of 'post-human' physical identity is characteristic of the broader cultural logic of SF.

Digital art

Given the manifold points where SF has crossed into the so-called mainstream in the last half-century it is, perhaps, surprising that the fertile, often extraordinary world of SF painting and illustrations has had relatively little impact on the 'official' art world. The 1960s phenomenon of Pop Art, for instance (which we might think ripe for SF imagery), is almost entirely empty of SF tropes: pop artists looted other forms of popular culture – war comics, love-stories, women's magazines, advertisements – but not SF. One possible exception is the work of the British pop artist Richard Hamilton (b. 1922), whose 1960 painting *Towards a Definitive Statement on the Coming Trends in Men's Wear and Accessories* portrays what appears to be a cyborg President Kennedy, possibly wearing a space suit, certainly looking towards the future.

But there is one suburb of the city of contemporary art that has been fruitfully cross-pollinated by SF: digital art. Perhaps it is the fact that a new generation of artists have been exploring the technical possibilities of a new idiom, which idiom is already associated with SF, that has led to this. Many digital artists can usefully be considered under the rubric of SF. Using the computer to doctor photographs enables Australian digital artist Patricia Piccinini (b. 1965) to create images such as *Last Days of the Holidays* (2001), in which a skateboarding boy encounters a weird, newt-like alien being in a sunny car park. The critic Christiane Paul calls Piccinini's work 'synthetic realism' (Paul, p. 37), which is not a bad descriptor for most of SF as a genre.

More characteristic of SF digital art for many are the iterated shapes and virtual-space fractal sculptures of the British artist William Latham (b. 1960s: www.artworks. co.uk/latham) or the twisted organic shapes and altered photographs of Austrian Dieter Huber (b. 1962: www.dieter-huber.com). Latham's work in particular epitomises an iconic look: the concretisation of alien forms whose spurious organic verisimilitude bypasses the question of surrealism altogether. These are creatures that are 'grown' in the computer according to certain algorithms, not painted or constructed in more conventional ways. Latham has said that his art was inspired by the Darwinist Richard Dawkins' own experiments in using the computer as an evolutionary environment.[1]

Many artists have been interested in the possibilities of using digital environments to create alien forms of life and allow them to interact. The American artist Karl Sims (www.genarts.com/karl/) has formed Darwinian virtual environments (one of his most famous works is named in honour of Darwin *Galápagos* (installed in Tokyo 1997–2000)) in which bizarre-looking creatures inhabit complex cyber-systems, influenced by the viewers, who are encouraged to interact with the system. By comparison Kenneth Rinaldo's interactive *Autopoesis* (2000: www.accad. ohio-state.edu/~rinaldo), in which a gallery space is festooned with robot limbs that jerk and move in response to viewers passing through, seems rather quaintly clumsy and dated; reminding us (as perhaps it is supposed to) that real space is far less malleable and aesthetically possible than virtual space.

Of course, with some digital art the viewer may be tempted to think that artists are doing things which, whilst interesting, are more effectively managed in the commercial world. Australian artist Jeffrey Shaw (b. 1944: www.jeffrey-shaw.net) created an installation called *The Legible City* (first exhibited in New York, 1989) in which viewers sat on an exercise bike and pedalled to move themselves through a virtual city-space in which buildings were replaced by words spelt from giant 3D blocky letters. There's a kind of rather dreary literalism in this conceptualisation of virtual reality as dominated by verbal hypertext (not unlike the literalism in Gibson's straightforwardly symbolic *Neuromancer* 'Matrix') and the messages spelled out by the 'buildings' lack either menace or depth (for instance, 'GET WORSE IN THE FUTURE'). Much more exciting work, aesthetically speaking, was being done in the many hundreds of navigable virtual spaces being created by commercial programmers, partly in the field of video and computer gaming (below) and partly in commercial applications of digital art. 'Industrial Light & Magic', the special effects company founded by George Lucas in 1975 to provide effects for Star Wars, has grown into one of the biggest SFX companies in movie-making. Although much of their work is the functional provision of effects of a variety of films, much of what they do is strikingly beautiful, and in many cases it is the visual iconography created by the 'ILM' team that is the best aspect of any given feature.[2] It might be argued that it is this commercial aspect of digital art – popular, imaginative, visually unconstrained – that represents one of the most significant developments of SF in the later century.

Arcade, video and computer games

There are three different hardware formats for SF screen games: arcade games (played on dedicated machines, usually in public arcades), computer games (played on home computers), and video games (played on consoles plugged into ordinary televisions). This last is by far the most popular, outselling computer games nearly fourfold.

The oldest of the three is the coin-operated arcade games: a bulky, five- or six-foot tall box with a television screen and controls at the front positioned either singly in bars, restaurants or gathered together in specialist arcades. These developed from coin-operated games and pinball tables popular in the West from the 1930s and 1940s (pinball is still popular in the West; in Japan so is a variant called *pachinko*). With the rapid development of computer technology in the 1970s a number of electronic arcade games became enormously popular. The first arcade game was created by the Atari Company (founded in 1972 by Nolan Bushnell and Ted Dabney): *Pong* (1972) reproduced in flat, white-on-black video format the logic of ping-pong. The success of this game encouraged a great many competitors to enter the market with variants. Taito, another Japanese company, released *Space Invaders* (1978), the first genuinely classic arcade game. Designed by Toshihiro Nishikado (who claimed to have been inspired by reading H. G. Wells) the game is a two-dimensional shooting gallery in which a laser-cannon slides along the bottom of the screen firing up at rank upon rank of slowly descending alien invaders. Lacking narrative, this game nevertheless manages to evoke with very minimal graphics and actions a genuine tension and a worthwhile sense of the relentlessness of the inhuman attackers. It is a significant SF text both in its own right and also in terms of its major influence on the development of the genre.

These first two notable arcade games emblematise the chief distinction gaming would go on to develop: on the one hand, games that reproduce existing sports or games (car racing, football and so on) in electronic form; and, on the other, games that created an environment marked by a novum or several nova. This latter sort has been the idiom for a number of significant SF texts. *Asteroids* (Atari, 1979) replaced the invaders with collision-course asteroids, and placed the laser-cannon on a fully rotating spaceship in the middle of the screen. *Galaxian* (Namco, 1979) was a souped-up plagiarism of *Space Invaders*, but moved the graphic component forward by being the first colour-screen arcade game. *Battlezone* (Atari, 1980) gave the player command of a futuristic tank, rendering the battlefield in three dimensions although with a primitive wire-frame visual algorithm. The next development was the creation of arcade games that tied in to cinema texts: notably *Tron* (Midway, 1982) which was actually released before the movie; and *Star Wars* (Atari, 1983), which was so successful that more sophisticated sequels (particularly *Return of the Jedi*: Atari, 1984) followed. These games piggybacked on the narrative of the film, mimicking the visual look of the film text as far as possible and including vocal recordings by some of the film's stars: after losing all one's lives, the arcade would hum with Alec Guinness's voice announcing that 'The force will be with you – always'.

By the mid-1980s personal computing had penetrated many homes, and a specialist market was growing up in computer games to be played at home. Several companies produced specialist consoles, and initially the biggest games were the child-fantasies *Super Mario Brothers* (Nintendo, 1985) and *Sonic the Hedgehog* (Sega, 1990). But SF soon followed. The first title of major importance was *Fainaru Fantajii* ('Final Fantasy', 1987) which took Japan (and, after 1990, the US) by storm, leading to a large number of sequels. The early games belonged to the Fantasy genre, although later games crossed the generic border between Fantasy and SF, with a group of heroic characters travelling through a complex world. A spin-off mid-budget movie, *Final Fantasy: the Spirits Within* (Hironobu Sakaguchi, 2001), was the first to feature realistically computer-animated humans, although its plot (soul-stealing aliens threaten a space station) had little to do with the original games. *Star Fox* (Nintendo, 1993), in which the player navigated space-fighters on various missions, used a new processing chip to produce very rapid, good-looking 3D visuals.

In *Doom* (id Software, 1993) the player is a space marine who has been banished to Mars for insubordination. A malfunction with a teleporter means that creatures from Hell start pouring into the Mars base, and the player must wander around the base shooting these. Hyper-violent and bloody, the game was extremely popular. A whole sub-genre of *Doom*-like games followed, now known as 'first-person shooters', among them the blackly humorous SF pastiche *Duke Nukem 3D* (3D Realms, 1996), and the sequel *Doom II: Hell on Earth* (1994), which was licensed by the US Marines and turned into a training tool, *Marine DOOM*.

Myst (Cyan, 1993) was the best-selling computer game for much of the 1990s. In this game the player wanders about an interactive landscape, the strange island world of Myst. Unusually for a computer game there are no enemies to fight, no lives for the player to lose: the enjoyment derives from solving the puzzles encountered and slowly unearthing the mysterious history of this place (the game's developers Robyn and Rand Miller claimed their inspiration was Jules Verne's *L'Île mystérieuse*, although the connection is one of mood rather than specifics). The success of Myst led to many game sequels (*Riven: the Sequel to Myst*, 1995; *Myst III: Exile*, 2001; *Uru: Ages Before Myst*, 2003; and *Myst IV: Revelation*, 2004) and a thriving online fan base. The mega-text also moved out of the electronic idiom, with three novels written by David Wingrove (b. 1954) and the games' creators (*Myst: the Book of Ti'ana*, 1996; *Myst: the Book of Atrus*, 1997; and *Myst: the Book of D'ni*, 1998) and a number of comic books published by Dark Horse. But it is as a visual artefact that the game is most significant; a large portion of its enormous success was due to its distinctive and often very beautiful graphics. Myst created a functioning and infinitely expandable SF universe into which fans could enter.

By the mid-1990s some video games had achieved considerable cultural cross-over. One notorious craze was centred on the Japanese game *Pokémon* (Nintendo, 1996) in which players collect various 'pocket monsters' (the meaning of the Japanese title) so as to fight them against one another. Developed by Satoshi Tajiri (b. 1965), this game gave rise to a rapidly expanding franchise in which Pokémon-inspired Manga comics, collectable 'trading cards' (introduced into the US and Europe in 1999, these became a fanatical craze amongst many children for several years), board games, more than one TV series and a number of motion pictures. By the late 1990s scores of impressively produced SF video games were being released every year, too many to list here. Both the *Star Trek* and *Star Wars* franchises have spawned a large number of spin-off games, to say nothing of plagiarised and derivative titles – such as the *Star Trek*-like *Star Ocean* (Square Enix, 1996) series. As with tie-ins and novelisations, some of these spin-offs have been very high quality. *Star Wars: Knights of the Old Republic* (LucasArts, 2003) combines absorbing game play with very good-looking cinematic visuals, adding material details to the imaginary universe of the film franchise. Other titles have been original developments drawing deeply on the traditions of SF. *Halo: Combat Evolves* (Microsoft, 2001) is set aboard 'Halo', a space habitat, through which space marines fight members of an alien race called The Covenant. Although essentially a Doom-style shooting game, the imagined universe of Halo represents a detailed and absorbing space-opera environment. By the late 1990s video game graphics were so detailed, so well-rendered and fluid, that they have begun to constitute a new form of art. The important feature they share with cinematic and televisual SF art is the extent to which they are immersive, allowing the participants to explore a visually imaginative and aesthetically interesting virtual world.

Audio science fiction

SF radio serials have a long pedigree, going back to the very popular American radio *Buck Rogers* (from 1932), or the BBC's fondly remembered *Journey into Space* (1953–55). Later the shift from a verbal to a visual logic left the purely audio SF experience rather stranded. The sole exception to this owed its success to the fact that, as comedy, it was premised on the logic of the joke rather than (as was increasingly the case for other SF) the logic of the spectacular or fantastic image. Two series of *The Hitch-Hiker's Guide to the Galaxy* (first series 1978, second 1980) by the British comic writer Douglas Adams (1952–2001) ran on the BBC, immediately establishing itself as a huge success. Adams' intimate knowledge of the tropes of SF and his ingenious comic imagination interacted wonderfully, with the result that the show worked as SF, albeit in an absurdist idiom, as well as comedy. Arthur Dent, an unexceptional Earthman of the sort of unredeemed ordinariness that only the English middle classes can produce, narrowly escapes the demolition of his world when his friend Ford Prefect beams him away. The two embark on an odyssey through a Pythonesque galaxy, in company with the two-headed barfly Zaphod Beeblebrox and the depressed robot Marvin 'the Paranoid Android'.

But in general, if we talk about 'audio SF', we are talking about the intersection of the genre and pop music. It is not until the 1960s that SF properly enters into popular music (nobody would mistake Sinatra's 'Fly Me to the Moon' for science fiction). The excitement of the ongoing space programme and what Harold Wilson, then UK prime minister, optimistically identified as 'the white heat of technology' could not fail to cross-pollinate with Pop, an idiom coming into sudden spectacular flower that same decade. Early examples are not inspiring: the satellite-inspired instrumental track 'Telstar' (1962) by the Tornadoes is a kitsch piece of backing music that owes its redeeming, chiming, almost metallic bounciness to its producer Joe Meek (1929–1967). But things quickly improved.

The American jazz composer Sun-Ra (born Herman Poole Blount, ?1914–1993) declared himself not human, but born on the planet Saturn as part of an 'angel race'. His inventively freeform music fleshed out this personal mythology with great potency and beauty, although without great precision; although many of his fans (who take the story literally) may not thank me for referring to it as a 'mythology'. Of his more than 100 discs many are excellent, including *We Travel the Spaceways* (mid-1960s), *The Heliocentric Worlds of Sun Ra, Volume One* (1965) and *Sun-Ra and his Solar Arkestra Visits Planet Earth* (1966). A low-budget film, *Space is the Place* (John Coney, 1974), dramatises Ra's arrival on Earth and his mission to save the Earth, and especially Black America.

The British songwriter and performer David Bowie (b. 1947) achieved his first significant fame in an SF idiom; first through the hippy-futurism of the song 'Space Oddity' (1969) about the modishly doomed astronaut Major Tom, and later with his 'Starman' (from *The Rise and Fall of Ziggy Stardust and the Spiders from Mars*, 1972) which steals and slightly morphs the melody from 'Somewhere over the Rainbow' to tell a similarly sentimental story about a benign alien who is waiting to visit us until he can be sure that he won't 'blow our minds'. Less gooey is 'Life on Mars' (from *Hunky Dory*, 1971), an almost Bradburyan song, in which the titular question 'Is there life on Mars?' focuses the alienation and triviality of Earthbound existence. But the most SF thing about Bowie was his stage persona, based on a deracinated alien-like strangeness that had something to do with his drug-taking and bisexuality (a brave lifestyle to be honest about in the 1970s). His role as the alien in *The Man Who Fell to Earth* (Nicolas Roeg, 1976) cemented this persona even as Bowie reinvented himself as the chilly 'Thin White Duke'. By the 1980s 'Ashes to Ashes' (*Scary Monsters … and Super Creeps*, 1980) had reworked the wide-eyed innocent Major Tom as a 'junkie / strung out in heaven high/reaching an all time low'; and Bowie's SF adventure had hit a permanent downer.

Other crossovers between 1960s pop psychedelia and SF followed a similar route. The British band Pink Floyd started their career with lengthy, throbbing hymns to outer space as a manifestation of inner space such as 'Astronomy Domine' (from *Piper at the Gates of Dawn*, 1967) or 'Set the Controls for the Heart of the Sun' (from *Saucerful of Secrets*, 1968); but they turned to symbolic dramas of schoolyard misery or suburban depression in the 1970s. The ambitious SF 'rock opera' *Lifehouse* was written by the British composer Peter Townshend for his band The Who in the late 1960s. This tale of a transcendental rock concert in a polluted near-future dystopia that makes its audience literally disappear into nirvana was unfinished, although many of the songs appeared on the Who's album *Who's Next* (1971), their greatest work and an obliquely SF text. The 1960s and 1970s vogue for themed 'concept albums', connected to 'Progressive' or 'Prog' rock music, was well suited to SF speculation. King Crimson's *In the Court of the Crimson King* (1969), especially its long first track '21st Century Schizoid Man', gave voice to a melo-dramatically glum vision of the future. Hawkwind, a British band that once included the SF author Michael Moorcock among its members, have returned many times in their extravagant, mind-expanding rock to SF, with albums such as *In Search of Space* (1971), *Space Ritual* (1973) and *Warrior at the Edge of Time* (1975). *Tarkus* (1971) by the British band Emerson Lake and Palmer, is a concept album about the battles of the titular cyborg creature, part-tank, part-animal. The American band The Grateful Dead produced a great many albums of shambolically inspiring guitar-based music, among them *Anthem of the Sun* (1968) a loosely SF psychedelic suite of music, and the lengthy single song 'Dark Star' (from *What a Long Strange Trip It's Been*, 1977), whose multilayered music underpins lyrics of rather gauchely hippy obliqueness:

> Dark star crashes pouring its light into ashes
> Reason tatters, the forces tear loose from the axis ...
> Spinning a set the stars through which the tattered tales of axis roll about the
> waxen wind ...

The Canadian rock-band Rush has found inspiration in SF on several occasions: the twenty-minute title track of *2112* (1976) is a dystopian fable about the rediscovery of music in the titular year, based on the Russian-born, American writer Ayn Rand's (1905–1982) novella *Anthem* (1938). Other Rush albums, such as *Fly by Night* (1975) and *Hemispheres* (1978), drew more on Fantasy than SF tropes. This taste for concept albums rooted in SF dystopia has not gone away; *OK Computer* (1997) by the British band Radiohead situated Douglas Adams' comic SF (the title, and the title of the main song 'Paranoid Android' are both quoted from *Hitch-Hiker's Guide*) in a much grimmer, less relenting vision of future hell. The American-born Jeff Wayne's (b. ?1946) *War of the Worlds* (1978) is an elaborate pop- (or pomp-) rock version of Wells's novel, narrated by Welsh actor Richard Burton, and with many of Wells's sentences set *stet* to music, rather absurdly (for instance: 'the chances of anything coming from Mars is a million-to-one he said'). And yet there is a certain gaudy pleasure to be had in the sheer naffness of the adaptation.

In a very different vein is Funkadelic, the musical collective organised by the American funk musician George Clinton (b. 1940), who invoked the idiom of

UFOs and 'mother ships', drawing some of their inspiration from Sun-Ra. The British band ELO ('the Electric Light Orchestra') fashioned a derivative 'mother ship' UFO to promote their double album *Out of the Blue* (1977). Although the music on that album had no SF content, their later album *Time* (1981) was a future-set rock opera. But this kind of project was the exception rather than the rule. Speaking personally, I've a soft spot for *Time* – and for the various other SF concept albums I own – because I was listening to this sort of music as a child. But, fond though I am of it, it cannot be denied that most people see this sort of music (and this sort of SF) as pretentious, offputting or even risible. It is not this mode in which SF music has made a significant cultural impact.

SF music was especially connected with the new instrumentation developed in the 1960s and 1970s. Where electric guitars (the spine of most pop and rock) have traditional overtones, looking back to the 'authentic' music of blues and R&B, electronic synthesisers from their first invention were seen as forward-looking 'futuristic' instruments. Robert Moog (b. 1934) invented the first playable synthesiser in 1963, from where the possibilities of the instrument were rapidly developed. The Japanese musician Isao Tomita (b. 1932) performed synthesiser-only versions of classical music, often managing – for instance, with his version of Gustav Holst's *Planets* (1976) – to give an otherworldly quasi-SF quality. The same is true of the original compositions for synthesiser by the Frenchman Jean-Michel Jarre (b. 1948). There is no explicit SF context to the instrumental suites *Oxygène* (1976) or *Equinoxe* (1978) but it is hard to escape the sense that these bleepy, throbbing, soaring soundscapes are aural SF. The Greek composer Vangelis (b. 1943) did something similar with his synthesiser albums *Earth* (1974), *Cosmos* (1974) and *Heaven and Hell* (1976). Tangerine Dream, a German synth-band, anchored their spiralling, pounding electronic compositions to SF with album titles like *Alpha Centauri* (1971) and *Stratosfear* (1976).

But it transpired that what we might call 'free composition' in electronic idioms was not the future of SF music either. Instead SF music found its popular forum as a dance idiom. The key band in this regard is the German four-piece Kraftwerk, who began in the 1970s establishing a cyborg musical aesthetic, using electronic synthesisers to which human vocals were only sometimes added. In 1991 one of the group's founders Ralf Hütter (b. 1946) explained that 'the soul of the machines has always been a part of our music. Trance always belongs to repetition, and everybody is looking for trance in life ... in sex, in the emotional, in pleasure, in anything ... so, the machines produce an absolutely perfect trance' (quoted in Savage, p. 310). Their breakthrough album, *Autobahn* (1974), found this blend of the human and machinic in the journey of a car along a motorway, the liberating monotony and repetition of the journey captured perfectly in the electronic pulses and sound-effects of the music. By the time of *Die Mensch-Machine* ('The Man Machine', 1978), however, the cyborg was explicitly robotic; the four personality-free members of the band were replaced in some stage shows with animated manikins. This properly SF robot music found expression in several subsequent albums, including *Computerwelt* ('Computerworld', 1981) and *Techno-pop* (recorded in 1983 although not released that year). 'Techno', a style of dance music that was either influenced by, or else directly 'sampled' (stolen) from Kraftwerk, became a major cultural phenomenon in the 1980s. Mass dances, or 'raves', took place to the accompaniment of a deliberately repetitive, machinic, cyborg music ('if there is one central idea in techno,' opines Jon Savage, 'it is of the harmony between man and machine'; Savage, p. 312). Dozens of bands and musicians worked in the style, among them The Orb, a British duo who combined Kraftwerk, Jarre and Pink Floyd in rambling, dancy electronic performances with whimsical titles such as 'A Huge Ever Growing Pulsating Brain that Rules from the Centre of the Ultraworld' (1990) and 'Toxygene' (1997). One offshoot from this somatic man-machine dance culture is the music of the British composer Gary Numan (b. 1958). Numan's songs, built from a more layered, denser electronic palate, have tended to be darker and bleaker than the more escapist 'techno' norm. He has also worked from specific SF templates: his 'Are "Friends" Electric?' is a doomy, synth-driven version of Dick's *Do Androids Dream of Electric Sheep?*; the inverted commas around 'friends' exactly capturing the teenage sulkiness of the track.

The success of Techno as an SF music depends less on the aural and more on the somatic possibilities: which is to say, it is as a musical landscape into which fans can enter, which they can explore via dance and style, that it succeeds. This is something akin to the possibilities of imaginative entry afforded by the most successful SF mega-texts (the fans' immersive relationship to 'the Star Wars universe' or 'the Matrix universe').

UFOs

Which leads this critical history of science fiction back to the actual world. There is something wrongheaded in discussing the crossovers between fiction and fact as if the two territories blurred into one another, since most of us are very good at distinguishing between them. Nevertheless, and especially in the field of SF, the extent to which fiction has insinuated itself into the assumptions of 'real life' speaks to the thoroughgoing penetration of SF into all aspects of western culture and society, as much as it does to the porosity (some call it gullibility) of that culture.

The most obvious point of crossover is the UFO, or 'unidentified flying object'. First sighted by the American private pilot Kenneth Arnold above Washington State in 1947, sightings of these supposedly extraterrestrial craft increased in frequency through the late 1940s and 1950s, becoming a cultural commonplace by the 1960s.[3] Many people today take it as axiomatic that aliens visit the Earth on a regular basis, and some that these beings abduct human beings and experiment on them. One such craft supposedly crashed in Roswell, New Mexico, in 1947: its remains and the body of its pilot are allegedly being held by the US government in a secret military installation known as 'Area 51' in Nevada. Government denials only calcify certainty in the minds of UFO believers, since one of the paranoid tenets of 'UFOlogy' is that the government is either covering up the truth to avoid mass panic or is actually complicit with the aliens.

It is, to say the least, a strange coincidence that aliens began visiting the Earth in so conspicuous a manner exactly as Golden Age SF, particularly in its cinematic form, was making its first significant impact on western culture. More level-headed commentators point to the close correlations between the cultural idiom with which people are familiar, and the explanations they give to unexplained phenomena. Howard E. McCurdy, after noting that pre-scientific society talked in culturally determined ways not of ETs but of demons and magic, points out that UFO sightings in the early 1950s owed more to the success of the alien saucer from 1951's *The Day the Earth Stood Still* than to objective reality. He adds:

> After an initial burst of interest, the rash of UFO sightings dropped off during the mid-1950s to an average of forty-six a month. This suddenly changed following the launch of Sputnik 1, when the number of reports rose sharply, with over six hundred sightings in the final three months of that year. Barring the unlikely possibility that aliens actually stepped up observations of Earth, one is left with the plausible explanation that the launch of the Soviet satellite excited fears that caused people to detect more unknown objects in the sky. (McCurdy, p. 74)

In the 1960s UFOlogy began to acquire its own self-sustaining internal logic. The first widely reported alien abduction (two Canadians, Betty and Barney Hill, claimed to have been kidnapped by aliens from New Hampshire in 1961) led to many copycat claims. Having been fed by the fictive logics of SF in the first place, UFOlogy soon began to cross-pollinate the genre. One Arizona resident, Travis Walton, claimed to have been abducted for several days in 1975; his account of the experience was filmed as *Fire in the Sky* (Robert Lieberman, 1993), providing a new generation of wannabe believers with an imaginative framework for their fantasies. The American horror writer Whitley Strieber (b. 1945) claimed to have recovered buried memories of alien abduction under hypnosis; his well-crafted book relating the hypnotic sessions *Communion* (1985) was a bestseller and a film was made of it (Philippe Mora, 1989). By the late 1980s the 'truth' of UFOs was so deeply embedded that enormously popular TV shows could be premised on it. Chris Carter's TV serial *The X-Files* (1993–2002) bestrode the puny world of TV culture like a crazy colossus throughout the 1990s: its two FBI agents, Fox Mulder (David Duchovny) and Dana Scully (Gillian Anderson), investigated all manner of bizarre and paranormal goings-on, but the main spine of the show was an ongoing story-arc about alien infiltration of Earth, which posited a breeding programme to cross humans with aliens, and an ever-imminent apocalypse of alien invasion and tyranny, all of which was being collaborated in by a quisling US government. What was really significant about this show was the way so many of its fans took it not as SF but as veiled fact. Other shows, such as *Dark Skies* (1996–97), or the mini-series *Taken* (Breck Eisner and others, 2002), plugged into similar belief structures.

It would be merely to break a butterfly upon a wheel to object to UFOlogy, which can be defended as a properly Feyerbendian approach to the possibilities of alien life (or at least as adding spice and potential to otherwise drab lives) were it not the case that SF as an idiom precisely mediates *religious* anxieties, and UFOlogy does this more explicitly than fictive SF. Some UFOlogists interpret the dialectic of their SF-determined beliefs in materialist terms (which is to say, they believe physical aliens have travelled across space in technologically-advanced space-craft). Some, however, interpret it in specifically religious terms, often with tragic results. In 1997 39 adults committed suicide in California at the behest of Marshall Herff Applewhite (1931–1997), the leader of the 'Heaven's Gate' cult. Applewhite had persuaded his followers that an alien spacecraft was following the Hale-Bopp comet being flown by a cult member called 'Ti' who ascended to a higher spiritual plane upon dying of cancer in 1985. He also preached that by killing themselves they would be rapturously transported there. They all tied plastic bags around their heads and took phenobarbital so as to suffocate in their sleep. Similar cults are legion in the religiously promiscuous United States.

The most famous SF-based religion is Scientology, created in the early 1950s by the SF author L. Ron Hubbard (1911–1986) as a development of his lucrative self-improvement programme, 'Dianetics'. Today this Church contains hundreds of thousands of members, in many countries, among them several high-profile celebrities. Hubbard taught that human beings were immortal spiritual beings (he called them 'thetans') who, while passing through many reincarnations, have accumulated all sorts of negative spiritual energy. Members of the Church can purge themselves of these so-called 'engrams' via a lengthy series of courses (called 'auditing') provided by their Church. This expensive procedure (a complete audit can cost between $300,000 and $500,000) transforms them from 'pre-clears' to 'clears'. Supposedly cleared out of the thetan's soul by auditing are the various traumas that Hubbard believed blocked the pathway of 'the bridge to total free-dom': instances of torture or cruelty from past lives, as well as encounters with unpleasant extraterrestrial races (the Macarb Confederacy and various Invader Forces). Auditing sometimes reveals past lives spent travelling around the galaxy.

Scientologists can be aggressively defensive about their faith (the Church has the reputation as one of the most litigious in the world); and from my personally atheist and non-spiritualist perspective there is nothing in this religion that seems to me *intrinsically* more absurd than is to be found in many other more main-stream faiths. But, that said, it's hard to deny that Scientology is, generally speak-ing, a malign manifestation of twentieth-century SF. Hubbard was a liar, a conman (or at best a man deeply self-deluded) and an exploiter of others. Nobody who reads the many exposés of the cult – the best of them is Russell Miller's *Bare-Faced Messiah: the True Story of L. Ron Hubbard* (1987) – can doubt the founder's primary motivation: 'If a man wanted to make a million dollars', Hubbard told a New Jersey convention of SF writers in the late 1940s, 'the best way to do it would be to start his own religion.' Tax-exempt and sheltered by social convention from much criticism, a religion is an ideal umbrella for the unscrupulous. Many people have been entrapped and fleeced by this Church. Of course, as is often the case with cult members, many of them do not consider that they *have* been exploited.

The idiom of the Church is deeply implicated in SF. One scientologist was told, during an audit, that he had first arrived on Earth 74,000 years ago to battle black magicians who were 'using electronics for evil purposes'. The account of the audit continues: 'he then goes to another planet by spaceship. A deception is accomplished by hypnosis and pleasure implants (rather like opium in their effects) whereby he is deceived into a love affair with a robot decked out as a beautiful red-haired girl' (quoted in Miller, p. 203). Hubbard himself announced that in one of his past lives he had lived on an alien planet, manufactured metal humanoids and offered them to the local thetans, sometimes selling them outright, sometimes by hire-purchase. It is the banality and cliché of these sub-Pulp adventures that is most interesting: to capture the hearts of so many thousands it is not even necessary, it seems, to write poetry of the calibre of the Koran or the Gospel of St John; all one need do is plunder the traditions of second-rate Pulp SF of the sort that Hubbard himself was writing (at one cent a word) in the days before he found a more remunerative income stream.

References

McCurdy, Howard E., *Space and the American Imagination* (Washington DC; Smithsonian Institute Press 1997)

Moore, Alan, Gary Leach and Alan Davis, *Miracleman. Book One: A Dream of Flying* (Eclipse Books 1988)

Miller, Russell, *Bare-Faced Messiah: the True Story of L. Ron Hubbard* (London: Michael Joseph 1987)

Moore, Alan and Dave Gibbons, *Watchmen* (New York: DC Comics 1986–87)

Paul, Christiane, *Digital Art* (London: Thames and Hudson 2003)

Savage, John, *Time Travel: Pop, Media and Sexuality 1976–96* (London: Chatto 1996)

Postscript: Twenty-First-Century Science Fiction

Let me restate the thesis of this Critical History by way of conclusion and to explain the rationale of my choice of texts, which perforce became more and more selective as it proceeded.

The first thing to say is that I am not arguing that 'all SF encodes religious myth'. Such a statement is patently untrue. Contemporary SF is a fantastically variegated and multifarious mode: any normative statement of the type 'SF is such-and-such' will be demolished by the myriad counter-examples that can be produced by anybody even slightly acquainted with the genre. Many writers have produced great SF which has little or nothing to do with religion. But one of my purposes in this long book has been to delineate a critical-historical approach to the business of defining SF; and one assertion I would (tentatively) stand by is that the SF that has had the greatest impact, particularly in the last 60 years or so, has done so because it articulates a dialectic that goes right back to the birth of the genre at around 1600. Modern SF is a genre, this book has argued, that developed out of a specific religious-ideological clash between Catholic and Protestant ways of viewing the world. I have sometimes used the terms 'Catholic' and 'Protestant' as roughly synonymous with 'Fantasy' and 'SF', by way of suggesting that SF owes its birth (though not its continuing development) to the latter movement. But I must concede that there is something far too narrowly sectarian in such labels, especially today. So permit me to suggest a different way of stating this dialectic for the twenty-first century. Some SF writers hold religious faith (of various sorts); some are atheists; the books and texts they produce articulate various positions on the scale from 'absolute materialism' to 'absolute mysticism or spiritualism'. It won't do to suggest that SF is a 'scientific' and 'materialist' idiom if by saying that we are trying to suggest that it is these things to the exclusion of religious/spiritual discourse.

But a way of approaching this, it seems to me, is to think of some of the assumptions behind religious belief as such. The German theologian Martin Buber (1878–1965) distinguished between two modes of human subjective relationships, the 'I–Thou' relationship of one subjectivity encountering another subjectivity (as when friends or lovers interact), and the 'I–It' relationship, in which a subjectivity encounters an objectivity (as when, say, you load and turn on the washing machine). It is often the case that humans relate to other humans as objects rather than as subjects, and according to Buber our relationship to other people fluctuates between 'I–Thou' and 'I–It'; he thinks that the only persistent 'Thou' on which an I–Thou relationship can permanently be predicated is God. More to the point is the way individual consciousnesses apprehend the cosmos as a whole. H. and H. A. Frankfort, in their 1946 study *The Intellectual Adventure of Ancient Man*, utilise Buber to talk about the difference between the modern 'scientific' view of the cosmos and that common to prehistoric man. Although they limit themselves to prehistory, it is I think the case that this latter circumstance underlies most latter-day religious worldviews as well. I find it very illuminating:

> The fundamental difference between the attitudes of modern and ancient man as regards the surrounding world is this: for modern, scientific man the phenomenal world is primarily an 'It'; for ancient – and also for primitive – man it is a 'Thou'. This formulation goes far beyond the usual 'animistic' or 'personalistic' interpretations ... for a relation between 'I' and 'Thou' is absolutely *sui generis* ... the knowledge that 'I' has of 'Thou' hovers between the active judgment and the passive 'undergoing of an impression', between the intellectual and emotional. (Frankfort and Frankfort, pp. 12–13)

One of the complaints often made by religiously- or spiritually-minded people against science is that it too cruelly 'objectifies' the world. Indeed I think the Frankforts are too sanguine in their assessment that 'modern man' takes an 'I–It' model for his relation to the cosmos. On the contrary, it seems to me that most people today prefer to think of their world as a Thou, a notion given pseudo-scientific plausibility with the (basically science fictional) 'Gaia' hypothesis of the ecologist James Lovelock (b. 1919), which argues that the world is literally a Thou. As Frankfort and Frankfort say, this positioning of one's relation to the cosmos as an 'I–Thou' relation goes beyond mere animism.

Where 'science' must in general treat the cosmos as an It (not as an inconsistent, cranky, personalised Thou), fiction has always been drawn to the I–Thou. What I mean by this is partly that the greatest writers have created not only 'Thou' characters but 'Thou' environments for those characters – Dickens' London, Hardy's Wessex or Faulkner's Yoknapatawpha County, for instance. Science fiction, on the other hand, and unlike most other fictions, mediates precisely an I–It and an I–Thou apperception. Individual texts may, of course, be instanced that explore one to the exclusion of the other; but the genre as a whole, in part because of the cultural dynamics of its birth, is involved in the dialectical inter-relation of both.

Another way of looking at this might be to invoke the work of the French philosopher Bruno Latour. Latour's definition of 'Modernity' sees it as a jamming together of two seemingly incommensurate discourses. On the one hand is 'the work of purification' which distinguishes carefully between 'human and thing', between scientific fact like gravity and social fact like astrology. This careful separation out into 'human beings on the one hand' and 'nonhumans on the other' is what Latour calls 'the modern critical stance': what we might want to identify with the I–It dichotomy. On the other hand, says Latour, is the 'work of translation', which 'creates mixtures between entirely new types of beings, hybrids of nature and culture'. Latour thinks the attempt to keep these two sorts of 'work' separate is distinctively modern:

> So long as we consider these two practices of translation and purification separately we are truly modern – that is, we willingly subscribe to the critical project, even though that project is developed only through the proliferation of hybrids down below. As soon as we direct our attention simultaneously to the work of purification and the work of hybridization, we immediately stop having been modern, because we become retrospectively aware that the two sets of practices have always already been at work in the historical period that is ending. (Latour, p. 11)

Latour's opinion on the question of whether we have ever actually been 'modern' is blazoned in the title of the famous study from which this quotation is taken: *Nous n'avons jamais été modernes* (We have never been modern). This, in a phrase of theoretical shorthand, can be taken as the overall thesis of this present critical history of one particular discourse that hybridizes 'science' and 'culture': SF.

The future

Some argue that SF is so closely wedded to the twentieth century that it faces desuetude in the twenty-first. This seems unlikely. Clearly SF in Europe, North America and to a certain extent Japan will change. In these areas it seems clear that today the most ground-breaking, the most exciting and also the most popular SF takes the form of visual narrative. Were I bold enough to hazard prediction, it would be that this will increasingly become what SF is: cinema, TV, computer games (with an increasingly narrative component) and graphic novels. The strength of these media is their ability to cater to a culture increasingly fascinated with visual art; not to say a culture that conceptualises and mediates its anxieties about technological change, its fearful uncertainties about its place in the cosmos and the extent to which redemption is possible.

This is not to say that written SF is in the doldrums. On the contrary, the number of good SF books being published each year is very large. But a culture cannot be maintained entire on fan interest alone. Fandom, as I have argued, has strengths, but it has weaknesses also. From the great mass of SF writers working today Fandom selects a few exemplars, some good, some frankly bad, and then cleaves to them with an aggressive Doberman loyalty that is tinged with cultural paranoia, in a manner best described as 'unbecoming'.

New SF novels are published in a world in which several hundred good, and several dozen great, genre books are published every year. The reader, or even the reviewer and professional critic, cannot read so many books. Choices have to be made. In the SF community, as in other fan communities, these choices are made partly on tribal grounds, those grounds on which SF fans allocate allegiance to authors and then act as advocates for that author's books, good or bad. This is a rather imprecise business, with many good authors effectively overlooked and some weaker authors over-praised; it also tends to militate against more general acceptance of these writers beyond the confines of genre. Even if I confine myself to the 'British SF Renaissance', a palpable if slightly nebulous phenomenon in literary SF at the start of the twenty-first century, the list of not merely good but great SF writers who lack the cultural dissemination to match their talent is enormous, and rather shaming: Geoff Ryman (b. 1951; Canadian-born but British resident), Roger Levy (b. 1955), Paul McAuley (b. 1955), Jon Courtenay Grimwood (n.d.), Eric Brown (b. 1960), Neil Gaiman (b. 1960), Peter F. Hamilton (b. 1960), Neal Asher (b. 1961), Michael Chabon (b. 1963), Richard Morgan (b. 1965), James Lovegrove (b. 1965), Liz Williams (b. 1965), Alistair Reynolds (b. 1966), Justina Robson (b. 1968) and Steph Swainston (b. 1975). Each of these authors deserves several pages'-worth of discussion of their output; some of them, I think, are the best writers working today in any idiom. Very few of them are familiar names outside the genre.

So much talent, producing such a wealth of literary art, may strike future critical-historians of literature as a sort of explosion, the biggest formally distinct revolution in letters since Modernism. But then again, it may not: the biases of culture may have already moved so solidly towards non-literary art that it becomes merely a footnote.

The situation is a little different, or seems to be, in other parts of the world. African SF was non-existent for most of the period covered by this book; but this fact may be changing. One reason for the absence may be the broad cultural bias in favour of 'spirits' or 'magic' as an explanatory discourse, something in conflict with the materialist emphases of contemporary science. In the words of Kwame Anthony Appiah:

> Most Africans cannot fully accept those scientific theories in the West that are inconsistent with [beliefs in invisible agents]. If modernisation is conceived of in part as the acceptance of science, we have to decide whether we think the evidence obliges us to give up the invisible world of the spirits. (quoted in Cooper, p. 222)

The rich traditions of African 'magic realist' and 'fantastic' writing grow from this culture, of course; and Appiah's question 'How much of the world of spirits must we intellectuals give up?' seems to express a reluctance to accept that the answer might be: 'All of it'. But another way of looking at this would be to note that it was precisely this cultural climate – this dialectic between new materialist-scientific discourses on the one hand, and magical-spiritual discourses on the other – that gave rise to European SF. Africa may prove to be one of the most important loci for twenty-first-century SF. As yet few African writers have made a mark in SF, although one notable exception is the challenging and brilliant Ghanaian writer B. Kojo Laing (b. 1946). His *Women of the Aeroplanes* (1988) is a utopian fantasy of sorts, set in Africa and Scotland; more ambitious still is *Major Gentl and the Achimoto Wars* (1992), a complex experimental fiction set in 2020 in 'Achimoto City'.

Abundance or lack?

Earlier in this study I mentioned Schopenhauer and Nietzsche, two philosophers who saw the Will as primary – which perspective (I suggested) shaped, through various channels, the near-ubiquitous fascination with Will that underpins so much SF produced in the late nineteenth century and throughout the twentieth. One difference between them is that Schopenhauer's vision was a pessimistic one, the cosmos a hungry Will that devours itself; whereas Nietzsche's was a triumphant one of a transcendent Will that revels in its abundance. People see the universe in different ways and whilst SF has taken a broader (and therefore more realistic) perspective on the cosmos than other art, it can still be constrained by these questions of passion. Damien Broderick quotes a section from *The Cosmic Connection* (1975) by the American scientist and occasional SF writer Carl Sagan (1934–1996). Broderick talks of 'this raw poetry of the scientific imagination':

> There is a place with four suns together in the sky – red, white, blue, and yellow; two of them are so close together that they touch, and star stuff flows between them.
>
> I know of a world with a million moons.
> I know of a sun the size of the Earth – and made of diamond.
> There are atomic nuclei a mile across that rotate thirty times a second.

Broderick declares 'that this is not fiction, but simple reality –' (Broderick, p. 11). And arguably this *is* a reality of the universe, a resource of wonder and poetry and truth that SF continues to excavate, and which illuminates the best of the genre with wonder. But it is hardly the *simple* reality of the cosmos. The simple reality of things is hydrogen and empty space; it is gravity, electromagnetism and some atomic forces acting on desolations of emptiness upon emptiness, leavened by only occasional near-emptiness. There are certain pockets of more complex matter, but as a proportion of the whole they are so small as to be insignificant. If we humans tend to spend our time looking at these more abundant scraps of creation, rather than contemplating the empty vastness that is the majority, it is because our brains prefer intricacy and complexity to nothingness. But this does not change the reality of things; the lack, rather than the abundance, of existence.

References

Broderick, Damien, *x, y, z, t: Dimensions of Science Fiction* (Holicong, PA: Borgo Press 2004; I. O. Evans Studies in the Philosophy and Criticism of Literature 20)

Cooper, Brenda, *Magical Realism in West African Fiction: Seeing with a Third Eye* (London: Routledge 1998)

Frankfort, H. and H. A. Frankfort (eds), *The Intellectual Adventure of Ancient Man: An Essay on Speculative Thought in the Ancient Near East* (1946: Harmondsworth: Penguin 1971)

Latour, Bruno, *We Have Never Been Modern* (New York: Prentice Hall 1993)

Serres, Michel, *We Have Never Been Modern*, transl. Catherine Porter (New York/London: Harvester 1993)

Chronology of Key Titles in Science Fiction and Developments in Science

*c.*80	Plutarch, *Peri tou prosôpou*
*c.*100	Antonius Diogenes, *Ta huper Thulên*
*c.*170	Lucian, *Alêthês Historia*
1516	Thomas More, *Utopia*
1543	Copernicus, *De revolutionibus orbium coelestium*
*c.*1600	Johannes Kepler's *Somnium* written (not published until 1634)
1622	Giovan Battista Marino, *L'Adone*
1638	William Godwin, *The Man in the Moone*
	John Wilkins, *The Discovery of a World in the Moone*
1644	René Descartes, *Principia Philosophiae*
1657	Savinien de Cyrano de Bergerac, *L'Autre Monde ou les Etats et Empires de la lune* (*Voyage dans la lune*)
1656	Athanasius Kircher, *Iter exstaticum coeleste*
1659	Jacques Guttin, *Epigone, histoire du siècle futur*
1665	Robert Hooke, *Micrographia*
1685	Isaac Newton, 'De Motu Corporum'
1686	Bernard de Fontenelle, *Entretiens sur la pluralité des mondes*
1687	Isaac Newton, *Principia Mathematica*
1690	Gabriel Daniel, *Voyage du Monde de Descartes*
1698	Christiaan Huygens, *Cosmotheoros*
1726	Jonathan Swift, *Travels into Several Remote Nations of the World* (*Gulliver's Travels*)
1730	Voltaire's *Micromégas* (published 1750)
1737	Thomas Gray, 'Luna habitabilis'
1741	Ludvig Holberg, *Nikolai Klimi iter subterraneum*
1750	Robert Paltock, *The Life and Adventures of Peter Wilkins*
1765	Marie-Anne de Roumier, *Les Voyages de Milord Ceton dans les sept planettes*
1771	Louis Sébastien Mercier, *L'An deux mille quatre cent quarante*
1781	Nicolas-Edme Restif de la Bretonne, *La découverte australe par un homme volant*
1798	Thomas Malthus, *An Essay on the Principle of Population, as it Affects the Future Improvement of Society*
1805	Jean-Baptiste Francois Xavier Cousin de Grainville, *Le dernier homme*
1813	Willem Bilderdijk, *Kort verhaal van eene aanmerklijke luchtreis en nieuwe planeetontdekking*
1816	Ernst Theodor Amadeus Hoffman, *Der Sandmann*
1818	Mary Shelley, *Frankenstein, or The Modern Prometheus*
1820	Adam Seaborn, *Symzonia: a Voyage of Discovery*
1826	Mary Shelley, *The Last Man*
1827	Jane Loudon, *The Mummy! A Tale of the Twenty-Second Century*
1830–33	Charles Lyell, *Principles of Geology*
1834	Félix Bodin, *Le Roman de l'avenir*
1835	Edgar Allan Poe, 'The Unparalleled Adventure of One Hans Pfaall'
1849	Edgar Allan Poe, *Eureka: a Prose Poem*
1858	Fitz-James O'Brien, 'The Diamond Lens'

1859	Charles Darwin, *On the Origin of Species by Means of Natural Selection*
1863	Nikolai Chernyshevsky, *Chto delat?*
1864	Jules Verne, *Voyage au centre de la terre*
1865	Jules Verne, *De la terre à la lune* Achille Eyraud, *Voyage à Vénus*
1868	Edward S. Ellis, *The Steam Man of the Prairies*
1869	Jules Verne, *Autour de la lune*
	Jules Verne, *Vingt mille lieues sous les mers*
1871	Edward Bulwer-Lytton, *The Coming Race*
	George Tomkyns Chesney, *The Battle of Dorking*
1872	Camille Flammarion, *Récits de l'infini*
1874	Jules Verne, *L'Île mystérieuse*
1880	Percy Greg, *Across the Zodiac: the Story of a Wrecked Record*
1882	Albert Robida, *Le vingtième siècle: Roman d'une parisienne d'après-demain*
1884	Edwin Abbott, *Flatland: a Romance of Many Dimensions* Mathias Villiers de l'Isle-Adam, *L'Eve future*
1887	Albert Robida, *La Guerre au vingtième siècle*
1888	Edward Bellamy, *Looking Backward 2000–1887*
1891	William Morris, *News from Nowhere, or An Epoch of Rest*
1893–94	Camille Flammarion, *La Fin du monde*
1895	H. G. Wells, *The Time Machine*
1896	H. G. Wells, *The Island of Doctor Moreau*
1897	H. G. Wells, *The Invisible Man*, Kurd Lasswitz, *Auf Zwei Planeten*
1898	Garrett P. Serviss, *Edison's Conquest of Mars*
	H. G. Wells, *The War of the Worlds*
1901	H. G. Wells, *Anticipations*
	H. G. Wells, *The First Men in the Moon*
1902	Georges Méliès' film *Le Voyage dans la lune*
1904	H. G. Wells, *The Food of the Gods and How it came to Earth*
1905	H. G. Wells, *A Modern Utopia*
1906	H. G. Wells, *In the Days of the Comet*
1908–11	165 Dime Novels appear in Germany under the general title *Der Luftpirat und Sein Lunkbares Luftschiff*
1909	Filippo Marinetti, *Il manifesto del futurismo*
1911–12	Hugo Gernsback, *Ralph 124C 41+: A Romance of the Year 2660*
1912	Edgar Rice Burroughs, *Under the Moons of Mars* (published in book form as *A Princess of Mars*, 1917)
1915	Charlotte Perkins Gilman, *Herland*
1920	Yevgeny Zamiatin, *We*
	David Lindsay, *A Voyage to Arcturus*
	Gustav Holst, *The Planets Suite*
1921	Karel Čapek, *R.U.R*
1923	René Clair's film *Paris qui dort*
1926	First issue of Gernsback's *Amazing Stories: the Magazine of Scientifiction*
	Fritz Lang's film *Metropolis*
1928	First publication of E. E. 'Doc' Smith's *The Skylark of Space*
1930	Olaf Stapledon, *Last and First Men*
	First issue of *Astounding Stories of Super-Science*
1931	James Whale's film version of *Frankenstein*
1932	Aldous Huxley, *Brave New World*
1933	Merian C. Cooper and Ernest B. Schoedsack's film *King Kong*
1936	William Cameron Menzies' film version of *Things to Come*
	Frederick Stephani's 13-episode film series *Flash Gordon*
1937	Olaf Stapledon, *Star Maker*
	Katherine Burdekin, *Swastika Night*

1938   C. S. Lewis, *Out of the Silent Planet*
           Jerry Siegel and artist Joe Shuster's 'Superman' character first appears in *Action Comics*
           Orson Welles' radio adaptation of *War of the Worlds*
           John W. Campbell becomes editor of *Astounding*
1940   'Captain Marvel' first appears in *Whiz Comics*
           Robert Heinlein, 'The Roads Must Roll'
1941   Isaac Asimov, 'Nightfall'
1943   René Barjavel, *Ravage*
1946   Hermann Kasack, *Die Stadt hinter dem Strom*
           A. E. Van Vogt, *Slan*
1947   Kenneth Arnold reports seeing a UFO flying above Washington State
1949   George Orwell's *Nineteen Eighty-four*
1950   Jack Vance, *The Dying Earth*
           Ray Bradbury, *The Martian Chronicles*
           Irving Pichel's film *Destination Moon*
1951   A. E. Van Vogt, *The Weapon Shops of Isher*
           John Wyndham, *The Day of the Triffids* Robert Wise's film *The Day the Earth Stood Still*
1953   Arthur C. Clarke, *Childhood's End*
           Arthur C. Clarke, 'The Nine Billion Names of God'
           Ray Bradbury, *Fahrenheit 451*
           Nigel Kneale's TV drama *The Quatermass Experiment*
1954   Isaac Asimov, *The Caves of Steel*
           Inoshiro Honda's film *Gojira*
1956   Alfred Bester, *Tiger! Tiger!* Fred McLeod Wilcox's film *Forbidden Planet*
           Don Siegel's film *Invasion of the Body Snatchers*
1957   Ernst Jünger's *Gläserne Bienen*
           John Wyndham, *The Midwich Cuckoos*
1958   James Blish, *A Case of Conscience*
1959   Robert Heinlein, *Starship Troopers*
           Walter Miller, *A Canticle for Leibowitz*
1960   George Pal's film of *The Time Machine*
1961   Robert Heinlein, *Stranger in a Strange Land*
           Stanisław Lem, *Solaris*
1962   Stan Lee and artist Steve Ditko's *Spider-Man* first appears
           Anthony Burgess, *A Clockwork Orange*
           Brian Aldiss, *Hothouse*
           La Jetée ('The Jetty') (dir. Chris Marker)
           Barbarella (dir. Jean-Claude-Forest)
1963   Pierre Boule, *La Planète des singes* ('The Planet of Apes')
           Dr Strangelove or: How I Learned to Stop Worrying and Love the Bomb (dir. Stanley Kubrick)
           First episode of *Doctor Who* broadcast
1965   Philip K. Dick, *The Three Stigmata of Palmer Eldritch*
           Frank Herbert, *Dune*
           Sun Ran, *The Heliocentric Worlds of Sun Ra, Volume One*
1966   Harry Harrison, *Make Room! Make Room!*
           J. G. Ballard, *The Crystal World*
           John Barth, *Giles Goat Boy*
1966–69   First series of TV's *Star Trek*
1967   Samuel Delany, *The Einstein Intersection*
           Harlan Ellison, *Dangerous Visions* (anthology)
1968   Michael Moorcock, *The Final Programme* (first of the *Jerry Cornelius* novels)

	Planet of the Apes (dir. Franklin J. Schaffner)
	Philip K. Dick, *Do Androids Dream of Electric Sheep?*
	2001: A Space Odyssey (dir. Stanley Kubrick)
	Keith Roberts, *Pavane*
1969	Philip K. Dick, *Ubik*
	David Bowie's hit single 'Space Oddity'
	Jack Vance, *Emphyrio*
	Ursula Le Guin, *The Left Hand of Darkness*
1970	Larry Niven, *Ringworld*
1971	Tangerine Dream, *Alpha Centauri*
1972	Robert Silverberg, *Dying Inside*
1973	Thomas Pynchon, *Gravity's Rainbow*
	Brian Aldiss, *Billion Year Spree: the History of Science Fiction*
1974	Ursula Le Guin, *The Dispossessed: An Ambiguous Utopia*
1975	Samuel Delany, *Dhalgren*
	Joanna Russ, *The Female Man*
	James Tiptree Jr, *Warm Worlds and Otherwise*
1977	Arkady Strugatski and Boris Strugatski, *Piknik na obochine* ('Picnic by the Roadside')
	Star Wars (dir. George Lucas)
	Close Encounters of the Third Kind (dir. Steven Spielberg)
	Hansruedi Giger, *H. R. Giger's Necronomicon*
1978	*Space Invaders* (designed by Toshihiro Nishikado)
	First radio series of *The Hitch-Hiker's Guide to the Galaxy* by Douglas Adams
	Jeff Wayne's pop-opera version of *War of the Worlds*
	Kraftwerk, *Die Mensch-Machine*
1979	*Alien* (dir. Ridley Scott)
	Mad Max (dir. George Miller)
1980	Russell Hoban, *Riddley Walker*
	Gene Wolfe, *The Island of Doctor Death and Other Stories and Other Stories*
1980–83	Gene Wolfe, *The Book of the New Sun*
1981	ELO, *Time*
1982	*Tron* (dir. Steven Lisberger)
	E.T. the Extra-Terrestrial (dir. Steven Spielberg)
	Blade Runner (dir. Ridley Scott)
	Brian Aldiss, *Helliconia Spring*
1984	William Gibson, *Neuromancer*
	Terminator (dir. James Cameron)
1985	Margaret Atwood, *The Handmaid's Tale*
	Orson Scott Card, *Ender's Game*
1986–87	Alan Moore and Dave Gibbons' graphic novel *Watchmen*
1987	Iain Banks, *Consider Phlebas*
	Octavia Butler, *Dawn* (first book of the *Xenogenesis* trilogy)
	RoboCop (dir. Paul Verhoeven)
1988	Sheri Tepper, *The Gate to Women's Country*
	Akira (dir. Katsuhiro Ôtomo)
1989	Dan Simmons, *Hyperion*
	Sheri Tepper, *Grass*
1991	Gwyneth Jones, *White Queen*
	Masamune Shirow's graphic novel *Kōkaku Kidōtai* ('Ghost in the Shell')
1992–95	Kim Stanley Robinson's 'Mars' trilogy (*Red Mars*, 1992; *Green Mars*,1993; and *Blue Mars*, 1995)

1993–2002	Chris Carter's TV serial *The X-Files*
1994	Neal Stephenson, *Snow Crash*
1996	David Foster Wallace, *Infinite Jest*
	Paul McAuley, *Fairyland*
1997	Greg Egan, *Diaspora*
1998	Michel Houllebeq, *Les Particules élémentaires* ('The Elementary Particles', translated into English as 'Atomised')
1999	Ken McLeod, *The Cassini Division*
	The Matrix (dir. Wachowski brothers)
2000	China Miéville, *Perdido Street Station*
2002	M. John Harrison, *Light*
	Spider-Man (dir. Sam Raimi)
	Minority Report (dir. Stephen Spielberg)
2003	Margaret Atwood, *Oryx and Crake*
	Stephen Baxter, *Coalescent*
2004	*Battlestar Galactica* (TV show, produced David Eick and Ronald D. Moore)
2005	*War of the Worlds* (dir. Stephen Spielberg)

Notes

Preface

1 See Darko Suvin, 'Science Fiction and Utopian Fiction: Degrees of Kinship', in *Positions and Presuppositions in Science Fiction* (Macmillan 1988). Chris Ferns, *Narrating Utopia: Ideology, Gender, Form in Utopian Literature* (Liverpool: Liverpool University Press 1999) is a very detailed and perceptive study that covers much of this ground.

2 The Italian novelist and critic Umberto Eco neatly dramatises this anxiety in what is his best book by a long chalk, *L'Isola Del Giorna Prima* ('The Island of the Day Before', 1994). A character called Saint-Savin is talking about the theological difficulty of a universe of infinite inhabited worlds: 'was Christ made flesh only once? Was Original Sin committed only once, and on this globe? What injustice! Both for the other worlds, deprived of the Incarnation, and for us, because in that case the people of all other worlds would be perfect, like our progenitors before the Fall ... Or else infinite Adams have infinitely committed the first error, tempted by infinite Eves with infinite apples, and Christ has been obliged to become incarnate, preach, and suffer Calvary infinite times, and perhaps He is still doing so, and if the worlds are infinite, His task will be infinite too. Infinite his task, then infinite the forms of His suffering: if beyond the Galaxy there were a land where men have six arms ... the Son of God would be nailed not to a cross but to a wooden construction shaped like a star – which seems to me worthy of an author of comedies.' At this point his interlocutor screams in rage at his blasphemy and attempts to stab him with a sword (Eco, pp. 139–40).

Chapter 1

1 Freedman's *Critical Theory and Science Fiction* (Hanover, NH and London: Wesleyan University Press 2000), pp. 24–30 discusses the question of literary canonisation, and SF's troubled relationship to 'the literary canon', in a lucidly convincing and non-paranoid manner.

2 Henry Jenkins, *Textual Poachers: Television Fans and Participatory Culture* (London: Routledge 1992). Also relevant here are John Tulloch and Henry Jenkins, *Science Fiction Audiences* (London: Routledge 1995); and Cheryl Harris and Alison Alexander (eds), *Theorizing Fandom: Fans, Subculture and Identity* (Cresskill, NJ: Hampton Press 1998).

Chapter 2

1 Even the advent of the space age has not purged SF of its nostalgic attachment to the 'sky' as the arena in which it takes place. An example: the X-Wing fighters of *Star Wars* (1977) screech through the soundless vacuum and explode loudly when shot. Of course, we know intellectually that a vacuum is a wholly soundless place, but we feel, somehow, that these space-battles are actually happening 'in the sky', like the Battle of Britain only higher up, and so we require 'aerial' sound effects to make the scene believable to us.

Chapter 3

1 C. S. Lewis points out that Kopernik's theory was not verified until the work of Kepler and Galileo in the early seventeenth century, and that 'general acceptance [came] later still.

Humanism, dominant in mid-sixteenth century England, tended to be on the whole indifferent, if not hostile, to science.' He goes on to argue that, while the Copernican revolution emptied the world of 'her occult sympathies', the result was 'dualism rather than materialism'. The present study does not share these conclusions.

2   Anthony Levi asserts 'that the phenomena we know as the renaissance and the reformation are connected, that they grew out of an impasse reached in the intellectual culture of Western Christendom, and that the intellectual positions to which they gave rise ... reflected their fundamental natures' (Levi, p. xi).

3   These numbers are cited to give a sense of relative proportions, not absolute numbers: the figure of 21 texts only includes works published in other languages if they were translated into English during this time; if we include continentally published works the number would double. Salzman's complete breakdown is as follows: 1. Elizabethan fiction (105 titles); 2. Sidnean Romance and additions to Sidney's *Arcadia* (16 titles); 3. 'Attenuated' Short Romance (6 titles); 4. Continental Romance, including adventure romance and pastoral romance (15 titles); 5. Didactic Romance (10 titles); 6. Political/Allegorical Romance and Religious Allegory (23 titles); 7. Didactic Fiction (21 titles); 8. French Heroic Fiction (15 titles); 9. Jest-Books (15 titles); 10. Criminal Biography (23 titles); 11. Imaginary Voyage/Utopia/Satire (20 titles); 12. Picaresque Fiction (24 titles); 13. Popular Chivalric Romance (26 titles); 14. Popular Compilations of History (9 titles); 15. Anti-Romance and 'Impure' Romance (18 titles); 16. The Novella (17 titles); 17. Memoirs (11 titles); 18. Scandal Chronicles/Secret Histories (17 titles); 19. Nouvelle Historique and Nouvelle Galante (72 titles); 20. The Political/Allegorical Novel (9 titles); 21. Oriental tales (15 titles); 22. Restoration Novel (58 titles); 23 Popular Non-Chivalric Fiction (34 titles).

4   It is worth dwelling for a moment on the probable reason why Kepler gave his 'privolvans' such a frantic, grotesque life. It represents, amongst other things, a complete break with the logic of the Ptolemaic solar system. A work such as the long poem *Le Premier des meteors* (1567) by Jean-Antoine de Baïf (1532–1589), set in a Ptolemaic solar system, sees only sublunary existence as capable of change, and dramatises the higher heavens as pure and incorruptible. Kepler's change-wracked privolvans live at that place where, according to the Ptolemaic way of thinking, the superlunary cosmos began. Kepler's gesture, in other words, is profoundly Copernican.

5   Arthur Lovejoy's influential study *The Great Chain of Being: a Study in the History of an Idea* (1933) does not accept that the Copernican revolution had any great impact on the development of the idea of a plurality of worlds, a concept that Lovejoy traces back to Neoplatonic thought. But as Ladina Bezzola Lambert observes, many critics have challenged Lovejoy's conclusions. In particular Karl Guthke 'disagrees with this position ... criticiz[es] Lovejoy for not clearly distinguishing between the idea of a plurality of *kosmoi* (which goes back to Aristotle) and a plurality of worlds within a single cosmos. He also points out that ... the Great Chain of Being represented a strictly vertical order which clearly adhered to the geocentric cosmos and the theological hierarchy to which the principle was applied. Such an order could not span the ubiquity of life in a post-Copernican universe of innumerable planetary systems' (Lambert, 106n.).

6   One antecedent for this work may have been *Les Hermaphrodites* (1605), a satirical dystopia by Thomas Artus (d. 1614) in which a hermaphroditic society is located on a floating island.

Chapter 4

1   Historians of science are cautious on this point: 'the idea that the reformed religion (its Calvanist sects especially) favoured the cultivation of natural science in a way that the Roman Catholic Church did not ... still attracts able exponents such as Christopher Hill, Reijer Hooykaas and Charles Webster. Historians from Catholic, Mediterranean Europe seem not to have espoused it. At a first inspection the correlation seems to have much to

commend it: the northern, Protestant half of Europe seems to have been rendered more fortunate in eminent scientific discovery than the south' (A. Rupert Hall, *The Revolution in Science 1500–1750* (London: Longman 1983), p. 23).

2   Marjorie Hope Nicholson found in this passage the title of her study of Newton-influenced poetry, *Newton Demands the Muse* (Princeton, NJ: Princeton University Press 1946).

3   Hence the first Minister of Brobdingnag carries 'a white staff, near as tall as the mainmast of the *Royal Sovereign*'; that Gulliver being carried overland in his box experiences 'agitation ... equal to the rising and falling of a ship in a great storm'; that Gulliver sleeps under a 'clean white handkerchief' that is 'larger and coarser than the mainsail of a man-of-war' and so on (Swift, *Travels*, pp. 146, 136, 131).

4   Aldiss's insistence on SF as a mode of the Gothic is shaped at least in part by his choice of Mary Shelley's Gothic novel *Frankenstein* (discussed in the next chapter) as the starting point for the whole genre. But *Frankenstein* is in many ways an uncharacteristic Gothic novel; appearing very late in the craze, it is a work which can be read with equal justice *either* as a novel founded on a supernatural occurrence (which is to say, as Gothic horror) *or* as a purely materialist science-based extravaganza (which makes it SF). There is a well-balanced discussion of 'Gothic SF' by Peter Nicholls, in Clute and Nicholls, pp. 510–12.

Chapter 5

1   Suvin discusses this possibility, and also advances the theory that increasing fascination with the future matched increasing ideological control of the bourgeois capitalist economy, 'with its salaries, profits, and progressive ideals always expected in a future clock-time' (Suvin, p. 73). Alkon, who discusses Suvin's theories intelligently and in some detail, also stresses the formal possibilities of experimental fiction as an enabling device for futurism. Neither critic discusses the publication, by Thomas Malthus (1766–1834), of *An Essay on the Principle of Population, as it Affects the Future Improvement of Society* in 1798. Malthus's theory was that increasing population would outrun the resources needed to support it with consequent famine, misery and mass death. His work was enormously influential and may have played some part in focusing European minds on the future as an actual, rather than merely a notional, entity.

2   Brian Stableford notes how common this comic-satiric use of the 'automaton' or 'robot' trope is in science fiction: 'William Wallace Cook's *A Round Trip to the Year 2000* (1903), which features robotic "mugwumps", and the anonymous skit *Mechanical Jane* (1903) are both comedies, as is J. Storer Clouston's *Button Brains* (1933), a novel in which its robot is continually mistaken for its human model' (Clute and Nicholls, p. 1018). More recently robots such as John Sladek's Roderick, the fussily comic C-3PO from *Star Wars* and 'Marvin the Paranoid Android' (from *The Hitch-Hiker's Guide to the Galaxy*) have continued this tradition.

3   John Sutherland's *Is Heathcliff a Murderer? Puzzles in 19th-Century Fiction* (Oxford: Oxford University Press 1996), inspired by Ellen Moers' analysis 'Female Gothic' (in G. Levine and U. Knoepflmacher (eds), *The Endurance of Frankenstein*, Berkeley: University of California Press 1979) suggests that the 'physical, eye-averting revulsion' implied in Mary's description of the process 'is a reflex and a rhetoric associated traditionally in Anglo-Saxon cultural discourse with ... sexual intercourse (and its variant, self-abuse) and childbirth (and its variant, abortion)' (p. 31). He goes on to suggest that Mary, drawing on her unfortunate experiences with childbirth and sex (her mother died ten days after she was born, she eloped with Shelley at 17 and gave premature birth to a daughter who died a few days later, all before writing *Frankenstein*) had Victor create the monster 'by a process analogous to fertilization and *in vitro* culture. The initial work "of his hands" which Victor refers to is, presumably, masturbatory. The resulting seed is mixed with a tissue, or soup composed of various tissues. The mixture is grown *ex utero* until ... it is

released into life. Victor Frankenstein ... is less the mad scientist than the reluctant parent, or semen donor. He does not make his monster, as one might manufacture a robot – he gives birth to him, as one might to an unwanted child, the sight of whom fills one with disgust' (p. 33).

4   For a much more sympathetic account of the epistemological ambitions of *Eureka*, see Peter Swirski's *Between Literature and Science: Poe, Lem, and Explorations in Aesthetics, Cognitive Science, and Literary Knowledge* (Liverpool: Liverpool University Press 2000).

Chapter 6

1   Darko Suvin is especially scathing about *The Romance of Two Worlds*: 'fraudulent ... proto-Fascist ... a narration based on ideology unchecked by any cognitive logic ... cobbled together from orts and scraps of esoteric metaphysics' (quoted in James, p. 30).
2   Another utopia from this period, although much less well known today, might be mentioned here as a challenger to Carey's laurels of 'most influential'. Nikolai Chernyshevsky (1828–1889) was a Russian political activist and radical. Imprisoned by the Tsar, he smuggled out a novel, *Chto delat?* ('What is to be Done?' 1863), which includes a utopian programme for a future society founded on the principles of socialist equality. Lenin (who published a book with the same title) was influenced by Chernyshevsky, and he of course had a far greater practical effect on changing the world than the Bellamyite Nationalist Party: 'as a blueprint for social and political change the novel exerted widespread influence and, through its effect on Lenin, contributed directly to changing the world' (Moser, p. 262).
3   A number of 'fairy' burlesques and theatrical extravaganzas light-heartedly adapted eighteenth-century SF about flying humanoids; for example, E. L. Blanchard's *Peter Wilkins; or, Harlequin and the Flying Women* (Drury Lane 1860) or James Planche's *The Invisible Prince* (1846). I am indebted to Jane Brockett for drawing these texts to my attention.
4   John Clute's expert article on 'Edisonade' (in Clute and Nicholls, pp. 368–70) discusses the iconic appeal of Edison during this period, as well as noting several other SF yarns featuring the great inventor, amongst them: J. S. Barney's *L.P.M.: The End of the Great War* (1915) in which advanced weapons made by an inventor called 'Edestone' end the First World War and establish a world state; and Cleveland Langston Moffett's *The Conquest of America* (1916), in which Edison himself appears as an inventor of super-weapons.
5   Comparing Nietzsche and Schopenhauer, Maudemarie Clark notes that 'both philosophers portray the world as will, and in doing so, rule out a goal or end-state of the world process that could be its justification ...[but] Schopenhauer sees it in terms of need or lack – his world is a hungry will devouring itself – whereas Nietzsche's is painted in tones of strength and abundance – his world is a superabundant will whose energy overflows. Schopenhauer idealizes the ascetic, one who turns against life and willing ... Nietzsche's ideal is opposite, the life-affirming person who does not find value in the process ... [who sees] the world under the aspect of abundance rather than lack, as the overflowing of energy without ultimate aim or goal, as play – in short, as will to power' (Clark, p. 144).

Chapter 7

1   'The preoccupations and techniques of Verne seem to me to link him with many other, more highly regarded writers (than SF writers). At the very least he resisted containment in the categories that had been allotted to him. When I came across Kurt Vonnegut's remark that "I have been a soreheaded occupant of a file drawer labelled 'science fiction' and I would like out, particularly since so many serious critics regularly mistake the drawer for a urinal," I seemed to hear the voice of Verne. The purpose of this book, then, is to deliver Verne from confinement in that drawer' (Martin, p. xi).

2 In John Clute's impassioned words, 'the reputation [Verne] long had in English-speaking countries for narrative clumsiness and ignorance of scientific matters was fundamentally due to his innumerate and illiterate translators who ... remained impenetrably of the conviction that he was a writer of overblown juveniles and that it was thus necessary to trim him down, to eliminate any inappropriately adult complexities, and to pare the confusing scientific material to an absolute minimum' (Clute, pp. 1276–7). It should be added that some more recent translations have been far more faithful to the original.

3 It was an emblem to which Verne returned several times in his writing career, most notably in *Les Cinq cents millions de la Bégum* ('The Bégum's Five Hundred Million', 1879) and *Sans dessus dessous* ('Anti-Topsy-Turvy', 1889).

4 The English doctor Benjamin Ward Richardson (1828–96) was so impressed by Verne's submarine that he even inserted one (made of wood) into his historical novel set in Roman times, *The Son of a Star: A Romance of the Second Century* (1889).

5 Andrew Martin discusses the difficulty of translating this title: the French phrase 'sens dessus dessous' means 'topsy turvy' or 'upside down', but Verne's title 'by replacing the *e* in *sens* with an *a*, signifies something like the opposite of these ... and might be loosely rendered as: "No More Ups and Downs" or "An End to Inversion"' (Martin, pp. 179–80).

6 There is one further point that must be made about Verne's enduring impact on the genre. In addition to being multi-modal works themselves, combining visual and textual elements, Verne's novels migrate promiscuously into a variety of other idioms. For instance, he became better known to most twentieth-century audiences through the many cinematic adaptations of his books. Even in the nineteenth century there were many theatrical and operatic versions of his books staged. Verne himself adapted some of his voyages extraordinaires to the stage in collaboration with Adolphe Dennery. *Le Tour du Monde en 80 Jours* (first staged in 1875) ran (in the words of Laurence Senelick) 'for a record-breaking 652 nights ... [and] set the style for the pièce à grand spectacle' (Senelick, p. 3). Offenbach's *opéra bouffe Le Voyage dans la lune* (1875) took from Verne's novel title and mode of travel (firing out of a large cannon), although the story develops in rather un-Vernean directions: the 23 scenes of this lengthy opera actually land the actors on a topsy-turvy satirical moon familiar from the seventeenth and eighteenth centuries. The work was a hit, running for 185 performances.

7 See Michel Foucault, *Surveiller et punir: Naissance de la prison* (1975).

Chapter 8

1 'Historians of ideas usually attribute the dream of a perfect society to the philosophers and jurists of the eighteenth century; but there was also a military dream of society; its fundamental reference was not to a state of nature, but to the meticulously subordinated cogs of a machine, not to primal social contract, but to permanent coercions' (Foucault, p. 169).

Chapter 9

1 Peter Nicholls has this evocative description of Pulps: 'printed on cheap paper manufactured from chemically treated wood pulp, a process invented in the early 1880s. The paper is coarse, absorbent and acidic, with a distinctive sharp smell much loved by magazine collectors. Pulp paper ages badly, largely because of its acid content, yellowing and becoming brittle. Because of the thickness of the paper, pulp magazines tended to be quite bulky, often ½ inch thick or more. They generally had ragged, untrimmed edges, and later in their history had notoriously garish, brightly coloured covers, many of the coal-tar dyes used to make cover inks being of the most lurid hues' (Clute and Nicholls, p. 979). 'Garish', 'brightly coloured' and 'lurid' also rather nicely describe the content and style of most Pulps.

2   Gary Westfahl, in the best and most detailed account of Gernsback's importance to the development of SF, admits the 'aesthetic failure' of *Ralph 124C 41+*, but insists that this novel 'is the one essential text for all studies of science fiction, a work which anticipates and contains the entire genre' (Westfahl, pp. 92–3). He makes an interesting case, although Westfahl's definition of 'science fiction', of which he sees Gernsback as both the inventor and neglected genius, is much more narrowly (he might say 'precisely') conceived than mine: see Westfahl, pp. 287–318.

3   The great historian of the SF magazine is Mike Ashley, from whose *The Time Machines* (2000) I derive the following (indicative not comprehensive) list of Pulp titles, with the most important titles (in terms of longevity and/or influence) in bold: *Air Wonder Stories* (1929–30); **Amazing Stories** (1926–present); *A. Merritt's Fantasy Magazine* (1949–50); *Astonishing Stories* (1941–3); **Astounding Science-Fiction** (1930–present, as *Analog*); *Captain Future* (1940–44); *Captain Hazard* (1938); *Captain Zero* (1949–50); *Comet* (1940–41); *Cosmic Stories* (1941); *Doc Savage* (1933–49); *Doctor Death* (1935); *Dynamic Science Stories* (1939); *Fanciful Tales of Space and Time* (1936); *Fantastic Adventures* (1939–53); *Future Fiction* (1939–43, later revived); *Futuristic Stories* (1946); *Marvel Science Stories* (1938–41; later revived); *Marvel Tales* (1934–5); *Miracle Science and Fantasy Stories* (1931); *Planet Stories* (1939–55); *Science Fiction* (1931–41); *Scientific Detective Monthly* (1930); *Science Wonder Stories* (1929–30); **Startling Stories** (1944–55); *Strange Tales* (1931–33); *Super Science Stories* (1940–43, later revived); *Tales of Wonder* (1937–42); *Uncanny Stories* (1941); *Uncanny Tales* (1940–43); *Unknown* (1939–43); **Weird Tales** (1923–54, later revived); **Wonder Stories** (1929–53; later *Thrilling Wonder Stories*).

4   Mark Bould disagrees with the 'more rational footing': 'domination is reinstituted with the promise of extraordinarily compromised arbitration, the workers having learned their lesson: it's hegemony in action!' (private correspondence).

5   The history of the different versions of *Metropolis* is complicated, with many different cuts being shown at different times; the longest was the 4,189 metres version premiered in Berlin; the US cut was 3,100 metres; recent restorations have produced cuts of 3,153 metres. For a more detailed discussion, see Thomas Elsaesser, *Metropolis* (London: BFI 2000).

Chapter 10

1   He remains 'the most influential figure in the history of American SF ... he had the capacity to make the wildest or the most personal, of visions, sound like common sense' (Clute, p. 128); 'Heinlein was indeed probably the most influential figure in the history of SF' (James, p. 65).

2   What is especially interesting in Gifford's article, I think, is the way in which, even as he points up the disparity between the actual novel and Heinlein's account of the novel in *Expanded Universe*, he strikes an apologetic and, frankly, subaltern tone. He argues that Heinlein *intended* to make 95 per cent of his Federal service in *Starship Troopers* Federal rather than military, but somehow forgot, and concludes, 'I wish I had had the chance to discuss this directly with Heinlein before his death ... but it is too late, too late – and I can only hope his shade can forgive me for saying he was wrong' (Gifford, p. 11). I cite this because it seems to me fairly representative of the extent to which SF fandom has colluded in Heinlein's construction of himself as a patriarchal authority figure; something indicative of at the least timidity, at the worst servility, and not, I think, a good thing.

Chapter 11

1   This is the explanation offered by James Blish for the outpouring of religiously inflected SF in the late 1950s and 1960s (an outpouring to which, as we have seen, he contributed some of the most powerful and eloquent texts). Such writing was, he argued, 'a chiliastic crisis, of a magnitude we have not seen since the chiliastic panic of 999 AD' (quoted in Clute and Nicholls, p. 1001).

2 Thomas Disch, who has a low opinion of the book, comments mordantly: 'it was this element of communes featuring dorm-style promiscuity under the auspices of [a] bullying alpha male that especially endeared the book to the counterculture of the '60s, including Charlie Manson, who made *Stranger* required reading for his followers' (Disch, p. 234). It can't be denied that the pitchy touch of Manson's endorsement has defiled Heinlein's novel (although the same cannot really be said of Manson's other enthusiasms – the Beatles' *White Album*, say); but it may be more to the point to see Manson's own homicidal messianism as one more symptom (an unusually baleful one) of a broader cultural logic.

3 We may wonder, for instance, how Dune's atmosphere is oxygenated in the absence of planetary vegetation. In later books Herbert suggests that the sandworms fart oxygen, which hardly addresses the problem.

4 Abraham Kawa disagrees with my judgement here, and locates Ellison's greatest achievements in other modes of SF: 'He is greatly influenced by superhero comics, particularly their hyperbolic mode of writing as exemplified by Stan Lee ('Repent' is, in many ways, a superhero story), and even made occasional plot contributions to the genre, like the Hulk story 'The Brute that Shouted Love at the Heart of the Atom!' (*Incredible Hulk* #140, 1971) in which the shrunken green behemoth becomes the king of a subatomic realm, and the grim Batman tale 'Night of the Reaper' (*Batman* #237, 1972). This willingness to 'play' with other creators' characters also served him to good stead in the screenplay for 'The City At the Edge of Forever' (1967), one of the most highly regarded episodes of the original *Star Trek* series' (Kawa, private correspondence).

5 The critic Sarah Lefanu quotes Silverberg's dotty certainty: 'it has been suggested that Tiptree is female, a theory that I find absurd, for there is to me something ineluctably masculine about Tiptree's writing ... lean, muscular, supple, Hemingwayesque', and then quotes his more elegant retraction ('she fooled me beautifully, along with everyone else, and called into question the entire notion of what is "masculine" or "feminine" in fiction'). But as Lefanu herself points out, 'there is something dangerous about seeing masculinity and femininity in such essentialist terms' (Lefanu, pp. 122–3).

Chapter 12

1 I don't want to draw this discussion out unnecessarily; but it might be objected that 'music' rather than poetry has always been the dominant form of 'lyric' or 'epiphanic' art. I disagree, although the case can be argued. It seems to me that the form of Romantic music that trades particularly in intensities (the choral movement of Beethoven's Ninth, say; or the Nimrod movement in Elgar's *Enigma Variations*) are fairly recent developments in art and relatively rare. Most music over the last thousand years – beautiful and endlessly supple as that art form has proved itself – was subordinated to other needs: church, court, ritual, and so on. Much pop music is similarly subordinated, particularly to the needs of dance (and of commerce). But a large portion of pop now functions culturally in the way poetry once did. Which is to say: very few people ever listened to music in the past the way that a rapt teenager will today listen to his favourite album, over and over again, milking the evanescent thrill, finding tremendous and even mystical significance in it, becoming extraordinarily intimate with it.

2 In the original series the Klingons are brutish creatures, dressed with a samurai Japanese styling, with darkened skin but otherwise humanoid. In the movies and *The Next Generation* the Klingons are more sympathetically rendered, and are given certain physical differences from humans, being taller and having a prominent cranial bone structure that looks like a giant Mars-bar melted into their foreheads.

3 Brian Aldiss is only one among many to lament the turning of SF into 'commercial coin' by the visual media. 'Much of the vitality of written SF lay in its conflict of ideas', he writes; accordingly, 'we are wise to have reservations about this sweeping success of [visual] SF. As Pamela Sargent has pointed out in a recent issue of *Science-Fiction Studies* (July 1997), "Visual science fiction is almost a virtual museum of the forms and ideas found in written SF, dumbed down to varying degrees" ' (Aldiss, p. 2). It is worth adding

that not all critics agree with the notion of SF as a literature of ideas. Mark Bould, for instance, makes two good points: 'first, the majority of sf stories have no more ideas than your average Mickey Spillane novel; second, what often passes for ideas in sf are more accurately described as conceits' (private correspondence).

4   *Solider* (Paul W. S. Anderson, 1998), a very bad SF film supposedly set in the same universe as *Blade Runner*, has almost no points of continuity with the earlier film. Talk of an actual sequel to *Blade Runner*, to be called *Metropolis*, has been circulating for a decade or more, but nothing has yet come of the project. The American writer K. W. Jeter (b. 1950) has published three quite good sequels in novel form: *Blade Runner 2: the Edge of Human* (1988); *Blade Runner 3: Replicant Night* (1996); *Blade Runner 4: Eye and Talon* (2000).

Chapter 13

1   An excellent discussion of this point can be found in Brian McHale's essay 'POSTmodern-CYBERpunkISM'.

2   This mildly interesting but unoriginal conception was prolonged into a number of sequels, each of which illustrates the thesis of diminishing returns: *Rama II* (with Gentry Lee, 1989), *The Garden of Rama* (with Gentry Lee, 1991), *Rama Revealed* (with Gentry Lee, 1993).

3   See www.locusmag.com/awards for a full list of SF awards currently active.

4   See, for instance, Rob Hansen's *Then: a History of UK Fandom* (www.dcs.gla.ac.uk/SF-Archives/Then/Index.html); Harry Warner, *A Wealth of Fable* (SCIFI Press 1992); Peter Weston, *With Stars in My Eyes: My Adventures in British Fandom* (NESFA Press 2004).

5   The comparison is presumably made because Wolfe is seen by some to be interested in 'memory'; but Severian has the inhuman ability to remember literally everything that happens to him; where the whole point of Proust's narrator is the intensely human, hit-and-miss manner in which his memory works. Wolfe substitutes a perfectly remembering but unreliable narrator (Severian does not tell the whole story, and sometimes lies, although it is hard to be sure when he is doing so) for the much more precise Proustian delineation by a narrator who is if anything too reliable, too dedicated to capturing the precise contours of his own imperfect memory.

6   My outrage has encouraged me to oversimplify the book's reception. Although many reviews did celebrate the book in these terms, and although it was shortlisted for the Booker Prize, several critics took issue with what they saw as its impertinent appropriation of the Holocaust by a tricksy game-playing gentile; and one of the Booker judges (Nicholas Mosley) resigned from the judging panel in protest at the shortlisting.

Chapter 14

1   See the book Latham wrote with Stephen Todd, *Evolutionary Art and Computers* (San Diego, CA: Academic Press 1992).

2   Leafing through a book such as Mark Cotta Vaz and Patricia Rose Duignan's *Industrial Light & Magic: Into the Digital Realm* (New York: Del Rey 1992) provides evidence for how strikingly beautiful these visual effects can be.

3   There is a vast literature about this phenomenon. A good place to start is Curtis Peebles, *Watch the Skies! A Chronicle of the Flying Saucer Myth* (Washington DC: Smithsonian Institute Press 1994).

Further Reading

Nobody seriously interested in studying science fiction can afford to be without John Clute and Peter Nicholls' *Encyclopedia of Science Fiction* (2nd edn., London: Orbit 1993); a virtually all-inclusive treasure chest of information about the genre and critical judgements upon key players.
   Three good introductory books about SF are:

James, Edward, *Science Fiction in the Twentieth Century* (Oxford: Oxford University Press 1994)
James, Edward and Farah Mendlesohn (eds), *The Cambridge Companion to Science Fiction* (Cambridge: Cambridge University Press 2003)
Luckhurst, Roger, *Science Fiction* (London: Polity 2005)

although the focus of all three is heavily twentieth century.
A good study of earlier forms of the genre is Paul Alkon's *Science Fiction before 1900: Imagination Discovers Technology* (1994; London: Routledge 2002).
The following books are all excellent, either as approaches to the genre as a whole, or as studies of important specific aspects of it:

Aldiss, Brian and David Wingrove, *Trillion Year Spree: the History of Science Fiction* (London: Gollancz 1986)
Ashley, Mike, *The Time Machines: the Story of the Science Fiction Pulp Magazines from the beginning to 1950* (Liverpool: Liverpool University Press 2000)
Broderick, Damien, *x, y, z, t: Dimensions of Science Fiction* (Holicong, PA: Borgo Press 2004; I. O. Evans Studies in the Philosophy and Criticism of Literature 20)
Disch, Thomas, *The Dreams our Stuff is Made of: How Science Fiction Conquered the World* (New York: Simon and Schuster 1998)
Gray, Chris Hables (ed.), *The Cyborg Handbook* (London and New York: Routledge 1995)
Jones, Gwyneth, *Deconstructing the Starships: Science, Fiction and Reality* (Liverpool: Liverpool University Press 1999)
Kneale, James and Rob Kitchin (eds), *Lost in Space: Geographies of Science Fiction* (London: Continuum 2002)
Kuhn, Annette (ed.), *Alien Zone: Cultural Theory and Contemporary Science Fiction Cinema* (London: Verso 1990)
Lefanu, Sarah, *In the Chinks of the World Machine: Feminism and Science Fiction* (London: Women's Press 1988)
Le Guin, Ursula K., *The Language of the Night: Essays on Fantasy and Science Fiction*, ed. with an intro. Susan Wood (London: Women's Press 1989)
Martin, Andrew, *The Mask of the Prophet: the Extraordinary Fictions of Jules Verne* (Oxford: Clarendon 1990)
Parrinder, Patrick, *Science Fiction: its Criticism and Teaching* (London and New York: Methuen 1980)
Parrinder, Patrick, *Shadows of the Future: H. G. Wells, Science Fiction and Prophecy* (Liverpool: Liverpool University Press 1995)
Suvin, Darko, *Metamorphoses of Science Fiction: On the Poetics and History of a Literary Genre* (New Haven, CT: Yale University Press 1979)
Suvin, Darko, *Positions and Suppositions in Science Fiction* (London: Macmillan 1988)
Westfahl, Gary, *The Mechanics of Wonder: the Creation of the Idea of Science Fiction* (Liverpool: Liverpool University Press 1998)
Wolmark, Jenny, *Aliens and Others: Science Fiction, Feminism and Postmodernism* (New York and London: Harvester Press 1994)

Index

Note: Page references in **bold** refer to the main discussion of the authors or texts highlighted. Films are given by title with director given in parenthesis afterwards; video games and similar texts follow the same format. All other entries are given by author, where known.